2007 Cleantech Conference and Trade Show

Cleantech 2007

T0186852

Cleantech 2007™

The Cleantech Conference, Venture Forum and Trade Show

May 23-24, 2007

Santa Clara Convention Center

Santa Clara, California, U.S.A.

www.techconnect.org

Transport IV (cover):
Transport IV renders electron flow paths in a "two dimensional electron gas". The research leading to the Transport series was inspired by the experiments of Mark Topinka, Brian Leroy, and Prof. Robert Westervelt at Harvard, which actually measured the paths taken by the electrons. The theory was performed with the help of Scot Shaw, a member of my research group at the time.

Transport IV is based on flow patterns for electrons riding over bumpy landscape, which is what electrons experience in the two-dimensional electron gas (2DEG) that they dwell in. A 2DEG is a sea of electrons confined to a sheet, i.e. two dimensions. The bumps they encounter are due to charged atoms lying above the sheet. The electrons have more than enough energy to ride over any bump, and the concentrations of electron flow into the branches seen here are recently discovered indirect effects of that bumpy ride. The channeling or branching was unexpected and has implications for small electronic devices of the future.

This image was made in a computer simulation by launching 100,000 electrons from the lower middle of the image, and following their tracks. Each electron had a slightly different launch angle. A color map was applied to turn the density of electron paths into color. Code written in Fortran computed the trajectories, and wrote information about them to a very large raster image. The algorithm used to write to pixels, which had already been accessed by previous electron paths gives the peculiar shading and form to this image.

The electron tracks in Transport IV are an excellent example of the wonderful way nature emulates herself in different contexts. The branching pattern is reminiscent of familiar natural forms.

2007 Cleantech Conference and Trade Show

Cleantech 2007

The Cleantech Conference, Venture Forum and Trade Show

May 23-24, 2007

Santa Clara, California, U.S.A.

www.nsti.org

NSTI – TechConnect 2007 Joint Meeting

2007 Cleantech Conference, Venture Forum and Trade Show, Cleantech 2007

The 2007 NSTI Nanotechnology Conference and Trade Show, Nanotech 2007

2007 NSTI Bio Nano Conference and Trade Show, Bio Nano 2007

10th International Conference on Modeling and Simulation of Microsystems, MSM 2007

7th International Conference on Computational Nanoscience and Technology, ICCN 2007

6th Workshop on Compact Modeling, WCM 2007

NSTI Nanotech Ventures 2007

Nanotech for Investors 4th Bi-Annual Summit

2007 TechConnect Summit

Cleantech 2007 Proceedings Editors:

Matthew Laudon
mlaudon@techconnect.org

Bart Romanowicz
bfr@nsti.org

TechConnect
Boston • Geneva • San Francisco

TechConnect
One Kendall Square, PMB 308
Cambridge, MA 02139
U.S.A.

The papers in this book comprise the proceedings of the 2007 Cleantech Conference and Trade Show, Cleantech 2007, Santa Clara, California, May 23-24 2007. They reflect the authors' opinions and, in the interests of timely dissemination, are published as presented and without change. Their inclusion in this publication does not necessarily constitute endorsement by the editors, TechConnect, or the sponsors. Dedicated to our knowledgeable network of willing and forgiving friends.

ISBN 1-4200-6382-0

Additional copies may be ordered from:

CRC Press
Taylor & Francis Group
an informa business
www.taylorandfrancisgroup.com

6000 Broken Sound Parkway, NW
Suite 300, Boca Raton, FL 33487

270 Madison Avenue
New York, NY 10016

2 Park Square, Milton Park
Abingdon, Oxon OX14 4RN, UK

Printed in the United States of America

Cleantech 2007 Proceeding Editors

VOLUME EDITORS

Wade Adams
Rice University
Daniel L. Laird
Sandia National Laboratories

Matthew Laudon
TechConnect
Bart Romanowicz
Nano Science and Technology Institute

Producing Sponsor

TechConnect

Co-Producing Sponsors

MIT·Stanford·UC Berkeley Nanotechnology Forum

Nano Science and Technology Institute

Press Sponsor

Business Wire

Commercialization Partner

BusinessWeek

Sponsors

Buchanan Ingersoll & Rooney, PC
Clean Technology International Corp.
Controlled Contamination Services, LLC
Economic and Workforce Development through the
California Community Colleges
Full Circle Solar
NanoSelect

Institute of Biophysics, Chinese Academy of
Sciences
Lawrence Berkeley National Laboratory
NanoDynamics, Inc.
Omron Advanced Systems, Inc
SOPOGY, Inc.
Tesla Motors

Media Sponsors

Nature
Science Magazine
Red Herring, Inc.
Fuel Cell Magazine
Battery Power Products & Technology

R&D Magazine
Sustainable Industries Journal
Inside Greentech
CLEAN — Soil, Air, Water

Table of Contents

Emissions, and Environmental

Enviroment, Health and Society

Novel Technologies

Industry and Policy

Photovoltaics and Fuel Cells

Nanoparticle Processes & Applications

Cleantech 2007 Program Committee

TECHNICAL PROGRAM CO-CHAIRS
Daniel L. Laird   *Sandia National Laboratories*
Wade Adams   *Rice University*

ORGANIZING CO-CHAIRS
Kitu Bindra   *Buchanan Ingersoll & Rooney*
Matthew Laudon   *Nano Science and Technology Institute*
Bart Romanowicz   *Nano Science and Technology Institute*
Shaym Venkatesh   *OMRON*

TECHNICAL PROGRAM COMMITTEE
Paul Alivisatos   *Lawrence Berkeley National Laboratory*
Lawrence Dubois   *SRI International*
Dan Kammen   *University of California, Berkeley*
Jerry McNerney   *US Congress*
Nitin Parekh   *PARC*
Wendy Pulling   *Pacific Gas & Electric Company*
Dan Rastler   *EPRI*
John Sylvia   *Texas Pacific Group*
Loucas Tsakalakos   *GE*
Case P. van Dam   *University of California Davis*

BUSINESS & INVESTMENT COMMITTEES
3i
Analog Devices
Arrowhead Research
Atomic Venture Partners
BASF
BASF Venture Capital
Battery Ventures
Boston University
Buchanan Ingersoll & Rooney PC
Cabot Corporation
Cargil Investments
Chevron
Cisco Systems
Council of Scientific & Industrial Research, India
DaimlerChrysler
Dorsey and Whitney
Draper Fisher Jurvetson
DuPont
Eastman Kodak Company
Emerald Technology Ventures
Ford Motor Company
Garage Ventures
General Catalyst Partners
Genzyme
Georgia Tech
Goodrich
Greenberg Traurig
Harris & Harris Group
Hewlett-Packard Company
Honda
Honeywell
IBM

Imperial College London
In-Q-Tel
Intel
Intel Capital
Massachusetts Technology Transfer Center
Medtronic
Merck
Mohr, Davidow Ventures
Motorola
Motorola Investments
MSBi Capital
Nanodimension
New York University
North Bridge Venture Partners
Northwestern University
Novartis
Pfizer
Procter & Gamble
Rice University
Sanyo
Stanford University
UBS
UC San Diego
UCLA
University of California
University of Florida
University of Geneva, CH
University of Illinois
University of Michigan
University of Minnesota
University of Pittsburgh
University of Washington
University of Wisconsin-Madison
Vista Ventures
Washington University
Wilmer Hale
Yale University

CONFERENCE OPERATIONS MANAGER
Jennifer Rocha   *TechConnect*

Energy Balance Development in a Cogeneration with Biogas for H₂ Production by Catalytic Reforming

Irma Paz Hernández Rosales[1], Arturo Fernández Madrigal[1], Luz García Serrano[2]

[1]Centro de investigación en Energía, UNAM, Apto. Postal 34, 62580 Temixco, Mor. MEXICO
[2]Universidad Autónoma Metropolitana – Av. San Pablo Núm. 180, Col. Reynosa Tamaulipas,C.P. 02200, México D.F.
[1]Tel. (0155)56229705, Fax (0155)56229742, iphr@cie.unam.mx

ABSTRACT

The main objective of this research was to carry out technically a mass and energy balances in an energy cogeneration plant with cattle excrement having biogas + natural gas as a fuel. It was used like raw material for hydrogen production.

Also to evaluate the economic pre-feasibility of the steam reforming plant using GN and Biogas + GN mixture fuels. An analysis of an electric energy cogeneration plant was carried out; this plant used biogas and GN. Based on these fuels, an industrial significant scale for this research was fixed, as a consequence an installed power of 75 MW was selected. Mass and energy balances were carried out over this installed power.

1 INTRODUCTION

Hydrogen can be produced in big amounts starting from primary energy sources such as fossil fuels (coal, petroleum or natural gas), different intermediaries (refinery products, ammonia and methanol) and alternative sources like biomass, biogas and waste materials. Water steam reform of natural gas represents approximately three quarter of the total hydrogen production. The process is based on the water steam reaction and the high- temperature methane over a catalyst. Other gases that have hydrocarbons are also suitable for the hydrogen production, such as different gases (biogas) coming from the biomass and waste anaerobe fermentation. [1,2,3]

A project that has the hydrogen production by the gas reforming, as a product of the biomass decomposition, is the example of the cogeneration plant in Tizayuca Hidalgo. The object of this plant is to obtain biogas, organic fertilizer and a water treatment plant. However it is also possible to obtain hydrogen by catalytic reform starting from biogas obtained from the cattle excrement wastes; as a consequence the biogas plant energy balance

will be perform, this balance will start from the plant process that will have the following stages: cattle excrement accumulation, anaerobe treatment digestores, liquid/solid separation, biogas production, motors and thermal recovery to give us electricity and heat. These results will permit us to determine by the biogas reforming the hydrogen quantity that was produced. It can be used as a fuel for the cogeneration equipment motors.

2 BIOMASS

In this research the following **Buswell** equation for the excrement biological reaction was used:

$$C_n H_a O_b + (n - a/4 - b/2) H_2O = (n/2 - a/8 +b/4) CO_2 + (n/2 + a/8 - b/4) CH_4 \qquad (1)$$

In the natural process the production is 60-70 by 100 CH_4 and 40-30 by 100 CO_2. It is constant basically for each type of biomass. This final gaseous product of the reaction is commonly named biogas. [7]

3 STEAM REFORMING

Steam reform is the common method to produce gases enriched in hydrogen. Methane, main component of the natural gas, reacts with steam according to the following equilibrium [8,9,10]:

$$CH_4 + H_2O \longrightarrow CO + 3H_2 \qquad (2)$$

3.1 Reforming by Natural Gas, Biogas (methane) and GN + Biogas mixture

This section shows that starting from choosing the steam reforming process for hydrogen production, the energy balances will be used starting from three possibilities of fuels to know about the hydrogen production. It will be possible with the characteristics of the plant and of the reformer.

Fig 3.1 shows three possibilities of fuels in the steam reformer using the Tizayuca plant diagram. In this figure the steam reformer is added to produce hydrogen starting from three possibilities of fuels to develop material and energy balances; table 3.1 gives the calculation of the percentages that will be used in the RV.

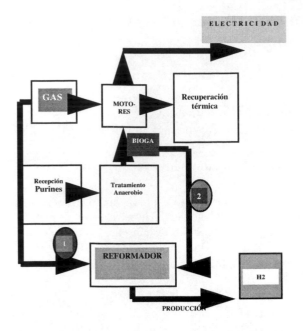

Fig. 3.1 Diagram of the different fuels for feeding RV to produce hydrogen in the Tizayuca Plant.

Table 3.1 Percentages of the compositions fuels.

	GN	Biogás	Mezcla Biogás + GN
Composición original (ton/d)	10.14	70.2	80.34
Composición original (m³/d)	13170	58,500	71670
%	18.38	81.62	100

4   MASS AND ENERGY BALANCE RESULTS

We consider a global mass and energy balance of the steam methane reforming plant in a stationary state.

1. Reforming (molar)

$$CH_4 + 2H_2O \longrightarrow CO_2 + 4H_2 \qquad (3)$$

$$\Delta H_R = 15.8 \left[\frac{MJ}{kg_{CH4}} \right]$$

2. Combustion (molar)

$$CH_4 + 2O_2 \longrightarrow CO_2 + 2H_2O \qquad (4)$$

$$\Delta H_R = -55 \left[\frac{MJ}{kg_{CH4}} \right]$$

When the mass and energy balance is performed for the steam methane reforming plant, the following reactions in a mass base are (kg).

1) $1CH_4 + 4O_2 \longrightarrow 2.75CO_2 + 2.25H_2O$
2) $3.48CH_4 + 7.83H_2O \longrightarrow 9.57CO_2 + 1.74H_2$

$4.48 CH_4 + 4O_2 + 5.58H_2O \longrightarrow 12.32CO_2 + 1.74H_2$
$\Longleftrightarrow 1CH_4 + 0.89O_2 + 1.25 H_2O \longrightarrow 2.75CO_2 + 0.39H_2$

The one estequiometricamente molar is to:

$$1CH_4 + O.44O_2 + 1.12H_2O \longrightarrow 1CO_2 + 3.12H_2$$

4.1   Balance analysis

When we had the composition of the excrement reaction, we developed the mass and energy balance where we obtained, after developing a mass and energy balance in a stationary state, the following global process result expressed in molar terms (volumetrical): the first and third hypotheses gave the same result, 3.120 mol of hydrogen and 2.184 mol of hydrogen with regard to the second one. In conclusion biogas can be a fuel for the steam reforming process that can produce the same hydrogen quantity; also it can be produced using 100% of GN.

According to the excrement quantity used in the plant, it will be required a double increase of the quantity. As a result the steam reforming plant will work with 100% of biogas as a fuel. It is also important to mention that this biogas can be richer with regard to its calorific heat if different excrement mixtures are used such as excrements of pigs, cows and organic wastes. This is an important fact that can be part of later researches.

5 ECONOMIC EVALUATION

It was possible to calculate the following economic indicators, VPN and TIR because of the balances.

Table 5.1 Economic indicators that were obtained for the steam methane reforming process.

	RV	RV
	100% GN	70%Biogás + 30%GN
VPN (miles de US$)	$885.292	924.557
TIR	152%	152%

The balanced generation cost is the quotient of VPN cost and VPN of electricity production.

Costo nivelado de generación US$/kW	RV	RV
	100% GN	70%Biogás + 30%GN
	4.066	1.539

The one that emerged for biogas + GN mixture is 2.64 times smaller, regarding to the generation balanced cost.

5.1 Economic evaluation analysis

By the economic evaluation we have that the net present value VPN is bigger than zero in both cases, this is because of the investments costs, therefore they are part of the same magnitude. In conclusion the project is profitable. The internal rate of rebate is bigger than the deduction rate (9%); it is totally possible to obtain great benefits using biogas (methane) as a fuel in the steam reformer.

The main difference in the project profitabilities is because of the changeable costs. Using biogas (methane) as a fuel only the water process costs, cooling water and electricity have an influence, the raw material costs are reduced in a 70% because only 30% of GN is used.

6 CONCLUSION

It is technically feasible to operate a combine biotechnological plant of hydrogen production with the biogas properties and the mass and energy balances under the used conditions.

The biogas is generated starting from a solid waste quantity (excrement); when it is used as a fuel in a steam reformer we get the following mass and energy balance percentages for the hypotheses used in this work:

- First hypothesis: 100% of GN we obtain 99% of hydrogen.
- Second hypothesis: 100% of biogas we obtain 70% of hydrogen.
- Third hypothesis: a mixture of biogas + 30% of GN we obtain 99% of hydrogen.

It is feasible to use a biogas + GN mixture with regard to the economic evaluation since the hydrogen production costs by the steam methane reforming technology using as a fuel the biogas + GN mixture is cheaper; from the order of **US$0.1994 kg H_2**, in comparison with the hydrogen production costs using GN of the order of **US$0.469 kg H_2** as a fuel. As a result we get in both cases an annual production of **29.683.500** kg/year of H_2.

The result obtained for the biogas + GN mixture regarding to the generation balance costs is = **US$1.539/kW** and the one of GN is =**US$4.066/kW**. The first one is 2.64 times smaller.

This result proves that it is attractive, for the businessman, to invest in this type of combined biotechnological plants.

REFERENCES

[1] Gaudemark, B. and Lynum, S. Hydrogen production from natural gas without release of CO_2 to the atmosphere. Florida, USA, Proceedings of the 11th World Hydrogen Energy Conference, 1996, p.511-523.

[2] Manuales sobre energía renovable: Biomasa/ Biomasa.Users Network (BUN-CA). -1 ed. -San José, C.R. : Biomass Users Network (BUN-CA), 2002. p. 42 - 54

[3] Houghton J. T. Climate Change 2001: The Scientific Basis. Contribution of Working Group I to the Third Assessment Report of the Intergovernmental Panel on Climate Change. Cambridge Univ. Pr., 2001, p.125-175.

[4] IPCC. Climate Change 1995: Scientific-Technical Analyses of Impacts, Adaptations, and Mitigation of Climate Change. Cambridge University Press, Cambridge, UK and New York, NY, USA, 1996, p. 235-267.

[5]   FAO-WETT. Wood energy information in Africa: Review of TCDC Wood Energy country reports and comparison with the regional WETT study. Rome, Food and Agriculture Organization, 2002 of the UN: 61.

[6]   OLADE (organización latinoamericana de energía).Biogás energía y fertilizantes a partir de desechos orgánicos. México 1994, p.7-65.

[7]   Antonio Alonso Concheiro y Luis Rodríguez Viqueira. Biomasa, Alternativas Energéticas CoNaCyT y el Fondo de Cultura Económica 1985. p.175

[8]   Biogás.
http://www.ingenieroambiental.com/Biogas
http://www.roseworthy.adelaide.edu.au/
        ~pharris/biogas/beginners.html
http://www.fao.org/ag/aga/agap/frg/Recycle/biodig/
        manual.htm
http://www.hcm.fpt.vn/inet/~recycle

[9]   Borda Bremen. Biogas plants, building instructions. German Appropriate Technology Exchange, Germany, 1998. p. 225-236

[10] Information and advisory service on appropriate technology. Biogas Digest. Biogas Basics. 2004. Volume I.
http://www5.gtz.de/gate/id/Download.afp?PubNam e=../publications/BiogasDigestVol1.pdf, 12 de abril de 2004.

[11] Evaluación de mezclas de estiércol de bovino y esquilmos vegetales para obtención de biogás por fermentación anaeróbica. Informe IIE/FE-A2/12. Instituto de investigaciones Eléctricas. Cuernavaca, Mor., México. Junio, 1979.

[12] Information and advisory service on appropriate technology. gtz project. *Biogas Digest. Volume II. Biogas – Application and Product Development.* http://www5.gtz.de/gate/id/Download.afp?PubNam e=../publications/BiogasDigestVol2.pdf, 12 de abril de 2004

[13] Gavaldá Oguiu E. Viabilidad técnico-económica del aprovechamiento energético del biogás producido por codigestión de purines de cerdo. Aplicación a una granja de 5000 cerdos situada en Catalunya. Proyecto de final de carrera, Escuela Técnica Superior de Ingenieros Industriales de Catalunya, Universidad Politécnica de Catalunya. 2000.

[14] German federal ministry for economic cooperation and development (gtz). Naturgerechte technologien, bau- und wirtschaftsberatung (tbw) GmbH : Anaerobic Processes for the treatment of Municipal and Industrial Wastewater and Waste. An Overview.

[15] Ministerio de Economía, Dirección General de Política Energética y Minas. Planificación y desarrollo de las redes de transporte eléctrico y gasista 2002-2011.
http://www.mineco.es/transporteelectricoygasista, 15 de julio de 2000

Cleantech 2007, ISBN 1-4200-6382-0

Extracting unburnt coal from black coal fly ash

M. Kusnierova[*], P. Fecko[**], V. Cablik[**], I. Pectova[**], N. Mucha[**]

[*] Institute of Geotechnics of Slovak Academy of Sciences, Kosice,
Watsonova 45, 043 53 Kosice, Slovak Republic, kusnier@saske.sk
[**] Faculty of Mining and Geology, VSB-Technical University of Ostrava, Czech Republic
17. listopadu 15, 708 33 Ostrava – Poruba, Czech Republic, e-mail: peter.fecko@vsb.cz

ABSTRACT

Black-coal fly ash from Kosice Power Plant in Slovakia contains between 22 and 25% of unburnt coal, which is, strictly speaking, a useful component. Studying optimal separation methods of unburnt coal from the examined sample of fly ash, three methods of dry mechanical screening on sieves, counterflow air classification, dry and wet gravity separation and flotation have been verified. The combination of dry mechanical screening and flotation appears to be the most efficient, through the application of which we retrieved coal concentrate with ash content below 10%, which is applicable in the process of power generation in power plants.

Keywords: fly ash, unburnt coal, separation methods

1 INTRODUCTION

Fly ash is the finest fraction of waste from combustion of fossil fuels, which gets intercepted in the thermal power plant separators. Out of the total volume of power-engineering waste from fossil fuel combustion it forms approximately 80%. In fact, fly ash is a heterogeneous material formed by morphologically different particles with various physical, chemical, mineralogical and technological properties, which are affected by both the quality of combusted coal and the technological process of own combustion.

Many types of fly ash are used untreated in a limited number of production technologies, especially in the building industry. On the other hand, fly ash often contains a range of useful components, as stated in Table 9, whose world-wide utilization does not exceed even 10 % of the overall production of this waste.

More economical and complex utilization of fly ash could be achieved through benefication methods focusing on the concentration and retrieval of the individual useful components. The objective of our research was to verify the possibility to separate the useful component of **unburnt coal** and thus increase the efficiency of utilization of the power-producing potential of the examined fossil fuel.

The research dealt with black coal fly ash from a heating plant supplying heat for the city of Kosice (Slovak Republic, Europe).

Useful component	Separation methods	Utilization methods
unburnt coal	sizing, flotation	power generation
cenospheres	sizing	filling agents, etc.
Fe-minerals: magnetite, maghemite, hematite	electromagnetic separation	Fe production, medium solids for gravity separation
Ti minerals: perovskite, rutile, ilmenite, ilmenorutile	electromagnetic separation, flotation, biohydrometallurgy	Ti production
Al minerals: boehmite, hydrargillite	bio and hydrometallurgy	Al production
Al-Si minerals: zeolite group	not stated	production of sorbents
Si minerals: quartz, cristobalite	not stated	Currently, separation and separate utilization is economically unacceptable
Ca minerals: anhydrite, gypsum	not stated	production of building materials
amorphous phase	partial concentration possible by gravity methods, hydrothermal alternation	production of synthetic zeolites, building materials
Trace elements	hydrometallurgy	e.g. production of Ge, etc.

Table 1. An overview of useful components contained in fly ash, methods of their separation and utilization.

2 MATERIAL AND METHODS

2.1 Characteristics of the examined fly ash

Grain-size and chemical composition in terms of distribution of majority constituents into the grain-size classes of fly ash.

Sample	Grain size [mm]	Mass yield [%]	Ignition loss [%]	SiO_2 [%]	Al_2O_3 [%]	Fe_2O_3 [%]
KOSICE	over 0,5	0,9	15,8	48,1	15,3	8,9
	0,1-0,5	13,7	35,5	38,0	11,4	5,7
	0,04-0,1	62,8	23,5	40,3	13,4	10,4
	-0,04	22,6	15,7	27,5	20,9	13,2

Table 2. Distribution of the majority elements in the grain-size classes of the examined sample of black coal fly ash.

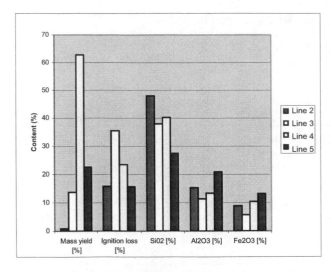

Figure 1. Distribution of the monitored elements in power-engineering fly ash of Kosice heating plant; Line 2: +0.5 mm, Line 3: 0.1-0.5mm, Line 4: 0.04 – 0.1mm, Line 5: -0.04 mm.

Morphological properties of the examined fly ash

The powder dispersion of fly ash contains mainly spherical particles (cenospheres) (B), allotriomorphic particles of crystalline constituents of fly ash (A) and porous particles of unburnt coal (C).

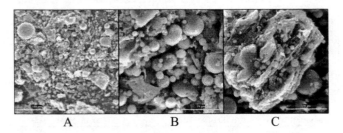

Figure 2. Morphology of fly ash particles.

Phase composition of the examined fly ash

The examined fly ash contained approximately 80% of amorphous material in which especially aluminosilicate vitrain phase prevailed and partly unburnt coal. From the crystalline constituents there were quartz, cristoballite, feldspar, graphite, hematite, magnetite, montmorillonite, bayerite, corundum, and mullite.

The easiest benefication operation is sizing, where in case of polymineral mixtures it is possible, in some cases, to obtain a certain concentrated constituent of fly ash. In the experiments we used the methods of dry mechanical screening on sieves and counterflow air classification on Alpine screens.

To concentrate the unburnt coal, classical froth flotation was also used, applying the collectors of Montanol, Flotalex and depressant (water glass). The experiments were carried out on a laboratory flotation machine VRF-1 with a flotation cell volume of 1l.

Results

Mechanical screening respects only one property of all the fly ash constituents, i.e. particle size. The results of the distribution of the monitored majority elements into the individual grain sizes are given in Table 2 and Figure 1. It is apparent that unburnt coal with approximate 20% mass yield gathers in the material with grain size 0.1-0.5mm, in which the ignition loss is 35.5%.

Air classification

Air classification of polymineral mixtures of various densities of the individual constituents is often accompanied by a partial classification effect. Air classification was applied on the sample of sized fly ash of –0.1mm grain size. The sample contained unburnt coal ($\rho = 1.2$ g.cm^{-3}), quartz, Fe minerals (maghemite and hematite) ($\rho = 4.2$-5.3 g.cm^{-3}), and amorphous phase, in which prevailed Si-Al vitrain material in the form of cenospheres ($\rho = 0.3$-0.8 g.cm^{-3}) and allotriomorphic particles ($\rho = 2.6$-3.2 g.cm^{-3}). The classification was done under the following conditions: ρ: 0.65 and 1.2 g.cm^{-3} and dividing size: 0.04 and 0.07 mm.

D	ρ	Product	Mass yield	Ignition loss	Fe₂O₃	SiO₂	Al₂O₃
(mm)	(g.cm⁻³)		(%)	(%)	(%)	(%)	(%)
0.04	0.65	1	18.00	19.25	8.97	47.28	14.00
		2	82.00	23.74	7.73	40.33	15.00
0.04	1.2	1	24.00	22.16	7.82	45.77	13.87
		2	76.00	22.13	8.61	40.97	15.77
0.07	0.65	1	9.40	16.75	9.04	47.22	14.46
		2	90.60	23.62	7.79	41.68	15.51
0.07	1.2	1	10.00	18.10	8.48	46.59	14.25
		2	90.00	24.30	7.80	40.86	14.88

Table 3. The results of air classification experiments with black coal fly ash.

It is apparent from the results given in Table 3 that in the process of air classification of fly ash, a minimal classification effect was manifested, which is documented by the results of unburnt coal accumulation. What are surprising are the results of Fe contents, where a significantly higher level of concentration was expected. As obvious from the following analyses, displayed in Figures 3 and 4, Fe occurs in the examined fly ash largely in the form of incrustation of various thicknesses on the spherical particles formed by vitrain aluminosilicate mass. The given results also imply that even if there were certain theoretical prerequisites for reaching the classification effect in the process of air classification, practical experiments did not confirm them.

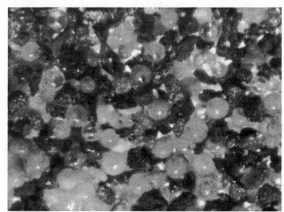

Figure 3. Oversize fraction of air classification with the division of d=0.04 mm and dividing density of 0.650 g.m^{-2}.

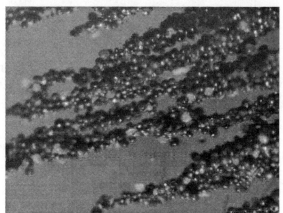

Figure 4. Agglomeration of magnetic particles in the magnetic lines of force from the product in Figure 3.

Flotation

The flotation experiments were carried out with the samples of examined fly ash with grain size of 0.1-0.5mm, -0.1mm and deslimed fraction of –0.02-0.1mm. For flotation we selected two classical flotation regimes making use of a collector and a depressant of aluminosilicate constituents of fly ash material. The flotation experiments were implemented under the following conditions: pulp density 100g/l, collector dose (Flotalex, Montanol) 500g/t, and an alternative dose of depressant (Na$_2$SiO$_3$) 2000g/t in the form of 1% solution. Froth products - concentrates of unburnt coal were fractionally sampled after 5 min (K1), 5-10 min (K2), and 10-15 min (K3). The final product after 15 minutes of flotation was final tailings.

The flotation process was as follows: 1/- agitation of water glass and the sample for 2 min, 2/- admixture of collector (agitation for 1 min) 3/- flotation for 5 minutes and concentrate sampling + admixture of collector in the dose of 500g/t, 4/- flotation for 5 minutes and concentrate sampling + admixture of collector in the dose of 500g/t, 5/- flotation for 5 minutes and concentrate sampling + admixture of collector in the dose of 500g/t, 6/- filtration of the froth and final product + drying + weighing + determination of ash content.

The results of the individual flotation exams in terms of ash yield are given in Figures 5, 6, and 7.

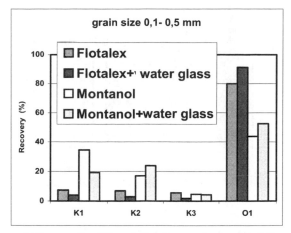

Figure 5. The results of fly ash flotation, grain size 0.1-0.5 mm.

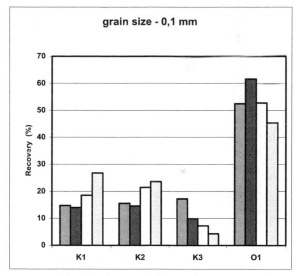

Figure 6. The results of fly ash flotations, grain size -0.1 mm.

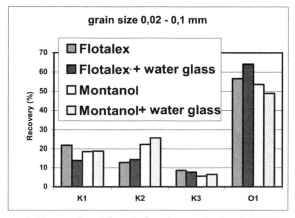

Figure 7. The results of fly ash flotation, grain size 0.02-0.1 mm.

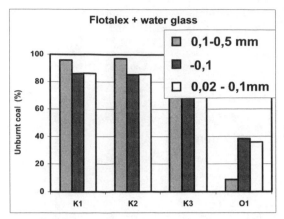

Figure 8. The best results of fly ash flotation, according to the individual fractions.

The results of the individual flotation tests in terms of ash yield are given in Figures 5, 6, and 7. The best results of flotation of the individual floated samples in terms of unburnt coal yield, See Table 4 and Figure 8, confirm that the best and also practically significant results were obtained in the flotation of material of 0.1-0.5mm grain size using Flotalex collector and depressant. The content of unburnt coal in tailings was 8.75 %, which makes the material useful for the building industry. However, the ash content in the concentrate was 16.3% and therefore, refining flotation was applied, which provided unburnt coal concentrate with ash content of 9.8%.

Floated material	Products of flotation			
	K1	K2	K3	O
0,1-0,5 mm				
ε unburnt coal (%)	95.92	97.09	98.23	8.75
− 0,1mm				
ε unburnt coal (%)	86.07	85.43	90.11	38.38
0,02-0,1 mm				
ε unburnt coal (%)	86.2	85.63	92.25	35.92

Table 4. The best results of flotation tests of the individual fly ash grain size fractions.

3 CONCLUSION

The presented results of experiments confirmed that there is a potential retrieval of coal concentrates utilizable as secondary power-producing material provided that the examined fly ash is treated applying combined processes of dry mechanical screening and following flotation of sizing product of 0.1-0.5 mm grain size, in which unburnt coal accumulates. The possible fly ash treatment shall increase the level of the power-producing potential of coal and thus decrease the volume of produced waste.

REFERENCES

[1] P. Fecko, M. Kusnierova et. al., Fly ash, VSB-Technical University of Ostrava, ISBN 80-248-0836-6, 2005
[2] F. Michalikova, L. Florekova and M. Benkova, Vlastnosti energetickeho odpadu – Popola, Vyuzitie technologii pre environmentalne nakladanie. TU Kosice, ISBN 80-8073-054-7, 2003
[3] V. Cablik, M. Kusnierova, et. al. Possibilities of Ti and Al recovery from Fly ash of heating plant Pisek a.s., Proceedings: 6th Conference on Environment and Mineral Processing, VSB-Technical University of Ostrava, p. 777-782, Part II, ISBN 80-248-0072-1, 2002
[4] P. Fecko, V. Cablik, M. Kusnierova, Possibility of Recovery Al and Ti from Fly Ash. Proceedings: International Mining and Environment Congres "Clean Technology": Third millennium challenge, p. 27-32, Lima – Peru, 1999

Bacterial Desulphurization of Brown Coals

P. Fecko, M. Kusnierova, M. Safarova, V. Cablik, I. Pectova

Faculty of Mining and Geology, VSB-Technical University of Ostrava,
17. listopadu 15, 708 33 Ostrava – Poruba, Czech Republic,
tel. +420 596993575, e-mail: peter.fecko@vsb.cz

ABSTRACT

The objective of the paper was application of bacterial leaching on 2 brown coal samples from bore S 187 (CV) from locality Mine CSA Most. Based on the results of bacterial leaching and petrologic analyses of the given samples, it is possible to state that the individual samples are very similar, they contain significant shares of clay materials and pyrite is predominantly represented in a framboidal form, which intergrowths into a massive form. Applying bacterial leaching it is possible to remove from 32 to 38% of total sulphur and from 30 to 32% of pyritic sulphur from the coal; better results are obtained eliminating sulphate sulphur, i.e. up to 50% desulphurization.

Keywords: Bacterial leaching, *Thiobacillus ferrooxidans*, Desulphurization, Coal macerates

1 INTRODUCTION

Desulphurization of the fuels is a problem, which even with big effort of humans beings it has not been solved so as to the stop SO_2 introduction to air. It is well known that the high amount of sulphur in coal has adverse influence on its utilization and that contributes to environment contamination as the acid rain (Fecko et al. 1991, Fecko et al.1994, Fecko 1997). There are more evidences that the Czech Republic occupies one of the first positions in Europe in environment contamination by sulphur oxides, because our main source of energy is the combustion of solids fuels. In the amount of industrial emissions per km^2 the Czech Republic is second in the world (25 t km^{-2}) and in the amount of emissions per inhabitant is third (0.2 t). Combustion depends on the concentration of sulphur which reaches somewere around 12%.

2 MATERIALS AND METHODS

2.1 Distribution of sulphur in coal

Sulphur is presented in its organic and inorganic forms in coal. Free sulphur is presented only sporadically. Pyrite and marcasite are presented in large quantities but their proportions vary. Sulphates, mainly gypsum, originated primarily during the carbonisation process and secondarily during the weathering of pyrites. Organic sulphur is mainly bound to the structures of dibenzenethiophene, benzenethiophene and thiols. Pyrite is presented in the epigenetic and syngenetic forms. Syngenetic pyrite was formed during the first phase of the coal forming process and that is, why it is interspersed within the coal substance. Epigenetic pyrite is geologically younger, therefore it acts as a filling material in the joints and fissures. It is less intergrown within the coal substance, it forms larger crystals and is easier to eliminate by suitable coal processing methods. A whole range of chemical techniques with the potential to separate pyrite from coal was reviewed, and also some microbiology techniques. The chemical techniques for desulphurization employ relatively non-specific reactions, functioning at high temperatures and pressures, and with relatively high consumption of chemicals. The methods of microbiology have the advantage of very specific reactions in a simple reactor, at ambient temperature and normal pressure, but they need a longer leaching time. The aim of this work is confirmation the viability of bacterial leaching applications on the samples of black coal from the different localities. [2, 3, 4]

2.2 *Thiobacillus ferrooxidans*

They are aerobic, chemoautotrophic organisms that require atmospheric oxygen and inorganic compounds with CO_2 for production of their new biomass. They are non-sporulating gram negative bacteria. In appearance they are sticks of average 0.5-0.8 μm with length 0.9-1.5 μm. *Thiobacillus ferrooxidans* can get energy by oxidation of sulphur components and by oxidation of Fe^{2+}. Optimal temperature for this bacteria is: 28-30 °C and optimal pH is n the range 1.8-2.2. [5]

2.3 Principle of pyrite oxidation by *Thiobacillus ferrooxidans* bacteria

Bacterial leaching can be either direct or indirect. In the direct interaction the surface of the minerals is occupied by bacteria and metal sulphides are attacked by enzymatic oxidation.

Direct leaching oxidation of pyrite is best described by the following equations:

$$FeS_2 + 3.5\,O_2 + H_2O \rightarrow FeSO_4 + H_2SO_4 \tag{1}$$

$$2\,FeSO_4 + 0.5\,O_2 + H_2SO_4 \rightarrow Fe_2(SO_4)_3 + H_2O \tag{2}$$

The bacteria create a leaching agent in indirect bacterial leaching. This agent oxidises sulphidic minerals. In acid solutions Fe^{3+} is the active agent. The solubility of pyrite can be written as:

$$Fe_2(SO_4)_3 + FeS_2 \rightarrow 3\ FeSO_4 + 2S^0 \qquad (3)$$

$$2S^0 + 3O_2 + 2\ H_2O \rightarrow 2\ H_2SO_4 \qquad (4)$$

2.4 Methods of bacterial leaching

For the bacterial leaching testwork a 10-litre airlift glass bioreactor patterned on the research of Deutsche Montan Technologie Company – Essen. (Bayer, 1988) was used. For bacterial leaching 2 brown coal samples from bore S 187 (CV) from Mine CSA Most (followed as S 23 and S 31) was used. After sterilization of the reactor, the prepared samples of coal were placed in it together with the medium 9K without $FeSO_4$. After one hour of mixing and homogenising of the suspension, 1,000 ml of the bacterial culture *Thiobacillus ferrooxidans* was introduced into the reactor. Clean bacterial cultures of *Thiobacillus ferrooxidans* from the Czech-Slovak Collection of Micro-organisms in Brno were used for the test programme. The concentration of introduced bacteria in the process was 10^9 in 1 ml bacterial solution. The bioreactor was connected to the aquarium water aerator, which supplied the reactor with air. The air was cleaned in washers in 1 M H_2SO_4 solution to have more moisture and to remove airborne bacteria. Mixing of 5% suspension was using air. pH was measured by laboratory pH-meter "RADELKIS" and the pH was kept at the optimal value 1.8 – 2 during the whole experiment (28 days) to prevent formation of unwanted jarosite. The temperature was kept in the range 26-30 °C during the whole experiment.

During the leaching, after 1, 2, 3, and 4 weeks, samples of approximately 50 ml were taken from the bioreactor for analysis, was filtered on a Buchner funnel where the filtrate and the filter cake were separated, the content of total sulphur and its separate forms were determined in the filter cake. The cake was washed in 100 ml of 1M HCl and in 200 ml distilled water before the determination. [1]

2.5 Petrographic analysis

Measurement condition

Maceral analysis was performed on the grains according to CSN ISO 7404-2 [6] and CSN ISO 7404-3 [7] using a Zeiss NU-2 microscope. Planimetric analysis was evaluated in oil immersion, with refractive index $n_D = 1,515$, and the length wave $\lambda = 546$ nm, temperature t = 20 °C, objective enlargement 32x.

Determination of sulphur

Sulphur was determined at the Research Coal Institute in Ostrava Radvanice on a LECO SC 132 instrument, directed by microprocessor with detection of SO_2 using infrared detector. Different forms of sulphur were determined by thermal phase analyses at temperatures 420 °C (organic sulphur) 820 °C (pyrite sulphur) and 1,370 °C (total sulphur) with constant programme conditions. The sulphate sulphur was calculated.

3 RESULTS OF BACTERIAL LEACHING AND MINERALOGICAL-PETROLOGICAL ANALYSIS AND DISCUSSION

3.1 Sample No. 23

Results of petrologic analysis of Sample No. 23 – input

The maceral subgroup of humotelinite was chiefly formed by ulminite and textoulminite. There was less textinite than ulminite (euulminite). In the grains of humotelinite, corpohuminite was quite frequent, mineralized by clay minerals, formed by fine stripes or they were finely scattered in the coal mass. Its total abundance was 20.2 %.

Humocolinite was quite abundant (34.2 %). Gelinite grains were often disrupted by fissures of retreat. In corpohuminite there were liptinite group macerals – especially sporinite, less of suberinite.

Humodetrinite was represented by 11.4 %.

The maceral group of liptinite was mainly made up by sporinite and cutinite, less by suberinite. There was very little alginite. The total abundance of this group was 5.3 %. However, in some grains it accumulated to such a degree that monomaceral microlithotype of liptite was formed.

The maceral group of inertinite was very rare – planimetrically mainly funginite and fragments of fusinite were traced. Inertinite was represented by 2.6 %.

Inorganic impurities were frequent and formed predominantly by pyrite (17.5 %) and clay minerals (8.8 %). Pyrite prevailed as framboidal, which accumulated in places (Figure 1) but was often finely scattered in the coal mass.

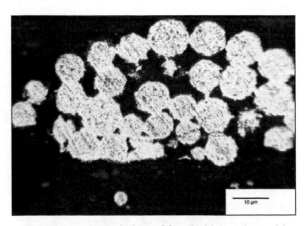

Figure 1. Accumulation of framboidal pyrite and its partial intergrowth with transition into massive pyrite.

Sulphur	Prior to leaching	Post leaching	Degree of desulphurization
	(%)	(%)	(%)
S_{total}	4.73	3.20	32.35
$S_{pyritic}$	2.93	1.99	32.08
$S_{organic}$	0.92	0.80	13.04
$S_{sulphate}$	0.88	0.41	53.41

Table 1. Results of bacterial leaching.

The results of bacterial desulphurization of the sample (See in the Table 1) imply that after one-month leaching it is possible to eliminate approximately 33 % of total sulphur, 32 % of pyritic sulphur and 53 % of sulphate sulphur from the sample. The elimination of organic sulphur is only about 14 %.

Results of petrologic analysis of Sample No. 23 post bacterial leaching

The maceral group pf huminite was represented by the subgroup of humotelinite, which made 24.2 %. Textinite was rarer than ulminite or textoulminite. There are heavily jellified cell walls.

The humocolinite subgroup was made up by gelinite and corpohuminite, in roughly comparable abundance; its total abundance was as high as 40 %.

The subgroup of humodetrinite was less frequent; its total abundance was 9 %. Attrinite and densite occurred in approximately same amounts. Some grains were mineralized by clay minerals or by finely intruded pyrite.

The liptinite group was primarily represented by sporinite and cutinite (Figure 2). In certain grains there was high occurrence of alginite. The total abundance of this group was 6.1 %.

Figure 2. Cutinite with framboidal pyrite, partly leached and corpohuminite.

Inertinite was represented by funginite and fragments of fusinite. It only totalled to 1 %.

Inorganic impurities were mainly represented by clay minerals, namely by 14.1 %. They formed separate grains or mineralized the individual macerals. Pyrite chiefly occurred as framboidal, it often mineralized textinite – or as finely scattered it mineralized other macerals. Its total abundance was 6.2 %.

3.2 Sample No. 31

Results of petrologic analysis of Sample No. 31 - input

The humotelinite maceral subgroup was predominantly formed by textoulmite and ulminite. There was less textinite and it was often mineralized by finely intruded pyrite. Its total abundance was 26.2 %.

Humocolinite made 17.6 % and it was represented especially by gelinite, which had frequent fissures of retreat and corpohuminite (flobafinite). Some grains were heavily jellified and they were labelled as colinite.

Humodetrinite was represented by attrinite and densinite. Its abundance was 14.2 %.

There was quite high occurrence of liptinite, namely as sporinite and cutinite. Alginite was rare. In places, liptinite transited into liptite. The sample contained 7.5 % of liptinite.

The macerals of inertinite were made up by fragments (fusinite) or funginite occurred. Both macerals were often mineralized by pyrite.

There was 14.0 % of pyrite. It was often massive or finely intruded. However, framboidal pyrite prevailed (Figure 3). Clay minerals mainly formed separate grains or mineralized humotelinite or humodetrinite. As a rule, humocolinite occurred without mineralization. The content of clay minerals in the sample was 15.8 %.

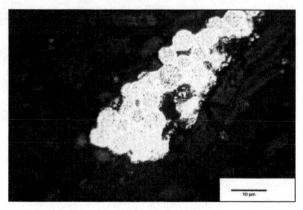

Figure 3. Ingrowths of framboidal pyrite with transition into massive pyrite.

Bacterial leaching results

It is apparent from the results of bacterial leaching (Table 2) that after one-month leaching there is an approximate 38 % desulphurization of coal, desulphurization of pyritic sulphur is about 29 % and sulphate sulphur desulphurization is about 53 %. Interestingly, there is high desulphurization of organic sulphur, i.e. as high as 54 %.

Sulphur	Prior to leaching (%)	Post leaching (%)	Degree of desulphurization (%)
S_{total}	8.15	5.08	37.67
$S_{pyritic}$	5.38	3.80	29.37
$S_{organic}$	1.83	0.84	54.10
$S_{sulphate}$	0.94	0.44	53.19

Table 2. Results of bacterial leaching.

Results of petrologic analysis of Sample No. 31 post leaching

The maceral group of humotelinite was represented by textinite, which was very often mineralized by pyrite and ulminite. Its total abundance was 17.6 %.

The group of humocolinite was characteristic for its abundance as high as 47.0 %. Gelinite slightly prevailed with typical fissures of retreat over corpohuminite.

Humodetrinite was represented by mere 5.2 %. Both macerals of attrinite and densinite were present in almost identical amounts.

The liptinite maceral group was frequent in sporinite and cutinite. Suberinite (it was overtopped) and alginite were rare. The total abundance of liptinite was 9.0 %.

Inertinite was formed by rare fragments of fusinite and funginite.

Inorganic impurities were abundant in pyrite – 15.0 % and clay minerals – 6.2 %. Pyrite was finely intruded in textinite; clay minerals filled the cell space of funginite. In places, the occurrence of hematite was traced (Figure 4).

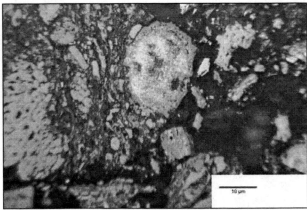

Figure 4. Secondary mineral post pyrite leaching – hematite.

4 CONCLUSION

The objective of the paper was application of bacterial leaching on 2 brown coal samples from bore S 187 (CV) from locality Mine CSA Most. Based on the results of bacterial leaching and petrologic analyses of the given samples. It is possible to state that the individual samples are very similar, they contain significant shares of clay materials and pyrite is predominantly represented in a framboidal form, which intergrowths into a massive form. Applying bacterial leaching it is possible to remove from 32 to 38 % of total sulphur and from 30 to 32 % of pyritic sulphur from the coal; better results are obtained eliminating sulphate sulphur, i.e. up to 50 % desulphurization. Desulphurization results could be improved under the following conditions:

- applying bacterial cultures of *Thiobacillus ferrooxidans* pre-adapted to a given coal type
- prolonging the leaching time
- applying mixed bacterial cultures of *Thiobacillus ferrooxidans* and *Thiobacillus thiooxidans*
- applying bacterial leaching post removal of clay materials, large shares of which in the sample complicate leaching, especially of pyritic grains.

ACKNOWLEDGEMENTS

This work were supported by research project No.1M06007 and project KONTAKT ME118 of Ministry of Education of the Czech republic.

REFERENCES

[1] M. Beyer, Mikrobielle Kohleentschwefelung, Forschung Bericht, Essen. 1988

[2] P. Fecko, Bacterial desulphurization of coal from Lupeni in Romania. VSB-Technical University of Ostrava, Department of Mineral Processing, Czech Republic. 1996

[3] P. Fecko, et al., Desulphurization of coal from Northern Bohemian brown coal basin by bacterial leaching, FUEL, Vol 70, pp.1187-1191. 1991

[4] P. Fecko, Bacterial Desulphurization of coal from open pit Mine Michal. 5th.Southern Hemisphere Meeting on Mineral technology, INTEMIN, Buenos Aires, pp.249- 252. 1994

[5] P. Fecko, Bacterial desulphurization of coal from Sokolov lignite basin, XX.IMPC, Vol.4., pp.573-584, Aachen. 1997

[6] CSN ISO 7404-2. Methods for the petrographic analysis of bituminous coal and anthracite. Part 2: Method of preparing coal samples. Cesky normalizacni institut, Praha, 16 p. 1995

[7] CSN ISO 7404-3. Methods for the petrographic analysis of bituminous coal and anthracite. Part 3: Method of determining group composition. Cesky normalizacni institut, Praha, 12 p. 1997

Amorphous Diamond Solar Cells

James C. Sung[*,1,2,3], Tun-Jen Hsiao[1], Ming-Chi Kan[1], Michael Sung[4]

Address: KINIK Company, 64, Chung-San Rd., Ying-Kuo, Taipei Hsien 239, Taiwan, R.O.C.
Tel: 886-2-2677-5490 ext.1150
Fax: 886-2-8677-2171
E-mail: sung@kinik.com.tw

[1] Kinik Company, 64, Chung-San Rd., Ying-Kuo, Taipei Hsien 239, Taiwan, R.O.C.
[2] National Taiwan University, Taipei 106, Taiwan, R.O.C.
[3] National Taipei University of Technology, Taipei 106, Taiwan, R.O.C.
[4] Advanced Diamond Solutions, Inc., 351 King Street Suite 813, San Francisco, CA 94158, U.S.A.

ABSTRACT

Silicon based photo voltaic panels are the mainstream of solar cells. However, there are intrinsic limitations of such devices for replacing power plants. The low conversion efficiency (<20%) would make it higher cost than generators run by fossil fuels. The repeated high temperature processes for extracting, purifying, and ingoting silicon actually wasted more electricity that may be recovered by solar cells. Additionally, silicon is not radiation hard so its crystalline structure will be gradually damaged by the bombardment of UV photons from the sun.

All these drawbacks can be overcome by using amorphous diamond as the electron generator in vacuum. Amorphous diamond contains the highest amount of atoms per unit volume (about $180/nm^3$). Most carbon atoms are tightly held by distorted tetrahedral (sp^3) bonds. Because these distortions are all different, all carbon atoms have unique electron energies. The presence of numerous discrete energy states allows valence electrons to be excited by absorbing minute energies. Amorphous diamond is the only blackbody material that can absorb and emit low energy photons (e.g. 10 microns wavelength IR) and phonons (e.g. 100 C heat). If amorphous diamond is exposed in vacuum, the highly excited electrons can be emitted readily. Thus, amorphous diamond can be the most efficient thermionic material with the capability to convert up to half of sunlight's energy into electricity.

Experimental data has confirmed that amorphous diamond can absorb more energy than silicon when exposed to sunlight. The conversion efficiency for generating electricity can be boosted by narrowing the vacuum gap between the energy input anode and electron output cathode. If the gap is reduced to about one micron, the conversion efficiency can be much higher than that of silicon solar cells.

Keywords: amorphous diamond, solar cell, thermionic emitter, super entropy material

1 THE POLLUTED SOLAR CELLS MADE OF CRYSTALLINE SILICON

The mainstream solar cells are made of crystalline silicon. Although the cost for generating one KwH is 5-10 times of conventional fuel fossil power plants, crystalline silicon solar cells are thought to be environmentally friendly. However, crystalline silicon is typically produced from highly pure quartz that may be extracted from beach sand. The fact that white beach sand can only be produced after tens of million years weathering and washing implies that quartz is very stable. Consequently, it would require very high temperature (e.g. 1700! to reduce quartz to form metallurgical grade silicon. While silicon is reduced, 1.5 time of carbon dioxide is released that contributes to the green house effect. Moreover, the metallurgical grade silicon must be purified, again at high temperature. Eventually, the pure silicon is melted above 1410!) to form either ingot for single crystal or casting for polycrystals. In all these high temperature process steps, high amount of electricity may be needed.

The crystalline silicon for making solar cells once get to a price tag of about $100/Kg. For making solar cells, about 0.5 mm thick, including cutting kerf, may be needed, so one kilogram of silicon may produce about 1 m^2 to exposed under sun. As the solar constant is about 1 Kw/m^2, and the average total efficiency one day of silicon solar cells as measured around the clock is less than 10%, we may expect that the power for one kilogram silicon material is about 0.1 Kw. Assuming that the power cost is about $0.05/KwH, the above 0.1 Kw solar panel would require about 2000 KwH or energy to produce. To pay back the bill, the silicon solar cells may have to run 20,000 hours or about 2.3 years! During this period, the governments have to subsidize the inefficiency of the electricity produced by such pollution causing silicon. Consequently, it is desirable to constructing solar cells by using more environmental friendly materials. One of such materials is amorphous diamond.

2 SUPER-ENTROPIC MATERIAL

Amorphous diamond appears to be contradictory term, like liquid crystal or glassy metal. Amorphous means non-crystalline and diamond implies crystalline. However, this terminology is meaningful because unlike silicon that forms only sp^3 bonds, i.e. diamond structure, carbon may form either sp^2 (graphitic) or sp^3 (diamond) bond. Although there is one form of amorphous silicon, there can be at least two forms of amorphous carbon, so amorphous diamond can be distinguished from amorphous graphite, and together they are amorphous carbon.

Amorphous diamond is formally known as tetrahedral amorphous carbon (tac), it is really a diamond-like carbon (DLC) that contains no non-carbon impurities (e.g. H). Amorphous diamond is essentially a chaotic carbon mixture with distorted sp^2 and sp^3 bonds. As such it possesses both metallic character of conductive graphite and semiconductor character of insulating diamond. Moreover, as each carbon atom is unique in its electronic state that is determined by the degree of distortion of its bonds. Hence, amorphous diamond contains numerous discrete potential energy for electrons. In fact, amorphous diamond may have the highest density of atomic occupancy (1.8×10^{23} per cubic centimeter) that is several times higher than ordinary materials (e.g. about four times of iron atoms or silicon atoms). Thus, amorphous diamond has the highest configuration entropy for both atoms and valence electrons.

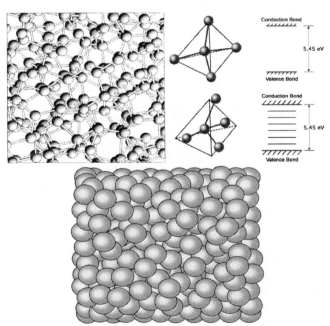

Fig. 1: The high atomic density and the unique way of distorting bonds for each atom makes amorphous diamond the material with the highest configurational entropy. As a result, amorphous diamond has the densest electron states that are discrete. This is in contrast of all materials that have either overlapped electron orbitals, as in the case of metal, or few discrete electron states, as in the case of semiconductors or insulators.

Amorphous diamond can be conveniently deposited by PVD methods, such as by sputtering or arc depositions. Due to the low temperature (<150 !) of deposition, amorphous diamond can be coated on most materials including metal, semiconductor, or even polymers. This flexibility makes amorphous diamond useful for many applications.

Due to such high configuration entropy of valence electrons, amorphous diamond is capable to advance electron energy by absorbing small increments of energy, such as by converting thermal energy (lattice vibration) to potential energy (electron state). If amorphous diamond is exposed in high vacuum (e.g. 10^{-6} torr), the energy state may be higher than vacuum state so amorphous diamond my emit electrons simply by heating. Because amorphous diamond has the highest discrete electronic states, it is the most thermionic material known.

3 THERMIONIC EMISSION

Even without high vacuum, amorphous diamond coated nickel electrodes of cold cathode fluorescent lamps (CCFL) used for back lighting can reduce significantly the turn-on voltage.

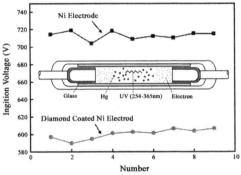

Fig. 2: The reduction of ignition voltage of CCFL by coating nickel electrodes with amorphous diamond.

Due to its exceptional ability to increase the potential energy of electrons by absorbing heat, amorphous diamond coated metal is highly thermionic.

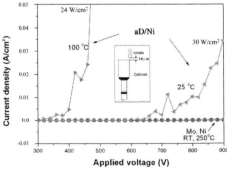

Fig. 3: The great enhancement of emitted current from amorphous diamond coated nickel electrode in CCFL by modest heating.

Based on the above thermionic effect, the effective work function, i.e. the activation energy for electron emission in vacuum, can be lower than 1 eV. This is the lowest of all materials that have effective work function higher than 2 eV. Due to this unique thermionic character, amorphous diamond can emit more current than even carbon nanotubes (CNT) that have a high work function, but with a nanometer radius to enhance the electrical field. Moreover, as amorphous diamond is solid in content, it can emit electrons at a much lower temperature than CNT that will concentrate electricity on the skin of the hallow structure. In fact, the skin of each CNT will burn out when the current exceeds 20 ! A. As a result, CNT devices are not reliable (e.g. Samsung's CNT front panel display or Iljin's CNT backlight). In contrast, amorphous diamond field emission can be highly robust. This is particularly suited for display or backlight applications.

4 AMORPHOUS DIAMOND SOLAR CELL

The merit of amorphous diamond to convert either light or heat to electricity can be applied to solar cell panels or thermal electrical generators. For example, amorphous diamond was over covered coated on indium tin oxide (ITO), the transparent electrical conductor that was coated on a glass substrate. This panel was separated from another ITO coated glass by glass bead spacer. The gap was sealed around and the space was pumped down to high vacuum (10^{-6} torr). This panel was exposed to a xenon light that irradiated a spectrum with an energy output similar to solar constant (AM1.0 or $0.1 W/cm^2$). An external bias was applied and the electric current was monitored. It was demonstrated that the current increased substantially when light shone through or when amorphous diamond was heated up.

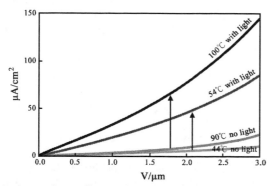

Fig. 4: Amorphous diamond can enhance electron emission in vacuum by light absorption and by thermal agitation.

When the applied bias was gradually reduced to zero, the current enhanced by xenon lamp was not dependent on the bias. Hence, the field emission could be triggered by sunshine directly without adding an external bias. However, the current density was too low to be useful as a solar panel unless the vacuum gap could be reduced further from 7 microns.

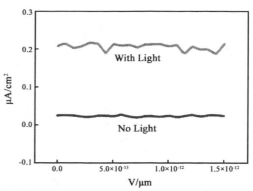

Fig. 5: The field emission became spontaneous when the external bias was reduced to zero.

When the vacuum was back filled with iodine, the current could be generated noticeably without applying the external field. This current was further increased by sensitizing amorphous diamond with a light absorbent dye. But even so, it was still low with the conversion efficiency.

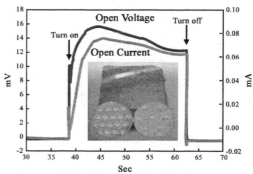

Fig. 6: The photo-electric effect of amorphous diamond when exposed to a xenon lamp (AM1.0) of about 0.1 W/cm^2. In the experiment, the vacuum gap was back filled with liquid electrolyte of iodine.

The amorphous diamond solar cell was also constructed with a silicon layer in a hybrid design. In this case, no vacuum was needed. In one example, the nitrogen doped amorphous diamond was coated on boron doped silicon substrate. This hybrid design showed a dramatic increase in photo electricity, much higher than using a vacuum gap or back filed with a liquid electrolyte.

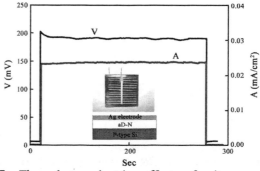

Fig. 7: The photo electric effect of nitrogen doped amorphous diamond coated on boron doped silicon layer.

When a monochrometer was used to filter the broad spectrum of the xenon lamp and the photocurrents of amorphous diamond coated silicon and silicon solar cell were measured and compared, the former exhibited a much higher value. Upon cooling the semiconductors to a cryogenic temperature of liquid nitrogen (70 K), the electrical current generated by light increased. Moreover, the increase was higher with shorter wavelengths (i.e. with higher energy). However, amorphous diamond coated silicon showed much higher cooling enhancement and also the blue shift of the peak wavelength.

The above observation demonstrated that amorphous diamond could absorb light and generate electricity more effectively than silicon. This is particularly attractive as amorphous diamond is radiation superhard and it would not be susceptible to UV damage. Amorphous silicon solar cells have the advantages of thin film and low cost, but the aging problem of UV damage makes it less useful so more costly crystalline silicon plates are used as solar panels. It would appear that amorphous diamond coating of amorphous silicon solar cells can boost both the energy conversion efficiency and the longevity of the service.

Due to the high electric resistance of amorphous diamond, the electric current it generated was dissipated as heat so the final output of electricity was significantly reduced. The dampening effect was greatly reduced by cooling the device at liquid nitrogen temperature. However, alternative methods by channeling out electricity rapidly once it is formed may also be effective to preserve the electricity generated. One example is to coat amorphous diamond on amorphous silicon layers and stack them together. Due to the thinness (e.g. 100 nm) of the amorphous diamond and amorphous silicon, the light absorbed by each layer can generate electricity independently. The electricity can then be channeled out readily due to the short distance of travel to reach the electrode. The combined electricity would retain most of the energy derived from sunlight.

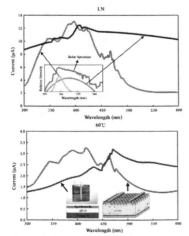

Fig. 8: The projected photo electrical current as a function of optical wavelength. Note that amorphous diamond could convert more current with IR irradiation than pure silicon. Moreover, the cooling enhancement of both energy and intensity was more obvious.

In summary, amorphous diamond has the highest density of discrete electronic states. This unique feature makes amorphous diamond particularly useful as energy converters, such as field emitters, solar cells, thermal generators, radiation coolers, and heat absorbers.

REFERENCES

[1] Ming-Chi Kan, Jow-Lay Huang, Chien-Min Sung, Ding-Fwu Lii, Kuei-Hsien Chen, "Field Emission Characteristics of Amorphous Diamond", Journal of the American Ceramic Society, 86, 9 (2003) p.1513-1517.

[2] Ming-Chi Kan, Jow-Lay Huang, Chien-Min Sung, Ding-Fwu Lii, Bao-Shun Yau, "Field Emission of Micro Aluminum Cones Coated by Nano-Tips of Amorphous Diamond", Diamond and Related Materials, 12 (2003) p.1610-1614.

[3] Ming-Chi Kan, Jow-Lay Huang, Chien-Min Sung, Kuei-Hsien Chen, Bao-Shun Yau, "Thermionic Emission of Amorphous Diamond and Field Emission of Carbon Nanotubes", Carbon, 41 (2003) p.2839-2845.

[4] Ming-Chi Kan, Jow-Lay Huang, Chien-Min Sung, Kuei-Hsien Chen, "Thermally Activated Electron Emission from Nano-Tips of Amorphous Diamond and Carbon Nano-Tubes", Thin Solid Films, 447-448 (2004) p.187-191.

[5] B. R. Huangal, C. S. Huang, J. T. Tan, Chien-Min Sung, R. J. Lin, "The Field Emission Properties of Amorphous Diamond Deposited on the Cu Nanowires", 2004 Asian CVD-III, The 3rd Asian Conference on Chemical Vapor Deposition.

[6] B. R. Huang, C. S. Huang, C. F. Hsieh, Chien-Min Sung, "The Field Emission Properties of Samarium/Amorphous Diamond Field Emitters", 2004 Asian CVD-III, The 3rd Asian Conference on Chemical Vapor Deposition.

[7] Ming-Chi Kan, Jow-Lay Huang, Chien-Min Sung, Kuei-Hsien Chen, Bao-Shun Yau, "Stability of Field Emission Characteristics of Nano-Structured Amorphous Diamond Deposited on Indium-Tin Oxide Glass Substrates", New Diamond and Frontier Carbon Technology, 14, 4 (2004) p.249-256.

[8] Chien-Min Sung, Kevin Kan, Michael Sung, Jow-Lay Huang, Emily Sung, Chi-Pong Chen, Kai-Hong Hsu, Ming-Fong Tai, "Amorphous Diamond Electron Emission for Thermal Generation of Electricity", NSTI-Nanotech 2005, Anaheim, California, U.S.A., p.193-196.

[9] Chien-Min Sung, Kevin Kan, Michael Sung, Jow-Lay Huang, Emily Sung, Chi-Pong Chen, Kai-Hong Hsu, "Amorphous Diamond Electron Emission Capabilities: Implications to Thermal Generators and Heat Spreaders", ADC/NanoCarbon 2005, Chicago, Illinois, U.S.A.

[10] Chien-Min Sung, "Amorphous Diamond Materials and Associated Methods for the Use and Manufacture Thereof", U. S. Patent 6,806,629.

[11] Chien-Min Sung, "Amorphous Diamond Materials and Associated Methods for the Use and Manufacture Thereof", U. S. Patent 6,949,873.

Integrated Multistage Supercritical Technology to Produce High Quality Vegetable Oils and Biofuels

G. Anitescu, A. Deshpande, P. A. Rice and L. L. Tavlarides

Department of Biomedical and Chemical Engineering, Syracuse University, Syracuse, NY, 13244, USA

ABSTRACT

A multi-step integrated technology to produce vegetable oils and biodiesel (BD) is proposed, documented for technical and economic feasibility, and preliminarily designed. The 1st step consists of soybean oil extraction with supercritical (SC) fluids. The 2nd step is designated to transform the soybean oil into BD by transesterification (TE) with SC methanol/ethanol. The degradation of glycerol to light fuel products makes this method simple and cost effective by eliminating costly glycerol separation steps. Part of the BD is in-situ consumed by a diesel engine which, in turn, provides the mechanical power to pressurize the system as well as the heat of the exhaust gases for the extraction and TE steps. Different versions of this system can be implemented based on the main target: oil and BD production or diesel engine applications. The efficiency of the combustion and cleaner emissions render the proposed technology attractive for the transportation sector with only TE step needed to provide fuels to engines.

Keywords: integrated supercritical technology, supercritical extraction of soybean oil, supercritical transesterification, biodiesel, economic analysis.

1   INTRODUCTION

Unfortunate developments in crude oil prices combined with disruptions in supply and refining capacity have given new urgency to the development of alternative, renewable fuels that not only reduce the reliance on petroleum feedstock, but also result in reduced emissions of airborne pollutants. Biodiesel (BD) derived from plant/animal fats is one such fuel that is under a great deal of consideration.

We propose to produce BD with a continuous integrated process, starting with soybean oil (SO) extraction in multiple extractors and cascaded separators. The extraction step should use three parallel batch extractors with supercritical (SC) carbon dioxide (CO_2), and CO_2 recovery stages. The BD production step will use the extracted oil or oil from other sources and SC methanol/ethanol in a continuous tubular reactor without catalysts and multi-step separations. The process will use a diesel generator to provide power and heat for the upstream processes.

If implemented, this integrated system has potentially a wide range of applications. First, it is very attractive for farmers by reducing their dependency of high petroleum fuel prices. Many farmers which crop soybeans already use SO as a fuel. However, SO is not an acceptable fuel because of the damage it renders to the engines, especially on long-term usage. Secondly, this system, without the SC extraction (SCE) step, should have great potential to be implemented on transportation vehicles since only a small reactor is required to be inserted between the injection pumps and common rails of diesel engines. All the needs for the TE reaction to proceed are already available with these engines, including the free heat of the exhaust gases.

The overall goal of this study is to develop an industrial technology for continuous SO extraction coupled to BD production by using a SC alcohol as shown in Figure 1. Specific objectives for this stage of the project were to: (i) develop a conceptual design with key steps established for the integrated process, (ii) determine optimal parameters for oil extraction and the reaction parameters and yields at lab-scale, and (iii) perform an economic analysis for easy comparison with current methods.

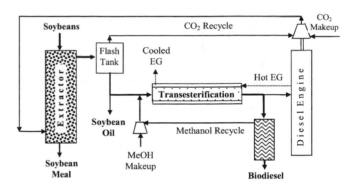

Figure 1: *Integrated Multistage SC Technology System* to produce vegetable oils and biodiesel (schematic).

2   CONCEPTUAL DESIGN

2.1   Supercritical Oil Extraction

For the *extraction step*, the reported data corroborated with our own experimental information show that SCE of SO requires high pressures (beyond 400 bar, with our results suggesting the lower limit of these ranges) and relatively high temperatures (>80°C, with our optimum yield at ~100°C) for acceptable high efficiencies of oil recovery when pure SC CO_2 is used. However, under either of these conditions, appropriate process optimization can render this method competitive with the conventional extraction with liquid hexane [1]. Also, we expect that a further suitable selection of a co-solvent such as propane/butane can lower the extraction conditions of *P-T* parameters with significant cost savings. Alcohols and pentanes/hexanes should be excluded due to their interaction and extraction of the proteins. Furthermore, a battery of three coupled extractors makes this step quasi continuous with positive results on cost cuttings.

2.2 Oil Transesterification

In the *transesterification step*, BD is prepared from SO and SC methanol/ethanol. The conditions at which bulk glycerol decomposed in our reaction system have been determined as 350-400°C and 100-300 bar. Here we assume that the reaction is pseudo first order and the reverse reactions are not significant. These assumptions were made as for the design proposed we are using a stoichiometric alcohol to oil molar ratio of 3:1. We will also carry out experiments to explore and gather more information about the aforesaid issues and refine, reevaluate, and reassess our design. However, when a TE process in an industrial scale flow reactor will be implemented, a possible separation of glycerol and BD phases will be less complicated than the acid/base catalyzed TE process.

In conclusion, a conceptual design for the integrated steps of extraction and TE, including a diesel engine as a source of energy, is schematically shown in Figure 1. Updated modifications will be implemented as research progresses in the next stages of this project.

3 EXPERIMENTAL RESULTS

Using a high pressure view cell as a diagnostic aid, it was found that the oil miscibility with methanol is crucial to attaining a high conversion to BDF. Experiments have been carried out with this apparatus to test its capabilities regarding SO-MeOH miscibility and phase transitions under different *P-T-composition* conditions. The results for equal volumes of SO and MeOH heated in the view cell in a batch mode are shown in Figure 2. In this case, SO and methanol are only partially miscible from room *P-T* conditions up to near 400°C, while beyond 400°C and 100 bar, one homogeneous phase has been obtained. These experiments are very important because complete TE reactions can be achieved, in a reasonable short time, only under complete miscibility conditions.

Figure 2 shows selected photographs of phase transitions associated with TE reactions between SO and methanol (3 mL each) heated from ambient T-P conditions to 420°C and 120 bar. The images (seen down the columns) are for SO(L)-MeOH(L) (#1), SO(L)-MeOH(L)-MeOH(V) (#2-3), SO-BDF-MeOH(L-SC) (#4-5), and BDF-MeOH(SC) (#6).

Figure 3 is a GC-MS comparison between commercial BD and our TE products which shows similar compositions with the latter containing fewer high molecular byproducts.

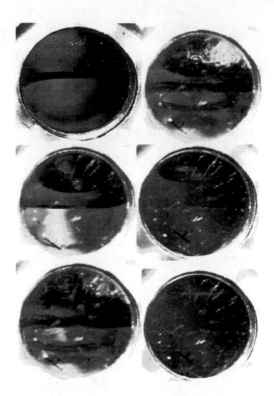

Figure 2: Photographs of SO-MeOH phase transitions with TE reactions.

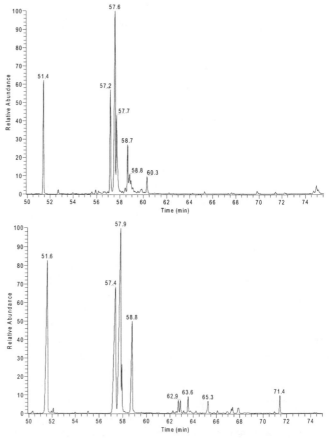

Figure 3: A GC-MS comparison between commercial BD (bottom) and TE products from our experiments (top).

4 ECONOMIC ANALYSIS

Economic estimates for different system configurations were performed to select the most desirable option based on maximum benefits.

4.1 Biodiesel Production

Experiments carried out in our laboratory have shown that reacting stoichiometric quantities of methanol or ethanol with SO at 400°C, 100 bar for about three minutes gives nearly complete conversion of the SO to BD (esterified fatty acids). Under these conditions, in contrast to the BD production under traditional conditions (1 bar, 65 °C, catalysts), the reaction rate is 60 times faster and very little glycerol is left in the final reaction products. Instead, other compounds are formed resulting from the breakdown of the glycerol. Not all of these products have been qualitatively and quantitatively determined, but it is expected that many of them can be used directly as part of the BD product. This would simplify the separation process greatly and additional BD fuel may be produced. In addition, under near stoichiometric SO:MeOH ratio, essentially all of the alcohol quantities will be consumed, eliminating the need for alcohol separation and recycling.

Thus, the process for producing BD from SO becomes quite simple: a reactor with a heat source to produce BD and one or two distillation columns to separate water and other impurities from the product. In our preliminary process design, the reaction product stream is assumed to contain only two glycerol degradation products, diglycerol and glycerol formal. These two products have been identified by GC/MS analysis of the product stream from the laboratory reactor. The process flow diagram is shown in Figure 4.

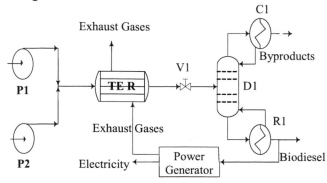

Figure 4: Process flow diagram for biodiesel production coupled with power generation. (P1, P2: oil and MeOH pumps; TE-R: transesterification reactor in a heat exchanger; V1: pressure reduction valve; D-1: distillation column; C1: Condenser; R1: Reboiler).

The process consists of pumping the methanol and SO to the TE reactor at 100 bar (the reactor operating at 400°C) and a distillation columns operating at 1 bar with steam driven reboilers and condensers cooled by cooling water. The distillation column separates the water and most of the reaction byproducts from the methyl esters. A 2nd column can be added, as needed, to separate byproducts from the water. The process is unique in that the exhaust gases from a diesel engine are used to provide heat for the reaction.

The basis for the base-case design was the TE of 2100 gal/day of SO with 269 gal/day of methanol to produce 2300 gal/day of BD. The plant was assumed to operate 24 hours/day for 346 days a year. The thermodynamic properties of SO, the methyl esters of the three main fatty acid chains of SO (palmitic, linoleic, and oleic) and the assumed byproducts (diglycerol and glycerol formal) were estimated from group contribution methods using CHEMCAD process simulation software. Vapor-liquid equilibrium compositions were estimated by CHEMCAD using the Soave-Redlich-Kwong equation of state.

CHEMCAD was used to design all of the process units except the TE reactor and diesel engine/generator set. The reactor was sized based on the residence time used in the laboratory experiments (3 minutes) to obtain nearly complete conversion of the SO, although extra volume was allowed to provide sufficient heat transfer area for heating the reaction mixture with the diesel engine exhaust gases. While designing distillation column D1, a small percentage of the reaction byproducts were allowed to remain in the bottoms product (BD) since we believe that some byproducts can be included into the BD. The percentage of the byproducts that can be included in the BD fuel is presently unknown and a somewhat arbitrary 1.62 % was included, which possibly could be higher. The remaining byproducts in overheads stream of D1 were separated from water in D2. Some of the byproducts may be valuable to be recovered and sold as separate products, although no credit is taken for these products in the preliminary estimate of the cost of manufacturing (COM). The diesel engine was sized to provide a sufficient amount of heat for the reaction and uses part of the produced BD (11.3 mol %) as fuel for the engine. The power from the engine is used to generate electricity that is to be sold to the local grid for $0.10/kWh.

The capital and total manufacturing costs for the process were estimated using the methods outlined in Turton et al. [2] and Ulrich [3]. All pieces of equipment with their bare module costs and the total fixed capital investment (FCI) have been considered in 2006 dollars.

The module cost for each piece of equipment includes the purchased cost of equipment and cost of piping, instrumentation, structural supports, etc. associated with the equipment as well as the labor cost for installation and indirect costs such as freight, insurance, engineering fees, and overhead. An additional 18% was added for contingency and for the contractor's fee. Carbon steel was specified for equipment exposed to temperatures less than 250°C and stainless steel was specified for equipment exposed to higher temperatures.

The raw materials costs are for the SO and the methanol. SO was assumed to be available for $2.10/gal [4] and methanol at \$0.33/gal [5]. The utilities cost was taken from Turton et al. [2]. Because no waste was generated by the proposed technology, no cost of waste treatment was considered. COM was calculated as [2]:

$$COM = 0.2860\ FCI + 2.5755 C_{OL} + 1.163(C_{UT} + C_{WT} + C_{RM})$$

The various other items in the COM, such as maintenance costs, supervisory and clerical labor, depreciation of the FCI, administration, distribution and selling, and R&D costs are estimated as fractions of the FCI, operating labor cost, utilities cost or raw materials cost. The cost of manufacture is given in Table 1, together with revenues from the sale of electricity, and the break-even cost of the net BD available for sale.

Table 1: Comparison of biodiesel cost for three options.

Option	BD Capacity (gal/day)	COM (\$/yr)	Electricity Revenue (\$/yr)	BD Cost (\$/gal)
Case 1[*]	2300	2,572,804	115,316	3.46
	4600	4,385,660	228,887	2.95
	9200	8,036,670	457,774	2.68
	12000	10,217,965	598,267	2.61
	15000	12,567,301	747,834	2.56
Case 2[*]	9200	7,908,132	-	2.59
Case 3[*]	9200	7,733,783	-	2.49

Case 1[*]: Byproducts separation and power generation;
Case 2[*]: Byproducts separation and fired heater;
Case 3[*]: Water separation and fired heater.

For the base case design of 2300 gal/day of BD, COM was estimated to be \$ 2,572,804/yr. The total electricity produced from power generator (with 40% efficiency) is 1,171,196 kWh/yr, but 18,036 kWh/yr are used to drive the methanol and SO feed pumps, leaving 1,153,160 kWh/yr available for sale. Assuming that the selling price of electricity is \$0.10/kWh, the revenue obtained from electricity would be \$ 115,316/yr. Hence the net COM was estimated to be \$ 2,457,488/yr. The total BD produced per year is 800,557 gal from which 11.4 % is used for generation of electricity by the diesel-generator; hence the remaining 88.6 % of BD available for sale is 709,774 gal/yr. Accordingly, the break-even cost of BD is (\$2,457,488/yr)/(709,774 gal/yr) or \$ 3.46/gal.

Most of the total cost of manufacturing comes from the raw materials and labor costs. Although little can be done to reduce the raw materials cost, the COM/gal of product would be significantly reduced for a larger sized plant since the number of operators would not change. Also some savings in capital cost would be realized with a larger capacity plant. Thus the COM for plant capacities of 4600 gal/day, 9200 gal/day, 12,000 gal/day, and 15,000 gal/day were calculated. As expected these led to reduced break even costs for the BD product of \$2.95/gal, \$2.68/gal, \$2.61/gal, and \$2.56/gal respectively.

A second alternative design taken into consideration was the use of a direct-fired *Dowtherm* heater fueled by the BD products to heat the TE reactor instead of exhaust gases from the diesel-generator. This option reduces the BD fuel needed for heating the reactants to only 3.0 % of the BD product, substantially reducing the cost of the BD. For a process producing 9200 gal/day of BD, the break-even cost is reduced from \$2.68/gal to \$2.59/gal.

Further reductions in cost also may be possible if more of the reaction byproducts can be used directly in the BD fuel. In the case that essentially all of the byproducts can be used and only the water must be separated from the reaction products, the quantity of BD fuel increases by 16.5 % and only a single distillation column is required. For producing 9200 gal/day of BD using this option, its cost is reduced to \$2.49/gal. This would be competitive with the current price of diesel fuel of ~\$2.60/gal. [6]

The options considered using the diesel-generator may be more attractive. This option is most suited for isolated locations when electricity is not readily available. Thus it would be competing with other diesel-generator options where the cost of power generated is much higher than \$0.10/kwh. A more appropriate number to calculate may be the cost of power generated by the diesel-generator if the BD can be sold for \$2.80/gal.

4.2 Soybean oil extraction

We are currently working on the optimized design and economics of SC-CO$_2$ extraction of SO. It is expected that the SCE cost will be significantly reduced by using a quasi-continuous extraction process with a battery of four extractors (one for loading/unloading) operated at 400 bar and 100°C. Compared to literature reported values of 600 bar and 80°C, the former conditions will lead to similar yield of SO but lower operating costs.

Acknowledgement. Funds for this study have been provided by NYSERDA (Judy Jarnefeld, Senior Project Manager) and Syracuse Center of Excellence in Environmental Systems (Edward Bogucz, Director).

REFERENCES

[1] E. Reverchon and L. Osseo, "Comparison of Processes for the Supercritical CO$_2$ Extraction of Oil from Soybeans," *JAOCS*, 71(9), 1007-1012, 1994.
[2] R. Turton, R. Bailie, W. Whiting and J. Shaewitz, "Analysis, Synthesis, and Design of Chemical Processes," Prentice Hall International Series, 1998.
[3] G. Ulrich, "A Guide to Chemical Engineering Process Design and Economics," John Wiley & Sons, 1984.
[4] http://www.cbot.com/cbot/pub/page/0,3181,1341,00.html
[5] http://www.thefeaturearchives.com/topic/Archive/ Fuel.Cells_More_Beyond_Pure_Hype.html
[6] http://tonto.eia.doe.gov/oog/info/gdu/gasdiesel.asp

8nm Cerium Oxide as a Fuel Additive for Fuel Consumption and Emissions Reduction

G. Wakefield and M. Gardener

Oxonica Materials Ltd, 7 Begbroke Science Park, Sandy Lane, Yarnton, Oxford OX5 1PF, UK,
gareth.wakefield@oxonica.com

ABSTRACT

8nm cerium oxide is surface functionalised for diesel compatibility and added to fuel at a level of 5-8ppm. Static engine tests and large scale field trials show that engine fuel consumption is reduced by 3.5-8% and emissions, particularly of black carbon particulates and unburnt hydrocarbons are reduced by >15%. Cerium oxide acts as an oxygen donating catalyst which, as a refractory ceramic, survives the fuel burn and acts to improve engine performance in two ways. Firstly, the activation energy of cerium oxide is size related and as a consequence 8nm particles act to burn off carbon deposits within the engine cylinder at the wall temperature. The engine is progressively cleaned up by cerium oxide and performance improves. Secondly, the cerium oxide acts in the gas phase during fuel burning to reduce the deposition of non-polar compounds on the cylinder wall. This indicates oxidation of fuel during the burn which in turn results in a reduction of engine deposit build up and carbon based emissions.

Keywords: catalysis, fuel, cerium, oxidation, emissions

1 INTRODUCTION

The industrialization of the Western World has resulted in a huge increase in the use of fossil fuel combustion, approximately 50% of which is vehicular based. The overall combustion reaction is superficially simple, however it is one of the least understood widely used reactions. The basic equation, the release of CO_2 and H_2O from combustion of hydrocarbons with release of thermal energy, is [1]:

$$C_mH_n + \left(m + \frac{n}{4}\right)O_2 \rightarrow m.CO_2 + \left(\frac{n}{2}\right)H_2O$$

It is generally accepted that this is a free radical mediated reaction, and that the complex series of reaction pathways may lead to incomplete fuel burn, which leads to non-optimum engine performance and a variety of environmentally harmful pollutants, included unburnt hydrocarbons, carbon monoxide and black carbon or soot. The latter is a particular problem with diesel engines [2].

The source of the 12 million tonnes of soot emitted into the atmosphere every year is primarily combustion, from both fuel and biomass burning, and although sources are localized to urban environments in the main, the distribution in the atmosphere may range up to thousands of kilometers from the source. However, the atmospheric mass is concentrated in urban environments, with up to 45% sub-100nm atmospheric particles being soot [3, 4]. There are a variety of direct and indirect effects on climate from this atmospheric soot load, which have been considered in terms of the global mean energy budget of the planet – the "radiative forcing" (positive forces a mean warming, negative forces a mean cooling). Generally, studies indicate that soot has a positive radiative forcing up to about a third that of CO_2, making soot the second most important atmospheric contributor to global warming [5]. It has been suggested that a reduction in the emission of soot may be a relatively quick "win" in the battle against global warming [6].

Rare earth elements have been used for some decades as catalysts for a variety of processes in the field of environmental treatment of exhaust gases from vehicles [7]. More specifically, cerium oxide (CeO_2) is used to increase oxygen storage/release properties of three way catalysts used in vehicular exhaust systems [8]. The efficacy of cerium oxide as a catalyst is related to its ability to undergo a transformation from the stoichiometric CeO_2 (+4) state to the non-stoichiometric Ce_2O_3 (+3) valence state via a relatively (at least in comparison with other oxides) low energy reaction. This is in turn related to the general property of fluorite oxide structures to deviate strongly from stoichiometry. Even at a loss of considerable amounts of oxygen from the crystal lattice, and the formation of a large number of oxygen vacancies as a result, the fluorite structure is retained. Such sub-oxides may readily be reoxidised to CeO_2 in an oxidising environment. Because no crystal structure phase change is involved in the supply and re-absorption of oxygen from the CeO_2 lattice, CeO_2 may be used as an oxygen storage material in catalysis via the following reaction [9].

$$2CeO_2 \leftrightarrow CeO_2 + 0.5O_2$$

Cerium oxide has found an important application in exhaust catalytic converters where harmful emissions from fuel burning are converted to harmless gases by the following series of reactions.

Hydrocarbon combustion:

$$(2x + y)CeO_2 + C_xH_y \rightarrow \left[\frac{(2x + y)}{2}\right]Ce_2O_3 + \frac{x}{2}CO_2 + \frac{y}{2}H_2O$$

Soot burning :

$$4CeO_2 + C_{soot} \rightarrow 2Ce_2O_3 + CO_2$$

NO_x reduction :

$$Ce_2O_3 + NO \rightarrow 2CeO_2 + \frac{1}{2}N_2$$

Unburnt fuel, particulates and harmful gases are reduced when cerium oxide is used as an exhaust catalyst [10].

2. CeO₂ NANOPARTICLES AS A DIESEL FUEL ADDITIVE

Cerium oxide may also be used as a catalyst by addition of the material into the fuel itself. In the case of a dense inorganic oxide such as cerium oxide the material may only be additised effectively into the fuel when the particle size is nanocrystalline (8-10nm), as shown in Figure 1, and the surface is suitably modified such that the particles remain stable in the fuel medium. If such properties are obtained then the catalyst acts to promote lower emissions and a reduction in engine fuel consumption by two principal mechanisms, more complete fuel burning and carbon deposition destruction.

In order for a catalyst to be effective when additised into the fuel three principal properties are required. Firstly, complete hydrocarbon oxidation should be promoted, secondly formation of NO_x should not be favoured, and finally the catalyst should remain thermally stable [11,12].

As has been discussed, the ability of cerium oxide to donate oxygen allows complete hydrocarbon and soot burning in principle. In practice the activation energy of the cerium oxide, i.e. the lowest temperature at which oxygen donation occurs, is a crucial factor. Although the gas temperature of a diesel engine is high (approximately 1700ºC) [13], a low catalyst 'switch-on' temperature will clearly promote a more complete fuel burn in the milliseconds during which combustion occurs [14]. The high thermal stability of cerium oxide results in CeO_2 nanoparticles re-oxidising and remaining active after enhancing the initial combustion cycle. This is demonstrated in Figure 2, in which a thermogravimetric analysis (TGA) of a 5% w/v 8nm CeO_2 package is shown in

comparison with two standard organic commercial additive packages. It is clear that all of the organic material is burnt by 400°C, whereas the CeO_2 based additive remains functional at temperatures >1000°C.

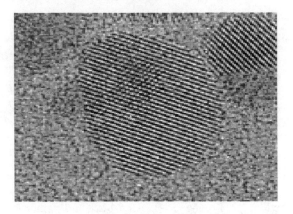

Figure 1. A TEM micrograph of 8nm CeO_2 fuel additive nanoparticles.

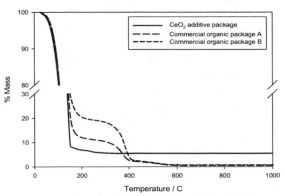

Figure 2.TGA analysis of 5% w/v CeO_2 additive package and two commercial organic additive packages.

3. EFFECT OF CeO₂ NANOPARTICLES ON INTERNAL CYLINDER DEPOSITS

CeO_2 nanoparticles are added to diesel fuel at between 5ppm and 8ppm in topical application. The effect on internal cylinder deposits is studied by analysis of deposits removed from the cylinder of a Kipor KDE2200E 2.8kW static diesel generator. The soluble fraction of the combustion chamber deposits (CCD) typically consists of a wide variety of low molecular weight hydrocarbons. Extraction of the soluble component into MeCN and subsequent separation by HPLC using a hydrophilic silica column allows a general distinction between polar and non-polar components.

A plot of the ratio of polar and non-polar soluble components is given in Figure 3. Addition of 5ppm CeO_2 to diesel fuel results in a reproducible decrease and increase of the non-polar/polar ratio as fuel is alternated between

Cleantech 2007, ISBN 1-4200-6382-0

additised and unadditised respectively. This variation indicates the effect of CeO_2 on the soluble fraction of the combustion chamber deposits is to convert non-polar to polar compounds. This results from the oxidative activity of ceria on non-polar compounds.

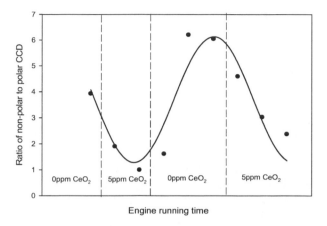

Figure 3. Variation of non-polar/polar ratio with CeO_2 additive level for 3 hour static engine cycles.

Studies of deposits formed prior to catalyst exposure show clear changes as a result of CeO_2 addition to fuel. All deposits in the cylinder are affected by the presence of ceria in the fuel. Deposits present on the head and piston surface area resulting from lubricant pyrolysis normally present in the engine as a shiny black material are converted to soft, matt black deposits easily removed by mild abrasion. Examination of areas close to the injector demonstrate that CeO_2 breaks up deposits with large grain structures being fragmented into smaller structures with some regions flaking away from the surface.

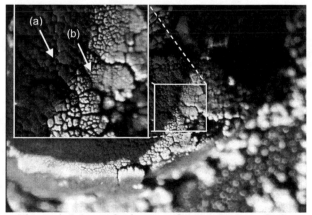

Figure 4. Injector body area after exposure to CeO_2 dosed fuel showing regions with (a) large grain deposit structures and (b) fragmented deposit structures exposed to flame

Deposits on intake and exhaust valves, shown in Figure 4, are typically present as a dense layer prior to CeO_2

exposure due to the rapid percussive movement of the valves removing loose deposits continuously during engine operation. The formation of deposits on injectors is well known to be detrimental to combustion characteristics.

4. EFFECT OF CeO_2 NANOPARTICLES ON FUEL CONSUMPTION

Accurate measurements of engine fuel consumption require an environmentally controlled engine test bed system. In these studies a Ricardo single cylinder E6 engine is used to monitor fuel consumption and emissions. Brake specific fuel consumption (BSFC) data from a single cylinder Ricardo E6 engine before and after CeO_2 addition is given in Figure 5. The engine has little carbonaceous deposit build up prior to exposure and hence the reduction in fuel consumption is primarily due to gas phase effects. Fuel consumption is reduced rapidly by 3.5% during the first two hours of exposure to 5ppm CeO_2 additive. Larger effects (7-8% fuel consumption reduction) have been demonstrated in engines following a suitable dirtying up period. Large scale field trails with up to 1000 additised public service vehicles have demonstrated fuel savings of 5-7%.

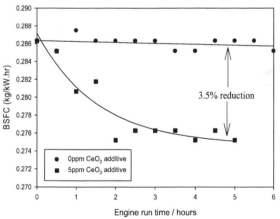

Figure 5. Fuel consumption reduction after 5ppm CeO_2 additised into diesel reference fuel

5. EFFECT OF CeO_2 NANOPARTICLES ON SOOT AND HYDROCARBON EMISSIONS

The fuel droplet is progressively vaporized once injected into the hot gases of the combustion chamber. As the fuel vaporizes from the droplet it mixes with air and spontaneously combusts. The rate of combustion during this stage – known as diffusive combustion – is limited by the rate of fuel vaporisation and oxygen mixing. If fuel vaporisation is not complete when the fuel temperature reaches a critical level then fuel pyrolysis (non-oxidation decomposition of fuel without oxygen) occurs resulting in soot formation. The particles agglomerate and, particularly as the combustion temperature is lowered at the end of combustion, hydrocarbon condensation onto the soot particles signficantly increases particle mass.

Emissions testing has been carried out on a number of engines under a variety of operating conditions. The results are summarised in Table 1, the numbers correspond to % reduction in emissions upon switching from unadditised to additised fuel.

	Peugeot 306	Peugeot 406	Cummins M11
Soot particulates	-15.5%	-19.0%	-16.5%
Hydrocarbons	-9.8%	-7.4%	-9.5%
Carbon monoxide	-6.6%	0	-9.8%

Table 1.Emissions data from fuel containing 5ppm CeO_2 nanoparticles.

The operating conditions for the engines are: Peugeot 306, standard EU urban/extra urban cycle; Peugeot 406, steady state 70kph; Cummins M11, standard speed and load conditions on an engine dynometer.

It is clear from the data that the incorporation of CeO_2 nanoparticles into diesel fuel results in a reduction in hydrocarbon, soot and CO emissions resulting from the ability of CeO_2 to act as an oxygen storage catalyst and assist in the oxidation of species resultant from an incomplete fuel burn.

6. CONCLUSIONS

The addition of CeO_2 nanoparticles into fuel at low loading results in two main effects within the engine cylinder, the net result of which is reduced fuel consumption and reduced emissions. CeO_2 particles act to enhance fuel burn and remove engine deposits thereby reducing fuel consumption of the engine. The improvement in fuel burn leads to a reduction in emissions associated with an incomplete burn.

REFERENCES

[1] G. Ertl, H. Knözinger and J Weitkamp, "Environmental Catalysis", Wiley-VCH (Weinheim), 1999

[2] A.C. Stern, R.W. Boubel, D.B. Turner and D.L. Fox, "Fundamentals of air pollution 2nd edition", Academic Press (London), 1984

[3] J. Houghton, Y. Ding, D. Griggs, M. Noguer, P Van der Linden, X. Dai, K. Maskell, C. Johnson, "IPCC (Intergovenmental Panel on Climate change), contribution of working group 1 to the third assessment report", Cambridge University Press (UK), 2001

[4] E. J. Highwood, R. P. Kinnersley, Environmental International, 32, 560-566, 2006

[5] M. Z. Jacobson, Nature, 409, 695-697 (2001)

[6] T.C. Bond and H. Sun, Environmental Science and Technology, 39, 5921-5926, 2005

[7] S. Colussi, C de Leitenburg, G. Dolcetti and A. Trovarelli, Journal of Alloys and Compounds, 374, 387-392 (2004)

[8] S. Bernal, J. Kasper, A. Trovarelli, Catalysis Today, 50, 173-443, 1999

[9] J. Kaspar, M. Grazini and P. Fornasiero, "Handbook on the Physics and Chemistry of Rare Earths", 29, 159pp, 2000

[10] A. Trovarelli, Catalysis Reviews: Science and Engineering, 38, 439pp, 1996

[11] R. Prasad, L.A. Kennedy, and E. Ruckenstein, Combustion Science and Technology, 22, 271, 1980

[12] R.L. Jones, Surface and Coatings Technology, 86-87, 127, 1996

[13] A.P. Kryukow, V.Y. Levashow and S.S. Sazhin, International Journal of Head and Mass Transfer, 47, 2541, 2004

[14] S. Logothetidis, P. Patsalas and C. Charitidis, Materials Science and Engineering C, 23, 803, 2003

Effect of Surface Hydration and Interfusion of Suspended Silica Nanoparticles on Heat Transfer

Denitsa Milanova, Xuan Wu and Ranganathan Kumar

Department of Mechanical Materials and Aerospace Engineering
4000 Central Florida Blvd.
University of Central Florida, Orlando, Florida 32816
407-823-2416 (Rnkumar@mail.ucf.edu)

ABSTRACT

Experimental results of silica nanofluids consisting of 10nm or 20nm silica particles have been performed. Particle size, zeta potential and the CHF values under different volume concentrations are provided, and agglomeration structures are seen to affect the critical heat flux of NiChrome wire immersed in a pool of water. The critical heat flux (CHF) of the wire does not increase monotonically with concentration. CHF decreases when particle concentration is increased depending on the particle shape and the hydroxylated surface of the nanoparticles.

Keywords: nanofluids, hydration layer, critical heat flux, zeta potential

In recent years, nanofluids, consisting of nanometer sized particles and fibers dispersed in base liquids, have been proven to be effective in enhancing the performance of future energy transport systems. The novel heat transfer fluids exhibit the anomalously enhanced thermal properties. In addition, they have the desirable characteristic of not settling down or clogging up the pores. All these characteristics make nanofluids promising for nanotechnology-based heat transfer applications [1-4]. Three features of nanofluids including anomalously high thermal conductivities at very low nanoparticle concentrations, strongly temperature dependent thermal conductivity and significant increases in critical heat flux (CHF) have been found till now. In addition, the Brownian motion of the nanoparticles, the microconvection around the nanoparticles, layering of liquid molecules at the particle-liquid interface, ballistic nature of heat transport in nanoparticles, and nanoparticle clustering are considered as possible mechanisms responsible for heat transfer enhancement [5-7].

Arguably, the Brownian motion of the nanoparticles and the microconvection around the nanoparticles only play a minor role [9], and the physical structure of layering of liquid molecules at the particle-liquid interface by itself can not contribute to the enhancement of heat transfer [10]. Thus, the mechanisms responsible for the heat transfer enhancement seem to be the chemical structure of solid and liquid interface and the nature of heat transfer inside the nanoparticles and between solid and liquid.

Vassallo et. al. [11]] showed as much as 200% increase in critical heat flux (CHF) over that for the base liquid in silica-water suspension. Here CHF is defined as the point at which the immersed silica wire broke. This tremendous improvement in CHF in water was partially attributed to the formation of wire coating and more nucleation sites for bubble initiation. A more recent paper by Milanova and Kumar [12] reports the effect of the negatively charged silica nanoparticle suspension in boiling conditions in water for different particle size, particle concentration, and under various pH conditions and ionic concentration. All these studies show that the chemical structure of the hydration layer on the particle-liquid interface plays a critical role in addition to the number density of the nanoparticles inside nanofluids and the surface area of nanoparticles. Since the hydration layer on the solid-liquid surface determines the shape of the nanoparticles, the aggregating structure of nanoparticles cluster and the ion density on the surface of the nanoparticle, it determines the way heat is transferred between the solid particles and the liquid.

In this paper, we will report our recent experimental results conducted by silica nanofluids consisting of 10nm or 20nm silica nanoparticles. The zeta potential, the particle size, and the CHF values under different volume concentrations are provided, and the aggregation and agglomeration structures and the effective diameters of silica nanoparticles under various volume concentrations will be discussed. In addition, the hydration layer on the silica-water interface will be presented, and its effect on heat transfer and viscosity of nanofluids will be discussed in detail.

Experimental Procedure

Silica nanofluids are prepared by diluting an aqueous nanosolution with de-ionized (DI) water at concentrations between 0.1% and 2%. An emphasis was put on lower concentrations (<0.5%) because they are of more practical importance. Higher concentrations change some physical properties of the fluid like color, density, viscosity, etc., which is undesirable and are examined only out of scientific interest. Silica nanoparticles are a commercial product of Alfa Aesar and were obtained in a suspension of two concentrations – 15% and 40%, at basic pH (10.5 and 9.5) corresponding to 10nm and 20nm particle sizes, respectively, which were confirmed by TEM analysis. The density of the as received solution is 1.3g/cm^3. The agglomeration characteristics and the diameter of silica nanoparticles at different volume concentrations are systematically determined by zeta potential and particle size analyzer. Electrophoretic Light Scattering (ELS) method and Dynamic Light Scattering (DLS) technique were adopted to characterize the potential and size of the nanoparticles in the fluid.

In addition to the particle size measurement, zeta potential also was measured. When the thickness of the diffuse layer declines, the particles exhibit near-zero zeta potential and a subsequent particle-particle contact occurs. Hence, zeta potential conveys information on the possible behavior of the particles within the solution.

Good dispersion of the solid phase was achieved by magnetic stirring for half an hour and ultra-sonicating for fifteen minutes. Viscosity readings of the prepared nanofluid were taken with a digital Brookfield viscometer before and after sonication in order to determine the effect of the ultrasound on the surface chemistry and agglomeration characteristics of nanosilica.

The heat transfer characteristics of silica nanofluid were systematically examined through pool boiling experiment, following the procedure given in [12]. A NiChrome wire (NiCr), serving as a thermocouple, was completely immersed in a glass container and current was applied at equal increments and time intervals. The temperature of the wire surface was maintained above the saturated temperature for water. The pool boiling curve was generated by plotting the heat flux q in kW/m^2 obtained from the current and resistance. The wall superheat was obtained from the resistance of the NiCr material [13]. CHF is the point at which the burning gradually progresses to a critical point when the wire breaks.

Results and Discussion

The surface chemistry and electrostatic potential of silica nanoparticles will be studied in detail because of their influence on agglomeration rate, distortion in shape, formation of an ordered second fixed layer (hydration layer). The high pH at which SiO$_2$ naturally exists implies high charge on the particles surface, mutual repulsion and attraction, and directed motion in the presence of electric current. A number of studies reveal a hydroxilated surface [14-16] with OH groups bond to SiO$_2$ skeleton. The ability of the sylanol groups to form hydrogen bonds with water molecules, i.e. adsorb water molecules at the interface will prove significant in viscosity effects and heat transfer enhancement, which will be discussed later. The size and curvature of the particles are aspects that determine the extent of hydration, dehydration, and recombining of the functional groups at the surface. The density distribution and the number of the sylanol groups determine the overall charge of the particles, their repulsion potentials, and the overall stability of the solution. Figure 2 shows a model of a charged silica nanoparticle with OH groups attached to surface silicon atoms. Bigger diameter particles intrinsically possess more sylanol groups which provide greater repulsive forces and less agglomeration. This study reveals

higher zeta potential value for 20nm-size particles (ζ_{20nm}= -27mV) compared to the one for 10nm-particles (ζ_{10nm}= -23.5mV), where both are at 0.2% concentration. At this loading, no agglomeration effect are observed based on the particle size measurement results, therefore the difference in the electrostatic potential is solely due to the particle size and curvature.

At concentrations as low as 0.1, 0.2, 0.3% the shape of nanosilica is spherical but with increase in the loading, the potential barriers weaken (fig.1A), which leads to increase in the effective diameter. The particle size with further increase in concentration doubles (from 10nm it goes up to 18nm, and 20nm for 1% and 2%). The dynamic light scattering technique gives us an accurate Gaussian distribution of the particles in the sample with a narrow variance. The shape of the particles plays an important role in the redistribution of the surface charge and TEM analysis was done in order to confirm the already obtained particle size and shed light to the change in the shape of the agglomerates (fig.1C.) As the sylanol groups are farther apart (small positive radius of curvature) they form fewer hydrogen bonds between them, which make them more readily removable at elevated temperatures. Therefore 10nm-size particles are generally expected to have lower zeta potential in absolute value. At concentrations greater than 0.4% the zeta potential is not high enough to provide repulsion and particles start to form agglomerates of two (as viewed by the TEM micrographs). At 0.5% a mutual interfusion (of 2-3particles) is observed which leads to a change in curvature from positive to negative at certain locations (fig.2A, B). The surface OH groups are brought together by the concave curvature, water is formed and retained in these crevices which results in an overall discharging. This statement is in accordance with the experimental values for the zeta potential, which was seen to absolutely increase asymptotically. This trend is followed by the particle size, which after a concentration of 0.5% is constant. Although the average diameter does not change, the shape undergoes distortions and reduction in the total surface area, which could impact the heat transfer properties of the nanofluid, as analyzed further.

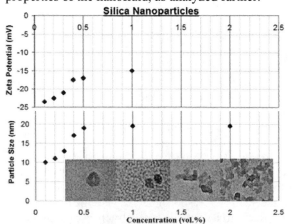

Figure 1: A. Zeta potential measurements; **B.** Particles size analysis; **C.** TEM micrographs.

Silica nanoparticles have been diluted in de-ionized water without the addition of any additives or stabilizers. The water molecules selectively orient towards the sylanol groups with their oxygen atoms towards the surface. The process of water adsorption takes place simultaneously with the redistribution of the OH groups at the interface, i.e. places where curvature effects contribute to the formation of water. It has been shown [17] that $Si_sOH:OH_2$ bond is weaker compared to $SiOH:OH_2(OH_2)$. The following interactions are given by the equations with the corresponding bonding energy:

$$Si_sOH + H_2O = Si_sOH:OH_2 \quad 6kcal\ mole^{-1} \quad (1)$$
$$Si_sOH:OH_2 + xH_2O = SiOH:OH_2(OH_2)\ 10.5\ kcal\ mole^{-1}, (2)$$

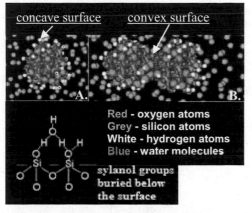

Figure 2: **A.** Silica nanoparticle built using the crystallographic method; **B.** Two silica nanoparticles simulating agglomeration just before a critical concentration.

Where Si_s represents silicon atom at the surface. The higher bonding energy between the water molecules within the ordered liquid layer compared to the one at the particle-liquid adsorption state is contributed to fact that water is intrinsically stable in a cluster of 4-5 molecules as it is in bulk. This ordering is approximately 3 molecular units and does not exceed 0.3nm. [18] The thickness is not expected to change with particle size, although it has greater influence on the smaller particle diameters due to the relative ratio of liquid and solid. Similar increase in the density profile at the solid interface was observed by [19] through Molecular Dynamics Simulations (MD) and numerically found to be in the order of 1-3nm.

The viscosity of 10nm- and 20nm- sized silica at varying concentrations is experimentally determined (fig. 3A) and shows an increase of up to 17% at concentration as high as 1% and 2%. However, if we apply ultrasound prior to the measurements, the viscosity drastically decreases by approximately 5 to 10%. It hasn't been determined if this is a direct result of removal of the hydration layer at some places on the surface, breaking of the formed clustered (as viewed by the TEM analysis), or both. These results conducted for 10nm and 20nm silica nanoparticles are representative of the extent of agglomeration, surface charging, and hydration and their combined effect. At lower

loadings the viscosity increases solely due to the fixed water layer, which tends to move along with the particle. Particles are not agglomerated at this stage and they exhibit non-Newtonian behavior, due to their small size and negligible weight. The applied ultrasound could not only serve as a source of good dispersion, but also as a continuous "breaker" of the hydrogen bonds between the solid surface and the liquid layer, due to the difference in the bonding energy between liquid-liquid and liquid solid interactions. The breakage of the immobilized layer is attributed to the smaller energy barriers for the sylanol – water bond.

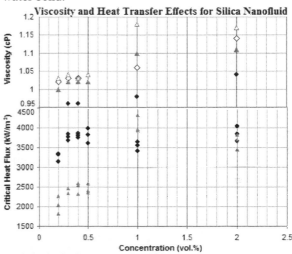

$\diamond$ viscosity before sonicating (10nm)
$\blacklozenge$ viscosity after sonicating (10nm)
$\triangle$ viscosity before sonicating (20nm)
$\blacktriangle$ viscosity after sonicating (20nm)
$\blacklozenge$ heat transfer for 10nm-silica after sonication
$\blacktriangle$ heat transfer for 20nm-silica after sonication

Figure 3: **A.** Viscosity; **B.** Critical Heat Transfer for Silica nanofluid.

At higher concentrations, agglomerations of 2-3 nanoparticles start to evolve and the viscosity increases drastically. After the process of ultra-sonication, these aggregates seem to be broken, because at 1% and 2% both 10nm and 20nm silica steadily shows a decrease in viscosity in the order of 7-8%. The effect of sonication on heat transfer and CHF of nanofluids is not known. Although the ultrasound waves break particle aggregates, they also break the hydration layer, which improves the heat transfer due to its ordered structure. The sites where desorption of water layer has taken place become discharged which may possibly lead to further agglomeration during the pool boiling experiment. So when current is applied, the particles tend to collapse onto the NiCr wire. It has been experimentally observed that the lower the zeta potential of the particles, the thicker the deposition on the thermocouple. The effect of thickness and porosity of the deposition on bubble size and intensity of growth on heat transfer has been discussed in another work. [21]

Although it has been shown that smaller particles – 10nm vs. 20nm generally exhibit better heat transfer characteristics [12] due to the smaller surface area to volume ratio for the later, clusters of a few particles are theoretically postulated [20] to increase the thermal transport due to the thermal resistance of the voids. Therefore not only the size of the formed agglomerations, but also the shape of the clusters determines the heat transfer properties of the nanofluid. The surface area of the particles considerably reduces when two particles fuse into one another, compared to the case when they merely touch.

Referring to Fig. 3, it has been established that the increase in nanoparticle size will decrease the heat transfer and the critical heat flux at the point of wire breaking. It is also known that the critical heat flux can also increase due to an increase in concentration. Thus, two opposing effects on CHF are taking place, one due to the change in the surface area and another due to the increase in the number of particles in the suspension, i.e., particle number density. The change in CHF at different ranges of concentrations can be explained using the following table.

Table 1: Effect of concentration on CHF for 10nm particles

Region	Conc, %	Particle size, nm	Surface area	Number density	CHF
1	0.1 - 0.2	10	Maintains	Increases	Increases
2	0.2 - 0.5	10 - 20	Increases	Decreases	Increases
3	0.5 - 1.0	20	Increases/ maintains	Decreases	Decreases
4	1.0 - 2.0	20	Maintains	Increases	Increases

Referring to Table 1, in Region 1, the particle size and the surface area stay constant, but the number density doubles due to the doubling of concentration and CHF slightly increases. In Region 2, the average particle size increases to twice the original diameter due to agglomeration, the surface area increases, and the number density decreases. The rate of increase in surface area is higher than the rate of decrease in number density, and hence the CHF increases first sharply up to 0.3% concentration, and stays approximately constant in the rest of the region. In Region 3, as more particles start agglomerating, although the surface area either maintains or increases, the number density drops drastically. The cumulative effect is that the CHF decreases in this region. Finally, in Region 4, the surface area does not increase any further, but the number density increases with further concentration, so the CHF increases.

In summary, the critical heat flux of the wire does not increase monotonically with concentration as theory predicts. CHF depends on the agglomeration characteristics, particle shape and the hydroxylated surface of the nanoparticles.

References

1. Choi SUS, "Enhancement Thermal Conductivity of Fluids with Nanoparticles". FED-Vol. 231/MD-. 66(1995), 99-103
2. Choi SUS, Zhang ZG, Yu W, Lockwood FE, Grulke EA. "Anomalous thermal conductivity enhancement in nano-tube suspensions". Appl. Phys. Lett. 79(2001), 2252–54
3. Wang XW, Xu XF and Choi SUS, "Thermal Conductivity of Nanoparticle-Fluid Mixture", J. of Thermo. Heat Trans, 13(1999), 474-480
4. P. Vassallo, R. Kumar, and S. D'Amico, "Pool boiling heat transfer experiments in silica–water nano-fluids", Int. J. Heat & Mass Trans. 47(2004), 407-411.
5. Xuan Y and Li Q, "Heat Transfer Enhancement of nanofluids", Int. J. of Heat & Flow 21(2000), 58-64
6. P. Keblinski, et al, "Mechanisms of heat flow in suspensions of nano-sized particles (nanofluids)" , Int. J. Heat Mass Transfer, 45(2002), 855-863
7. Xue Q-Z. "Model for effective thermal conductivity of nanofluids". Phys. Lett. A 307(2003), 313–17
8. Yu W, Choi SUS. "The role of interfacial layers in the enhanced thermal conductivity of nanofluids: a renovated Maxwell model". J. Nanopart. Res. 5(2003), 167–71
9. Gupta A., Wu X. and Kumar R. " Possible mechanisms for thermal conductivity enhancement in nanofluids", 4th International Conference on Nanochannels, Microchannels and Minichannels, June 19-21, 2006, Limerick, Ireland
10. Xue L,Keblinski P, Phillpot SR, Choi SUS, Eastman JA. "Effect of liquid layering at the liquid-solid interface on thermal transport". Int. J. Heat Mass Trans. 47(2004), 4277-4284
11. Vassallo P, Kumar R, D'Amico S. "Pool boiling heat transfer experiments in silica-water nano-fluids". Int. J. Heat Mass Trans. 47(2004), 407
12. Milanova D. and Kumar R, "Role of ions in pool boiling heat transfer of pure and silica nanofluids". Appl. Phys. Lett. 87(2005), 233107
13. www.resistancewire.com
14. A.V. Kiselev, Kolloidn. Zh., (1936), 2, 17
15. L.D. Belyakova, O.M. Dzhigit, A.V. Kiselev, G.G.Muttk, K.D. Shcherbakova, (1959) Russ. J. Phys. Chem. Engl. Transl., 33, 551
16. P.F. Kane and G.B. Larrabee, Characterization of Solid Surfaces, (1994), Plenum, New York.
17. R.L. Dalton and R.K. Iler, (1956) J. Phys. Chem. 60, 955
18. R. K. Iler, in E. Matijevic, Ed. Surface and Colloid Science, Vol. 6, NY, 1973
19. X. Wu, R. Kumar, P. Sachdeva, (2005) Calculation of thermal conductivity in nanofluids from atomic-scale simulations, ASME, IMECE2005-80849.
20. P. Keblinski, S.R. Phillpot, S.U.S. Choi, and J.A. Eastman, (2002) Mechanisms of heat flow in suspensions of nano-sized particles (nanofluids), Int. J. Heat Mass Trans., **45**, p. 855.
21. Milanova, D. and Kumar, R. 2006 ICNMM2006, Limerick, Ireland [also submitted to J Heat Transfer].

Evaluation of Nanostructured Polymeric Coatings for Steel Corrosion Protection

A. Aglan*, A. Ludwick** and M. Reeves**

*Chemical Engineering Department, Michigan Technological University, Houghton, Michigan, USA, aaaglan@mtu.edu

** Chemistry Department, Tuskegee University, Tuskegee, Alabama, USA, aludwick@tuskegee.edu

ABSTRACT

The effect of the addition of multi-walled carbon nanotubes (MWCNT) to epoxy and vinyl chloride/vinyl acetate copolymer (VYHH) coatings on their ability to protect the substrates was studied. Coatings were formulated from these resins with and without MWCNT reinforcement. Steel substrates were prepared and coated with each formulated coating and submerged in 5% NaCl solution to study effectiveness by means of Electrochemical Impedance Spectroscopy (EIS). Optical microscopy was used to capture the progress of sample corrosion. EIS measurements showed that the addition of MWCNTs to epoxy and VYHH coatings increased their charge transfer resistance in comparison with the neat coatings. This is an indication of the enhanced corrosion protection of the nanocoatings.

Keywords: vinyl chloride/vinyl acetate, corrosion, multi wall carbon nanotubes (MWCNT), electrostatic impedance

1 INTRODUCTION

Two main mechanisms are responsible for the breakdown of coating protection; diffusion of water through the coating and disbond propagation between the coating and the substrate. One technique that has recently been used to increase the adhesive and anticorrosive properties of polymers is the addition of nanoparticles [1]. Nanoparticles in the form of silicates, single wall carbon nano tubes (SWCNT) and multi wall carbon nano tubes (MWCNT) have been the most popular nano phases either as reinforcements or for specific functionalization [2-5]

Yang et al. studied the effect of pigment to binder ratio on the corrosion resistance of polyurethane coatings applied to carbon steel substrates [6]. The pigment to binder ratio affects the corrosion resistance. When they used nano zinc oxide, the optimum pigment to binder ratio was 0.3. The same ratio when they used conventional zinc oxide was 1.0. They concluded that nano zinc oxide particles have improved the anti corrosion resistance of polyurethane, resulting from the enhancement of the density of the coatings; this lead to the reduction in the transport paths, which blocked the penetration of the corrosive electrolyte to the steel substrates. The EIS measurements showed that there is an increase of three fold in the charge transfer resistance of the nano coatings compared with the conventional zinc oxide coating. The three-fold increase experienced with the addition of the nano coatings was observed through the 1000 hours of testing.

Chen et al. studied the effect of a nickel coating with carbon nano tubes on the corrosion resistance of carbon steel [7]. The coating was applied by electrochemical deposition. It was found that in 3.5% wt. NaCl solution the nano tube coating performed superior to pure nickel coating. The explanation for this was due to the fact that the nano tubes acted as a physical barrier more completely filling in the micro holes and flaws on the surface of the nickel coating. This explanation is similar to that given by Yang et al.

Characterization of various coatings using EIS requires an equivalent circuit, which consists of a combination of resistance (R) and capacitance (C) in various arrangements [8]. When a coated steel substrate is submerged in an electrolyte solution the first resistance R_1 represents the electrolyte resistance. The capacitance (C) represents the coating/electrolyte interface in the cell. The resistance R_{ct} in the Randle circuit is the polarization resistance or the resistance of the surface of the steel substrate to corrosion. The summation of R_1 and R_{ct} on a Nyquest plot represents the Charge Transfer Resistance, which can be used as a parameter to characterize the resistance of coatings to corrosion. Bard and Faulkner indicate that the real current voltage relationship in the electrochemical theory is nonlinear [9]. Loveday et al. state that as long as 10mV or less is used to measure the EIS current the current-voltage curve can be assumed to be linear. However, in some coating applications larger amplitudes can be applied [8].

In the present work, the effect of the addition of MWCNTs to VYHH (previously formulated [10]) and epoxy (commercial) coatings on their corrosion resistance is studied using EIS.

2 MATERIALS AND EXPERIMENTAL

The polymer resins used were an epoxy; *Valspar* Dura Build TM High Build Epoxy Finish comprised of resin A and hardener B in a 2:1 volumetric ratio. Resin A is a alkyd glycidyl ether and hardener B is isophorone diamine. Union Carbide's VYHH, which is a vinyl chloride/vinyl acetate copolymer, was also used. Multiwalled carbon nanotubes supplied by Ahwahnee Technology (Dia: 2-15 nm, Length: 1-10 μm, Layers: 5-20) were used as reinforcement in some systems. The VYHH and epoxy coatings were formulated with and without nanoreinforcements via sonication and asymmetric mixing techniques to achieve uniform dispersion. The coatings were applied to steel substrates (25mm X 101.6mm X 1.25mm). The steel substrates were prepared for coating by polishing and rinsing with acetone. Dried substrates were then dipped in the coatings with care taken to ensure an even coat with few defects. After 3 days of hang drying, samples were submerged in a 5% NaCl solution.

EIS measurements were conducted on unimmersed coated substrates through the use of the corrosion cell (or flat cell), shown in Figure 1. The flat cell was connected to a potentiostat (PARSTAT® 2273), supplied by Princeton Applied Research. The frequency range used in this experiment ranged from 1×10^6 Hz to 1×10^{-2} Hz while the amplitude used was 10mV.

Optical microscopic observations, on the tested samples were performed. Measurements were taken of the 1cm^2 portion of the sample that was exposed to the NaCl solution in the flat cell.

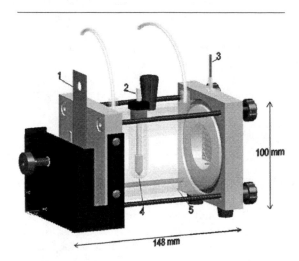

Figure 1. Flat Cell for EIS Measurements. (1) Working Electrode (Sample), (2) Reference Electrode, (3) Counter Electrode, (4) Reference Well, (5) Glass Chamber.

3 RESULTS AND DISCUSSION
3.1 EIS Measurements

In order to calibrate the set up which includes the PARSTAT® 2273 and the flat cell, the bare metal was inserted in the flat cell before and after 20 days immersion in 5% NaCl. The EIS measurement was carried out before immersion and a Bode plot was generated; after 20 days it was retested and a second Bode plot was produced. Figure 2 shows the comparison between no immersion and 20 days of immersion. The result of the charge transfer resistance (Rct) can be found by subtracting the minimum impedance from the maximum impedance. The value of Rct for no immersion was found to be 1250 Ω, compared to 550 Ω for that of 20 days immersion. As illustrated in Figure 2 the EIS measurement of the bare steel after 20 days immersion in 5% NaCl showed a decrease of more than 50% in the charge transfer resistance.

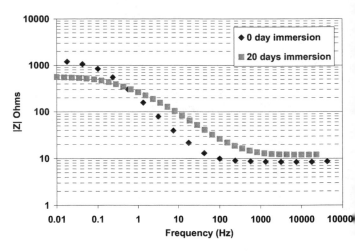

Figure 2. Bode plot of both 0 day and 20 days immersion of bare steel.

Results of EIS measurements are shown in the form of Bode plots in Figure 3 for both the neat and the nanoreinforced epoxy coatings. It can be seen that the nano epoxy coated sample displays higher impedance over the entire range of frequencies tested. The Rct from Figure 3 is about 2.08E9 for the nano-coated sample while that for the neat epoxy coating is only about 0.679E9. This indicates that the nano epoxy coating offers better corrosion protection.

Cleantech 2007, ISBN 1-4200-6382-0

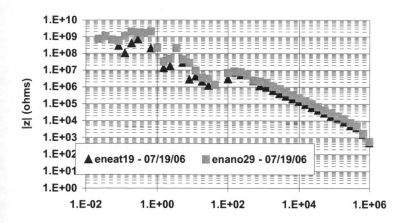

Figure 3. Bode plot of both neat and nano epoxy samples.

EIS measurements of the VYHH coated samples with and without MWCNT produced similar results, as seen in Figure 4. Again the impedance was greater for the nano specimen over the entire range of frequencies tested. The R_{ct} from Figure 4 is about 1.38E9 for the nano coated sample while that for the neat VYHH coating is only about 0.345E9.

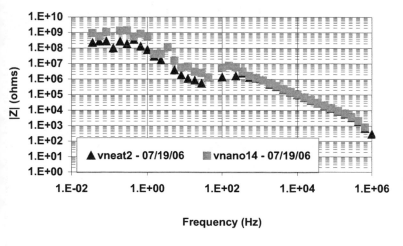

Figure 4. Bode plot of both neat and nano VYHH samples.

3.2 Immersion Test Observations

Visual and optical microscopy of the samples after 20 days of immersion in 5% NaCl solution was performed. For the bare steel it can be seen from Figure 5 that the surface has been adversely damaged; this was because no form of coating was added to aid in the prevention of corrosion.

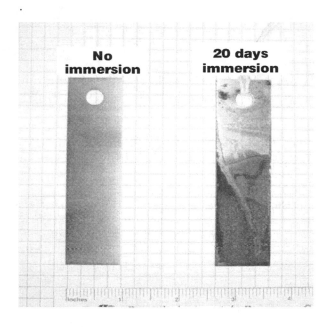

Figure 5. Corrosion characteristics of bare steel.

After 20 days of immersion the corrosion started on the edges of the neat epoxy coated samples, but the nano epoxy sample remained almost undamaged. This can be seen in Figure 6. Microscopic observation of the edge of samples with (right in Figure 6) and without MWCNT (left in Figure 6) after 20 days of immersion in 5% NaCl solution show the increased protection that the MWCNT has provided in the epoxy coating.

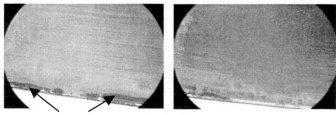

Severe edge corrosion on
neat epoxy coated specimen

Figure 6. Optical micrographs of the neat and nano epoxy samples after 20 days of exposure to 5% NaCl solution.

Optical micrographs of the neat and nano VYHH samples after 20 days of exposure to 5% NaCl solution are shown in Figure 7. Blisters are seen in the neat coating on the left compared to the completely unmarred nano-coating on the right.

blisters

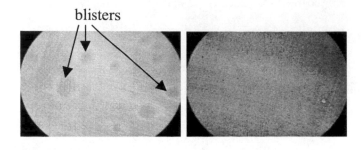

Figure 7. Optical micrographs of the neat and nano VYHH samples after 20 days of exposure to 5% NaCl solution.

Thus, the addition of a very small percent of MWCNT to the epoxy and the VYHH resins has improved the protection power of coating formulated from these systems. This finding has been verified from both the EIS measurements and the microscopic observations

4 CONCLUSIONS

Coatings have been successfully formulated containing MWCNT. EIS measurements have shown that the charge transfer resistance is higher for the nano coatings than the neat coatings for the epoxy and VYHH systems. Visual and microscopic observations have shown, based on 20 days 5% NaCl immersion, that nano reinforced epoxy and VYHH resins are better with respect to blistering and discoloration.

ACKNOWLEDGEMENTS

The NSF-REU at Tuskegee University supported this work. Dr. L Koons is gratefully acknowledged for his technical help. A. Allie, L. Allie, Maria Calhoun and Adam Charlton are gratefully acknowledged for their assistance.

REFERENCES

[1] http://www.inframat.com/hydro2.htm

[2] Hedia, H., Allie, L., Ganguli, S. and Aglan, H., The influence of nanoadhesives on the tensile properties and Mode-I fracture toughness of bonded joints. *Eng. Fracure Mech.* 73 (2006) 1826-1832.

[3] Ganguli, S., Aglan. H., and Dennig, P., Effect of loading and surface modification of MWCNT on the fracture behavior of epoxy nanocomposites. *J. Reinforced Plastics*, 25 (2006) 175-188.

[4] Ganguli,S., Bhuyan, M., Allie, L. and Aglan, H., Effect of multi-walled carbon nanotube reinforcement on the fracture behavior of a tetrafunctional epoxy. *J. Matrl. Sci. Letters,* 40(13) (2005) 3593-3595.

[5] Ganguli, S., Aglan, H. and Dean, D. Microstructural origin of strength and toughness of epoxy nanocomposites. *J. Elastomers and Plastics*, 37 (2005) 19-35.

[6] Yang, L.H., Liu, F.C., Han, E.H., Effects of P/B on the properties of anticorrosive coatings with different particle size. *Progress in Organic Coatings,* 53 (2005) 91-98.

[7] Chen, X.H., Chen, C.S., Xiao, H.N., Cheng, F.Q., Zhang, Yi, G.J., Corrosion behavior of carbon nanotubes–Ni composite coating. *Surface & Coatings Technology,* 191 (2005) 351–356.

[8] Loveday, D., Peterson, P., Rodgers, B., Evaluation of org coatings with electrochemical impedance spectroscopy. *J Coatings Tech.*, (August 2004) 46-52.

[9] Bard, A.J., Faulkner, L.R., Electrochemical Methods; Fundamentals and Applications, Wiley Interscience, 2000

[10] Mansour, E.M., A.-Gaber, B. Nabey, Tadros, A., Aglan, H., Ludwick, A., Solvent effect on the protection efficiency of vinyl resin varnish for preventing the corrosion of steel using electrochemical impedance spectroscopy. *Corrosion,* 58 (2002) 113-118.

Preparation and Characterization of Nafion/Microporous Titanosilicate Composite Membranes as Ion-Conducting Materials

S.-Y. Ryu[*], J.-D. Jeon[**] and S.-Y. Kwak[***]

School of Materials Science and Engineering, Seoul National University, San 56-1,
Sillim-dong, Gwanak-gu, Seoul 151-744, Korea
[*]suyol@daum.net, [**]jdjun74@snu.ac.kr, [***]sykwak@snu.ac.kr

ABSTRACT

Microporous ETS-4 particles were synthesized by the hydrothermal method and characterized by XRD and FE-SEM/EDX. Nafion/ETS-4 composite membranes were prepared by mixing 5 wt% Nafion solution with microporous titanosilicate ETS-4. The water uptake increased with an increase in the ETS-4 content. However, the IEC decreased with an increase in the ETS-4 content due to the addition of ETS-4 with no reactive sites ($-SO_3H$). From the FE-SEM/EDX images, it was confirmed that Nafion/ETS-4 composite membranes had homogenous distribution over the surface. The methanol permeability of composite membranes decreased with an increase in ETS-4 content up to 15 wt% of ETS-4 and increased with further addition. The proton conductivity decreased with an increase in the ETS-4 content. From methanol permeability and proton conductivity results, it was confirmed that the selectivity parameter of the NE10 composite membrane containing 10 wt% ETS-4 was the highest value.

Keywords: Nafion, ETS-4, methanol permeability, proton conductivity, fuel cell

1 INTRODUCTION

One of the most severe environmental issues is nowadays the pollution level due to internal combustion motor devices [1]. Fuel cells, characterized by near-zero emission levels, represent a promising alternative for on-board power generation [2]. Among the different types of such systems, direct methanol fuel cells (DMFCs) are suited for portable devices or transportation applications owing to their high energy density at low operating temperature and the ease of handling a liquid fuel. However, they have a major drawback such as slow oxidation kinetics of methanol at low temperature below 100 °C and high methanol crossover from anode to cathode. In particular, methanol crossover through the membrane as a result of electro-osmotic drag and the concentration gradient decreases the performance of the fuel cell because of the resulting mixed potential.

There have been many attempts to reduce the methanol permeability through the polymer electrolyte membranes: (1) to modify the surface of the membranes to block the methanol transport, (2) to control the size of the proton transport channels using different block copolymers and cross linkages, (3) to introduce a winding pathway (also called tortuous pathway) for a methanol molecule by making a composite with inorganic materials, and (4) to develop new types of electrolyte polymers. Among these attempts, Nafion has recently been modified with a variety of organic/inorganic materials, yielding Nafion-based composite membranes.

We chose the microporous Engelhard titanosilicate-4 (ETS-4) particles as an inorganic particle because of their regular pore size selective separations based on molecular size and shape. ETS-4 can also separate components based on preferential adsorption, in which the strong adsorption of one components [3,4]. Thus ETS-4 will effectively block the methanol molecules by tortuous pathway effect while maintaining the transport of the water molecules because it plays a role as size-slective adsorbent. That is, opening pore size (0.3 – 0.4 nm) of ETS-4 has smaller than that of methanol and bigger than that of water. However, membranes made of pure ETS-4 are plagued by defects such as cracks or gaps and exhibit poor mechanical properties such as brittleness and fragility. Therefore, Nafion/ETS-4 composite membranes represents a compromise between the non-selective polymeric films (against methanol/water mixture) and the brittle ETS-4 films. The objective of this study is to prepare and characterize the Nafion/ETS-4 composite membranes with various ETS-4 contents.

2 EXPREMENTAL

2.1 Materials

The ETS-4 was synthesized by the hydrothermal method using gels of the following molar compositions; $5H_2O_2:0.5TiO_2:10SiO_2:18NaOH:675H_2O$. Titanium(III) chloride and sodium silicate solution were used as titanium and silicon sources, respectively. The synthesis was carried out in Teflon-lined stainless steel autoclave at 200 °C for 3 days. The as-synthesized ETS-4 particles were ground using a mortar. Ground ETS-4 particle/water solution was sonicated for 30 min in a water-bath sonicator (Branson 3510, Branson Ultrasonic Corp.) and then filtered through a 0.45 μm pore size membrane filter (cellulose acetate, DISMIC-25cs, Toyo Roshi Kaisya Corp.). Nafion perfluorinated ion-exchange resin (5 wt% solution in a mixture of lower aliphatic alcohols and water) was purchased from Aldrich Chemicals and used as the membrane material; it has an equivalent weight (EW) of 1100 g for each sulfonic acid group.

2.2 Fabrication of Composite Membranes

The composite membranes were prepared by the solution casting method. The desired amount of ETS-4 was added into 5 wt% Nafion solution, and then stirred at room temperature and degassed by ultrasonication. The ETS-4 content of the mixture was 5, 10, 15, and 20 wt% with respect to Nafion. The resulting mixture was slowly poured into a glass dish in an amount that would produce a formed membrane thickness of approximately 80 μm. The filled glass dish was evaporated at 40 °C for 2 days and then annealed at 120 °C in a convection oven for 2 h. After cooling, the membrane was peeled off the glass dish by the addition of water. The membranes were stored in deionized water so that they were water-saturated. In this study, NEx denotes a Nafion/ETS-4 composite membrane containing x wt% of ETS-4.

2.3 Characterization

X-ray diffraction (XRD) was used to identify the phase of the ETS-4 by using a MAC Science MXP 18A-HF X-ray diffractometer with CuK$_\alpha$ radiation ($\lambda = 1.5406$ Å), which was generated at a voltage of 40 kV and a current of 40 mA. The diffraction angle was scanned at a 2° min^{-1} from 5 to 40°. Field emission scanning electron microscopy (FE-SEM, JEOL, JSM-6330F) equipped with energy dispersive X-ray (EDX) detector was used to observe the morphology of the ETS-4 crystals. The swelling characteristics of the membranes were determined with water uptake measurements. The samples were completely dried under vacuum for 3 days at 30 °C and then weighed. They were then placed in deionized water for a week at 25 °C. Water on the surfaces of the wet samples was removed with filter paper, and then the samples were immediately transferred to a weighing dish and weighed. The water uptake was calculated according to the equation

$$\text{water uptake (\%)} = \frac{W_{\text{wet}} - W_{\text{dry}}}{W_{\text{dry}}} \times 100 \qquad (1)$$

where W_{wet} and W_{dry} are the weights of the wet and dried membranes, respectively.

The ion exchange capacity (IEC) indicates the number of milliequivalents of ions in 1 g of the dry polymer. The samples of similar weight were soaked in 50 mL of 0.1 N NaCl solution for 24 h at 25 °C in order to achieve complete H$^+$ to Na$^+$ ion exchange. 10 mL of the released H$^+$ was then titrated with 0.1 N NaOH solution. The IEC was calculated from the titration data with the following equation:

$$\text{IEC (mequiv/g)} = \frac{V_{\text{NaOH}} \times N_{\text{NaOH}} \times 5}{W_{\text{dry}}} \qquad (2)$$

where V_{NaOH} is the amount of NaOH required to neutralize a blank sample, N_{NaOH} is the normality of the NaOH solution, 5 is the ratio of the amount of NaOH required to dissolve the sample to the amount used for titration, and W_{dry} is the weight of the dried sample.

Methanol permeability measurements were carried out using a home-made, glass, diffusion cell that consisted of two cylindrical glass compartments (A for feed and B for permeate) separated by a composite membrane with an effective area of 3.8 cm^2. In order to determine the methanol permeability of each membrane, liquid samples of 1 mL were taken from the feed and the permeate using a syringe at prescribed time intervals (20, 40, 60, 80, 100, and 120 min). The liquid samples extracted from permeate were analyzed with a calibrated gas chromatograph (HP 5980, Hewlett-Packard, USA) equipped with a capillary column (14% cyano propyl phenyl methyl polysiloxane, 30 m × 0.25 mm × 1.0 μm) and a flame ionization detection (FID). The uncertainty of the obtained values was less than 2%.

The proton conductivity measurements on fully hydrated membrane samples were carried out with the cell immersed in liquid water. The installed cell was placed in a chamber with controlled temperature. The impedance measurements were carried out in the frequency region from 1 Hz to 10^5 Hz and in the ac current amplitude of 1 mA using a Solartron 1255 electrochemical impedance analyzer with ZPLOT software.

3 RESULTS AND DISCUSSION

3.1 General Characterization of ETS-4

Figure 1 shows the X-ray diffraction peaks of the ETS-4 sample, synthesized by the procedures mentioned in the experimental section. Based on the two contributions by Kuznicki [3] and Chapman and Roe [4], this product was identified as pure ETS-4. As shown in Figure 1 (inset image), the ETS-4 produced intergrowth polycrystalline spherulitic particles composed of two semispheres, each with a shape that resembles a maple leaf. In addition, each semisphere was consisted of many needle-like crystals.

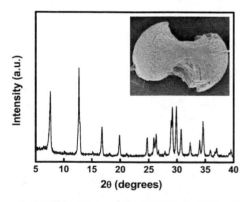

Figure 1: XRD patterns of the ETS-4 particles. The FE-SEM image in the inset shows the ETS-4 morphology.

The Si/Ti ratio in the ETS-4 particles is characterized by Si and Ti mapping images employing Kα peak of Si and Ti with FE-SEM/EDX. ETS-4 particles were found to have a Si/Ti ratio value of ca. 2.51. It was reported [5] that Si/Ti ratio of as-synthesized ETS-4 was 2.7, which is in relatively good agreement with our result. The ETS-4 crystals reported so far were large aggregations of highly intergrowth plates. Neither small ETS-4 crystals in the nanometer scale for the use as as-synthesized particles were reported, nor synthesis conditions for the crystallization of large, uniform crystals, were available. Therefore, grind procedures of ETS-4 particles were needed prior to the preparation of Nafion/ETS-4 composite membranes. The as-synthesized ETS-4 particles were ground using a mortar and filtered through a 0.45 μm pore size membrane filter. From the DLS measurements, it was confirmed that one narrow peak of the ETS-4 particle size distribution was observed and its average size was found to be ca. 217 nm. This narrow size is due to selectively permeative characteristic of a 0.45 μm pore size membrane filter.

3.2   General Characterization of Composite Membranes

Figure 2 shows the equilibrium percentage sorption of water, obtained by soaking the membranes in water at 25 °C. The water uptake of the membranes was found to increase from 21.4 for NE0 to 32.2% for NE20 with an increase in the ETS-4 content. This result shows that the water uptake of the membranes can be controlled by varying their ETS-4 content.

The ion exchange capacity (IEC) provides an indication of the number of ion-exchangeable groups present in an ion-conducting polymer membrane; these groups are responsible for the conduction of protons and thus the IEC is a reliable measure of the proton conductivity. The IEC of each membrane was determined by using the acid-base titration method. The results are shown in Figure 2 and demonstrate that the IECs of the membranes decrease from 0.89 to 0.72 mequiv/g with an increase in the water uptake. Thus the introduction of ETS-4 with no reactive sulfonic acid sites into the membranes contributes to a decrease in the IEC.

Figure 2: Water uptake and IEC for the composite membranes as a function of ETS-4 content.

The FE-SEM/EDX surface mapping images of silicon for the Nafion/ETS-4 composite membranes are shown in Figure 3. From FE-SEM results, it was demonstrated that Nafion/ETS-4 composite membranes had homogenous distribution of ETS-4 particles over the surface. These results are evidence that the ground ETS-4 particles are well dispersed in Nafion membranes by solution casting method.

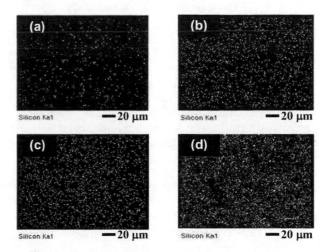

Figure 3: FE-SEM/EDX surface images of silicon for (a) NE5, (b) NE10, (c) NE15, (d) NE20 membranes.

3.3   Methanol Permeability

High methanol crossover is an enormous obstacle for membranes in fuel cell applications such as DMFC. Therefore, methanol permeability of the membranes needs to be reduced, while maintaining proton conductivity. The methanol permeability of NE series composite membranes decreased with an increase in ETS-4 content up to 15 wt% of ETS-4 and increased with further addition of ETS-4 as shown in Figure 4. The decrease in the methanol permeability can be explained in terms of internal changes in the Nafion structures depending on the ETS-4 content. That is, methanol molecules have a more tortuous path around the ETS-4 particles. Similar trends have been reported for other composite membranes, such as Nafion/montmorillonite composite membranes [6] and polyvinylalcohol (PVA)/zeolite composite membranes [7]. However, inspection of the NE20 containing the 20 wt% ETS-4 shows a slight drop in methanol permeability. One can assume that at high loading i.e. higher than 20 wt%, methanol can easily diffuse through the composite membranes due to ETS-4 particle's aggregate, while at low loading the crossover is hindered by the ETS-4 particles which are still finely subdivided in the Nafion matrix. From these results, it was demonstrated the critical role of ETS-4 loading in controlling the methanol crossover phenomena.

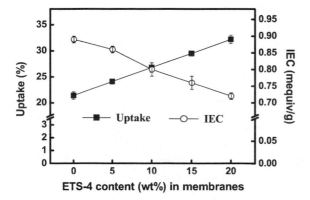

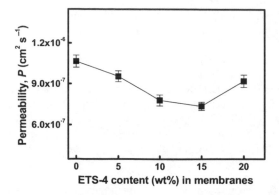

Figure 4: Methanol permeability of NE series composite membranes according to ETS-4 content in membranes.

3.4 Proton Conductivity and Selectivity

The proton conductivity of the composite membranes was measured using an ac impedance analyzer at various temperatures. Protons travel through ionic channels formed inside the Nafion matrix as well as through the matrix itself, provided the matrix possesses ionic sites. The proton conductivity of composite membranes showed behavior similar to that observed for their IEC. That is, as shown in Figure 5, the proton conductivity decreased with an increase in ETS-4 content at all different temperatures. In addition, all of the composite membranes exhibited a marked increase in proton conductivity with increasing temperature, indicating that temperature played a main role in the kinetics of proton motion in the polymer membrane and mobility of polymer chains. It was tentatively concluded that pristine Nafion membrane could be the highest level in the composite membrane for DMFC.

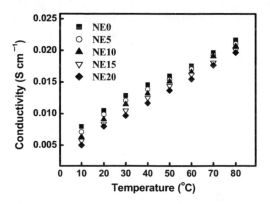

Figure 5: Proton conductivity of composite membranes fabricated with different ETS-4 content.

One of the methods used to elucidate membrane performance is the selectivity parameter, Φ (S s cm^{-3}), calculated using, $\Phi = \sigma/P_{\text{methanol}}$, where σ is the proton conductivity (S cm^{-1}) and P_{methanol} (cm^2 s^{-1}) is the methanol permeability through the membrane. Table 1 lists the selectivity parameters of NE series composite membranes

as a function of ETS-4 content for methanol permeability and proton conductivity measured at 20 °C. The maximum selectivity was found at 10 wt ETS-4 content, i.e., NE10 membrane, implying the optimum membrane.

Membrane	Methanol permeability (cm^2 s^{-1})	Proton conductivity (S cm^{-1})	Selectivity parameter (S s cm^{-3})
NE0	1.065×10^{-6}	1.05×10^{-2}	0.986×10^4
NE5	0.954×10^{-6}	0.99×10^{-2}	1.037×10^4
NE10	0.779×10^{-6}	0.91×10^{-2}	1.168×10^4
NE15	0.737×10^{-6}	0.84×10^{-2}	1.132×10^4
NE20	0.921×10^{-6}	0.80×10^{-2}	0.869×10^4

Table 1: Experimental parameters of composite membranes.

4 CONCLUSIONS

The ETS-4 was successfully synthesized by the hydrothermal method using gels of the following molar compositions; 5 H$_2$O$_2$: 0.5 TiO$_2$: 10 SiO$_2$: 18 NaOH : 675 H$_2$O. Nafion/ETS-4 composite membranes were prepared with the solution casting method. The water uptake values were found to increase with an increase in the ETS-4 content. The acid-base titration results show that the IECs of the membranes decrease from 0.89 to 0.72 mequiv/g with an increase in the ETS-4 content. From the FE-SEM/EDX mesurement, it was demonstrated that Nafion/ETS-4 composite membranes had homogenous distribution of ETS-4 particles over the surface. The methanol permeability of composite membranes decreased with an increase in ETS-4 content up to 15 wt% of ETS-4 and increased with further addition of ETS-4. The proton conductivity decreased with an increase in the ETS-4 content at all different temperatures due to the decrease of IEC values. From methanol permeability and proton conductivity results, it was confirmed that the selectivity parameter of the NE10 composite membrane is the highest value, implying the optimum membrane.

REFERENCES

[1] P. L. Antonucci, A. S. Aricò, P. Cretì, E. Ramunni and V. Antonucci, Solid State Ionics, 125, 431, 1999.

[2] A. K. Shukla, P. A. Christensen, A. Hamnett and M. P. Hogarth, J. Power Sources, 55, 87, 1995.

[3] S. M. Kuznicki, US Patent 4,853,202, 1989.

[4] D. M. Chapman and A. L. Roe, Zeolites, 10, 730, 1990.

[5] W. J. Kim, M. C. Lee, J. C. Yoo and D. T. Hayhurst, Micropor. Mesopor. Mater., 41, 79, 2000.

[6] C. H. Rhee, H. K. Kim, H. Chang and J. S. Lee, Chem. Mater., 17, 1691, 2005.

[7] B. Libby, W. H. Smryl and E. L. Cussler, AIChE Journal, 49, 991, 2003.

Amorphous Diamond as a Thermionic Material

James C. Sung[*,1,2,3], Ming-Chi Kan[1], Tun-Jen Hsiao[1], Ying-Tung Chen[4], Michael Sung[5]

Address: KINIK Company, 64, Chung-San Rd., Ying-Kuo, Taipei Hsien 239, Taiwan, R.O.C.
Tel: 886-2-2677-5490 ext.1150
Fax: 886-2-8677-2171
E-mail: sung@kinik.com.tw

[1] Kinik Company, 64, Chung-San Rd., Ying-Kuo, Taipei Hsien 239, Taiwan, R.O.C.
[2] National Taiwan University, Taipei 106, Taiwan, R.O.C.
[3] National Taipei University of Technology, Taipei 106, Taiwan, R.O.C.
[4] Department of Mechanical Engineering, Chung Cheng Institute of Technology, Tahsi, Taoyuan 33509, Taiwan, R.O.C.
[5] Advanced Diamond Solutions, Inc., 334 6th Street, Suite 4, San Francisco, CA 94103, U.S.A.

ABSTRACT

Amorphous diamond is essentially a chaotic carbon mixture with distorted sp^2 and sp^3 bonds. As such it possesses both metallic character of conductive graphite and semiconductor character of insulating diamond. Moreover, as each carbon atom is unique in its electronic state that is determined by the degree of distortion of its bonds, amorphous diamond contains numerous discrete potential energies for electrons. In fact, amorphous diamond may have the highest density of atoms (1.8×10^{23} per cubic centimeter) that is several times higher than ordinary materials (e.g. about four times of iron atoms or silicon atoms). Thus, amorphous diamond has the highest configuration entropy for both atoms and valence electrons.

Due to the distribution of discrete electronic energies with high density, amorphous diamond is uniquely capable to generate electricity and emit radiation. It has been demonstrated that amorphous diamond can be made as silicon free solar cells, front panel display field emission source, sensitive thermal sensing by IR detection, and perfect black body for energy conversion. Various amorphous diamond devices are being fabricated to exploit the superb properties of amorphous diamond.

Keywords: amorphous diamond, field emission, black body, front panel display, infrared detection

Amorphous diamond appears to be contradictory term, like liquid crystal or glassy metal. Amorphous means non-crystalline and diamond implies crystalline. However, this terminology is meaningful because unlike silicon that forms only sp^3 bonds, i.e. diamond structure, carbon may form either sp^2 (graphitic) or sp^3 (diamond) bond. Although there is one form of amorphous silicon, there can be at least two forms of amorphous carbon, so amorphous diamond can be distinguished from amorphous graphite, and together they are amorphous carbon.

Amorphous diamond is formally known as tetrahedral amorphous carbon (tac), it is really a diamond-like carbon (DLC) that contains no non-carbon impurities (e.g. H). Amorphous diamond is essentially a chaotic carbon mixture with distorted sp^2 and sp^3 bonds. As such it possesses both metallic character of conductive graphite and semiconductor character of insulating diamond. Moreover, as each carbon atom is unique in its electronic state that is determined by the degree of distortion of its bonds. Hence, amorphous diamond contains numerous discrete potential energy for electrons. In fact, amorphous diamond may have the highest density of atoms (1.8×10^{23} per cubic centimeter) that is several times higher than ordinary materials (e.g. about four times of iron atoms or silicon atoms). Thus, amorphous diamond has the highest configuration entropy for both atoms and valence electrons.

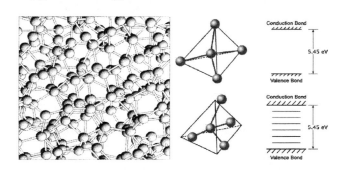

Fig. 1: The high atomic density and the unique way of distorting carbon bonds for each atom makes amorphous diamond the highest entropy material with the densest discrete atomic positions deviated from a crystalline lattice, and the most discrete electronic states of all materials.

Due to such a high configuration entropy of valence electrons, amorphous diamond is capable to advance electron energy by absorbing small increments of energy, such as by converting thermal energy (lattice vibration) to potential energy (electron state).   If amorphous diamond is exposed in high vacuum (e.g. 10^{-6} torr), the energy state may be higher than vacuum state so amorphous diamond my emit electrons simply by heating.   Because amorphous diamond has the highest discrete electronic states, it is the most thermionic material known.

In general, materials fall in three camps, conductor, semiconductor and insulator, amorphous diamond is an atomistic mixture of all, so electrons can pass through it and be emitted in vacuum.   By contrast, all other materials will stop electrons either inside the crystal lattice (e.g. an insulator) or on the surface (e.g. a conductor).   The unique ability for amorphous diamond to emit electrons in vacuum by receiving low levels of energy makes it the excellent of field emitter with very low apparent work function. Amorphous diamond can be coated on metallic substrate by cathodic arc process to become a useful field emitter in vacuum.

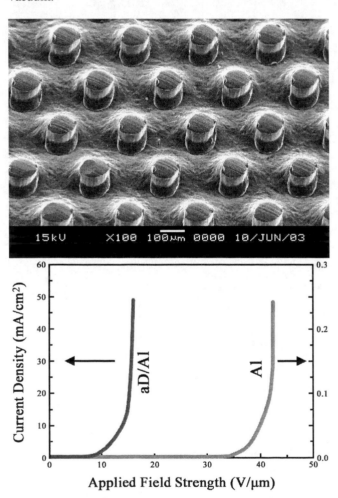

Fig. 2: The dramatic enhancement of emission current by coating amorphous diamond on aluminum cathode with bumps.

Even without high vacuum, amorphous diamond coated nickel electrodes of cold cathode fluorescent lamps (CCFL)

used for back lighting can reduce significantly the turn-on voltage.

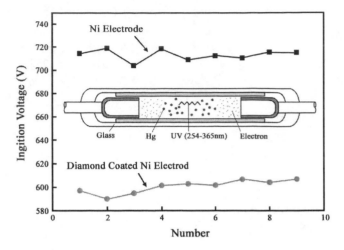

Fig. 3: The reduction of ignition voltage of CCFL by coating nickel electrodes with amorphous diamond.

Due to its exceptional ability to increase the potential energy of electrons by absorbing heat, amorphous diamond coated metal is highly thermionic.

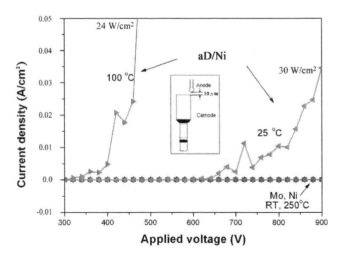

Fig. 4: The great enhancement of emitted current from amorphous diamond coated nickel electrode in CCFL by modest heating.

Materials fall under two camps: electrical conductors (metals) can conduct heat (phonon), but not emit heat (IR), and electrical insulators (ceramics) are just the opposite. However, amorphous diamond is both the thermal conductor and thermal emitter.   Amorphous diamond is not only thermionic, it is also the perfect black body. Typically a metal has a low emissivity (e.g. 2%), but an insulator has a poor thermal conductance (e.g. 10 W/mK), so both of them cannot sustain the emission far infrared from a warm surface.   However, amorphous diamond has a thermal conductivity (about 500 W/mK) that is even higher than the best metal (420 W/mK for silver), and its emissivity is nearly 100%. It was measured that at a temperature as low as 70℃, the sustained heat emission was 0.088 W/cm², this equals to what predicted by Stefan-Boltzmann's equation (5.67×10^{-8} T[K]⁴/m²) for black body. This implies that amorphous diamond has an emissivity of about 100% and the emission is not limited by its thermal conductivity. The exceptional ability to emit heat makes amorphous diamond an excellent thermal radiator for cooling high-powered LED.

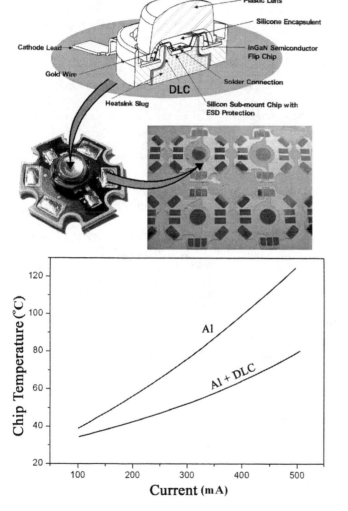

Fig. 5: The cooling of LED junction temperature by coating amorphous diamond on the surface of its aluminum substrate heat spreader.

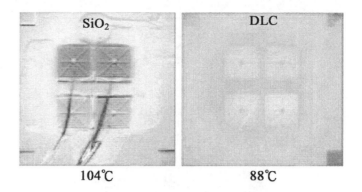

SiO₂   DLC

104℃   88℃

Fig. 6: The dramatic cooling effect by replacing silicon with an oxide coating with DLC coating for the submount of LED that was operated at 350 mA.

As the black body is reversible, amorphous diamond can be used as the heat absorber for applications related to advanced thermal imaging of infrared source or mundane water heating by absorbing sunlight.

The attribute of amorphous diamond to convert either light or heat to electricity may be used to make solar cells or thermal electrical generators.

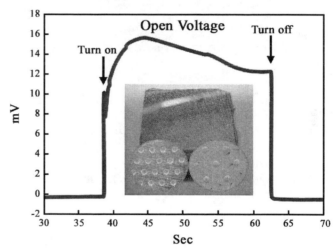

Fig. 7: The opto-electric effect of amorphous diamond when exposed to a xenon lamp (AM1.0) of about 0.1 W/cm². In the experiment, amorphous diamond coated ITO glass was separated by 7 microns from an uncoated ITO glass with isolated glassy spacer (shown as dots).

The sensitive field emission and thermionic emission of electrons by amorphous diamond would make it an ideal coating material for Spindt (metal spikes) array that may be used for front panel displays. Such field emission displays will have the lowest operational power and panel temperature.

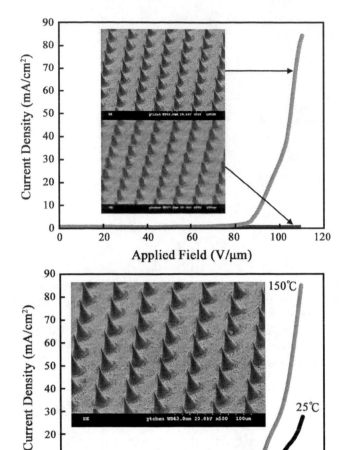

Fig. 8: The field emission of amorphous diamond coated nickel alloy cones can be enhanced by either sharpening the cone tips (upper diagram) or by modest heating (lower diagram).

In summary, amorphous diamond has the highest configuration entropy of electronic states, as a result, it is uniquely capable to emit electrons in vacuum upon receiving a low bias or simply by heating. Such unique properties makes amorphous diamond useful for a variety of applications such as field emitters, solar cells, thermal generators, radiation cooler, and heat absorbers.

REFERENCES

[1] Ming-Chi Kan, Jow-Lay Huang, Chien-Min Sung, Ding-Fwu Lii, Kuei-Hsien Chen, "Field Emission Characteristics of Amorphous Diamond", Journal of the American Ceramic Society, 86, 9 (2003) p.1513-1517.

[2] Ming-Chi Kan, Jow-Lay Huang, Chien-Min Sung, Ding-Fwu Lii, Bao-Shun Yau, "Field Emission of Micro Aluminum Cones Coated by Nano-Tips of Amorphous Diamond", Diamond and Related Materials, 12 (2003) p.1610-1614.

[3] Ming-Chi Kan, Jow-Lay Huang, Chien-Min Sung, Kuei-Hsien Chen, Bao-Shun Yau, "Thermionic Emission of Amorphous Diamond and Field Emission of Carbon Nanotubes", Carbon, 41 (2003) p.2839-2845.

[4] Ming-Chi Kan, Jow-Lay Huang, Chien-Min Sung, Kuei-Hsien Chen, "Thermally Activated Electron Emission from Nano-Tips of Amorphous Diamond and Carbon Nano-Tubes", Thin Solid Films, 447-448 (2004) p.187-191.

[5] B. R. Huangal, C. S. Huang, J. T. Tan, Chien-Min Sung, R. J. Lin, "The Field Emission Properties of Amorphous Diamond Deposited on the Cu Nanowires", 2004 Asian CVD-III, The 3rd Asian Conference on Chemical Vapor Deposition.

[6] B. R. Huang, C. S. Huang, C. F. Hsieh, Chien-Min Sung, "The Field Emission Properties of Samarium/Amorphous Diamond Field Emitters", 2004 Asian CVD-III, The 3rd Asian Conference on Chemical Vapor Deposition.

[7] Ming-Chi Kan, Jow-Lay Huang, Chien-Min Sung, Kuei-Hsien Chen, Bao-Shun Yau, "Stability of Field Emission Characteristics of Nano-Structured Amorphous Diamond Deposited on Indium-Tin Oxide Glass Substrates", New Diamond and Frontier Carbon Technology, 14, 4 (2004) p.249-256.

[8] Chien-Min Sung, Kevin Kan, Michael Sung, Jow-Lay Huang, Emily Sung, Chi-Pong Chen, Kai-Hong Hsu, Ming-Fong Tai, "Amorphous Diamond Electron Emission for Thermal Generation of Electricity", NSTI-Nanotech 2005, Anaheim, California, U.S.A., p.193-196.

[9] Chien-Min Sung, Kevin Kan, Michael Sung, Jow-Lay Huang, Emily Sung, Chi-Pong Chen, Kai-Hong Hsu, "Amorphous Diamond Electron Emission Capabilities: Implications to Thermal Generators and Heat Spreaders", ADC/NanoCarbon 2005, Chicago, Illinois, U.S.A.

[10] Chien-Min Sung, "Amorphous Diamond Materials and Associated Methods for the Use and Manufacture Thereof", U. S. Patent 6,806,629.

[11] Chien-Min Sung, "Amorphous Diamond Materials and Associated Methods for the Use and Manufacture Thereof", U. S. Patent 6,949,873.

Ultra-short pulsed laser for nano-texturization associated to plasma immersion implantation for 3D shallow doping:

Application to silicon photovoltaic structures.

M. Halbwax[*], T. Sarnet[*], Ph. Delaporte[*], M. Sentis[*]
H. Etienne[**], F. Torregrosa[**], V. Vervisch[**], I. Perichaud[***], S. Martinuzzi[***]

[*]Laboratoire LP3 CNRS UMR 6182 Luminy Marseille, France
[**]Ion Beam Services IBS, Peynier, France
[***]Laboratoire TECSEN CNRS UMR 6122 Marseille, France

ABSTRACT

It has been recently shown (Mazur *et al*) [1-7] that a simple way to improve the photocurrent of a silicon-based solar cell is to irradiate the silicon surface with a series of femtosecond laser pulses, in the presence of a sulfur containing gas. This improves the formation of micro-spikes on the silicon surface that strongly reduces the reflectivity of the illuminated surface for the incident solar light (Black Silicon).

We have prepared photovoltaic structures with different nano-texturization obtained by means of a femtosecond laser, without the use of corrosive gas (under vacuum). To take in account the 3D structured front surface, the emitter doping has been realized by using Plasma Immersion Ion Implantation (so-called PULSION). The results show a photocurrent increase of about 30 % in the laser textured zones.

Keywords: nanotexturization, black silicon, photocurrent, femtosecond laser, plasma immersion doping.

1 INTRODUCTION

In order to increase the efficiency of actual solar cells many different ways are currently being developed by researchers. They include the nano and micro-structuration of the surface [1-8], the use of antireflection (AR) coatings [9], rear totally diffused structures [10], or the use of absorbing nanoparticles [11].

A simple way to improve the silicon-based solar cell efficiency is to irradiate the silicon surface with a series of femtosecond laser pulses, in the presence of a sulfur containing gas [1-7]. This produces micro-spikes on the silicon surface that strongly reduces the incident solar light reflection (Black Silicon). We have in this study created a photovoltaic structure in a silicon wafer, which the illuminated surface was locally nanostructured (squares of 1mm^2) using a femtosecond laser, before the formation of

p-n junction. Various parameters, like polarization, spot size, energy density, number of shots, scanning parameters were chosen to make an appropriate nanotexturization without SF_6, i.e. the laser treatment was performed under vacuum (10^{-5} mbar).

The p-n junction was obtained by counterdoping the wafer surface by means of the Plasma Immersion Technique (PULSION tool, developed by IBS [11]) followed by Rapid Thermal Annealing (RTA).

It will be shown that the photocurrent increases by 25 to 30 % in the texturized areas.

2 EXPERIMENTS

2.1 Processing steps

Samples were n-type silicon doped phosphorus to 10^{15} cm^{-3} (5-20 ohms.cm), cleaned by means of the conventional RCA treatment.

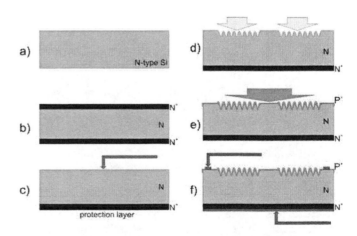

Figure 1: Processing steps of the photocell: a) RCA cleaning, b) creation of n$^+$ layer by diffusion, c) removal of the front n$^+$ layer, d) femtosecond laser structuring, e) plasma immersion doping, f) metallization of contacts

The surfaces were first phosphorus diffused from a POCl$_3$ source in order to create a n+ layer which helps the formation of a back ohmic contact , while the n+ front layer was chemically removed (CP$_4$ etch) before the laser treatment.

After the laser structuring of the surface, the samples have been boron implanted by Plasma Immersion (PULSION, BF$_3$, 2 kV, 900 °C, 30 mn) and RTA annealed. The junction depth in that case should be about 150 nm, which is much shallower than the 3D laser structures: therefore the junction follows the topography of the structures.

After realizing the p-n junction, the electrical contacts have been deposited by electron gun evaporation: a silver layer on the back side and one aluminum pad on the front non texturized surface, as shown in Fig. 1. A light-beam induced current LBIC mapping tool is used to detect the increase of the photocurrent in the texturized areas by comparison with the standard surface. This mapping tool was conceived to detect a photocurrent contrast in polycrystalline silicon between defect containing and defect free regions, giving rise to electrical image of defects [12]. In our samples it indicates directly if the photocurrent is increased below the nanotexturized areas. The photocurrent produced by a light spot (λ>800 nm), 20 m in diameter, is measured outside and inside the texturized regions, in order to evaluate the enhancement really due to the increase of the light transmission in the wafer bulk.

2.2 Laser Set-up

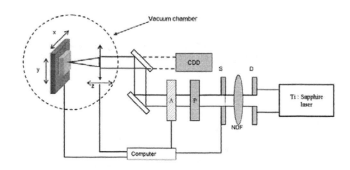

Figure 2: diagram of the femtosecond laser experimental set-up

The engraving of Si (001) was carried out in a vacuum system with a pressure of 5x10^{-5} to 1x10^{-5} mbar. This low pressure considerably reduces the redeposition of unwanted debris from the laser ablation process.

The optical set-up that was used to deliver the laser beam to the sample surface is presented in Figure 2. The micromachining experiments was performed using a Ti:sapphire laser (Hurricane model, Spectra-Physics) at 800 nm, 500 μJ energy, 1 kHz repetition rate and a laser pulse duration of 100 fs. To get a more uniform laser energy distribution, only the center part of the gaussian laser beam

was selected using a square mask (D) of 2 x 2 mm^2. A spot of about 35 x 35 m^2 area was obtained by projecting the mask image onto the sample surface with a planoconvex lens (f' 50 mm). The laser beam was perpendicular to the sample surface. A computer-controlled XY-stage (for the sample) and Z-stage (for the objective lens) has allowed precise positioning of the spot on the surface sample. The laser energy that was delivered to the sample surface could be attenuated by using a combination of analyzer (A) and polarizer (P) and completed by a set of neutral density filters (NDF). A PC controlled the analyzer rotation, the opening and closing of shutter (S) placed in front of the polarizer, and the XYZ stages. The engraving results are *in situ* monitored by a CCD camera.

Different square surfaces have been irradiated, ranging from 100x100 μm^2 to 1x1 mm^2. The experiment were carried out at two laser fluences: 140 and 185 mJ/cm^2. The laser-induced structuring of the sample surfaces was produced by scanning a simple straight line (30 μm width) at a speed-velocity of 150 m/s (maximum speed of our stages), with a d shift between the scans to treat the whole surface. The role of beam overlaps on the formed structures was analyzed by varying the distance d between the scans. For each laser fluence, the d value was equal to 1, 2, 5 and 15 m.

The laser spot displacement was perpendicular to the polarization.

2.3 Experimental Results

Nano and microstructuring of a silicon surface with a pulsed laser (from nanosecond to femtosecond) generally induces the formation of typical wave-like structures (LIPSS, capillary waves, ripples etc...) which have been studied for decades [13].

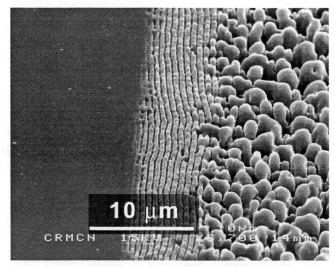

Figure 3: SEM photo of Laser Induded Periodic Surface Structures (LIPSS): capillary waves (periodicity 800 nm, center) and beads (about 2 μm, right)

In this study, the surface topography of these structures was characterized using Scanning Electron Microscopy (SEM) and Optical Microscopy (OM) in dark field mode.

Figure 3 shows different types of structures: the left part is the original Si surface, the center part is formed by capillary waves and the right part features "beads" that are formed with higher number of pulses and energy density. The periodicity of the capillary waves (center) is usually close to the laser wavelength (800 nm), they are formed by the interference between the incident beam and light scattered by minor surface defects. Ablation and melt formation occur at non-uniform depths: after resolidification, the ripple structure is frozen in place and acts as a precursor for the formation of beads, cones and spikes.

For higher energy densities and number of pulses those capillary waves tend to collapse to form a more hydrodynamically stable structure, like the beads observed in Figure 3 (right part). The absorption of light on these beads is not uniform: the ablation is maximized in the valley between the beads which tends to amplify the phenomena and creates more erected structures ("penguin-like" structures), when increasing the energy density and the number of laser pulses.

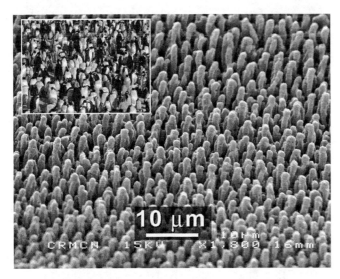

Figure 4: SEM photo of penguin-like structures created by femtosecond laser (top left corner is a picture of a real penguin colony in Antartica, *photo by G. DARGAUD www.gdargaud.net*).

Without using SF_6 gas we did not observe spikes, but rather rounded structures, like those shown in Figure 4: these "penguin-like" structures have been named from their similarity with colonies of penguins *(Aptenodytes forsteri)* when they spread over the antarctic ice, standing next to each other.

This type of structure tends to trap the incident light by decreasing the amount of specular (mirror-like) and diffuse reflections. Simple optical simulations have been performed using a commercial ray-tracing renderer (Autodesk 3ds Max) to observe the amount of light reflection on the structured surfaces, using spikes, penguin-like structures and pyramids like the one obtained by KOH anisotropic etching. The best structures seem to be the spikes, immediately followed by the penguin-like structures, in terms of light absorption. The pyramids like the one created by anistotropic etching, which are used nowadays on "high efficient" commercial photocells do not appear to be as absorbant as spikes or penguin-like structures. Moreover, these structures need to be coated (ARC) and are very dependent of the crystal orientation, which is a disadvantage on multi-cristalline silicons, in comparison with the laser structuration which is not sensitive to the different grains orientation.

Figure 5: SEM photo of a laser micro-structured area of a photo cell. Size of the treated zone is 300 μm^2, spot size 30 μm, F= 185 mJ/cm^2, d=1μm, scanning speed v=150 $\mu m/s$.

Figure 5 is a general view of a laser structured area (SEM) which shows a good homogeneity of the topography, for F= 185 mJ/cm^2 and d=1μm. When increasing d to higher values (d>5 μm) the homogeneity is not as good: the lack of overlaps (i.e. number of pulses) tends to create a mixture of capillary waves and beads. Also, working with smaller fluence or number of pulses advantages the formation of capillary waves instead of beads and cones.

The beneficial effect of the texturizations is demonstrated by the LBIC scan maps in Figure 6, which shows that the LBIC signal related to the photocurrent intensity is clearly increased in the laser treated regions.

It is also interesting to notice the high absorption around the laser treated area: this is attributed to the re-deposited nanoparticles that cover the surroundings of the treated zones.

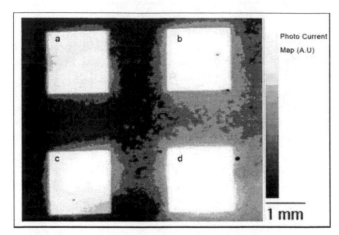

Figure 6: LBIC scan maps showing the increase in the photocurrent in the laser treated zones. spot size 30 μm, v=150 μm/s, a) F= 140 mJ/cm^2 d=1μm, b) F= 140 mJ/cm^2 d=2 μm, c) F= 185 mJ/cm^2 d=1μm, F= 185 mJ/cm^2, d=2μm

These nanoparticles were measured by SEM: they range in size from 10 to 100 nm, they are also clearly visible using dark field optical microscopy (Fig. 7).

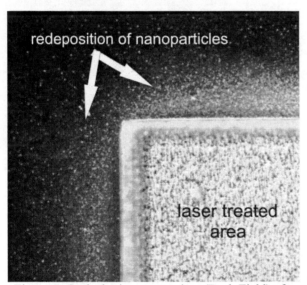

Figure 7: Optical microscope view (Dark Field) of a laser treated area showing the nanoparticle redeposition outside the spot. F= 185 mJ/cm^2, v=150 μm/s.

We measured an average photocurrent outside the spots in the order of 15 nA and smaller, whereas the current in the treated zones is in the range 19 to 21 nA: this implies an improvement in the photocurrent of at least 25 to 30 %. This improvement is probably even better because the photocurrent measured outside the spots have been overestimated by the nanoparticle-increased absorption. The real photocurrent of the non-laser treated surface should be smaller (<< 15 nA) and therefore the gain higher than 30 %. Additional measurements are still in progress and will give better quantitative results.

3 CONCLUSIONS

We have prepared laser-microstexturized Si structures which reduce the reflection of a silicon surface. 1 mm^2 areas were structured on n-type silicon with a femtosecond laser, under vacuum. The boron doped p+ regions were obtained using a Plasma Ion Immersion technique. An increase in the photocurrent was detected by LBIC and reaches more than 30 % in the treated zones.

Notice that this study has been done in a framework of renewable energy and sustainable development promotion: therefore it was important not to use SF$_6$ which has been identified by the Kyoto Protocol as one of the main greenhouse gas that contribute to climate change and global warming.

REFERENCES

[1] C. H. Crouch, J. E. Carey, J. M. Warrender, M. J. Aziz, and E. Mazur, Appl. Phys. Lett. 84, 1850, 2004.

[2] M. Y. Shen, C. H. Crouch, J. E. Carey, and E. Mazur, Appl. Phys. Lett. 85, 5694, 2004.

[3] J. E. Carey, C. H. Crouch, M. Shen, and E. Mazur, Opt. Lett. 30, 1773, 2005.

[4] J. E. Carey, Ph.D. dissertation, Harvard Univ., 2004.

[5] C. H. Crouch, J. Carey, M. Shen, E. Mazur, and F. Y. Genin, Appl. Phys. Lett. 79, 1635 2004.

[6] R. J. Younkin, J. E. Carey, E. Mazur, J. A. Levinson, and C. M. Friend, J. Appl. Phys. 93, 2626, 2003.

[7] B. R. Tull, J. E. Carey, E. Mazur, J. P. McDonald, and S. M. Yalisove "Silicon Surface Morphologies after Femtosecond Laser Irradiation" MRS Bulletin, Vol. 31, 626-633 2006

[8] J. Zhao and A. Wang, Appl. Phys. Lett. **88**, 242102, 2006

[9] A.E Mann,."Optical Coatings For Solar Cells", Spectrolab Sylmar Calif. Report AD0271358, 1960

[10] M Law et al,. "Nanowire dye-sensitized solar cells" *Nature Materials* 4:455-459, 2005

[11] F.Torregrosa, H.Etienne, G.Matthieu, L.Roux, Proceedings of the 16th International Conference on Ion implantation Technology. AIP conference pp609-613, 2006

[12] M. Stemmer, Appl. Surf. Sci., 63, 213, 1993.

[13] M. Birnbaum, "Semiconductor Surface Damage Produced by Ruby Lasers" Journal of Applied Physics, 36, 3688, 1965.

Dye-sensitized solar cell using a TiO$_2$ nanocrystalline film electrode prepared by solution combustion synthesis

C.M. Wang and S. L. Chung*

Department of Chemical Engineering, Nation Cheng Kung University, Tainan, 70101, Taiwan, ROC.
n3895115@mai.ncku.edu.tw; slchung@mail.ncku.edu.tw*

Abstract

This investigation was aimed at preparing well performance TiO$_2$ for dye-sensitized solar cell by solution combustion method. The solution combustion synthesis of TiO$_2$ powders was investigated over a wide range of synthesis conditions by using metal nitrates (oxidizer)-urea and glycine (fuel) system. Titanium (！) n-butoxide was hydrolyzed to obtain titanl hydroxide [TiO(OH)$_2$], and titanyl nitrate [TiO(NO$_3$)$_2$] was obtained by reaction of TiO(OH)$_2$ with nitric acid. Finally, the aqueous solution containing titanyl nitrate [TiO(NO$_3$)$_2$] and a fuel,i.e. urea or glycine was combusted to obtain the TiO$_2$ nanoparticles. The TiO$_2$ nanoparticles thus obtained were used to fabricate electrodes for DSSC. The performance of the DSSC was measured and compared with those fabricated with commercial TiO$_2$ powders.

Keywords: dye-sensitized solar cell, solution combustion synthesis, TiO$_2$

Introduction

Since Honda and Fuijishima found effects of photosensitization of titanium oxide electrode on the electrolysis of water into H$_2$ and O$_2$ in 1972, much attention has been focused on the use of the titanium oxide nanoparticles for various technical applications such as electrochromic materials, sterilization, self-cleaning surface, antifogging devices and dye-sensitized solar cells (DSSCs). Dye-sensitized solar cells as a new innovative technology have been developed very quickly during the past decade since Gratzal made a great breakthrough in 1991. In all these devices, the photoelectrode is made of a thin film of nanoparticulate TiO$_2$. The properties of these films certainly depend on the crystalline phase, morphology and preparation methods that were used, and it is of interest to investigate what kind of film is the most suitable for a specific application and the properties of TiO$_2$ for making the film.

TiO$_2$ films can be directly synthesized by a wide variety of techniques such as chemical vapor deposition, sol gel method, hydrothermal processing, microemulsion method, and flame pyrolysis process. Of these methods, the flame pyrolysis processing is a promising technique to synthesize nanometer-sized particles. The flame pyrolysis processing can be completed within a few seconds. This process has many potential advantages such as products obtained directly, low processing cost, energy-efficiency, and high production rate. The flame pyrolysis processing and self-propagating high-temperature synthesis (SHS) are different in approach but equally satisfactory in result. Solution combustion synthesis (SCS) was developed from self-propagating high-temperature synthesis (SHS) with combination of wet chemical techniques for the synthesis of metal oxide based materials. The process possesses both the advantages of SHS, which utilizes the in-built exothermicity of combustion of the reaction system to directly synthesize the required materials, and the advantages of wet chemical routes, which can produce a compositionally homogeneous mixture making both the ignition temperature and combustion temperature lower than those traditional SHS.

SCS is a flexible technique where oxidizing and reducing precursors are mixed on the molecular level and, under unique conditions of rapid high-temperature reactions, nanoscale powders of desired compositions can be synthesized in one step. The oxidizer (typically nitrates or oxalates) and fuel (e.g., hydrazine, hydrazide, glycine, or urea) are mixed in an aqueous solution to reach molecular level homogenization of the reaction medium. The solution is then heated until self-ignition, yielding a large volume of gas and converting the initial mixture to fine well-crystalline powders of desired compositions. Amomg the advantages of this process are low energy requirements, simple reactor setup, short reaction times, and molecular-scale mixing of precursors, leading to products of desired composition and microstructure.

We have an ongoing research program, which is aimed at investigating mechanisms of combustion synthesis, a novel technique for the synthesis of TiO$_2$ with a high specific surface area and small band gap that can function under visible light irradiation. The TiO$_2$ was used to fabricate electrodes for DSSC and was tested for performance.

Experimental

Titanyl nitrate was used as a precursor to prepare the nano-sized TiO$_2$ by solution combustion method. Titanium (III) n-butoxide was hydrolyzed to obtain titanyl hydroxide. Titanyl nitrate was then obtained by reaction of titanyl hydroxide with nitric acid. Urea or glycine(used as fuels) was added to the aqueous solution of titanyl nitrate and the solution was combusted to obtain TiO$_2$ particles. For preparation of TiO$_2$ thin-film, TiO$_2$ colloical suspensions were prepared by dispersing the TiO$_2$ particles in distilled water with polyethelene glycol, (PEG, average MW of 20000) as binder. The films were fabricated by dropping the suspension on a transparent conductive oxide glass (TOC) by using a doctor blade technique. The films were then dried at room temperature and then sintered in air at 450！ for 30min.

The sandwich-type solar cell was assembled by placing a platinum-coated TOC glass (counter electrode)beside the N719 dye-sensitized photoelectrode (working electrode) and the edges of the cell were sealed with a hot-melt Surlyn spacer (SX 1170-60, Solaronix) by heating it at ~105 ! . The redox electrolyte was composed of 0.1M LiI, 0.05M I_2 and 0.6M 1-propyl-2.3-dinethylimidazolium, and was introduced into the cell through two small holes drilled on the counter electrode. The holes were then covered and sealed with a small square sealing sheet of microscope objective glass. The resulting cell had an active area of about $0.25cm^2$. The current-voltage characteristics of the cells were measured with a dc voltage current source/monitor (Keithley, 2400). An AM1.5 solar simulator (Oriel, 66983 with a 300W Xenon lamp and an AM1.5 globle filter)was employed as the light source.

The crystalline size and phase of TiO_2 were characterized with X-ray diffraction (XRD, Rigakn, Cu-K " radiation). High resolution, field-emission scanning electron microscope (HEFE-SEM, JSM-6700F, JEOL) was used to examin the film thickness and surface morphology. The surface area and porosity of the nanoporous films were measured by a nitrogen absorption apparatus (Quantachrome NOVA 1000e).

Result and discussion

Combustion synthesis phenomena

A considerable amount of gas was generated during the combustion reaction and thus a certain amount of reaction heat is carried away through convective heat loss. The reaction can proceed in different modes. Smoldering combustion synthesis mode, is characterized by a relatively slow and essentially flameless reaction, leading to slower reaction rates as manifestes in the smoldering combustion behavior. Extremely fast reaction characterized the volume combustion synthesis mode. In this case, the reaction occurs essentially simultaneously in the whole reaction volume. With optimum fuel-to-oxidant ratios, the oxygen contained in the precursor is the main source of oxygen required for combustion reaction. Once the oxygen coming from NO_3^-, is generated, it immediately reacts with urea and oxidizes/consumes most of the fuel, and thus resulting in volume combustion synthesis reaction.

Typical time-temperature profiles of the solution combustion synthesis process when using glycine or urea as fuel are shown in Figure 1and 2, respectively.

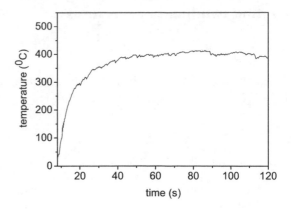

Figure 1. Temperature-time profile when using as fuel glycine

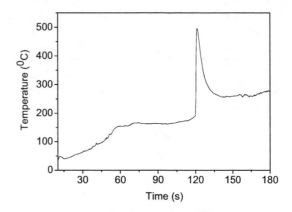

Figure 2. Temperature-time profile when using as fuel urea

The temperature first increased slowly and followed by either a sudden (at ignition temperature, T_{ig}) and uniform temperature rise to a maximum value, T_m(Fig.2), or by essentially constant profile(Fig.1). The former is a typical case of volume combustion synthesis (VCS) mode and the latter the smoldering combustion modes.

Characterization of TiO_2 powders

The variation trend of the BET surface area vs. fuel is listed in Table 1. The solution combustion method would result in gas formation since there is fuel to burn. The fuel acted merely as a space-filling template that dictated the porous structure of the product material, therefore, this combustion synthesis process yielded nanocrystalline powder with a high specific surface area.

Table 1 Effects of fuel on characteristics of TiO_2

Fuel	BET surface area (m^2/g)	crystalline phase	average particle size (nm)
Urea	78.56	Mixed-phase	38#3
Glycine	155.65	anatase	22#3

XRD pattern of solution combustion synthesized TiO_2 was recorded in 2$ range from 20 to 70˚ . Figure 3. illustrates the XRD patterns of the samples with different fuels. The pattern of the product when using glycine as fuel can be indexed to pure anatase phase of TiO_2, the rutile phase began to appear when using urea as fuel. The rutile phase of TiO_2 was formed over 600 ! and completely transformed to rutile phase at 800 ! . In this study, the rutile phase appears when the maximum combustion temperature reached 500 ! , suggesting that the synthesized TiO_2 in this process is unstable perhaps due to existence of carbon in the crystal. The carbon might be introduced from the alkoxide group, urea and improves the transformation of cyystallinity. The carbonaceous species of TiO_2 was detected with XPS, as shown in Figure 4.

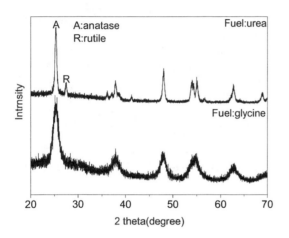

Figure 3. XRD patterns of the synthesized TiO_2

Figure 5 shows C(1s) spectra of combustion synthesized TiO_2. A peak at 285.5eV can be assigned to graphitic carbon and a lower binding energy peak at 284.3eV can be assigned to a carbidic species. Figure 4 shows Ti(2p) core level spectra. Peaks at 459 and 464.8eV indicate that Ti is in 4+ state. Therefore, we concluded that the TiO_2 indeed contains substitution of carbon in the of $TiO_{2-2x}C_xV_{O2}$, where V_{O2} represents the oxide ion vacancy created for charge balance.

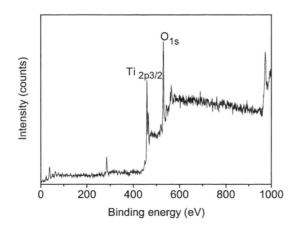

Figure 4. Ti(2p) core level spectra of urea-based TiO_2

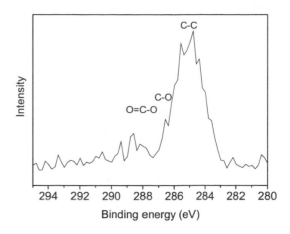

Figure 5. C(1s) spectra of urea-based TiO_2

The microstructure of solution combustion synthesized powder is presented in Figure 6, 7.

Figure 6. The FE-SEM images of urea-based TiO_2 thin film

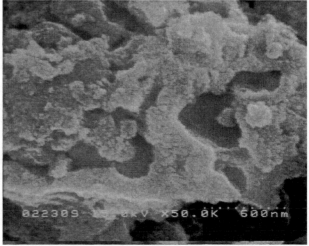

Figure 7. The FE-SEM images of glycine-based TiO_2 thin film

Figure 6 and Figure7 exhibits the FE-SEM images of TiO_2

thin film. Lots of pores are observed on the surface of the film prepared with the TiO_2 when using urea (The maximum combustion temperature is about 500 !) as fuel. (Fig.6) When small primary TiO_2 particles are employed as a starting material for preparing the thin film, the resulting TiO_2 paste is more viscous than that of large primary particles. It indicates that the severe aggregation of primary particles has already occurred before coating, resulting in the larger aggregated particles and pores. It was found that using TiO_2 powder ignition maximum temperature could significantly reduce the aggregation of primary particles during the preparation of TiO_2 slurry. When using glycine as fuel, maximum combustion temperature is about 400 ! , the sample results in a uniform thin film. There are no cracks and large aggregated particles as shown in Figure 7.

Photovoltaic performance of DSSC

The ability to control the surface structure of the TiO_2 colloids is important in all the applications that are based on processing related to the colloid surface. One such application is the dye-sensitized solar cell that consists of nanoporous TiO_2 electrode. A comparison between the photovoltaic performance of two cells with electrodes that were fabricated from the two types of colloids shows that the colloids made with glycine as fuel is preferable for dye-sensitized solar cells (Table 2). The compared cells were fabricated using similar procedures and measured under similar conditions in order to extract the contribution of the effect of surface structure. Commercial TiO_2 nanoparticles of P25 and ST01 were also used to fabricate electrodes and the performance of DSSCs were compared. Table 2 summarizes the cell performance data obtained from the I-V curve measurements.

Table 2 The cell performance data

Sample	V_{oc}(V)	J_{sc}(mA/cm^2)	Fill factor(%)	" (%)
P25	0.73	$7.92*10^{-2}$	0.62	0.36
ST01	0.69	$1.83*10^{-1}$	0.69	0.86
Glycine	0.76	$2.15*10^{-2}$	0.60	0.98
Urea	0.66	$1.88*10^{-3}$	0.67	0.82

Glycine-based DSSC exhibits the highest photovoltaic performance compared to the other DSSCs. The overall conversion efficiency(")of 0.98 for the glycine based DSSC is significantly higher than other based DSSCs. The high crystallinity and uniformity of glycine-based substrate along with the low charge transfer resistance may give rise to the overall conversion efficiency in the glycine-based DSSC. These results imply that the morphology and crystallinity of TiO_2 substrate are essential factors determining the charge transfer characteristics and overall conversion efficiency of the resulting DSSC.

Conclusion

In this work, well performance TiO_2 was synthesized by using solution combustion synthesis. To study the optimum properties in terms of morphological feature, crystallinity, specific area and photovoltaic performance,

the TiO_2 nano-particles with high specific area were synthesized by using different fuels. A DSSC fabricated with glycine-based TiO_2 showed the highest photovoltaic performance. This was attributed to the superior morphological features of glycine-based TiO_2, causing the high adsorption of dye and low interface resistance between morphological feature and specific surface area thus photovoltaic performance of TiO_2 particle was found to give rise to the significant enhancement of overall conversion efficiency of DSSC.

References

1. K. Honda, and A. Fujishima, Nature 37, 238 (1972)
2. A. Fujishima, T. N. Rao, and D. A. Tryk, J. Photochemistry and Photobiology C: Photochemistry Reviews 1, 1 (2000)
3. H. Furube, T. Asahi, H. Masuhara, H. Yamashita, and M. Anpo, Chem. Phys. Lett. 336, 424 (2001)
4. R. Asahi, T. Morikawa, T. Ohwaki, K. Aoki, and Y. Taga, Science 269, 293 (2001)
5. S. U. M. Khan, M. Al-Shahry, and W. B. Ingler, Science 2243, 297(2002)
6. Peter Erri, Pavol Pranda and Arvind Varma, Ind. Eng. Chem. Res. 43, 3092 (2004)
7. O'Regan, B.;Grätzel, M. Nature, 353, 737 (1991)

Synthesis and characterization of a highly cross linked PEGME and PEG for solid electrolyte and its application in dye-sensitized solar cells

M. Shaheer Akhtar, Ji-min Chun, Hyun-Chul Lee, Ki-Ju Kim and O-Bong Yang[*]
School of the Environmental and Chemical Engineering,
Chonbuk National University, Jeon-Ju

Abstract

In this paper, we were prepared a highly cross linked polymer with poly (ethylene glycol) methyl ether (PEGME) and poly ethylene glycol (PEG) in presence of inorganic salts by chemical method and used as polymer electrolyte with the addition of iodide couple. This electrolyte is showing ambient ionic conductivity of 2.35 mS/cm, it is comparably well high to reported electrolyte. Solid state dye sensitized solar were fabricated with this polymer electrolyte and achieved an open circuit voltage of 0.616 volt, short circuit current of 8.96 mA/cm^2 and over all conversion efficiency of about 3% under light intensity of 100mW/cm^2.

Keywords: dye sensitization, gel electrolyte, solid state dye sensitized solar cells, PEG, PEGME

Introduction

There is continuing interest in the development of high-quality and high ionic conductivity solid polymer electrolytes for the use in lithium batteries, sensors and solar cells [1-3]. The main advantage of polymer electrolytes is favorable mechanical properties, ease of fabrication of thin film of desirable size, and an ability to form effective electrode-electrolyte contacts. Generally, Polymer based electrolytes have displayed a low ionic conductivity at room temperature transport. Highly cross linked polymers have been appeared to be effective in enhancing ion conductivity, better mechanical stability and also good water absorbent, and improving the interfacial contact with electrode.

Recently, many researchers are searching an alternative candidate to replace liquid electrolyte in dye sensitized solar cells [4-5] because the use of liquid electrolyte has some limitation such evaporation and leakage of solvent in long term operation of device. A solid electrolyte has ability to solve the leakage and evaporation of liquid increase long-term stability with high performance. In spite of extensive researches, the performance of solid DSSCs is not satisfactory. This has been partially explained by imperfect contact between solid electrolytes and nanoporous TiO2 layers [6]. Many efforts have already been developed to increase the interfacial contact between electrolyte and nanoporous layers by using inorganic gel, polymer gel and organic gel electrolytes [7-9]. In this addition, we attempted to synthesize highly cross linked polymer with poly (ethylene glycol) methyl ether (PEGME) and poly ethylene glycol (PEG) in presence of inorganic salts by chemical method. In this paper, we were fabricated quasi-solid state dye sensitized solar cells by synthesized highly cross-linked polymers with iodide couple and achieved a high ionic conductivity.

Experimental

For the preparation of mesoporous TiO_2 nanocrystalline film, TiO_2 (P-25, Degussa) slurry was prepared by the incremental addition of aqueous polyethylene glycol (Fluka, average MW of 20,000) solution as binder to prevent cracking of film and control the porosity during preparation of film. Thus prepared uniform slurry was coated on FTO glass (by a doctor blade technique. After natural drying at room temperature, the thin film was calcined in static air at $450^{\circ}C$ for 30 min

The synthesis of acid of PEGME (A) was reported in literature [10]. Synthesized polymer (A, 5%) was dissolved in deionized water and adding different ratios of polyethylene glycol (PEG) under stirring and LiI (appropriate amount), after that the whole mixture heated up to $70^{0}C$ under vacuum condition for 10 hours. Finally a solid material was obtained. For Gel electrolyte, synthesized polymer mixed LiI 0.1M, I_2 0.010M in acetonitrile and then whole mixture was placed on stirrer and stirred over a period of 24 hours.

To fabricate the DSSCs, thus prepared TiO_2 thin film electrodes were immersed in the dye solution of 0.3 mM ruthenium dye (N-719) in dry ethanol at room temperature for 24 hrs. The dye-adsorbed electrodes were then rinsed with ethanol and dried under a nitrogen stream. Pt counter electrodes were prepared by electron beam deposition of Pt (60 nm thickness) on ITO glass. The resulting dye adsorbed film was seal with a Pt-sputtered conducting glass by a spacer (surlyn) and the gel electrolyte was introduced into the cell through one of two small holes drilled in the counter electrode.

The photoelectrochemical properties of the solar cell were studied by recording the current–voltage characteristics of the cell under an illumination of 1 Sun (100 mW/cm^2) using a solar simulator (Yamashita Denso, YSS-80). The area of the dye-coated TiO_2 electrode was 0.25 cm^2.

The photochemical characterization of DSSCs including photocurrent density was measured by using a scanning potentiostat (EG&G 273). The device was connected in a two-electrode configuration: the dye adsorbed TiO_2 film on TCO glass was connected as the working electrode and the Pt-coated TCO glass was used as the pseudo-reference (circuited with the counter electrode). Photocurrent–Voltage (I–V) curve was measured by using two computerized digital multimeters (Model 2000, Keithley) and a variable load. The light source was a 1000-W halogen lamp (Philips lighting) and its radiant power was adjusted with respect to Si reference solar cell (NERL, USA for Solar Energy System; Mono-Si + KG filter) to about one-sun light intensity (100 mW/cm^2).

Results and discussion

From the DSC thermograms (fig. 1) the glass transition temperature (T_g), the melting temperature (T_m) and the melting enthalpy (H_m) were determined. The obtained results for the pure PEG (figure not shown) and for the PEGME/PEG/I^-/I_3^- composite electrolyte are differed. Before the measurements, the materials were placed in a desiccator for 2 days. Then, they were heated from -80 to $100^{0}C$ with a rate of $10^{0}C$/min under nitrogen atmosphere. A slight decrease of melting temperature observed for the synthesized hybrid polymer which is around $57^{0}C$ and lower than the pure PEG ($64^{0}C$). In fact, according to the literature, the addition of inorganic salts and organic molecules in polymer matrix causes the lowering in crystallinity and prevents the recrystallization of polymer. Synthesized polymer electrolyte shows a weak signal of T_g which corresponds to high flexibility of polymer.

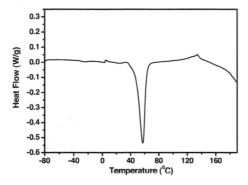

Fig.1 DSC of PEGME/PEG electrolyte

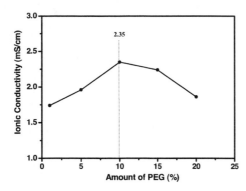

Fig. 2 Ionic Conductivity of synthesized polymer electrolyte with I^-/I_3^-.

Variation of the ionic conductivity at 25^0C as a result of changing amount of PEG is shown in fig.2. Originally, the ionic conductivity increases as increasing the amount of PEG. With 10% PEG, it attained maximum conductivity of 2.35 mS/cm after that again decrease as increasing the PEG; it is due to overloading of ether/acid groups in the hybrid polymer. Finally, we picked the best one hybrid polymer for the fabrication of solid state dye sensitized solar cells because higher conductivity of electrolyte leads the higher photocurrent.

The current-voltage characteristics of polymer gel electrolyte based DSSCs were observed under 1 sun illumination (fig.-3). Table-1 shown the averaged data extracted from I-V curve measurements of different types of electrolyte based DSSCs.

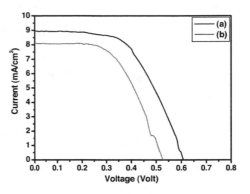

Fig. 3 IVs curve of DSSCs with (a) synthesized hybrid polymer and (b) only PEG based electrolyte.

In case of PEGME/PEG (10% w/w) based gel electrolyte cell shows the maximum efficiency (3.08%) with an I_{SC} of 8.96 mA/cm^2, a V_{OC} of 0.609 volt and fill factor 0.57 which are comparatively high in compare to only PEG gel electrolyte based cells. In the other words, the conversion efficiency of DSSCs with modified polymer based gel electrolyte increased because of the interference of acid of PEGME in the PEG matrix which provides the better interfacial contact between the electrolyte and TiO$_2$ surface. Also the high gelation supports in easy electron transport during the operation and improves the thermal stability, it gives long term life. The use of hybrid polymer materials is very effective to enhance the photocurrent and overall efficiency.

Table 1 Summary of IV data of PEGME/PEG and PEG electrolyte based DSSCs

Electrolyte	V_{OC}(volt)	I_{SC} (mA/cm²)	FF	η(%)
PEG	0.526	8.01	0.54	2.39
PEGME/PEG	0.609	8.96	0.57	3.08

Conclusion

A novel polymer (PEGME/PEG) is prepared by simple chemical method. The ionic conductivity of synthesized polymer is depended on the ratio of acid of PEGME and PEG. The polymer gel electrolyte shows high stability and a moderately high value of ionic conductivity about 2.35 mS/cm. Improved conductivity affects the photocurrent and performance of DSSCs with this gel electrolyte.

References

[1] M. Gratzel, Nature 414, 338, 2001.

[2] M. Gratzel, J. Photochem. Photobiol. A: Chem. 164, 3, 2004.

[3] A. F. Nogueira, J.R. Durrant, and M-A De Paoli, Adv. Mater. 13, 826, 2001.

[4] B. O'Regan and M. Gratzel., Nature 353, 737, 1991.

[5] M.K. Nazeeruddin, A. Kay, I. Rodicio, R. Humphry-Baker, E. Mueller, P. Liska, N. Vlachopoulos and M. Gratzel, J.Am.Chem.Soc., 115, 1263, 1993.

[6] A. Stashans, S. Lunll, R. Bergstrom, and A. Hsgfeldt, L. S. Eric, Phys. Rev. B 53, 159, 1996.

[7] N. Papageorgiou, Y. Athanassov, M. Armand, P. Bonhote, H. Pettersson, A. Azam and M. Gr̈atzel, J. Electrochem. Soc. 143, 3099, 1996.

[8] B. O'Regan, D.T. Schwartz, S.M. Zakeeruddin and M. Gr̈atzel, Adv. Mater. 12, 1263, 2000.

[9] T. Stergiopoulos, I. M. Arabatis, G. Katsaros and P. Falaras 2 (11), 1259, 2002.

[10] H. M.Xiong, Z.D.Wang, D.P.Liu, J.S.Chen, Y.G.Wang and Y.Y.Xia, Adv. Funct.Mater. 15 (2005), 1715.

Nano-Hetero Structure for direct energy conversion

L. Popa-Simil, I.L. Popa-Simil

LAVM LLC, Los Alamos, NM 87544

ABSTRACT

The direct conversion of nuclear energy becomes possible in hetero-nanostructures. To convert the kinetic energy of the fission product into electricity it has to be created a nanostructure formed from a repetitive conductor-insulator structure generically called "CIci". Its operation is based on the difference of electron shower intensity between the two conducting materials which makes that the electrons generated in the first conductor to pass through insulator and absorb in the second conductor, while this one producing a very small shower to pass in the next conductor so the conductors are polarizing. The conversion efficiency of such structures may be higher than 80%, and can be improved by quantum effects. To deliver the harvested energy outside the reactor zone a cascade of DC/AC and AC/AC micro converters have to be added at short distances inside. Because the fission energy transforms into electricity there is less energy left for heating the structure so the reactor will run cold. If superconductor structures is used the DC/AC converter may be achieved by using a SQUID Josephson junction. The direct nuclear energy conversion removes the actual thermo-mechanic devices with associated heat exchangers with higher efficiency, transforming the nuclear reactor into a solid-state electricity generator.

Keywords: nuclear fuel, hetero-structure, direct conversion, nano-structure,

1 INTRODUCTION

The direct energy conversion was a continuous subject of thinking starting from 1946, when the first patent [1] on this subject has been filed. In fact, the idea is much older, direct conversion being related to the radiation detection. The actual development of the beta batteries have used various versions of devices with ionizing radiation and charged particles [2].

The use of beta radioactive sources inside a capacitor like device to accumulate electricity were first proposed by 1956 [3] to produce low currents at high voltages, and recently retested [4].

The use of gamma rays to charge a capacitive structure by the photoelectric effect [5] was proposed by 1979 and has the advantage of higher power.

Another method of directly producing relatively low-voltage electrical power using a relatively high-energy radio-nuclide source is to irradiate a semiconductor device comprising one or a plurality of p-n semiconductor junctions connected in series or parallel [6, 7]. The p-n junction has high sensitivity to radiation damage and that is why they are using low-energy beta sources ^{147}Pr instead of ^{90}Sr. This concept is further developed in special applications of DoD [8].

The improved charged-particle powered electrical source [9] creating an improved battery for continuously-powered low-energy applications e.g., integrated microcircuits and/or sensors have been developed by 1997. The improved battery is powered by charged particles having kinetic energy that is transformed in electricity into a plurality of plate pairs or cells.

In the improved battery, the (relatively higher) kinetic energies of (relatively few) intercepted primary charged particles are incrementally converted to (relatively lower) kinetic energies of (relatively many) secondary electrons. These incremental kinetic energy conversions take place as the primary charged particles each pass through a plurality of cells comprising relatively thin plates. This relatively higher secondary electron yield in emitter plates will preferably be obtained by appropriate choices of plate materials, plate coatings, and/or plate geometry.

Differential secondary electron emission from secondary emitter [9] plates and collector plates can also be attained through emitter plate coatings (such as magnesium oxide over platinum or carbon) which increases secondary electron emission relative to that of a collector plate comprising, for example, a thin (for example, about 100 nm thickness) carbon film. Still, another method to achieve a desired cell plate differential in secondary electron emission is through control of plate geometry to maximize the probability of interaction with primary charged particles and minimize self-absorption of secondary electrons in emitter plates. Additionally or alternatively, collector plate geometry may be controlled to minimize the probability of interaction with primary charged particles and maximize self-absorption of secondary electrons.

The preferred methods of making an improved battery [2] may also comprise an additional step of choosing materials for each collector plate and each emitter plate so that, cell collector Fermi energy levels exceed cell emitter Fermi energy levels for each cell. By this procedure the chosen cell potential is about 3-10 V [9].

2 THEORY

The most nuclear reaction, fusion or fission ends by transferring the mass defect into kinetic energy of the resultant particles. These particles interact with the matter stopping and transferring all their energy to lattices warming them.

A brief description of the stopping power [10] mainly characterized by the formula:

$$S = \frac{4\pi r_0^2 m\, c^2 Z_2 Z_1^2}{\beta^2} \left[\left(\ln \frac{2mc\,\beta^2}{1-\beta^2} - \beta^2 \right) - \ln\langle I \rangle - \frac{C}{Z_2} - \frac{\delta}{2} - Z_1 L_1(\beta) + Z_2^2 L_2(\beta) \right] \quad \text{Eq.1}$$

where r_0 is the Bohr's electron radius, the first parenthesis in the second term is the relative effect on stopping, I is the mean ionization energy, C/Z_2 is a shell correction, $\delta/2$ is the dielectric polarization correction , $Z_1 L_1$ is the Barks effect for charge type corrections, $Z_2 L_2$ Bloch effect.

Finally, the energy is mainly transferred to the electrons in the lattice and nuclei and after secondary interactions dominated by the electrons, the whole energy becomes heat. For exemplification purposes a multi-layers sandwich was developed, formed from the following layers: 1st U 100 nm; (2nd S 30 nm, 3rd Al 30 nm; 4th Li 40nm; 5th Pb 30 nm; 6th SiO 40 nm; 7th Mg 30nm; 8th I 100 nm; 9th CF$_2$ 100 nm; 10th Th 100 nm; 11th Au 100 nm.

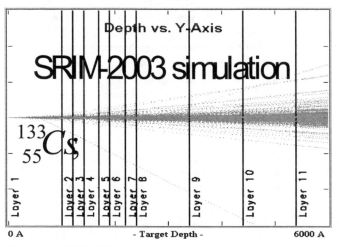

Fig. 1 – 100 MeV Cs stopping in a multi-material

The recoil implantation experiments [11] showed the capability of beams to deposit the kinetic energy into the recoil of other particles, nuclei and electrons opening various reaction channels.

The Monte Carlo simulations [12] presented in Fig. 1 shows the trajectories of Cs atoms one of the fission products in the material sandwich. The main stopping energy of moving particles in matter is due to the interactions with the electrons as Fig. 2 shows. During this process energy of 10 keV/nm is released. Considering that in 1 nm^3 there are less than 30 atoms and about 1200 electrons while only 3-4 molecules have been in the direct path of the particle. It turns that this energy is transferred to very few electrons removed by knockout from the molecular or atomic orbital.

All the range the ionization produces knock-on electrons that discharge their energy in showers down to energies of several eV. The free path in material is of several nm, comparable with the "Debye length", and is material dependent. Over the lengths of this size the electron

trajectory has ballistic behavior if does not resonates with transitory quantum states due to the fact that the structure is far from equilibrium.

The simulation showed that the ionization is a complex dynamic function of energy, ion type, and material type. Fig. 2 shows a good differential ionization rate for all the type of materials, and its dependencies, which drives to case sensitive material optimization. It also shows that using a smart material combination is possible to create a nano-hetero-structure similar to those proposed for charged particles energy harvesting [2, 5, 9] which to harvest the energy of any kind of moving particle or radiation and to be used as nuclear fuel tile.

3 DISCUSSION
3.1 The electricity harvesting structures

The previous calculations showed that a structure created of high electronic density materials, where the ionization rate is high, and produces electronic showers, followed by an insulator, with a thickness comparable with the ballistic range through electrons to pass with no absorption, if their energy exceeds a level, of few eV. After this layer, a low electronic generator material which captures the electrons and drives to an external negative plot has to have a thickness such as to stop effectively all the electronic shower and emit no or minimal shower followed by an insulator. It is created a repetitive nano-structure composed of a Conductor – Insulator-Conductor Insulator (CIci) module. Between the two conductors of the

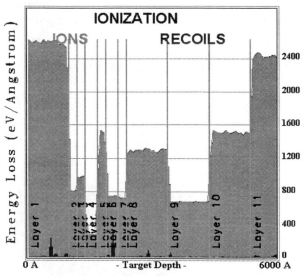

Fig. 2 – Ionization energy of 100 MeV Cs

"CIci" structure the electric charge accumulates. There are two basic potential electric connection types possible of being used: the parallel connection, where all the conductor of a kind are connected together, and this delivers lower voltages in the mV domain, given by the breakdown voltage admitted by the insulators. The serial connection, for domains where the moving particles field is constant at micrometric level, characterized by bi-polar conductive particles oriented with the high electron yield conductor towards the source of moving particles deepen into insulator material and having two conductive layers, constituting the converter module plots. This material is forming a super lattice. The actual progress is made in binary super lattice building [13] with constituents included nanoparticles of gold, lead selenide (PbSe), palladium, lead sulfide, iron oxide, and silver, as well as triangular nano-plates of lanthanum fluoride. The resulting super lattices had a range of crystal structures to produce novel materials and generate novel properties by engineering material composition at the nanometer scales and by employing natural self-assembly phenomena to control self-assembly of the lattices by tailoring the shape of the nanoparticles and using different proportions of the two nanoparticle constituents.

3.2 Conductive and super conductive materials

The analysis of the heating process in nuclear fuel driven to the conclusion that the fast moving fission products are inducing in the insulator Uranium Dioxide ceramics electron showers of few micron range. The moving electrons interact with the lattice heating it. To prevent heating it is necessary to cut the electron showers circuits and drive them through conductive materials in circuits where the total resistance to be significantly smaller.

The volume of the harvesting voxel is given by the conductivities of the two conductive layers, which have to

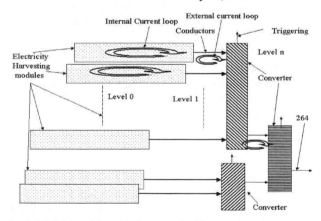

Fig. 3 Multi level based electric circuit

create an internal resistance much smaller than the previous one of the ceramics. This electrical structure reasons drives to the dimensional optimization of the direct conversion voxels.

Fig. 3 shows that each conversion voxel have to be accompanied by an electric converter DC/AC type realized in MEMS structure or micro-electronics having its impedance matched with the voxel.

The "CIci" structure has nanometer thick insulator, the maximum voltages will be less than 10 mV.

The level 1 converter has to bring this in the volts range, and another 2-3 stages are needed to get the power grid compatibility. The levels 1 output are connected into a level 2 adder, which has an AC/AC converter that increases the voltage and reduces the current in order to be applied to the next level. The final output level is compatible with the load requirements.

These modulus devices are connected in parallel at the tile level, in order to create the necessary redundancy. Due to nano-structure the controlled Josephson effect may be used to convert the electric DC power into alternate current $K_J = (483\ 597.879 \pm 0.041)$GHz/V [14]. The Andrew effect [15] may be applied to increase the conversion efficiency by quantum reflections in bi-materials, created at the minus pole, or inside bimetal polarized nano-beads.

There are three types of hetero-structures suitable for fission energy harvesting:

a) Having the fissile material U, Pu, Am not included in the structure but being surrounded by the harvesting structure thick as the range of the fission products. The ratio between the fissile material and the harvesting structure being dictated by the neutron criticality reasons.

b) Having the fissionable material mixed in the structure, mainly in the high density conductor structure by homogeneous or heterogeneous mixture

c) Having no fissile material in the structure, and shaped in tiles designed to stop by harvesting the energy of an external radiation like alpha particles from fission.

The usage of normal conductive materials drives to dimensions of the harvesting voxel in the range of mm^3 to cm^3, while the usage of superconductors may drive this volume up to dm^3.

To be used inside of a nuclear reactor volume the superconductor have first to exhibit a low absorption cross-section and to present low radiation damage. Another condition difficult to be achieved is the energetic balance, which simply means that the electric harvested energy to be smaller than the energy consumed to remove the thermal

Table 1 – Absorption cross-section [barn] of PuCoGa$_5$

	238Pu	239Pu	240Pu	242Pu
Pu	558	1017.3	289.6	18.5
Co	37.18	37.18	37.18	37.18
Ga	13.75	13.75	13.75	13.75
Amount	611.68	1068.23	340.53	69.43
Absorption decrease	91%	95%	85%	27%

effects inside, given by the particles direct heating and electric current heating effects. There are two super-conductive Plutonium based materials PuCoGa$_5$ and PuCoRh$_5$ with Tc around 18.5 K and 8.7K respectively is higher than any 5f electron configuration [16] for both materials, and equivalents based on Uranium, Americium, etc. The neutron absorption cross-section of Rh = 144 barns makes it unusable in the nuclear reactor, while Ga is exhibiting a total absorption cross-section of only 2.5 barns, over ten times bigger than that of Oxygen and Carbon, but less than ½ of that offered by W and Ti. Table 1 shows the potential superconductor material using isotopic Plutonium combinations. The stability over time of Pu structures is analyzed because of Pu's 5 MeV alpha decay into U, which under recoil becomes a Frenkel defect in the lattice inducing vortexes in the superconductor and loosing the superconductivity [17]. Compared with other superconductors e.g. $Y (Ni_{0.8} Pt_{0.2})_2 B_2C$, $CeRu_2$, or $Yba_2Cu_3O_{6.95}$, it shows about the same sensitivity to radiation damage as layered superconductors or even to anisotropic ones like BSCCO ($Bi_2Sr_2CaCu_2O_{8+\delta}$) it exhibits same properties. What matters is the fact that an aging process is active driving to the modification of the material properties in time. The difference of usage of a superconductor structure inside a nuclear reactor will become a dynamic equilibrium between the rate of destruction of the superconductor structure under radiation and the rate of repairing it under the nucleation and growth of the interstitial loops and cavities [18]. To define a radiation resilient micro structure we have to further understand the role of 5f electrons [19] and superconductivity of the nano-structures. Hybrid materials and hetero-superconductive structures are possible if the cross-sections and radiation specific behavior are optimal. A similar process as quenching and re-crystallization may be used to create a dynamic superconductive layer floating inside the fissionable product in excess, until the nuclear reactor life is over.

4 CONCLUSIONS

The fabrication of nano-metric hetero-structures for the direct conversion of the radiation energy into electricity is an important development for the future power-sources.

The usage of the Actinides superconductors opens the way to the cryogenic nuclear reactors acting like ultra compact batteries with huge powers based on fission.

The development of energy harvesting panels for outer space applications and fusion energy harvesting using superconductors or normal conductors

The development of self-healable super-conductive or normal nano-hetero-structures will provide the longevity of the nuclear structure until all nuclear fuel is wasted.

The usage of the nano-hetero structure is pushing up the limits of power density in the nuclear fuel from actual about 0.2-0.5 [kw/cm^3] to 5000 [Kw/cm^3] theoretically predicted.

REFERENCES

1. LINDER E. G., Method and means for collecting electric energy of the nuclear reactions. US2517120, 1946.
2. YOUNG R.D., H.J.P., LIGHT G.M., SEALE S.W. Jr.,, Charged-Particle Powered Battery. WO9748105A1, 1997.
3. WILLSON V.C., Generator of Power. US2728867, 1955.
4. Polansky Gary, Direct Energy Conversion Fission Reactor. NERI, 1999. **Proposal No.: 99-0199**(1): p. 2.
5. RITTER J.C., Radioisotope photoelectric generator. US4178524, 1979.
6. RAPPAPOR P., Radioactive batteries. US3094634, 1953.
7. OLSEN K., Semiconductor nuclear battery. U.S. Pat. No. 3,706,893, 1972.
8. Tsang T.F., The Liquid Electronics Advanced Power Systems (LEAPS). GTI -News/DARPA, 2004. **web**(1): p. 1-3.
9. YOUNG R.D., H.J.P., LIGHT G.M., SEALE S.W. Jr.,, CHARGED-PARTICLE POWERED BATTERY. US5861701, 1999.
10. ZIGLER J.F., Stopping of energetic light ions in Elemental Matter. Journal of Applied Physics, 1999. **85**: p. 1249-1272.
11. MUNTELE C.I. et al., The recoil implantation technique developed at the U-120 cyclotron in Bucharest. AIP Conference Proceedings, 1999. **475**(1): p. 558-560.
12. ZIGLER J.F., SRIM. IBM-web, 2003.
13. Talapin D., Nanoparticle superlattices offer new properties. nanotechweb.org, 2006. **web**.
14. (CODATA), C.o.D.f.S.a.T., Joshepson constant. NIST References, 2002. **Constants**.
15. Andreev A.F., Particle Scattering at Interfaces Between Superconductors or Superconductor-Normal Metal. Sov. Phys. JETP, 1964. **19**: p. 1228.
16. Thompson J. D., T.P., Curro N. J., Sarrao J. L.,, Progress and Puzzles in Plutonium Superconductors. J. Phys. Soc. Jpn., 2006. **75**(Suppl.): p. 1–3.
17. Kazuki Ohishi, T.U.I., Wataru Higemoto, Robert H. Heffner,, Influence of Self-Irradiation Damage on the Pu-Based Superconductor PuCoGa5 Probed by Muon Spin Rotation. J. Phys. Soc. Jpn., 2006. **75**: p. 53–55.
18. Rest J., H.G.L., Irradiation-Induced Recrystallization of Cellular Dislocation Networks in Uranium-Molybdenum Alloys. Mat. Res. Soc. Symp. Proc., 2001. **650**.
19. Gouder T., W.F., Rebizant J., Lander G.H.,, Understanding Actinides through the Role of 5f Electrons. www.mrs.org/publications/bulletin, 2001. **September**: p. 688.

Nanotechnology – Disruptive Technologies for Electric Utility Systems. Challenges and Opportunities.

John Stringer* and Roger H. Richman**

*Izambard, Redwood City, California, USA,
jstringer@izambard.com
**Daedalus Associates, Inc., Mountain View,
California, USA.

ABSTRACT

Within the next 25 years, the electric power utility system in the U.S. will face a number of serious challenges. These will include issues related to diminishing supplies and increased costs of fossil fuels, the demands for a reduction in the emissions of greenhouse gases, an increasing requirement for distributed generation and its integration into the grid system; and an increased demand for "digital quality power". Digital devices are highly sensitive to the slightest fluctuations in power supplies, and it is expected that 30% or more of the demand in 2025 will be for power capable of meeting this requirement. Nanotechnology offers possible solutions to these challenges; and some of these are discussed here. While some of these developments are evolutionary in character, based on the improvement of technologies that are currently in research and early development, the magnitude of the problems suggest a need for what have been called "disruptive technologies" and these are also discussed here.

1 INTRODUCTION

The U.S. electricity enterprise is one of the largest industries in the U.S.; it is approximately twice the size of telecommunications and nearly 30% larger than the automobile industry in terms of annual sales. Demand for electricity is projected to more than triple by 2050. Coincident with (and related to) the need for increased generating capacity is a number of serious challenges that face the power industry:

- Diminishing supplies and increasing cost of fossil fuels;
- Mandatory reduction of greenhouse gases released to the environment;
- Need for distributed generation and its seamless integration with the electricity grid;
- Demand for "digital quality" power.

This last challenge represents a major change in the demand pattern for electricity.

Digital devices are highly sensitive to even the slightest interruption in power: an outage of even a fraction of a single cycle can compromise performance. Likewise, variations of power quality caused by transients, harmonics, and voltage surges must be avoided. Power with sufficient quality and reliability to serve digital loads now constitutes about 10% of total electrical load in the U.S.; it is expected to reach 30% by 2020.

The immense size of the electricity enterprise would appear at first to be incompatible with nanotechnologies; several studies, however, have identified numerous possible directions [1,2]. Some of these nanotechnologies are evolutionary; that is they are extensions and improvements of conventional technologies. However, others fall into another class where they represent a radical change from the existing practice: this is called 'Disruptive Technology'.

The term 'disruptive technology' is generally credited to Clayton M. Christensen, in his 1995 article *Disruptive Technologies: Catching the Wave*, co-authored with Joseph Bower [3]. He further developed the concept in his book *The Innovator's Dilemma* [4] (Harvard Business School Press, 1997). He remarked "Disruptive technologies typically enable new markets to emerge." It is interesting that even in the 2002 edition of his book, he makes no mention of nanotechnologies. The concept met with considerable criticism; for example in 2004 John C. Dvorak in *The Myth of Disruptive Technology* [5] wrote "There is no such thing as a disruptive technology. There are inventions and new ideas, many of which fail while others succeed. That's it." At almost the same time Christensen, with Michael E. Raynor, wrote a sequel to his earlier book entitled *The Innovator's Solution: Creating and Sustaining Successful Growth* [6] (2003; Harvard Business School Publishing Corporation). In this, he replaced the term 'disruptive technology' with 'disruptive innovation' because they recognized that few technologies are intrinsically disruptive: it is strategy that creates the disruptive impact. The general idea is that a new technology first targets customers at the low end of the market who do not need the full performance capability. The performance of the product is lower than the incumbent, but exceeds the requirements of certain segments at a lower cost, thereby gaining a foothold in the market.

This is called 'low-end disruption'. To go further, the disruptor has to innovate, and the effect is to squeeze the incumbent into the higher-end markets. There is also the situation where the new product is inferior by most measures of performance, but fits a new or emerging market segment: Christensen calls this 'New Market Disruption'. There comes a point where the new technology outperforms the older technology, but the existing player may be unable to afford to move into the new area, for example because of the level of the investment in the older technology:

Joab Jackson, in *WashingtonTechnology,*[7] (1/27/2003) identified nanotechnology as a "coming disruptor", noting the president's FY 2003 budget request for $710 million for nanoscale science, engineering and technology. John Taylor, Director-General of the Research Councils in the UK, in a report entitled "New Dimensions for Manufacturing: A UK Strategy for Nanotechnology" [8] discusses the implications of nanotechnology under the heading "Nanotechnology is Disruptive – What this Means for Manufacturing Sectors with Reference to the UK". Specifically, he writes "A key issue therefore that could disadvantage the UK, compared to other advanced industrial nations, would be a failure of its companies to appreciate that nanotechnology is really disruptive – that it will generate major paradigm shifts in how things are manufactured. Nanotechnology could lead to changes that equal the revolutions ushered in by semiconductor technology and biotechnology."

In this paper, we will discuss four technologies of importance to the future of the electric power industry that may be regarded as 'disruptive nanotechnologies'.

2 PHOTOVOLTAIC SOLAR CELLS

A prime example of non-polluting distributed generation is the use of solar cells to supply electricity. In fact, if solar cells were price-competitive with grid electricity, the way that utility customers obtain power would be changed radically. Several possible routes to inexpensive solar cells are outlined next.

A promising approach is the dye-sensitized solar cell (DSSC), also called the Grätzel cell [9]. This consists of a nanocrystalline mesoporous network of a wide band gap semiconductor (usually TiO_2) that is covered a monolayer of dye molecules (usually a Ru dye). The semiconductor is deposited onto a transparent conductive oxide, through which the cell is illuminated. The TiO_2 pores are filled with a redox electrolyte that acts as a conductor to a platinum electrode.

A different approach to nanostructured PV uses a "bulk heterojunction" design; the idea is to develop a structure of two interpenetrating continuous polymer phases, one of which is composed of donor molecules, the other of acceptor molecules, with each phase attached to a different electrode [10]. The structure evolves because the selected polymers have a low entropy of mixing and separate on the nanoscale. At this time, structures produced by this strategy are not sufficiently

regular to achieve good performance. However, the introduction of time-resolved electrostatic force microscopy [11] should be a powerful tool for attacking this problem.

An embodiment of a quantum dot solar cell is a nanocomposite consisting of a porous oxide and a conjugated polymer. SnO_2 films with pore diameters of ~100 nm have been fabricated. Intercalation of polymers into the pores by absorption from solution yields structures with ~75% of the free volume filled with polymer [12].

Another route to nanostructured solar cells is the Hybrid Nanorod–Polymer Solar Cell [13] in which the bandgap is tuned by altering the nanorod diameter. A device has been fabricated, by solution processing, that consists of 7 nm x 60 nm CdSe nanorods in a conjugated polymer [poly-3(hexylthiophene)]. A power-conversion efficiency of 6.9% was obtained under 0.1 mW/cm^2 illumination at 515 nm.

One of the approaches currently being investigated might enable the disruptive technology needed to make solar cells a primary power source.

3 SENSORS

Integration of widespread distributed generation with an electricity grid that delivers digital-quality power will require current and voltage information on a continuous basis from many locations simultaneously. Although conventional sensors cannot provide that capability because they are not miniaturized [14], nanosensors based on magnetoresistance effects can [15]. For example, giant magnetoresistance devices (also called spin valves) consist of two or more layers of ferromagnetic metal separated by nonferromagnetic spacer layers. With a total thickness of 30 nm or less providing magnetoresistance ratios of 10-15%, they are the basis for 250 million magnetic read heads manufactured each year for the hard-disk industry. Tunneling magnetoresistance sensors are similar to spin valves, except that an ultra-thin insulting layer separates two ferromagnetic layers. Only when the magnetization directions are aligned in the magnetic layers is there a high probability of electrons tunneling quantum-mechanically through the insulator. Magnetoresistance ratios of >100% have been measured in prototype devices. Spintronic nanosensors with sensitivities of 1nT (1 nanoTesla) or better could be deployed to the grid in just a few years.

4 THERMOELECTRICS

Thermoelectric systems convert thermal gradients to electricity or electricity to thermal gradients. Although they are quiet, rugged, stable, and reliable, thermoelectrics have been regulated to niche applications because they are inefficient (≤5%). Efficiency improvements by factors of seven or eight are needed to make thermoelectrics competitive for distributed generation or for refrigeration [16]. In conventional materials the parameters of thermoelectric efficiency cannot be separately optimized because they are not independent: changing one parameter also changes the others. However, reducing dimensions to the nanometer scale can uncouple the efficiency parameters. 2-D quantum wells [17], short-period superlattices [18], and quantum dots [19] all show significantly higher efficiencies than their bulk counterparts.

Multilayer heterostructures probably cannot be made large enough for grid-connected electricity generation. Two approaches for developing the requisite high-efficiency bulk materials are being studied: (1) formulation of alloys in which nanoscale composition modulations can be induced in the solid state [20]; (2) self-assembling nanocrystal superlattices in which separately optimized nanoparticles interact synergistically [21]. Although truly high-efficiency thermoelectrics are still in the future, early results are very promising.

5 CATALYSIS

Catalysts touch directly several aspects of the electricity enterprise, among them the Polymer Electrolyte Membrane Fuel Cell (PEMFC) and photochemical splitting of water. Both applications could become disruptive technologies. PEMFC's low operating temperature (~80°C), and consequent quick start-up time, make it an obvious candidate for distributed generation. Platinum-alloy catalysts are used at both the cathodes and the anodes of PEMFCs. At the cathode, the oxygen-reduction reaction is still too sluggish. At the anode, Pt-based catalysts are susceptible to poisoning by CO and sulfur species remnant in the hydrogen produced from hydrocarbon feedstocks. Nanoscale strategies have offered some improvement: submonolayer clusters of Pt atoms on Ru nanoparticles are more effective catalysts than current Pt-Ru alloys and they are more resistant to CO poisoning [20]. Mitigation of sulfur poisoning remains to be addressed.

Photocatalytic splitting of water by sunlight is an appealing route to a "green" hydrogen economy, because carbon dioxide is not produced. It has been known since 1972 that n-type TiO_2 is a photocatalyst that can split water [21]. However, the 3.0 eV bandgap of TiO_2 allows absorption of only the UV portion of the solar spectrum, which is but 2-4% of the available energy. There has been some progress in reducing the bandgap by doping TiO_2 with carbon [22]: the absorption threshold is shifted from 414 nm to 535 nm, that is, into the visible range, but much of the spectrum is still not utilized.

An increase in our capability to create and manipulate nanoscale structures might lead to new, more potent catalysts for low-temperature reactions. The discovery that nanoparticulate gold is an excellent catalyst for some reactions, whereas bulk gold is not [23], encapsulates that hope.

REFERENCES

[1] *Energy and Nanotechnology: Strategy for the Future,* held at the Baker Institute, Rice Univ., May-4, 2003.

[2] *Program for Technology Innovation: Nanotechnology Opportunities for the Electric Utility Enterprise,* EPRI, Palo Alto, CA: 2005. 1012933.

[3] Clayton M. Christensen, in his 1995 article *Disruptive Technologies: Catching the Wave*, co-authored with Joseph Bower

[4] Clayton M. Christensen, *The Innovator's Dilemma* (Harvard Business School Press, 1997).

[5] John C. Dvorak, *The Myth of Disruptive Technology*, www.pcmag.com, article date 08/17/2004.

[6] Clayton M. Christensen and Michael E. Raynor, *The Innovator's Solution: Creating and Sustaining Successful Growth* (2003; Harvard Business School Publishing Corporation).

[7] Joab Jackson, *WashingtonTechnology,* (1/27/2003)

[8] John Taylor, (U.K. Department of Trade and Industry Report, June 2002; see www.dti.gov.uk)

[9] B. O'Regan and M Grätzel, *Nature,* **353** (1991) 737.

[10] G. Yu, J. Gao, J. C. Hummelen, F. Wudl, and A. J. Heeger, *Science,* **270** (1995) 1789.

[11] D. C. Coffey and D. S. Ginger, *Nature Mater.,* **5** (2006) 735.

[12] S. Shaheen, K. Brown, A. Miedaner, *et al,* "Polymer Based Nanostructured Donor-Acceptor Heterojunction Photovoltaic Devices," presented at the Solar Program Review Meeting, 24-26 March, 2003, Denver, CO.

[13] A. P. Alvisatos, "Hybrid Nanorod–Polymer Solar Cell," Final Report, 19 July-19 September 2002. NREL/SR-520-34567, August 2003.

[14] J. Stringer and R. H. Richman, in *Encyclopedia of Smart Materials, Vol. 2,* M. Schwartz, ed., John Wiley & Sons, Inc., New York, 2002, pp. 873-890.

[15] A. Jander, C. Smith, and R. Schneider, in *Advanced Sensor Technologies for Nondestructive Evaluation and Structural Health Monitoring,* Proc. of SPIE, Vol. 5770, 2005, pp. 1-13.

[16] R. H. Richman and J. Stringer, in *Mass and Charge Transport in Inorganic Materials: Fundamentals to Devices,* P. Vincenzini and V. Buscaglia, eds., Techna Srl, Faenza, Italy, 2000, pp. 1355-1362.

[17] S. Ghamaty and N. B. Elsner, in *Thermoelectric Materials 2003—Research and Applications,* G. S. Nolas, *et al.,* eds., MRS Symp. Proc. Vol. 793, Materials Research Soc., Warrendale, PA, 2004, pp. 225-228.

[18] R. Venkatasubramanian, E. Siivola, T. Colpitts, and B. O'Quinn, *Nature,* **413** (2001) 597.

[19] T. C. Harman, P. J. Taylor, M. P. Walsh, and B. E. LaForge, *Science,* **297** (2002) 2229.

[20] K. F. Hsu, S. Loo, F. Guo, *et al., Science,* **303** (2004) 818.

[21] J. J. Urban, D. V. Talapin, E. V. Shevchenko, C. R. Kagan, and C. B. Murray, *Nature Mater.,* **6** (2007) 115.

[22] S. M. Haile, *Acta Mater.,* **51** (2003) 5981.

[23] A. Fujishima and K. Honda, *Nature,* **238** (1972) 37.

[24] S. U. M. Khan, *Science,* **297** (2002) 2243.

[25] M. Haruta, *Catalysis Today,* **36** (1997) 153.

The Use of Microfluidics to Combat Fuel Crime
(Catching Mr Big with the Small)

I. M. Eastwood*, M. Al Jafari**, E. T. Bergstrom***, E. Dorland*, D. M. Goodall*** and A. Taylor*

*Authentix Ltd, 7 Chessingham Park, York, YO19 5SE, UK
*** Department of Chemistry, University of York, York, YO10 5DD, UK

ABSTRACT

Taxation on fuels is an incentive for crime syndicates to set up smuggling and laundering groups that extract billions of dollars from the global economy every year. This crime falls into three categories, adulteration, grade swapping and tax evasion e.g. smuggling. In order to combat these crimes Authentix has developed and installed advanced microfluidic systems to test the fuel. Our system allows efficient marker testing in-field and we have demonstrated consistent extractions under a variety of conditions. Although microfluidics drives the technical engine of the testing program, our multi-million dollar success in these areas is also related to our ability to bring about legal enforcement. This is done through wide-scale testing and control programs.

Keywords: fuel marking, microfluidics, revenue recovery laundering, smuggling

1 INTRODUCTION

The significant taxation on fuels is a big incentive for criminals and terrorist groups to set up smuggling and laundering syndicates that cost legitimate businesses and governments billions of dollars every year. In our experience this varies from between 7 to 30% of sales, depending on the measures in place within the country (or company). This can be reduced significantly with an effective marking program. In some cases these syndicates undermine the stability of the government and reduce a government's ability to attract investment. Moreover, the lost revenues can fund groups that have an active interest in destabilising and undermining governments. Although losses are large due to the volumes of fuel used, in fact this is a greater loss to the country than first appears. In addition to funding organised crime and thus taking physical resources in terms of policing etc., the company loses the immediate revenue that would have resulted from a sale and the government its tax. Furthermore, there is less legitimate money running through fuel stations and consequently distribution companies reduce their investment as they see very little return from these loss-making franchises. This ultimately leads to the loss of jobs and people turning out of

necessity to the very crime syndicates that have reduced their income. From the government standpoint the income from personal taxation and company taxes is reduced. As GDP is reduced locally, it interferes with the ability to invest in infrastructure (which would ordinarily produce more income/GDP). Thus there is a crossover point where organised crime can become so profitable that the government becomes starved of finances and resources to such an extent that it can never gain effective control without taking extraordinary measures.

Authentix is heavily involved in a variety of nanotechnology programs in fuels management; the company has successfully used Nanotag™ in Malaysia and has used nano and microfluidic systems to combat many of these crimes. It has applied this in fuel systems from heavy crude through to Liquid Petroleum Gas (LPG).

Authentix has developed three signature systems for meeting the needs of different client groups such as:

1.1 Those wishing to identify even tiny quantities of adulterants in fuels (such as low tax kerosene, tax exempt diesel in road fuel or even LPG). For this system, we use markers that are recognised by antibody receptor molecules in a lock-and-key fashion. The exquisite selectivity and sensitivity of the antibodies renders them particularly useful for identifying and quantifying a few parts per billion of the adulterant marker. These markers must be resistant to extraction and attempts at laundering.

1.2 Those wishing to determine if the fuel has been tampered with or is below specification (grade swapping).
In these situations, each of the fuels is coded with a different marker. Any dilution with the wrong or cheaper grades will be flagged with a value of less than 100%. A different technology from that used in 1.1 needs to be employed, and preferably one that can read a marker for each grade

1.3 Those wishing to identify dilution of fuels with smuggled fuel from, for example, neighbouring states. In this case (as for 1.2) normally the fuel is marked and results below 100% imply that smuggled fuel or an adulterant has been added; either way, this is illegal.

Although microfluidics provides the engine of the testing program, our success in these areas is also related to our ability to test the fuel and bring about legal enforcement. This is done through wide-scale testing and control programs. One drawback from field testing is the need to have highly skilled employees doing the testing and this can be time consuming, costly and produces some chemical waste that needs to be disposed of appropriately.

Authentix has been actively involved in developing microfluidic systems based on lab-on-a-chip technologies. The goal was to efficiently and quantitatively remove the marker from the fuels without the need of a full scale laboratory. The use of microfluidics lends itself to our needs. The ability to produce a lab-on-a-chip system that will perform chemical reactions and extractions in one device takes away much of the need for highly skilled staff. Samples can be introduced into a black box, a button is pressed and after a few minutes a reading is given. Very little waste is produced during the process and because the volumes are so very small, diffusion can usually take the place of mixers etc..

Most investments in microfluidics have supported development of uses with biological fluids – which are almost invariably water-based systems. It is much more difficult to do this work with two immiscible fluids. It is also a challenge to find pumps and materials that are compatible with the extremes of pH and the solvents we use. So initially glass was our preferred substrate.

2 MICROFLUIDICS DEVELOPMENT

2.1 The Reactor

This consists of a 2 layer glass microchip. A D shaped groove is etched into the first layer at 35 micrometers deep and 60 micrometers wide. The channel is sealed with the second glass layer. Both glass surfaces are optically polished before sealing. (See Figure 1)

Figure 1. Picture of the re-usable fuel chip prior to connection into a reader.

2.2 Extraction

This is done using marked diesel and an aqueous extractant. The two fluids flow side by side for a few seconds (channel is ~11 cm long). The two flows are then split such that 100% of the fuel and some 10-20% of the extractant moves down one capillary whilst the remaining 80-90% extractant is passed to the detector

2.3 Detection

Detection of the fluorescent markers was done by epifluorescence using a sapphire ball lens and suitable beam splitting mirrors, a blue LED light source and a silicon photodiode as a sensor (see Figure 2). This has the benefit of making the detector very robust, small and simple to situate at the extractant outlet of the chip. Synchronous detection using a laboratory lock-in-amplifier is used in order to keep the electronic and light bleed noise to a minimum. With this setup, the noise level is not limited by the detector as the chemical background of the system' is higher than the noise of the detector.

Figure 2. The epifluorescent detector – note the capllary running along the front edge of the device beside the sapphire ball lens.

3 RESULTS

Figure 3 shows the effect of running fuel and extractant together, producing stable laminar flow. It was necessary to vary the flow rate to achieve stable laminar flow as this is dependent upon the viscosities of the fuel and the extractants, which in turn are dependent upon temperature. Under the right conditions, it was possible to get $100 \pm 3\%$ extraction of the marker. Figure 4 shows the eluates from the separation system.   In order to prevent any fuel contamination it was found that it was best to have a ratio of 1.1–1.2 : 1 of extractant to fuel volume flow rates, resulting in pure extractant at the other outlet (is my reword correct?). The system was optimised for consistency and speed rather than purely extraction efficiency. At flow rates of 4   l min⁻¹, extractions of 88 - 90% efficiency were generally achieved using diesel fuel.   These results could be generated within 30 seconds of starting the test. More importantly, much less than 500   l of fuel and reagent were used in the test, so that there was very little disposable waste.

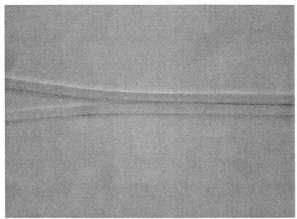

Figure 3. Laminar flow of two phase systems containing diesel and an aqueous extractant

Figure 4. Excellent separation of markers from fuel and aqueous extractant.

4 DISCUSSION

The results of this study have shown that Authentix can produce a simple lab-on-a-chip device capable of doing the normal extractions required for a typical laboratory-based system in a fraction of the time required to do the same work at a macro scale (30 seconds *vs* 10 – 20 minutes). It also gives us the concurrent benefit of using minute quantities of reagents and producing a fraction of the waste. The system is almost solid state with few moving parts and can be easily reduced to a small hand-held device. Automating the process has reduced the possibilities of user error and initial field tests on the system have demonstrated good reliability. However, we have seen problems of 'tube flow' under certain conditions where the extractant coats the walls of the glass, with the fuel flowing as a tube within the extractant.   Some levels of temperature control and a redesign of the chip to reduce the chance of this happening are under investigation.

Authentix recognises the help of two DTI funded Knowledge Transfer Partnerships between Authentix, The University of York and The University of Hull in order to accelerate this work.

TiN/GaN Metal/Semiconductor Multilayers for Thermionic Energy Conversion

V. Rawat* and T. Sands**

School of Materials Engineering, School of Electrical & Computer Engineering, Purdue University,
501 Northwestern Avenue, West Lafayette, IN-47906
*vrawat@purdue.edu, **tsands@purdue.edu

ABSTRACT

TiN-GaN multilayers were grown for potential application as solid-state thermionic direct energy conversion devices using reactive pulsed laser deposition in an ammonia ambient. The crystallographic analysis of the multilayers by high-resolution x-ray diffraction and cross-sectional TEM revealed that, despite the difference in crystal structures of TiN and GaN, it was possible to grow thick uniaxially textured columnar-grained multilayers. In-plane electronic transport was assessed using Hall effect and Seebeck coefficient measurements. Thermal conductivity measurements have shown that by increasing the interface density, the cross-plane thermal conductivity of the multilayers can be reduced to 3.6 W/m-K, compared to 135 W/mK for bulk GaN and 38 W/mK for bulk TiN.

Keywords: thermionic energy conversion, reactive pulsed laser deposition, TiN, GaN.

INTRODUCTION

The prospects for a compact, solid-state thermal-to-electrical energy conversion device with efficiency above 20% for hot-side temperatures in the range of 300-700°C has motivated research in nanostructured thermoelectric (TE) materials for more than a decade. Such a device would find early application in vehicle waste heat recovery systems, steam-free powering for naval vessels, and terrestrial generators employing concentrated solar energy. Although recent laboratory-scale research on bulk and thin-film nanostructured materials based on PbTe show promise for enhanced efficiencies, the best such material does not yet meet the requirements for 20% device efficiency [1,2]. Recent theoretical investigations [3,4] have suggested that metal/semiconductor multilayer-based thermionic (TI) energy converters can achieve much higher efficiencies than conventional thermoelectric devices. Such multilayers with nanoscale periods employ metal/semiconductor barriers to enhance the asymmetry of the differential conductivity about the Fermi energy (i.e. high hot electron concentrations) [3]. For applications involving moderate to high hot-side temperatures (~300-700 C), the multilayers must be stable against corrosion, decomposition, and interdiffusion. The nitrides meet these criteria, and offer potential materials combinations for metal-semiconductor multilayers. Nitrides such as TiN, ZrN, VN, and TaN are metals whereas GaN, InN, ScN and their alloys are semiconductors. Furthermore, the electrical and thermal properties of the nitride multilayers and their interfaces can be tuned by alloying. In this work, we describe our efforts to evaluate the potential of TiN-GaN multilayers for direct thermal energy conversion.

EXPERIMENTAL DETAILS

The nitride films and multilayers were deposited on sapphire and MgO substrates in a high vacuum pulsed laser deposition system (PVD Products, Inc.) with a base pressure of 8×10^{-8} torr. The targets were a TiN disk of 2" diameter and a liquid gallium target contained in a stainless steel dish. A 248 nm KrF excimer laser (Lambda Physik 305i) was used to generate 25 ns pulses at 5 Hz with a pulse energy of 650 mJ and a fluence of 8 J/cm^2 at the target. The process gas was ammonia at a pressure of 20 mtorr and a flow rate of 55 sccm. To ensure uniform film deposition, the targets and the substrate were rotated and the laser beam was rastered over the target surface.

Prior to deposition, both sapphire and MgO substrates were ultrasonically cleaned in acetone and isopropanol and then rinsed in deionized water. The above process was repeated three times. Sapphire substrates were then chemically etched in a 3:1 solution of sulphuric acid: phosphoric acid (H_2SO_4:H_3PO_4) at a temperature of 100°C for 15 minutes. After etching, the sapphire substrates were rinsed in DI water for 3 minutes. The substrates were mounted with indium on a Mo disk and loaded into the deposition chamber. Both of these substrates were then annealed in vacuum at 585°C for 30 minutes to allow for surface reconstruction [5]. Using the TiN/GaN multilayers with different multilayer periods were grown on sapphire and MgO substrates.

RESULTS & DISCUSSION

X-ray diffraction results obtained from TiN/GaN multilayers grown on sapphire (0001) substrates confirmed the presence of uniaxial texture within the multilayers as shown in figure 1. The orientation relationship between the TiN layers, GaN layers and the sapphire substrate was determined using asymmetric θ-2θ scan and Φ-scan about GaN (10-12) peak and it was found to be TiN (111)[1-10] || GaN (0001)[11-20] || sapphire (0001)[11-20]. The lattice mismatch between a TiN (a=0.424 nm) film and a GaN film (a=0.319 nm, c=0.518 nm) grown in the crystallographic orientation state above is 8.55%.

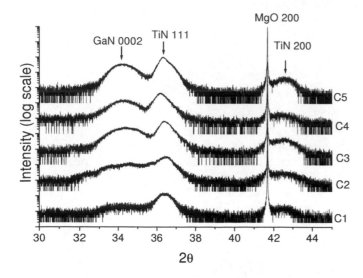

Figure 1: X-ray diffraction patterns obtained from five different TiN-GaN multilayers grown on sapphire with following periods, a) C1, λ=3.4 nm, b) C2, λ=6.8 nm, c) C3, λ=15.9 nm, d) C4, λ=19.1 nm and e) C5, λ=22.5 nm.

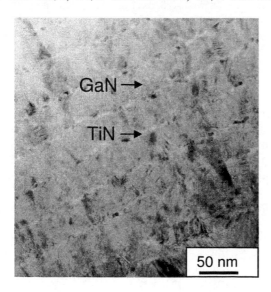

Figure 2: Cross-sectional TEM image of a TiN(5nm)-GaN(30nm) multilayer grown on a sapphire (0001) substrate.

The cross-sectional TEM image shown in figure 2 confirmed that, despite the differences in crystal structure and lattice parameter, the layered structure was preserved during growth. Similarly, TiN/GaN multilayers with thin layers of GaN sandwiched between TiN layers were grown on a cubic substrate (i.e. MgO) in an attempt to pseudomorphically stabilize the metastable rocksalt phase of GaN. Such pseudomorphic stabilization of the rocksalt phase has been achieved in the case of AlN, which exhibits a T-P phase diagram that is similar to that of GaN with wurtzite, zincblende and rocksalt phases[6]. X-ray

diffraction patterns obtained from such short-period multilayers grown on MgO are shown in figure 3. Presence of evenly spaced satellite peaks and the absence of wurtzite GaN peak, confirms of the stabilization of rocksalt phase of GaN. This was seen only in the multilayers where the GaN layer thickness was less than 2nm. With the increase in GaN layer thickness, the wurtzite phase of GaN started stabilizing. The [010] zone axis electron diffraction pattern from one such multilayer grown on MgO with thick GaN layers is shown in figure 4. The electron diffraction pattern confirms the presence of the wurtzite phase of GaN.

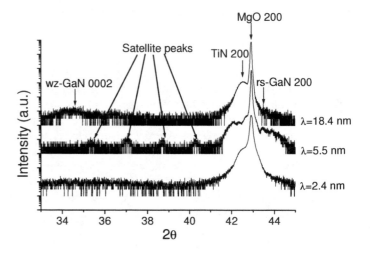

Figure 3: X-ray diffraction patterns obtained from TiN/GaN multilayers with short periods, grown on MgO substrates.

In-plane electrical characteristics were determined using Hall effect measurements. In-plane Hall mobility, carrier concentration and electrical conductivity of TiN/GaN multilayers grown on sapphire are shown in figure 5. In these samples the thickness of the TiN layer was varied from 0.5 nm to 5nm while the GaN layer thickness was kept constant at 10 nm. It can be inferred from the figure that with increasing TiN layer thickness, the electrical characteristics of the multilayers can be tuned from that of being a pure semiconductor to that of a metal. Similarly, the Seebeck coefficients of the multilayers vary from values characteristic of a degenerate semiconductor to those expected from a metal. The effective power factor "$S^2\sigma$" for these multilayers was calculated from the experimentally determined electrical conductivity and Seebeck coefficient values. The maximum value of in-plane power factor was found to be 1.5×10^{-3} W/m-K^2 for a TiN/GaN multilayer with a TiN layer thickness of 1 nm and a GaN layer thickness of 10 nm. The cross-plane power factor is expected to be much higher than the in-plane power factor due to the energy filtering effect.

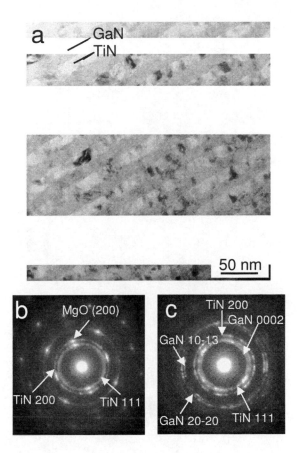

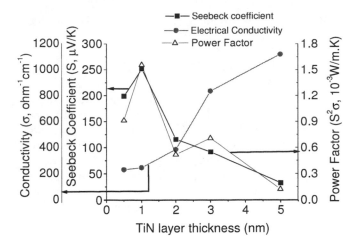

Figure 4: (a) Cross-sectional TEM image of a TiN(10nm)-GaN(variable thickness) multilayer grown on MgO(100). (b) [010] zone axis diffraction pattern obtained from the MgO substrate and the multilayer. (c) Selected area diffraction pattern obtained from the multilayer only, showing presence of uniaxial texture.

Figure 6: Left axis: In-plane electrical conductivity and Seebeck coefficient, Right axis: Calculated power factor as a function of TiN layer thickness. GaN layer thickness was kept constant at 10nm while the TiN layer thickness was changed.

Preliminary measurements of the room temperature cross-plane thermal conductivity of multilayers comprising of TiN(bulk κ=38 W/mK) and GaN(bulk κ=135 W/mK) layers on sapphire using time-domain photoreflectance show that short period multilayers have thermal conductivities as low as 3.6 W/m-K Measurements as a function of temperature are in progress. Further reduction in thermal conductivity is expected with solid-solution alloying of the metallic and semiconducting layers.

CONCLUSIONS

TiN/GaN multilayers with uniaxial texture were grown on cubic (MgO) and rhombohedral (sapphire) substrates using reactive PLD in an ammonia ambient. In-plane electrical measurements have shown that, depending on the relative thicknesses of the layers, the multilayer's electrical behavior can be tuned to yield semiconductor or semimetal like behavior. Thus, by varying the multilayer period, it is possible to obtain a multilayer with an optimum in-plane power factor($S^2\sigma$). The cross-plane thermal measurements revealed that by increasing the interface density, the thermal conductivity of these otherwise thermally conducting nitrides can be reduced drastically.

The TiN/GaN multilayers prepared in this study may represent the first crystalline metal/semiconductor multilayers with the nanoscale periods necessary to investigate thermal and electrical transport phenomena at characteristic length scales that are comparable to electron and phonon wavelengths. Preliminary x-ray scattering data suggests that rocksalt-structured metal-semiconductor superlattices are possible in this system for GaN layers thinner than 2 nm. Although measurements of in-plane electronic transport properties are promising, cross-plane thermal and electronic transport measurements will be

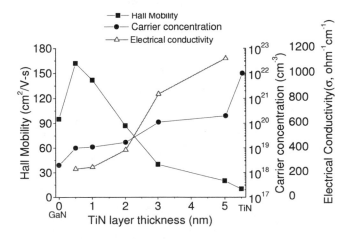

Figure 5: Left axis: In-plane Hall mobility, Right axis: In-plane electrical conductivity and carrier concentration as a function of TiN layer thickness. GaN layer thickness was kept constant at 10nm.

necessary to test the solid-state thermionic concept for direct thermal energy conversion.

ACKNOWLEDGEMENTS

The authors would like to acknowledge Rajeev Singh and Prof. Ali Shakouri (UCSC) for Seebeck coefficient measurement, and Kee Yon Koh and Prof. David Cahill (UIUC) for thermal conductivity measurements.. This work is funded by ONR/DoD through a MURI grant. A part of this work was carried out in the Center for Microanalysis of Materials, University of Illinois, which is partially supported by the U.S. Department of Energy under grant DEFG02-91-ER45439.

REFERENCES

[1] R. Venkatasubramanian, E. Siivola, T. Colpitts and B. O'Quinn, Nature **413**, 597(2001).

[2] T. C. Harman, P. J. Taylor, M. P. Walsh and B. E. LaForge, Science **297**, 2229(2002).

[3] D. Vashaee and A. Shakouri, Phys. Rev. Lett. **92**, 106103(2004).

[4] G. D. Mahan and L. M. Woods, Phys. Rev. Lett. **80**, 4016(1998).

[5] K.G. Saw, J. Mat. Sci. **39**, 2911(2004).

[6] A. Madan, I. W. Kim, S. C. Cheng, P. Yashar, V. P. Dravid, S. A. Barnett, Phy. Rev. Letters **78**(9), 1743 (1997).

Pt and Pt-Ru/Carbon Nanotube Nanocomposites Synthesized in Supercritical Fluid as Electrocatalysts for Low-Temperature Fuel Cells

Yuehe Lin[1,*], Xiaoli Cui[1], Jun Wang[1], Clive. H. Yen[2] and C. M. Wai[2]

[1]Pacific Northwest National Laboratory, 902 Battelle Blvd, P. O. Richland, WA 99352, Tel: 1-509-376-0529, Email: yuehe.lin@pnl.gov
[2] University of Idaho, Moscow, ID 83844, USA

ABSTRACT

In recent years, the use of supercritical fluids (SCFs) for the synthesis and processing of nanomaterials has proven to be a rapid, direct, and clean approach to develop nanomaterials and nanocomposites. The application of supercritical fluid technology can result in products (and processes) that are cleaner, less expensive, and of higher quality than those that are produced using conventional technologies and solvents. In this work, carbon nanotube (CNT)-supported Pt and Pt-Ru nanoparticles catalysts have been synthesized in supercritical carbon dioxide (scCO$_2$). The experimental results demonstrate that Pt, Pt-Ru/CNT nanocomposites synthesized in supercritical carbon dioxide are effective electrocatalysts for low-temperature fuel cells.

Keywords: carbon nanotubes, fuel cell, supercritical fluid

1 INTRODUCTION

Direct methanol fuel cells (DMFC) is considered as one of the most promising options to solve the future energy problem because of its high energy conversion efficiency, low pollutant emission, low operating temperature, and simplicity of handling and processing of liquid fuel.[1-7] Electrocatalysts with higher acitivity for methanol oxidation at room temperature are critically needed to enhance its performance for commercial device applications. It is well known that platinum is the only single-component catalyst that shows a significant activity for methanol oxidation. Considerable efforts have been devoted to design and synthesize Pt-based alloy catalysts with higher poison tolerance and greater methanol oxidation activity.   PtRu is a promising catalyst for methanol oxidation in DMFC. It is well known that the preparation technique is one of the key factors to determine its catalytic activity. Numerous approaches including impregnation and chemical reduction electrodeposition sputtering method have been developed in order to generate clusters on the nanoscale and with greater uniformity on the carbon supports. The PtRu nanoparticles can be prepared by chemical reduction with formic acid or impregnation method, a microwave-assisted polyol process using metal precursors H$_2$PtCl$_6$ and RuCl$_3$. Chemical methods are the most widely used in the synthesis of metal or mixed metal nanoparticles. However, conventional preparation techniques based on wet impregnation and chemical reduction of the metal precursors are often time-consuming and labor-intensive. In addition, these procedures often do not provide adequate control of particle shape and size. The physical methods mainly proceed in a vacuum through atomization of metals by thermal evaporation or sputtering.

The use of supercritical fluids (SCF) for the synthesis and processing of nanomaterials has gained considerable interest in recent years. SCF exhibits an attractive combination of the solvent properties of a gas and a liquid. It can dissolve solutes like a liquid, and yet possess low viscosity, high diffusivity, and zero surface extension like a gas. Furthermore, the solvent strength of SCF can be varied by manipulating fluid temperature and pressure, thus allowing a degree of control and rapid separation of products, which is impossible using conventional solvents. It provides a rapid, direct and clean approach to preparing nanomaterials and nanocomposites. These special and unique features make SCF an attractive medium for delivering reactant molecules to areas with high aspect ratios, complicated surfaces, and poorly wettable substrates. The supercritical carbon dioxide (scCO$_2$) allows reactive components to penetrate inside the porous materials themselves, partitioning into the inner regions of the porous supports. Through hydrogen reduction of metal-β-diketone complexes in scCO$_2$, multiwalled carbon nanotubes can be decorated by metal nanoparticles with uniformity to achieve nanocomposites. In previously papers, we have demonstrated high activity for oxygen reduction reactions of Pd/CNT and Pt/CNT processed in SCF. Recently, platinum and ruthenium naoparticles were successfully loaded on carbon aerogel in scCO$_2$. In principle, a number of metal precursors can be used as starting materials in SCF and metal alloys can be coated on CNT to form nanocomposites. Cu-Pd alloy nanoparticles attached to SiC nanowires through hydrogen reduction of a mixture of Cu(hfa)$_2$·xH$_2$O and Pd(hfa)$_2$·xH$_2$O [hfa= hexafluoroacetylacetonate] in scCO$_2$ have been obtained. In this paper, platinum/ruthenium alloy nanoparticles were decorated on carbon nanotubes in scCO$_2$ and the nanocomposites were characterized by transmission electron microscopy (TEM) and X-ray diffraction (XRD). The PtRu/CNT powder was loaded on the glassy carbon electrode through a casting process, and the electrocatalytic activity for methanol oxidation was investigated in 1 M H$_2$SO$_4$ at room temperature using electrochemical methods such as cyclic voltammetry (CV), linear sweep voltammetry (LSV), chronoamperometry (CA), and electrochemical impedance spectroscopy (EIS). Its catalytic performance was compared with that of Pt/CNT synthesized in scCO$_2$.

Experiments

2.1 Decorating PtRu Nanoparticles on Carbon Nanotubes

The PtRu/CNT catalyst was synthesized by using the following procedures: The CNT (20 mg) and the metal precursors Pt(acac)$_2$ (20 mg) and Ru(acac)$_2$ (20 mg) [acac = acetylacetonate] with a small amount of methanol (3 mL) as a modifier were all loaded in a high pressure reaction cell (10 mL) located in a oven at 200°C. Pt(acac)$_2$ and Ru(acac)$_2$ have a low solubility in supercritical CO$_2$. Addition of methanol modifies the polarity of CO$_2$ and enables dissolution of the Pt and Ru precursors in the fluid phase. Carbon dioxide gas was introduced into the reaction cell and was pressurized to 80 bar in order to make the gas become a supercritical fluid. Hydrogen gas at 10 bar was initially filled in the H$_2$+CO$_2$ mixer cell and CO$_2$ gas of 120 bar was then added to the cell. After one hour of waiting for the precursors to completely dissolve in the supercritical CO$_2$, the H$_2$+CO$_2$ gas was introduced into the reaction cell by pressurizing it to 160 bar. The reductions (Pt^{2+} to Pt^0 and Ru^{2+} to Ru^0) were fast and took 15 min only. After depressurizing the reaction cell, PtRu/CNT powder could be recovered and then was washed 5 times using methanol and ultrasonication for 30 minutes each time. The detailed procedures for the preparation of Pt/CNT were described in our previous report.

2.2 Electrode preparation and modification

A 0.5 wt% Nafion solution was prepared by diluting the 5 wt% Nafion solution with water. Catalyst powder was dispersed ultrasonically in 0.5% Nafion solution to obtain a homogeneous black suspension solution with 1 mg/mL PtRu/CNT or Pt/CNT, and a 5 L aliquot of this solution was pipetted onto glassy carbon (GC, 3 mm in diameter, BAS, West Lafayette, IN) electrode surface. Before the surface modification, the GC electrode was polished with 0.3 μm and 0.05 μm alumina slurries, washed with water and acetone, and finally subjected to ultrasonic agitation for 1 minute in ultrapure water and dried under an air stream. The coating was dried at room temperature in the air for 1 hour. The catalyst should be homogeneously dispersed and the same procedures for experiments should be controlled in order to obtain reproducible results. The modified electrode surface was then washed carefully with ultra pure water before measurement.

2.4 Apparatus

Cyclic voltammetry, linear sweep voltammetry, and chronoamperometry experiments were performed with a CHI 660 electrochemical workstation (CH Instruments Inc, Austin, Texas). All electrochemical experiments were carried out with a conventional three-electrode system. The working electrode was glassy carbon coated with Pt/CNT

or PtRu/CNT composite films. An Ag/AgCl (saturated by KCl solution) reference electrode was used for all electrochemical measurements, and all the potentials were reported versus this reference electrode. A platinum wire was used as a counter electrode. In order to obtain reproducible and reliable results, a fresh methanol solution was used for every measurement. All the electrochemical experiments were carried out at room temperature.

RESULTS AND DISCUSION

Figure 1 (left) shows a typical TEM image of the carbon nanotube-supported platinum nanoparticles. The little dark spots which represent the Pt nanoparticles are closely attached to the CNT surfaces. The Pt nanoparticles have a size distribution range of approximately 5 to 15 nm. This size distribution of Pt nanoparticles is considered suitable for fuel cell applications since Peuckert et al. showed that once the Pt particle size was below 4 nm, the oxygen reduction peak would drop significantly.

It has been suggested that the curvature of the carbon nanotubes was a controlling factor for attachment of a certain range of metal nanoparticles to the nanotube surfaces during the supercritical fluid deposition.[1,5] Larger particles probably would fall off from the nanotube surfaces during the deposition process. There are two possible mechanisms for the formation of

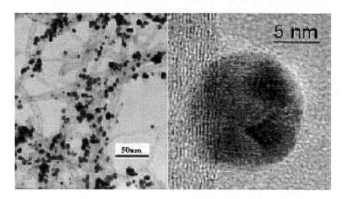

Figure 1. TEM (left) and HRTEM (right) images of a Pt-CNT sample prepared by the supercritical fluid deposition method

nanoparticles on CNT surfaces. The first one is that if the metal precursor molecules are close to the CNT surface when hydrogen is introduced, reduced metal atoms may grow directly on the functional groups of the substrate surface to develop uniformly distributed metal nanoparticles that strongly adhere to the surface of the CNT. The second mechanism is the aggregation of the reduced metal, forming nanoparticles in the fluid phase far away from the CNT surface, followed by attachment of the particles to the functional groups of the substrate surface by collision. If the size of the metal nanoparticles is too large for attachment to the CNT surface, these nanoparticles will fall off because of the curvature limitation. If the size of the metal nanoparticles is moderate and suitable for attachment to the curved surface, they may still adhere to the CNT surfaces. The second mechanism probably is a main cause of the variation in particle size distribution of the metal particles observed on CNT surfaces after washing and sonication. A HRTEM micrograph shown in Figure 1 (right) indicates that Pt nanoparticles are crystallites with visible lattice fringes. The Pt-CNT sample was washed and ultrasonicated several times to remove the metal particles that are not attached or adhered to the CNT' surfaces before characterization. The remaining platinum nanoparticles were attached strongly to the surfaces of CNT and showed no obvious changes from TEM images taken after catalysis experiments. Methanol Oxidation Reaction is also a very important indication of the electrocatalytic activity of a catalyst especially for direct methanol fuel cell.

The Oxygen Reduction Reaction (ORR) is especially important for the realization of highly efficient fuel cells, batteries, and many other electrode applications. Platinum particles on a variety of carbon supports are the most widely used and efficient catalysts for the cathode of fuel cells. For the ORR experiments at the GC/Pt-CNT electrode, a solution of 0.1 M H_2SO_4 was purged with ultrapure oxygen for 15 min. The solution became completely saturated with oxygen. The electrode was scanned over a potential range from 0.7 V to 0 V for 5 cycles to ensure reproducibility and the last cycle is shown in Figure 2. Figure 2 illustrates the cyclic voltammograms of GC/CNT (a) and GC/Pt-CNT (b) electrodes for the reduction of oxygen reaction. For the

oxygen reduction proceeded in a relatively positive potential region. A cathodic catalytic peak current occurred at about 0.42 V. The oxygen-reduction potential showed a significant shift anodically in the presence of platinum on the carbon nanotubes. The peak current increases linearly with the square root of the scan rates which indicates that the ORR process on Pt-CNT is controlled by the diffusion of oxygen to the electrode surface. This observation reveals a high activity to oxygen reduction by Pt-CNT synthesized in supercritical CO_2.

The electrocatalytic activity for methanol oxidation of Pt-CNT prepared in supercritical CO_2 was characterized by cyclic voltammetry in an electrolyte of 1 M H_2SO_4 and 2 M CH_3OH at 50 mV/s. Figure 3 shows the cyclic voltammograms of GC/CNT (a) and GC/Pt-CNT (b) electrodes for the oxidation of methanol reaction. This feature of CV curve is in agreement with the literature report for Pt/C catalysts. The ratio of I_f to I_b can be used to describe the catalyst tolerance to carbonaceous species accumulation. High I_f/I_b value implies good oxidation of methanol to carbon dioxide. In our experiments, the ratio was estimated to be 1.4 for the Pt-CNT electrode. Such a high value indicates that most of intermediate carbonaceous species were oxidized to carbon dioxide in the forward scan. For comparison, a ratio of 0.87 was reported with a nanosized Pt on XC-72 synthesized by a microwave-assisted polyol process. The experimental results highlighted the high activity of the Pt-CNT prepared from supercritical CO_2 for methanol oxidation. The high activity may result from the high surface area of CNT and the nanostructure of the platinum particles.

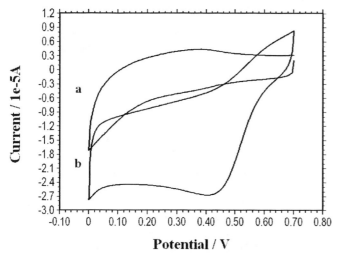

Figure 2. Cyclic voltammograms at GC/CNT (a) and GC/Pt-CNT (b) electrodes for oxygen reduction reaction at scan rate of 0.02 V/s. The electrolyte is 0.1 M H_2SO_4 saturated with oxygen The scan involves 5 cycles and the fifth cycle (last cycle) is shown here.

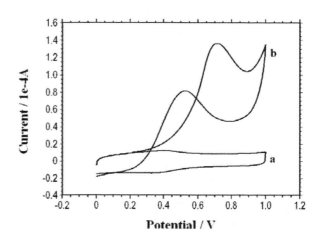

Figure 3. Cyclic voltammograms of room-temperature methanol oxidation on GC/CNT (a) and GC/Pt-CNT (b) electrodes cycled from 0 V to 1.0 V vs Ag/AgCl at 50 mV/s in 1 M H_2SO_4, 2 M CH_3OH.

electrode of GC/CNT, oxygen reduction current increased with the potential decrease, suggesting that the catalytic reduction of oxygen at GC/CNT is a kinetics-controlled process. For the electrode of GC/Pt-CNT

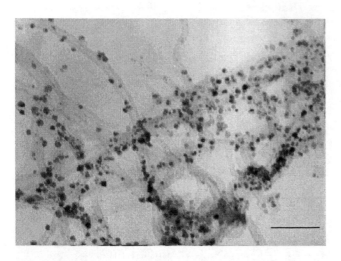

Figure 4. A typical TEM image of CNT decorated by platinum/ruthenium nanoparticles synthesized in supercritical carbon dioxide. The scale bar is 50 nm.

Figure 4 shows the typical transmission electron microscopy micrograph of PtRu/CNT electrocatalysts prepared in $scCO_2$. The nanoparticles show a good distribution on the surface of CNT and the particles with an average size of 5~10 nm. The EDS results indicates that the Pt and Ru content in the nanocomposite is 4.1% and 2.3 % by weight and molar ratio approximately Pt:Ru = 45:55, respectively.

The electrocatalytic activity for methanol oxidation of PtRu/CNT prepared in $scCO_2$ was characterized by cyclic voltammetry in an electrolyte of 1 M H_2SO_4 and 2 M CH_3OH at 50mV/s, and the resulting voltammograms are shown in Figure 5. Figure 5 shows the results of Pt/CNT electrodes for the electrooxidation of methanol. Before recording cyclic voltammograms, the electrode was soaked in the test solution for 10 minutes to allow the system reaching a stable state. The potential scan starts from its open circuit potential positively to 1.0 V then to -0.1 V and then it cycles between -0.1 V and 1.0 V for PtRu/CNT electrode. The cyclic voltammograms of PtRu/CNT are different from those obtained on Pt/CNT electrode. The onset of methanol oxidation occurs at about 0.05 V at the electrode of PtRu/CNT. The lower onset potential indicates clear evidence for superior electrocatalytic activity for methanol oxidation. There are large changes in the potential cycles over time. The early cycles exhibit a low peak current and lower peak potential in the forward scan and there is almost no peak during the reverse potential scan. As the number of cycles increases, the potential for the peak current in the forward scan shifts to higher values and the peak potential increases greatly. The forward anodic peak potential shifts more positively from cycle to cycle. By the sixth cycle, the forward peak potential is located at about

0.57 V and a broadened peak centered at 0.45 V appeared in the reverse scan. Then, anodic peaks appeared in both the forward and backward scans. Significant changes can be observed between the sixth and twenty-five cycles as shown in Figure 4. The peak potential in the reverse scan moves to higher voltages and the peak current increases. This phenomenon may result from the leaching of ruthenium during the cycling of the electrode between oxidizing and reducing potentials. At last, the cyclic voltammogram of the catalyst was found to have become similar to that of pure Pt/CNT, indicating excessive loss of Ru. Previous results have been shown that similar performance was observed on a commercial E-TEK catalyst PtRu/C on carbon paper as reported in a recent paper. The loss of ruthenium was confirmed by TEM and X-ray microanalysis in the commercial catalyst.

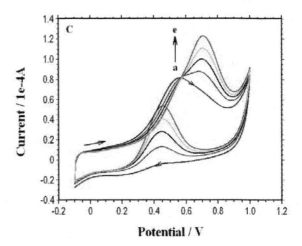

Figure 5. Cyclic voltammograms of room-temperature methanol oxidation on PtRu/CNT (C) electrodes cycle cycles 5, 10, 15, 20, 25 from a to e (C) respectively at 50 mV/s in 1 M H_2SO_4, 2 M CH_3OH.

REFERENCES

1. X.-R. Ye, Y. Lin, C. Wang, M. H. Engelhard, Y. Wang, and C. M. Wai, *J. Mater. Chem.* 14, 908 (2004).
2. Y. Lin, X.-R. Ye, and C. M. Wai, in *Dekker Encyclopedia of Nanoscience and Nanotechnology*, ed. J. A. Schwarz, C. Contescu, K. Putye, Marcel Dekker, New York (2004), p. 2595-2607.
3. X.-R. Ye, Y. Lin, C. Wang, and C. M. Wai, *Adv. Mater.* 15, 316 (2003).
4. X.-R. Ye, Y. Lin, C. M. Wai, J. B. Talbot, and S. Jin, *J. Nanosci. Nanotech.* 5, 964 (2005).
5. X. R. Ye, Y. Lin, and C. M. Wai, *Chem. Commun.* 642 (2003).
6. Y. Lin, X. Cui, and X. Ye, *Electrochem. Commun.* 7, 267 (2005).
7. Y. Lin, X. Cui, C. Yen, C.M. Wai. *J. Physical Chem. B,* 109, 14410-14415, 2005

Nanoporous Silicon Membrane Based Micro Fuel Cells for Portable Power Sources Applications

K. L. Chu[*], M. A. Shannon[**], R. I. Masel[***]

[*]Department of Chemical and Biomolecular Engineering, University of Illinois at Urbana-Champaign
600 South Mathews Avenue, Urbana, Illinois, 61801, USA, kchu1@uiuc.edu
[**]Department of Mechanical and Industrial Engineering, University of Illinois at Urbana-Champaign
1206 West Green Street, Urbana, Illinois, 61801, USA, mshannon@uiuc.edu
[***]Department of Chemical and Biomolecular Engineering, University of Illinois at Urbana-Champaign
r-masel@uiuc.edu

ABSTRACT

In this paper the preparation of nanoporous silicon membranes and their usage for the solid electrolyte in micro fuel cells compatible with silicon micro-fabrication technology is presented. The effects of different membrane structures and fuel concentrations were studied. And the micro fuel cell design for improved performances is discussed.

Keywords: nanoporous silicon, formic acid, fuel cell

1 INTRODUCTION

Silicon-based fuel cells are under active development for electrical power supply to micro systems by some research groups using perfluorinated polymer materials, for example Nafion membranes, as solid electrolyte [1-4]. But polymer materials are not suitable for silicon micro-fabrication technology due to its volumetric variation with changes in hydration level and its incompatibility with photolithography process. For solid electrolyte in micro-scale fuel cells, the proton conducting material reported in this work is based on a nanoporous silicon membrane, which was formed by electrochemically etching silicon in hydrofluoric acid solution and is compatible with current silicon micro-fabrication technology.

2 EXPERIMENTAL

Nanoporous silicon membrane fabrication procedure is shown in Figure 1. N-type antimony-doped silicon wafers with resistivity range from 0.005 to 0.02 Ω-cm were used. (Figure 1 A) Wafers were patterned by silicon nitride deposition and photolithography to obtain circular windows with diameter of 5.3 mm. (Figure 1 B and C) Silicon membranes with thickness of 50, 100, or 150 μm were obtained by wet-etching silicon-nitride-patterned wafer using KOH solution. (Figure 1 D) Nanoporous silicon was formed by electrochemically etching silicon membrane in an electrolyte solution consisted of wt. 49% hydrofluoric acid and ethanol (volume ration of 1:1), with anodic current density of 20, 40, or 80 mA/cm^2 applied to the silicon membrane. (Figure 1 E) The electrochemical etching was carried out in an AMMT etching system. Nanopores grew from one side of the silicon membrane to the other, and a thin silicon substrate layer beneath the nanoporous silicon layer was removed by reactive ion etching. (Figure 1 F)

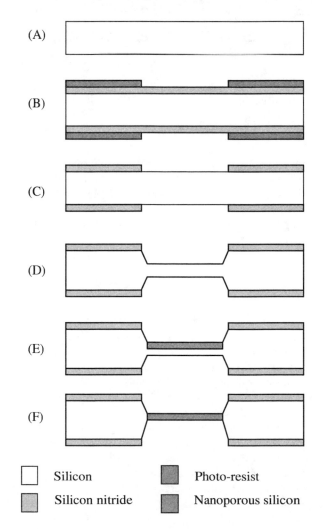

Figure 1. Nanoporous silicon membrane fabrication process

Membrane-electrode-assembly (MEA), the majoe part in micro fuel cell, was made by direct ink-painting technique. Ink of platinum nanoparticles, Millipore water, and Nafion solution was mixed and painted on one side of membrane as cathode, and ink of same solvent but with palladium nanoparticles was painted on the other side as anode. The catalyst ink was dried by heating samples on a hot-plate. After catalyst films were formed, a 5 nm thick gold-palladium alloy film was sputtered on top of catalyst layers on both sides. Then current collector was formed by painting gold ink on top of the sputtered thin film. The final structure of MEA is shown in Figure 2.

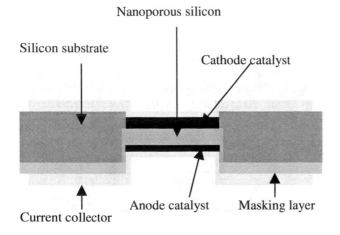

Figure 2. Membrane-electrode-assembly (MEA)

The microstructures of nanoporous silicon membranes were characterized by scanning electron microscopy (SEM). A Teflon cell was used to hold the MEA for fuel cell performance tests. Polarization curves were obtained using different formic acid concentrations of 1M, 5M, or 9M as fuel for the anode. The cathode was air-breathing. No pumping was done for either anode or cathode.

3 RESULTS AND DISCUSSION

The SEM image of nanoporous silicon membrane made by 40 mA/cm^2 and 80 mA/cm^2 are shown in Figure 3, 4, and 5, respectively. The nanopore diameter in membranes made by 20 mA/cm^2 (not shown here) and 40 mA/cm^2 are almost uniform throughout the membranes. But in the case of 80 mA/cm^2, it is observed that nanopore diameter increases with depth, as shown in Figure 4 and 5. It is proposed that nanopores started growing from the substrate surface with relatively small diameter. As nanopores grew deeper into the substrate, electrolyte concentration decreased due to mass transport limit, and thus lead to increased nanopore diameter. This phenomenon is more significant when nanopores were formed by higher current density.

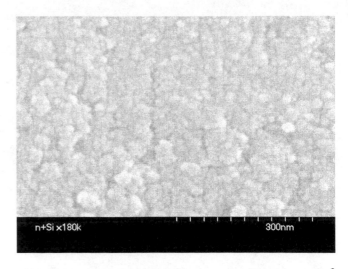

Figure 3. SEM image of nanopores formed by 40 mA/cm^2

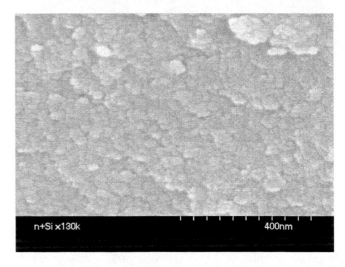

Figure 4. SEM image of nanopores formed by 80 mA/cm^2, close to the substrate surface

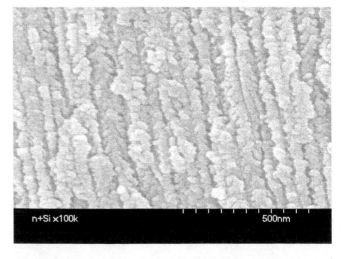

Figure 5. SEM image of nanopores formed by 80 mA/cm^2, deep into the substrate

Using 5 M formic acid as fuel plus 0.5 M sulfuric acid to increase fuel solution conductivity, polarization curves in Figure 6 show that micro fuel cell with 150 μm thick nanoporous silicon membrane gives higher open cell voltage of 0.695 V than the other two thickness (50 μm and 100 μm). It can be seen that the thinner membrane in a miniature fuel cell, the lower open cell voltage it produces. Figure 7 shows the power density curves of miceo fuel cells with these three membrane thickness.

Comparing polarization curves and power density curves in Fig. 6 and Fig. 7, it can be seen that micro fuel cell with 100 μm thick membrane achieves higher current density than 150 μm thick membrane in the high current region, and thus its maximum power density, which is 93.99 mW/cm^2, is higher than that of 150 μm thick membrane, which is 89.44 mW/cm^2.

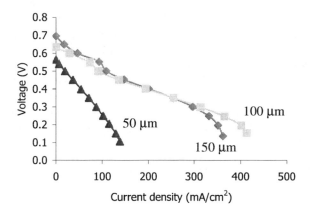

Figure 6. Polarization curves of micro fuel cells with three different thickness of nanoporous silicon membrane

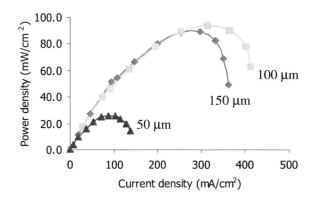

Figure 7. Power density curves of micro fuel cells with three different thickness of nanoporous silicon membrane

Polarization curves and power density curves of micro fuel cells with nanoporous silicon membrane with same thickness of 100 μm but made by three different current

densities of 20, 40, and 80 mA/cm^2, are shown in Fig. 8 and 9, respectively. It has been reported that for nanoporous silicon produced from n-type silicon substrate, nanopore diameter increases with current density used for nanopore formation [5]. Micro fuel ccll with membrane produced using 80 mA/cm^2 gives higher open cell voltage than the other two. Under the same operation voltage, micro fuel cell with membrane produced using 80 mA/cm^2 also gives higher current density and maximum power density.

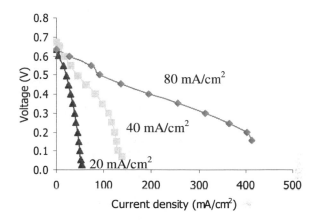

Figure 8. Polarization curves of micro fuel cells with three different nanopore formation current densities

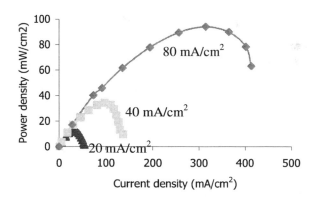

Figure 9. Power density curves of micro fuel cells with three different nanopore formation current densities

For micro fuel cell with 100 μm thick membrane, three fuel solutions with different formic acid concentrations of 1 M, 5 M, and 9 M were used to test the fuel cell performance. The results are shown in Figure 10 and 11. Open cell voltage decreases with formic acid concentration. Micro fuel cell performance suffers transport limit more significantly when using 1 M formic acid than when using 5 M and 9 M, as can be seen from the more rapid drop of current density along forward scan of polarization curves. And this rapid drop of current density contributes to lower power density output when using 1 M formic acid than using the other two fuel concentrations. Micro fuel cell performance was better when using 5 M formic acid than 9 M in terms of open cell voltage, current density, and power density output.

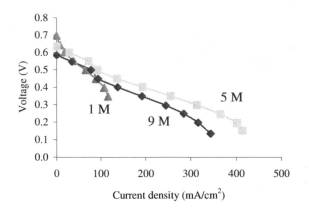

Figure 10. Polarization curves of micro fuel cells with three different formic acid concentrations

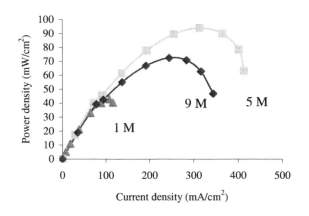

Figure 11. Power density curves of micro fuel cells with three different formic acid concentrations

Summarizing the above results, it can be seen that fuel cell performance improves as nanopore size in the membrane increases. And this is proposed mainly due to proton conductivity improvement. It can also be seen that fuel cell performance degrades as membrane thickness decreases. The reason is proposed to be increase in fuel crossover through the membrane as thickness decreases. And when different fuel concentrations were used for performance test, effects of fuel transport limit and fuel crossover can be seen in the cases of using 1 M and 9 M formic acid, respectively. And thus using 5 M formic acid for miniature fuel cells with 100μm thick membrane gave the best performance.

4 CONCLUSIONS

In this study nanoporous silicon membrane based micro formic acid fuel cells were demonstrated to be potentially promising for power generation for portable electronic devices. The fuel cell peak power density reached 93.99 mW/cm^2 at current density level of 314.37 mA/cm^2 when fuel cell voltage being 0.3 V. The effects of nanoporous silicon membrane thickness, nanopore size, and formic acid concentration on micro fuel cell performance were studied as well. Fuel crossover, proton conductivity, and fuel transport limit were proposed to explain the observation. Using 5 M formic acid with 0.5 M sulfuric acid as fuel, 100 μm thick nanoporous silicon membrane made using the highest current density in this study (80 mA/cm^2) gave the best micro fuel cell performance.

5 ACKNOWLEDGEMENT

This research is funded by the Defense Advanced Research Projects Agency (DARPA) under U.S. Air Force grant F33615-01-C-2172. Any opinions, findings, and conclusions or recommendations expressed in this manuscript are those of the authors and do not necessarily reflect the views of the Defense Advanced Projects Research Agency, the U.S. Air Force, or the Department of Energy. Micro and Nanotechnology Laboratory and Material Research Laboratory in University of Illinois at Urbana-Champaign also provided uses of microfabrication facility.

REFERENCES

[1] G. Q. Lu, C. Y. Wang, T. J. Yen, and X. Zhang, Electrochimica Acta, 49, 821, 2004.
[2] K. Shah, W. C. Shin, and R. S. Besser, Sensors and Actuators B, 97, 157, 2004.
[3] T. J. Yen, N. Fang, and X. Zhang, Applied Physics Letters, 83, 4056, 2003.
[4] J. S. Wainright, R. F. Savinell, C. C. Liu, and M. Litt, Electrochimica Acta, 48, 2869, 2003.
[5] V. Lehmann, R. Stengl, A. Luigart, Material Science and Engineering B, 69, 11, 2000.

Novel Electrolyte Membrane for Fuel Cell Utilizing Nano Composite

K.C. Nguyen*, H.V. Nguyen*, S.T. Do*, T.T. Doan*, A.T. Nguyen*, T.V.Le*, T.T. Nguyen*

*Saigon Hi Tech Park Research Laboratories, HoChiMinh City, Viet Nam

ABSTRACT

In the previous study [1], we reported the achievement of nano scale particles of carbon black by multiple diazo coupling of specific anchor groups having electrolytic functionality onto the same material. As a result, the pseudo products called as the "liquid" nano carbon (LNC) having average particle size down to the range between 20-30nm, can be isolated by water flush.

It is discovered in the present study that the electrolytic groups is the main cause for the significant reduction of the electrical conductivity of the carbon black raw material and believed due to the enhanced proton transport efficiency associated with the electrolytic chemistry. We reported here, a novel type of proton transport material comprised of LNC embedded in a polymer matrix to form proton conducting nano composite. The LNC proves stable nano scale [1] in aqueous environment and the nano composit demonstrated excellent film forming properties with a wide range of binders, especially, aqueous emulsion polymers and highly cross linking polymers. The nano composite also exhibits the excellent uniformity of membrane with well known coating process such as dip coating, spin coating, spray coating, roll coating, blade coating, rod coating, brush painting , inkjet printing....

It is found that the nano composite shows stabilization of DTA/TG data curve over that of the single LNC itself. The nano composite also exhibits remarkable thermal stability over wide range of temperature up to 350C in the ambient environment compared to the Nafion 117, a well known proton exchange membrane (PEM) material in the PEM fuel cell [2] market. In fact, in a PEMFC configuration, the nano composite PEM shows higher current density and 2X higher power efficiency than that of Nafion 117. The PEMFC using LNC nano composite PEM also exhibits superior shelf life exceeding 3.5 months when an aqueous methanol liquid fuel system is continuously fed.

1. INTRODUCTION

USP 5554739 and 5922118 [3] , demonstrated unique carbon black material for inkjet colorant application by attaching water soluble anchor group onto it to form black particles stably suspending in water. The attachment occurs via a diazo coupling reaction using primary amine precursor containing desired functional groups wanted to be on black carbon ring system. In aqueous solvents, the commercial products; Cabojet 200 and Cabojet 300 is reported to exhibit aggregate having average particle size in the vicinity of 130nm and they are not quite nano material yet [4]. In the previous study [1] , we selected the anchor group to be electrolytic which can form ionized particles in strong polar solvents or even in an electrically biased environment and then, repeated the coupling process on the same material for multiple times expecting that the multiple coupling process could increase the concentration of electrolytic groups on the invidual carbon particle surface and the particles carrying the same sign charge could rebel more effectively each other to stabilize the nano scale . It is already observed that the more diazo coupling cycle occurred; the coupling product shows more water solubility. The term "liquid" nano carbon (LNC) is originated here to determine nano scale particles of carbon which look "soluble "but not really a liquidified product.

It could be noted right here that the multiple diazo coupling process does not reduce the primary individual particle size but rather forms aggregate more easily broken down into nano scale in suitable solvents .

In this report, for the convenience of the terminology expression, Dn term indicates the product from the nth diazo coupling cycle , thus, D0 is the carbon black raw material which has not been exposed to any diazo coupling process yet, D1 is the product of the 1st diazo coupling reaction, D4 is the product of the 4th diazo coupling reaction etc . . . Actually, the nano scale is only achieved after the 2nd diazo coupling process as the D1 product is still showing the aggregate form with average particle size above 100nm, well agreed with what is observed for Cabojet 200 and 300 products . So we decided to choose D4 as standard nano material for the entire study in this report.

In the present study, we investigated the electrical properties of surface modified carbon black as functions of the number of diazo coupling cycle, confirming that electrical measurement can give some insights for the effect of the multiple diazo coupling processes. We also investigated the polymeric binder effect on the electrical properties of LNC, leading to a suggestion that a nano composite of LNC and aqueous emulsion polymers can be used as electrolyte membrane material.

The principle of proton exchange membrane fuel cell (PEM FC) has been described in details somewhere [2] in which the fuel source which can be either hydrogen gas or an aqueous solution of low alkyl chain alcohol can generate a proton when in touch with Pt catalyst sandwiched between an electron transport molecule and a proton transport molecule. It is assumed that the effective proton transport membrane in a fuel cell system must be able to transport the geminate proton out of the generation area as much efficient as it could in order to avoid the geminate recombination between electron and proton [5] causing the electron lost in the outside loop . In reality, Nafion [6] product from Dupont is a well known PEM product in the market. However, Nafion is sulfonated Teflon polymer and exhibits poor adhesion properties against any substrate due to the low surface energy associated with fluoro chemistry [7] and thus, the film casting from Nafion solution is much harder to be successful [8]. Several efforts [9] had been made to overcome these issues by changing the polymer backbone as reported for poly sulfone (PS), poly benzimidazole (PBI), poly ether ether ketone (PEEK). Even though, the challenges maintained in the balance between the density of proton transport functionality $-SO_3H$ and the hydrophobicity in the same film, leaving behind the major issues related to high fuel crossover, high water cross over, low methanol tolerance, ignoring the issues related to the material and manufacturing cost.

In this report, we presented a novel type of electrolyte membrane material using a nano composite of LNC [10] and emulsion polymer. The nano composite looks promising in terms of high heat resistance and more choices for the binder design and selection. In this case, LNC is acting as proton transporter and polymeric binders tend to provide protection for LNC to meet the requirements of operating condition.

2. EXPERIMENTAL PROCEDURE

2.1) Preparation of carbon black D0

Carbon black was prepared by burning acetylene gas coming out from the reaction of water with calcium carbide. The fume product is quenched with cooled jacket water and collected in a magnetic stirrer water vessel. The primary aggregate of carbon black was respectively washed with acid, base, organic solvents (toluene, acetone), rinsed with water and baked at 140C at least for 4 hrs in a convection oven.

2.2) Preparation of D4

The carbon black starting material D0 prepared in 2.1) was exposed to 4 times of diazo coupling process referring the previous report [1] [3]. In this case, sulfanilic acid was chosen as coupling precursor providing electrolytic group -SO$_3$H and p-amino benzoic acid to provide –COOH. The forming of D4 product was confirmed with the significant increase of water solubility and FtIR spectroscopic measurement.

2.3) Testing procedure

I) Ft IR measurement

FtIR spectroscopy measurement of LNC was carried out using FtIR Tensor 37 (Bruker) equipment.

ii) Thermal analysis

Thermal analysis of LNC was measured by DSC (Differential Scanning Colorimetry) technique using STA 409 PC-Netzsch equipment.

iii) Electrical resistivity

A pair of silver paste electrode spacing 1cm from each other was screen-printed on the top of 10 mil thick plain paper and air dried for several hours.

LNC was milled without milling media in a glass jar containing distilled water for 2 hours to form black slurry having solid content of 7.5% wt. The slurry was coasted on the pair of silver electrode by wound wire bar to form 10 um thickness (measured by a stylus profiler (KLA Tencor)) after being baked at 100C for 2 hours in a convection Blue M oven (VWR). The surface resistivity of LNC was directly measured using in-house 4 point probe tester.

For the testing of the nano composite, selected aqueous emulsion polymers were added into the milling jar with suitable weight content and the test film was prepared by the same way above described.

3. RESULTS AND DISCUSSION

3.1) Electrical properties of LNC

3.1.1) Electrical resistivity as functions of number of diazo coupling cycle

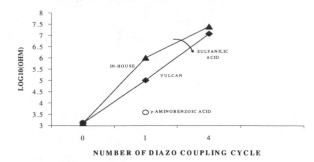

Fig.1 Effect of number of coupling cycle on electrical resistivity

Fig. 1 shows the electrical resistivity of various types of Dn products made out of D0 (in-house) using sulfanilic acid as diazo coupling agent. For a comparison purpose, Dn products were also prepared with commercial carbon black Vulcan XR 72C (Cabot Corporation). The measured data of the D1 (-COOH) product prepared with p-amino benzoic acid and in-house carbon black D0 was also plotted. It should be noted that all electrical measurement was done in the ambient condion (room temperature, 72% Rh).

It is observed that the electrical resistivity of D1, D4 is much higher than that of D0 in a magnification of 10^3 and 10^4 X, respectively, no matter what the raw carbon black material is. On the other hand, the D1(-COOH) product shows less electrical resistivity than other Dn products made out of sulfanilic acid, agreeing well with the fact that –SO$_3$H group is known to exhibit higher protonation efficiency than the –COOH group . If we assumed that each diazo coupling cycle can add certain amount of electrolytic group to the carbon black until all of the coupling sites run out, then the electrolytic group should be the cause for the electrical conductivity reduction in the carbon black due to the increased proton transport sites. The strong dependence of electrical resistivity on the number of diazo coupling cycle suggests that electrical resistivity can be an effective measure of the effect of coupling processes than other measurement procedures such as SEM or AFM.

3.1.2) Binder effects

As LNC product is easily and well dispersed in aqueous environment, the aqueous emulsion polymers or water soluble polymers are preferred to give a uniform dry film in which the nano scale of LNC is still maintained. However, the dried film of LNC/polymer requires the hydrophobicity to insure the performance stability with the use of liquid fuel, the aqueous emulsion polymer give better choices. In a hydrophobic solvent system, the LNC tends to reaggregate out of nano scale domain, thus hydrophobic solvent soluble polymers are not the first choice to form ideal nano composite.

Several different kinds of aqueous emulsion polymers and hydrophobic cross linking poly urethane (PU) were tested .The in-house emulsion copolymers exhibit average molecular weight in the range of 30,000- 40,000 and the Tg = 100C .

Fig. 2 exhibits the effect of binder content on the electrical resistivity of LNC (D4) for various kinds of polymer. All of the measurement was carried out at room temperature and

72 % RH condition. From this result, it is able to classify the behavior of polymer into 3 groups

Group A: The polymer exhibits the increased electrical

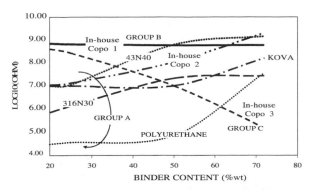

Fig. 2 Binder effect on electrical resistivity of LNC

resistivity of nano composite with binder content. Group A comprised of cross linking poly urethane (PU), emulsion polymers (43N40, 316N30 from CHEMCOR), emulsion polymer products from KOVA, in-house emulsion copolymer 2 (in-house Copo2).

Group B: The polymer shows no effect on resistivity such as the in-house emulsion copolymer 1 (in-house Copo1)

Group C: The polymer exhibits the decreased resistivity with binder content such as the in-house copolymer 3 (In-house Copo3).

At binder content < 50% wt, the electrical resistivity of nano composite is dominated by LNC. However at the binder content greater than 50% wt, the electrical resistivity is believed more dominated by the binder. Based on these

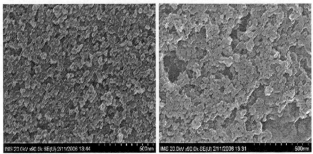

Fig.3 FE-SEM image of D4LNC (left) and the nano composite (D4/Copo3) in the scale of 500nm

data, it is speculated that the Group A and Group B is composed of conventional insulative polymers. The unique properties of Group C suggest that the polymer is a more conductive polymer, compared to other polymers of Group A and B. However, it should be noted that most of emulsion polymers containing a large quantity of ionic surfactants to stabilize the monomer emulsion before polymerization. And the electrical properties measured in the present study could be a mixed effect of electron transport and ionic transport properties. The detailed study of the nature of electrical transport mechanism in the nano composite will be discussed in another report. Fig. 3 is FE-SEM (Field Emission Scanning Electron Microscope, Hitachi S4800) image of the D4 LNC and the nano composite of D4 in in-house Copo3, in the scale of 500nm. It is recognized the nano scale with average size of 20-30nm of D4 in a dried stack (left). This scale is still maintained in a nano composite structure (right) in which the nano scale particles

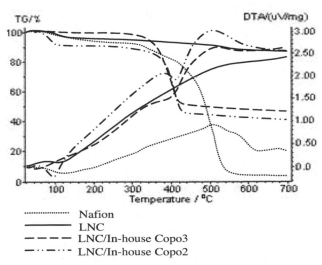

.... Nafion
———— LNC
– – – – LNC/In-house Copo3
–·–··–· LNC/In-house Copo2

Fig.4 DTA/TG of nano composite

are nicely surrounded with thin layer of emulsion polymer; in-house Copo3.

Next, the binder effect on thermal properties of D4 was investigated with DTA/TG data measurement and the results are illustrated in Fig. 4. In this case, the nano composite was formulated with D4/ polymer ratio = 1/1 by wt and the study was carried out with two types of emulsion copolymers; in-house Copo2 and in-house Copo3. D4 exhibits exothermic peak at 110.2C and it could be due to the cleavage of azo bond –N=N- but the confirmation is needed in a separated study. However, it is noteworthy that this peak is completely disappeared when D4 is blended with in-house Copo3 and the mass loss of the composite film also becomes flat between 0 and 350C. The flatness of the DTA/TG curve may reflect a recompansation in mass loss between LNC and polymeric binder and it might stabilize the azo bond against heating effect. On the other hand, the in-house Copo2 reversely, didn't show any improvement on thermal stabilization effect but further a

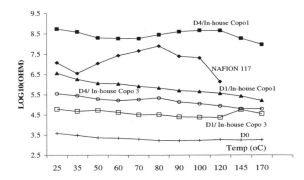

Fig. 5 Electrical resistivity of the nano composite as functions of temperature

deterioration of thermal stability and greater mass loss. In the same Fig. 4, DTA/TG data of Nafion (from Nafion DE2201 solution) was also shown. Nafion shows exothermic peaks in the vicinity of 100-150C and starts gradually loosing mass even at low temperature zone between 100C and 450C; the mass loss in Nafion is almost completed at about 550C. As previously mentioned, the aqueous hydrophobic emulsion polymers maintain the nano scale of LNC particles in the film thus could perform strongly the interaction between the chemistry of LNC and polymer, which might reflect in the thermal stabilization effect of LNC by polymeric binder.

Fig.5 summarizes the electrical resistivity of the nano composites, measured at various temperatures. In this case, 4 composites were formulated. These are D1, D4 blended in in-house Copo1 and in-house Copo3. It is observed that the nano composite comprised of D4 blended in the in-house Copo1 shows the highest resistivity with stable performance until the limitation of temperature measurement (180C). On the other hand Nafion 117 shows the second high resistivity. It should be noted that the Nafion 117 starts deform after 80C; the film becomes dark brown and completely destroyed at 120C, making the measurement unavailable. Other nano composite films show relatively stable electrical properties and mechanical strength during the temperature range of measurement between room temperature and 180C without any signals of physical changes or damage of the film. Actually, further heating test was done up to 350C and shows no physical deterioration on the composite film .

Next, a PEM FC was formulated to test out the PEM performance of the nano composite. In this case, a PEM fuel cell structure is composed of liquid fuel diffusion layer (LDL) using Toray Carbon Paper TGPH-060 (E-Tek product) over coated with electro catalyst layer composed of 20% Pt, 10% Ru , 70% Vulcan XR72C (E-Tek product) in anode and 30% Pt/70% Vulcan XR72C (E-Tek product) in cathode . The electro catalyst layer was prepared by mixing the electro catalysts with MEK solvent using ultrasonic bath (Branson 2510) for 5 minutes then brush painted on the LDL to achieve a weight loading of 4 mg/cm^2 after being baked at 80C in a convection oven set at 80C for 30 minutes. The active electrode area is about 4 cm^2. The membrane is composed of 50% D4 embedded in 50% in-house Copo3 was ultrasonically mixed for 5 minutes , then brush painted on the top of the electro-catalyst layer and baked in the same oven for 30 minutes at 80C . The PEM thickness was detected to be around 168 um and is equivalent to that of the Nafion 117 membrane. Actually, Fig. 6 shows the V-I characteristic curves of the

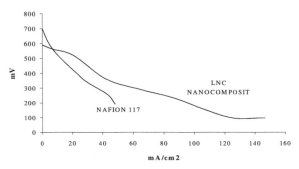

Fig. 6 V-I characteristic curves of the nano composite compared to the Nafion 117 PEM

PEM made out of D4/Copo3 composite compared to commercial Nafion 117 product when an aqueous fuel containing 3% methanol was fed . One can recognize that the nano composite PEM shows higher current density than that of Nafion 117. Fig.7 shows power efficiency curve of the nano composite PEM compared to that of the Nafion

117 PEM in the same PEMFC configuration. The nano composite PEM shows 2X higher power efficiency than that of Nafion PEM.

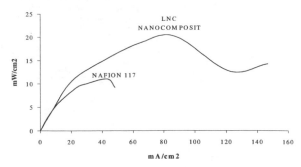

Fig. 7 Power efficiency curve of PEM fuel cell utilizing LNC composite membrane and Nafion membrane

Besides the above electrical performance, the PEM FC utilizing LNC nano composite exhibits life time exceeding 3.5 months with continuous feeding of aqueous liquid fuel to light up the white LED light bulb . The cell just looks like "sleeping" whenever the liquid fuel is ran out and starts working again when the membrane is wetted with a few drops of liquid fuel.

4. CONCLUSION

It can be concluded that

a) the electrolytic groups –SO3H, -COOH strongly reduce the electron conductivity of the carbon black , and enhance the proton transport properties

b) The linking of electrolytic group onto the carbon system occurs via azo bond –N=N- which is thermally unstable. However, the interaction between specific emulsion polymers when maintaining the nano scale of LNC could improve the thermal stability of the nano composite.

c) Actually, the nano composite PEM exhibits suitable PEMFC performance, at least equal to or superior than that of the Nafion 117 in terms of current density and power efficiency.

d) The nano composite also exhibits superior heat stability over the Nafion 117 as this PEM starts detoriated at 80C

e) The nano composite PEM exhibits a shelf life exceeding 3.5 months with continuous feeding of aqueous methanol fuel source.

5. REFERENCES
1. Khe et al, Proceeding of Digital Fabrication 2005 , p.209
2. Fuel Cell Technology Handbook, edited by Gregor Hooger, 2003 CRC Press, p. 7-1
3. US Pat 5554739 and 5922118
4. Nanotech Conference 2004, Boston , Opening Session
5. Rodrigues R. et al, Langmuir, 22 (3), 933 -940, 2006
6. www.mitstanfordberkeleynano.org/.../
7. US Pat 5,834,564, " Photoconductor coating having perfluoro copolymer and composition for making same"
8. US Pat 6902839, and also see S.H.Kwak et al, J. New Mat. Electrochem. Systems, 4, 25-29(2001)
9. G. Alberti, M. Casciola, Solid State Ionics, 145 (2001) 3 And also see K.D. Kreuer, J.Membr. Sci., 185(2001) 29
10. US Patent application pending.

Heat Transfer Cost-Effectiveness of Nanofluids

Diao Xu, Lunsheng Pan and Qiang Yao

Institute of High Performance Computing,
1 Science Park Road, #01-01 The Capricorn, Singapore Science Park 2, Singapore 117528
xud@ihpc.a-star.edu.sg

ABSTRACT

When metal or oxide nano particles are dispersed in liquids to form nanofluids, the particles improve thermal conductivity of the liquids. Therefore, it is suggested to use nanofluids as coolants to improve heat-exchanger efficiency. However, the nano particles also cause the increase of fluid viscosity. The present paper has numerically studied the flow and heat transfer of the nanofluids in a 2-D microchannel by using Computational Fluid Dynamics method. It is found that although the nano particles enhance the heat transfer rate of the fluids about certain percentage, the nano particles also cause an increase of viscous shear stress, and further causes an increase of the power consumption to deliver the nanofluids through the microchannels.

Keywords: Nanofluids; Computational fluid dynamics; Micro channels, Heat transfer, Coolants.

1. INTRODUCTION

The nanofluids are produced by dispersing metal or oxide nano particles in liquids. It is found that the metal and oxide nano particles have positive effects to enhance thermal conductivity of the liquids (base liquids). Some small amount of the nano particles dispersed in the base liquids will greatly increase thermal conductivity of the fluids [1-3]. Keblinski *et al.*[1]. proposed some potential mechanisms to explain why the small amount of the nano particles can greatly affect the solution's thermal conductivity.

The nano particles are very stable. They do not settle, and do not clog the components of a flow system, even in a micro fluidic system [1-3].

To explore their advantage, nanofluids are suggested to be used as coolants to improve the thermal efficiency and to reduce the size of heat exchangers. However, the nanofluids also enlarge fluid shear stresses on solid interfaces. This is because that the nano particles increase the viscosity of the fluids. The enlarged shear stresses will increase the fluid drags. This makes it difficult for the nanofluids to flow through the fluidic systems comparing with those base liquids [2, 3]. Therefore, a big pressure difference is required to drive the nanofluids to flow through the fluidic systems. This in turn will cause more power consumption. So, one has to carefully analyze the gain and the loss or cost-effectiveness, before adopting the nanofluids as coolants.

To investigate the cost-effectiveness of using nanofluids as coolants, Computational Fluid Dynamics method is employed to directly simulate the flow and heat transfer of the nanofluids in a 2-dimensional micro channel in the present paper. Basically there are two different numerical methods for doing these. One is based on molecular dynamics which directly focuses on the molecular behaviors of the nano particles. This method needs more *CPU* time and computer memory. The other is based on Navier-Stokes questions with introducing the thermal and dynamic parameters of the nanofluids obtained from the mixture fluid theory and experimental measurements. The latter provides useful information for researchers and engineers to understand the flow and heat transfer profiles of the fluidic devices with less *CPU* time and computer memory [2, 3]. Therefore, it is employed in the present paper to study the cost-effectiveness of the nanofluids in a 2-dimensional micro channel.

2. MATHEMATIC MODEL

In the present study, the nanofluids are regarded as incompressible, well-mixed and uniform single phase solutions. The flows are considered as steady laminar flows. The governing equations are Navier-Stokes and energy equations as below.

$$\nabla \cdot (\rho_{nf} \vec{V}) = 0 \tag{1}$$

$$\nabla \cdot (\rho_{nf} \vec{V}\vec{V}) = -\nabla P + \nabla \cdot (\mu_{nf} \nabla \vec{V}) \tag{2}$$

$$\nabla \cdot (\rho_{nf} \vec{V} C p_{nf} T) = \nabla \cdot (k_{nf} \nabla T) \tag{3}$$

where ρ_{nf} is nanofluid density, $\vec{V}$ velocity vector, P pressure, μ viscosity, Cp_{nf} nanofluid specific heat capacity, k_{nf} nanofluid thermal conductivity, and T temperature.

The density and specific heat capacity of the nanofluids are calculated using following formulas [2, 3].

$$\rho_{nf} = (1-\varphi)\rho_{bf} + \varphi\rho_p \tag{4}$$

$$Cp_{nf} = (1-\varphi)Cp_{bf} + \varphi Cp_p \tag{5}$$

For viscosity and thermal conductivity, the most widely acceptable data are from experiment measurements. For example, the viscosity and the thermal conductivity of water-γAl_2O_3 solutions are [2, 3].

$$\mu_{nf} = \mu_{bf}(123\varphi^2 + 7.3\varphi + 1) \tag{6}$$

$$k_{nf} = k_{bf}(4.97\varphi^2 + 2.72\varphi + 1) \tag{7}$$

where the subscripts p, bf and nf refer to the particles, the base fluid and the nanofluid respectively; φ is volume concentration of the nano particles. Eq.(4) and Eq.(5) are used for classic mixtures [4]. Eq.(6) is directly adopted from [5], which used a least-squares curve fitting from experimental data. The thermal conductivity expressed in Eq.(7) is from Hamilton and Crosser model [6]. Eq.(4) to Eq.(7) have been successfully employed in [2] and [3]. For ethylene glycol-γAl_2O_3 solutions, similar with Eq.(6) and Eq.(7), the viscosity and thermal conductivity are [2]

$$\mu_{nf} = \mu_{bf}(306\varphi^2 - 0.19\varphi + 1) \tag{8}$$

$$k_{nf} = k_{bf}(28.905\varphi^2 + 2.8273\varphi + 1) \tag{9}$$

There are also some models on viscosity from theoretical analysis. However, these models are less accurate. Here, some recommended models are given below [7]:

Eistein model

$$\mu_{nf} = \mu_{bf}(1 + 1.25\varphi) \tag{10}$$

Brinkman model

$$\mu_{nf} = \frac{\mu_{bf}}{(1-\varphi)^{2.5}} \tag{11}$$

Hamilton and Crosser proposed a model to present colloidal suspensions [6].

$$k_{nf} = k_{bf}\frac{\gamma + (n-1) - (n-1)\varphi(1-\gamma)}{\gamma + (n-1) + \varphi(1-\gamma)} \tag{12}$$

where $\gamma = \dfrac{k_p}{k_{bf}}$, and n is the shape factor respectively. For spherical particles, n has a value of 3.

Bruggeman model [8]

$$k_{nf} = k_{bf}\frac{(3\varphi-1)\gamma + \{3(1-\varphi)-1\} + \sqrt{\Delta_B}}{4} \tag{13}$$

$$\Delta_B = [(3\varphi-1)\gamma + \{3(1-\varphi)-1\}]^2 + 8\gamma \tag{14}$$

3 RESULTS AND DISCUSSIONS

A 2-dimensional microchannel with its width of 200 μm and length of 200 mm (length vs. width ratio is 1000) is employed in the present study.

After grid-independent verifications, it is found that a 11×1001 grid system can provide numerical results with enough accuracy. Therefore this grid system is used for the further calculations. Fig.1 shows the part of the grid system.

Figure 1: Grid system of the 2-dimensional microchannel, 11×1001

For all numerical simulations, the temperature at the walls is set to be 353K, the temperature of fluid at inlet is fixed to be 293K, and the inlet velocity is given according to Reynolds number based on the microchannel width and water properties. Three different fluids are simulated: pure water, water + 5% Al_2O_3 particles, and water + 10% Al_2O_3 particles. The thermal properties of the nanofluids are considered by introducing Eq.(4) to Eq.(7) into the calculations. The control volume numerical method is used, and numerical iteration is performed till the maximum numerical residual is less than 10^{-7}.

Cleantech 2007, ISBN 1-4200-6382-0

Fig.2 shows the velocity distribution of nanofluid (water + 5% Al_2O_3 particles) at inlet section with $Re=10$, and Fig.3 gives its temperature distribution.

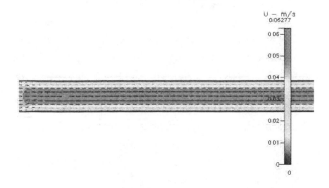

Figure 2: Velocity distribution of nanofluid (water + 5% Al_2O_3 particles) at $Re=10$

Figure 3: Temperature distribution of nanofluid (water + 5% Al_2O_3 particles) at $Re=10$

Comparing the velocity and the temperature profiles, it can be seen that the temperature of nanofluids soon becomes a uniformed distribution with same temperature as the hot wall. It only takes a very short distance from the inlet to become a uniform distribution.

The heat transfer rate and drag force are obtained from the computational results. Fig.4 presents the heat transfer ratio of the nanofluid to the pure water (base fluid).

$$R_q = \frac{q_{nf}}{q_{bf}} \qquad (8)$$

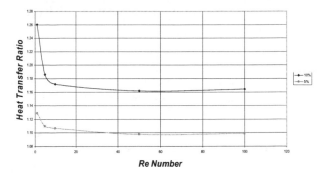

Figure 4: Heat transfer ratio of nanofluid vs. pure water

From Fig.4, it can be seen that the nano particles enhance the heat transferring ability of the base fluid. There is about 10% increase for water + 5% Al_2O_3 particles, while about 16% increase for water + 10% Al_2O_3 particles. After $Re>50$, the heat transfer ratio is almost a constant.

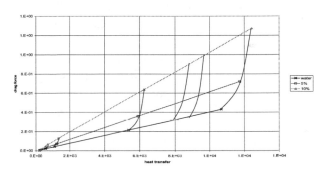

Figure 5: Heat transfer rate vs. drag force of nanofluids

Fig.5 shows the curves of the heat transfer versus the drag force of the microchannel, with various volume concentrations of the nano particles. The horizontal axis is the heat transfer, and the vertical axis presents the drag force. The marks on the curves stand for Reynolds numbers. From left to right, Reynolds numbers are 1, 5, 10, 50, and 100 respectively, which can be regarded as flow rate here. It is clearly shown that at the same flow rate, the heat transfer rate increases with the presence of the nano particles in fluids. The more nano particles, the more increase of the heat transfer rate. However, it also can be seen that there is an increase of the drag force. The increase of the drag force is very significant with the increase of the volume concentration of the nano particles.

As we know that the drag force is directly linked with the power consumption. In order to drive nanofluids through microchannels, the power has to be consumed to overcome the drag force. The large drag force means that more power is required to pump the nanofluids, or in other words, a more powerful pump is required for the purpose of delivering nanofluids.

If examining Fig.5 carefully, one can find that at same flow rate the increase of the drag force is much faster than the increase of heat transfer. For the same amount of heat transfer, the nanofluids will consume more power than the water does. In other words, if one uses the water as the coolant, one can achieve same heat transfer target by simply increasing the flow rate of the water. As a result, one can use less power by using the water comparing with by using the nanofluids. Here, it should be emphasized that the manufacturing cost of the nano particles and the nano fluids have not been included.

As shown in Fig.3, the temperature of the incoming cold nanofluids soon increases to the hot wall temperature, and becomes a uniform distribution. This means at the end of the channel, the outgoing fluids will be of the same temperature of the hot wall. Therefore, the factors which determine the heat flux out of the channel are specific heat capacity and fluid flow rate, not the thermal conductivity and the heat transfer rate of the nanofluids. So, in our present case, as the outgoing fluids have the same temperature of the hot wall, we need only to see fluids' specific heat capacity. Now we turn to discuss Eq.(4) and Eq.(5). We rewrite Eq.(5) by dividing Cp_{bf} both sides of Eq.(5).

$$\frac{Cp_{nf}}{Cp_{bf}} = 1 + \varphi(\frac{Cp_p}{Cp_{bf}} - 1) \qquad (9)$$

where Cp_{nf}/Cp_{bf} is the ratio of specific heat capacity of nanofluids to that of base fluid, and Cp_p/Cp_{bf} is the ratio of specific heat capacity of nano particles to that of base fluid. As we know, the base liquid (the water at the present case) usually has much higher specific heat capacity than those of solid metals. Therefore, Cp_p/Cp_{bf} is less than 1. So, from the right hand side of Eq.(9), we can conclude $Cp_{nf}/Cp_{bf} < 1$ for water base nanofluids. This implies that the nano particles improve heat transfer rate of nanofluids, but decrease the specific heat capacity of nanofluids.

For a channel or a tube used for heat exchanger, the liquid coolants will reach temperature of hot wall at certain length from the inlet section. Because of enhanced heat transfer rate, nanofluids will reduce this length comparing pure water. If the channel or tube has a length longer than the length, the nanofluids will have no effect to improve efficiency of a heat exchanger.

4. CONCLUSIONS

Flow and heat transfer of nanofluids in a 2-dimensional microchannel are numerically studied in the present paper using CFD method through introducing the thermal and dynamic parameters of the nanofluids. It is found that:

- Nano particles in fluids enhance heat transfer rate of fluids and increase viscosity as well.
- Increase of heat transfer rate is less significant than increase of viscous drag.
- Nanofluids have lower specific heat capacity than pure water.
- For long channel or tube, nanofluids will have no improvement for heat exchanging.
- Using the pure water as coolant will be more cost-effective in the present model.

REFERENCES

[1]. P. Keblimski, et al, "Mechanisms of Heat Flow in Suspensions of Nano-sized Particles (Nanofluids)", *International Journal of Heat and Mass Transfer*, 45 (2002) 855-863.

[2]. S.E.B. Maiga, et al, "Heat Transfer Behaviours of Nanofluids in a Uniformly Heated Tube", *Superlattices and Microstructures*, 35 (2004) 543-557.

[3]. G. Roy, C.T. Nguyen, and P.R. Lajoie, "Numerical Investigation of Laminar Flow and Heat Transfer in a Radial Flow Cooling System with the Use of Nanofluids", *Superlattices and Microstructures*, 35 (2004) 497-511.

[4]. B.C. Pak, Y.I. Cho, "Hydrodynamic and heat transfer study of dispersed fluids with Submicro metal oxide particles", *Exp. Heat Transfer* 11 (2) (1998) 151-170

[5]. X. Wang, X. Xu, and S.U.S. Choi, "Thermal Conductivity of Nanoparticles – fluid mixture", *J. Thermophys, Heat Transfer* 13 (4) (1999) 474 - 480

[6]. R.L. Hamilton, O.K. Crosser, "Thermal Conductivity of Heterogeneous Two-components System", *I & EC Fundamentals* 1 (3) (1962) 187-191

[7]. Y. Xuan and W. Roetzel, "Conceptions for heat transfer correction of nanofluids", Int. J. Heat Mass Transfer 43, 3701 (2000)

[8]. B.X. Wang, L.R. Zhou, and X.F. Peng, "A fractal model for predicting the effective thermal conductivity of liquid with suspension of nanoparticles", Int. J. Heat Mass Transfer 46, 2665 (2003)

Influence of His-195 on the nitrogenase FeMo-cofactor activity

Cavallotti, C., D. Moscatelli, S. Carrà

Dipartimento di Chimica, Materiali e Ingegneria Chimica "Giulio Natta",
Politecnico di Milano, via Mancinelli 7
20131 Milano, Italy

ABSTRACT

Mo-nitrogenase is the enzyme responsible for the conversion of atmospheric nitrogen to ammonia. Its active site is the iron-molibdenum cofactor, FeMo-co, which is composed by two incomplete cubes, Fe_4S_3 and $MoFe_3S_3$, connected through three bridging S. N_2 adsorption is expected to take place on the FeMo-co, and to be catalytically converted to two NH_3 molecules through the successive addition of protons and electrons. Mutant studies have evidenced that the activity of the nitrogenase FeMo-co is significantly influenced by the environment. In particular the mutation of HIS-195 to glutamine inhibits its ability to reduce nitrogen, though leaving intact its capability to adsorb N_2. In this work we investigated theoretically the adsorption of N_2 and of one hydrogen atom in presence of HIS-195. We performed our simulations with both C and N as central atoms, as it has been recently shown experimentally that N might not be the FeMo-co interstitial atom and C is a reasonable candidate We report energetic results for four different N_2 adsorption mechanisms. In particular we identify a favorable adsorption mechanism characterized by a partial opening of the FeMo-co which is here proposed for the first time.

Keywords: Biochemistry, Nitrgenase, Enzyme, Quantum Chemistry, Simulation.

1   INTRODUCTION

The enzyme is composed by two metallo-proteins, the iron protein, that has the function of transferring electrons, and the molybdenum iron protein, on which the nitrogen fixation takes place. The enzyme active is the iron-molibdenum cofactor, FeMo-co, which can be considered as composed of two incomplete cubes, $Fe_4(\mu_3\text{-}S)_3$ and $MoFe_3(\mu_3\text{-}S)_3$, connected through three bridging μ_2-S. The structure of the FeMo-co has been the subject of extensive experimental[1-3] and theoretical investigations,[4-8] which showed that an interstitial atom is present at the center of the MoFe-co. The theoretical attribution of the central atom identity to Nitrogen[4,9] has been recently questioned on the basis of experimental studies.[3] N_2 adsorption is expected to take place on the FeMo-co, and to be catalytically converted to two NH_3 molecules through the successive addition of protons and electrons. Mutant studies have evidenced that the activity of the nitrogenase MoFe-co is significantly influenced by the environment. In particular the mutation of

αHIS-195 to glutamine and αGLN-191 to lysine inhibits the ability of the MoFe-co to reduce nitrogen, though living intact its capability to adsorb N_2. While GLN-191 is relatively far from the candidates N_2 adsorption sites of the FeMo-co, HIS-195 is positioned in proximity of a bridging μ_2-S, with which it can interact to form an N-H-S bond. This suggests that N_2 adsorption is likely to take place in proximity of a μ_2-S, and that one or both of the two Fe atoms to which it is connected are directly involved. It has been proposed that the action of HIS-195 is either that of assisting directly the N_2 adsorption, or to play an active role in transposing of adsorbed hydrogen atoms from the FeMo-co and N_2.

2   METHOD

In this work we investigated theoretically the adsorption of N_2 and of the first of the hydrogen atoms necessary for the Thornely-Lowe scheme, in presence of HIS-195. Simulations were performed with density functional theory using the BLYP functionals. The Stuttgart-Dresden effective core potential basis sets were used for the Fe and Mo atoms while for all the other atoms we adopted the Dunning–Huzinaga double ζ basis set. All simulations were performed with the Gaussian 03 computational suite.[10] The molecular model used for the simulations is shown in Fig. 1. It has been obtained truncating the protein structure and replacing CYS-275 by SH and homocitrate with OCH_2CO_2.

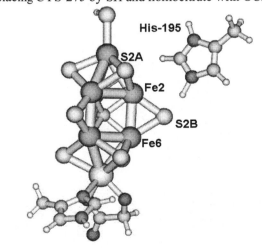

Figure 1. Site model

Only the imidazole ring of HIS-195 and HIS-442 is considered, while a CH_3 group is adopted to truncate the connection with the protein. The protonated form of HIS-195 was used in all the simulations, as it gave the results, in term of structural configuration, in best agreement with experimental data. The C atoms of the CH_3 groups and the S atom of the Cys residues are held fixed in the X-Ray positions of the 1M1N pdb structure during all optimizations. Given the uncertainty about the identity of the FeMo-co interstitial atom, two different sets of simulations, with N and C as central atoms, were performed. It seems in fact reasonable that, if N is not the central atom, than the second best candidate, on the basis of previous computational studies and electronic properties, is C. The total molecular charge and spin used in all the simulations were those of minimum energy. Interestingly the energy of the system as a function of the charge has a minimum at –2, of which –2.71 (N central atom) and –2.75 (C central atom) localized on the FeMo-co. This is in agreement with previous computational studies,[8] which predicted that, if the interaction of the FeMo-co with its protein environment is considered, the FeMo-co charge should be –3.

3   RESULTS AND DISCUSSION

An ample conformational search was performed with the aim of identifying possible interaction structures of HIS-195 with the FeMo-co. It has in fact been proposed that the adsorption of N_2 on the nitrogenase active site might lead to a significant structural change of the FeMo-co itself.[6] Our calculations showed that a possible interaction mechanism for HIS-195 with the FeMo-co is to exchange a H atom with a μ_2-S and bind the vicinal Fe atom of the $Fe_4(\mu_3$-S$)_3$ cluster through a N-Fe bond (Fig. 2).

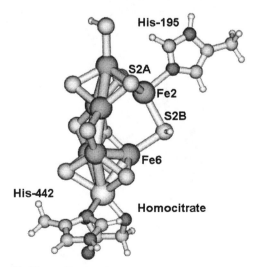

Figure 2. Open site structure.

This structure, which we will refer to as open, is slightly unstable with respect to the closed structure of Fig. 1 by

about 1.2 kcal/mol if N is the central atom and by 7.7 kcal/mol if it is C, which is consistent with the fact that the observed structure is in the closed form. However the transition from closed to open structure might take place either upon the adsorption of hydrogen atoms, or in concomitance with the N_2 adsorption, provided that the open structure is capable of interacting more efficiently with the adsorbing molecules. Two different FeMo-co structures are thus possible for the adsorption of N_2, the closed (Fig. 1) and open (Fig. 2) structures.

Another possibility for the adsorption of N_2, is to strongly modify the FeMo-co structure and interact with more than one Fe or Mo atoms contemporarily, which would thus compensate the energy required to modify the original cluster, as originally proposed by some authors. The interaction with vicinal aminoacids, among which HIS-195, is sometimes invoked to support this proposal. To account for all these possibilities, we investigated the N_2 adsorption for four different configurations. The first is the adsorption of N_2 on the FeMo-co without HIS-195, and in particular on one of the Fe atoms of the $Fe_4(\mu_3$-S$)_3$ cluster, that we found, as other authors[9], to interact more strongly with N_2 than the Fe atoms of the $MoFe_3(\mu_3$-S$)_3$ cluster and is better positioned, from a sterical point of view, to justify a significant contribution of HIS-195. The second and the third are the closed and open structures of Figs. 1 and 2. The candidate Fe atom is the same as that of the first conformation. The fourth is the results of an ample conformational search started from the structures proposed by Blochl et al. and performed under the hypothesis that HIS-195 must be somehow involved and that N_2 must interact with more than 1 Fe atom. We refer to this structure as 'bridged'. The minimum energy structures so determined are sketched in Fig. 3-6, while the computed adsorption energies are reported in Table 1 (In Figure 7 is reported the Global process of formation of adsorbed N_2H specie adopting the open structure of the FeMo-co).

Table 1. N_2 and H adsorption energies (kcal/mol) and energy required to transfer a H atom from a Fe site to adsorbed N_2 for C and N central atoms.

	Central Atom	FeMo	FeMo His195 Closed	FeMo His195 Open	FeMo His195 Bridge
N_2 ads	N	-5.6	4.0	-5.4	0.8
H ads on Fe6	N	-25.7	-23.2	-20.6	-38.3
FeH → N_2H	N	-6.3	-14.5	-2.1	1.0
N_2H ads	N	-37.6	-33.7	-28.1	-36.3
N_2 ads	C	-12.2	4.9	-10.6	3.0
H ads on Fe6	C	-30.8	-36.3	-24.8	-30.4
FeH → N_2H	C	-2.0	-7.5	-10.3	-12.5
N_2H ads	C	-41.1	-38.9	-45.7	-40.0

Cleantech 2007, ISBN 1-4200-6382-0

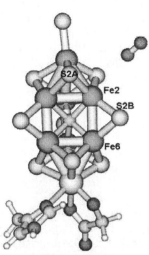

Figure 3. Structure of the the FeMo-co without HIS-195 with N$_2$ adsorbed.

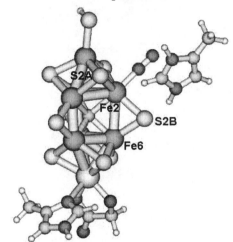

Figure 4. Closed structure of the FeMo-co with N$_2$ adsorbed.

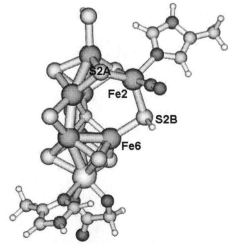

Figure 5. Open structure of the FeMo-co with N$_2$ adsorbed.

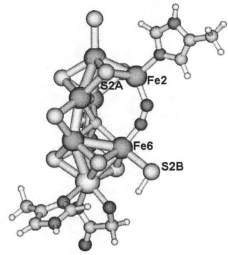

Figure 6. Bridged structure of the FeMo-co with N$_2$ adsorbed.

The calculated asdorption energies indicate that HIS-195 can significantly influence the adsorption process. In particular the comparison between N$_2$ adsorption on the FeMo-co with and without HIS-195 show that, in the absence of a major conformational change, such as the transition from closed to open site, the adsorption energy is positive. The N$_2$ adsorption energy for the open site conformation reported in Table 1 is calculated with reference to Fig.1a structure. Thus it does not comprise the energy necessary to reach the open site conformation (+1.2 and 7.7 kcal/mol for N and C as central atoms), despite of which the process is however overall exothermic. The nature of the central atom can influence the adsorption process. In particular FeMo-co clusters with C as central atom are more reactive towards N$_2$ than those with N, probably because of the lower extent of interaction of C with Fe than N. This is confirmed by the fact that the N$_2$ adsorption in the bridge conformation is highly unstable if N is the central atom, while the bond energy increases significantly, though not becoming negative, with N as central atom.

As the reduction of N$_2$ to NH$_3$ requires the adsorption of several H atoms, we determined the adsorption energy of H on one of the Fe atoms of the MoFe$_3$(μ_3-S)$_3$ cluster, and used it as a reference value to determine the energy required to transfer the adsorbed H to N$_2$ to form N$_2$H. The Fe2 atom of the MoFe$_3$(μ_3-S)$_3$ cluster is in fact likely to be one of the entry sites for H on the FeMo-co cluster[11], and a favorable energetic profile for transferring it from Fe2 to N$_2$ is required to let the N$_2$ reduction process proceed. H adsorption energies were computed with reference to gas phase molecular hydrogen. H transposition is significantly exothermic for almost all the examined configurations, with the notable exception of the bridge site with central N, which allows to rule out this conformation as a possible intermediate of the N$_2$ reduction mechanism. It is interesting to observe that the nature of the central atom can influence significantly hydrogen adsorption energies, and in particular the protonation of S2B in the open form greatly decreases the Fe-H bond energy. It is also interesting to observe that some preliminary studies of the stable conformations of two adsorption H atoms revealed that the open site conformation can adsorb both of them in a pseudo-octahedral structure on Fe2, which is very similar to one recently proposed to explain experimental observations.[2] The energy of this structure is among those of lowest energy we identified through an ample conformational search.

The main results of this study can be summarized as follows. The closed site conformation might be a candidate for the N_2 adsorption, provided that a sufficient amount of H is already adsorbed on the FeMo-co site and that the hydrogen transposition kinetics to form N_2H is sufficiently fast, as the energy gained through this reaction is sufficient to energetically stabilize adsorbed N_2. In this case the N_2 reduction would appear as dynamic process in which it is the high flux of H on the FeMo-co to create the feasible conditions for the reaction to proceed. The role of HIS-195

in this case would thus be that of transferring protons of N_2. A similar consideration holds for the bridge site with central C, though, given the large conformational change necessary to reach the N_2 adsorbed structure, it seems reasonable that this pathway might be kinetically limited. An interesting possibility for N_2 adsorption is that represented by the transition from closed to open site. This conformation is in fact only slightly unstable with respect to the closed site and the energy released by N_2 adsorption would make the structural change energetically accessible. A PES scan of the reaction of transition from closed to open site structure showed that this conformational change does not require overcoming any energetic barrier. In particular, if C is the central atom, the transposition of H to S2B and migration of HIS-195 towards Fe2 is helped by the formation of an intermediate structure, similar to Fig1b but with the FeMo-co in the 'closed' form, which has a similar energy to that in which HIS-195 is protonated (Fig. 1). Such structure would rapidly lead to the formation of the 1e structure in presence of N_2.

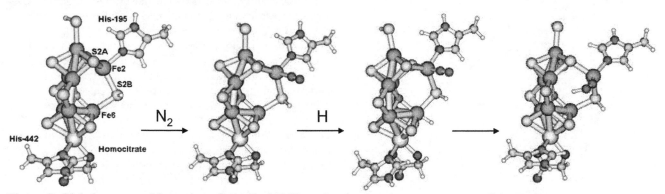

Figure 7. Global process of formation of adsorbed N_2H specie adopting the open structure of the FeMo-co.

References

(1) Einsle, O.; Tezcan, F. A.; Andrade, S. L. A.; Schmid, B.; Yoshida, M.; Howard, J. B.; Rees, D. C. *Science* **2002**, *297*, 1696-1700.

(2) Igarashi, R. Y.; Laryukhin, M.; Dos Santos, P. C.; Lee, H. I.; Dean, D. R.; Seefeldt, L. C.; Hoffman, B. M. *J. Am. Chem. Soc.* **2005**, *127*, 6231-6241.

(3) Yang, T. C.; Maeser, N. K.; Laryukhin, M.; Lee, H. I.; Dean, D. R.; Seefeldt, L. C.; Hoffman, B. M. *J. Am. Chem. Soc.* **2005**, *127*, 12804-12805.

(4) Dance, I. *Chem. Comm.* **2003**, 324-325.

(5) Hinnemann, B.; Norskov, J. K. *J. Am. Chem. Soc.* **2004**, *126*, 3920-3927.

(6) Huniar, U.; Ahlrichs, R.; Coucouvanis, D. *J. Am. Chem. Soc.* **2004**, *126*, 2588-2601.

(7) Kastner, J.; Hemmen, S.; Blochl, P. E. *J. Chem. Phys.* **2005**, *123*.

(8) Lovell, T.; Li, J.; Liu, T. Q.; Case, D. A.; Noodleman, L. *J. Am. Chem. Soc.* **2001**, *123*, 12392-12410.

(9) Hinnemann, B.; Norskov, J. K. *J. Am. Chem. Soc.* **2003**, *125*, 1466-1467.

(10) Gaussian03 *Revision C. 01, see supplementary informations for full reference* Wallingford CT, 2003.

(11) Dance, I. *J. Am. Chem. Soc.* **2005**, *127*, 10925-10942.

Development of Functionally Graded SiC-Based Diesel Engine Exhaust Gas Filter

Thorsten Gerdes, Monika Willert-Porada,
Chair of Material Processing, University of Bayreuth, D-95440 Bayreuth, Germany

ABSTRACT

The development of a SiC-carbide based filter that combines a low pressure drop due to a graded pore structure with high filtration performance ascribed to different filtration mechanism is presented.

The processing route is based on carbon paper/carbon black composites with graded porosity. By microwave assisted chemical vapor infiltration (CVI) the composite is coated with silicon and partially carbidized in a way, that the oxidation resistivity of the obtained SiC is combined with the good mechanical strength of the remaining carbon fiber core.

To reduce the soot removal temperature during operation, the material is coated with a Platinum/TiO$_2$-catalyst.

Keywords: particle filter, diesel soot, microwave, chemical vapour infiltration

1 INTRODUCTION

The incomplete combustion of diesel fuel in diesel engines as well as the recombination in the exhaust gas leads to carbon particle, inorganic oxide, and hydrocarbon emissions. The particle size varies between 5–20 nm up to 50 – 150 nm for agglomerates [1].

Because of the extreme pressure drop filter material with bulk pore sizes in the range of the soot particles are not useful. Therefore the filter materials consist of large pores and the particles are collected by "Deep-bed-filtration" (impaction and diffusion) and "Cake-filtration" instead of "Sieve-filtration".

Hot gas filtration using ceramic filters for diesel engine exhaust gas cleaning has gained considerable attention and growth over the last 15 years now.

State of the art diesel particle filters (DPF) utilize mainly cordierite or silicon carbide wall-flow monoliths with a honeycomb structure to trap the soot produced by diesel engines. Exhaust gases escape through the pores in the wall material and most of the soot particles are trapped in the filter walls.

The requirements of the removal efficiency, especially for extreme fine particles, leads to a high pressure drop in the filter cartridge and a significant increase of fuel consumption.

Therefore the development goal was a SiC-carbide based filter that combines a low pressure drop due to a graded pore structure with high filtration performance ascribed to different filtration mechanism.

To achieve the different requirement of a modern DPF, different performance parameters has to be taken into account (see figure 1). The filter wall material development has to consider conflictive requirements like high permeability and high particle collection efficiency.

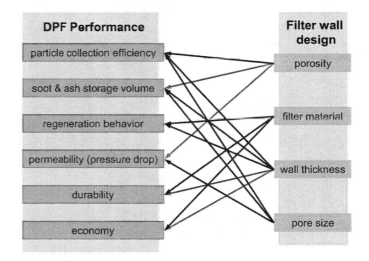

Figure 1: Diesel particle filter (DPF) development.

During exhaust gas filtration from time to time it is necessary to clean the DPF by a thermal treatment, because the collected particles enhance pressure drop.

Soot particle start burning at temperatures higher than 600°C but the burn off can be promoted by catalysts down to temperatures of 300–400°C [2].

2 EXPERIMENTAL

An overview of the DPF processing steps are shown in figure 2.

The material processing starts from thin commercial carbon papers with low pressure drop but insufficient removal efficiency for nano-scale soot particles. To achieve a high particle removal without a strong increase of pressure drop, from one side a carbon black slurry is partially infiltrated into the carbon paper. At this stage the composite can be folded into complex shapes to increase the filtration area.

To achieve the necessary oxidation resistivity at high temperatures, the material is heated in a 2.45 GHz microwave cavity while trichlorosilane is poured though the porous material (see figure 3).

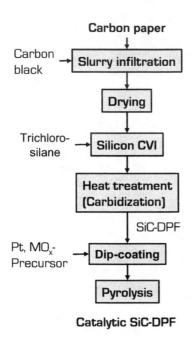

Figure 2: Diesel particle filters (DPF)-material preparation process

The thermal decomposition of the trichlorosilane can take place either homogeneous in the gas phase followed by an impaction and/or diffusion of the nano scale silicon particles at the carbon surface or by heterogeneous decomposition of the precursor at the carbon surface (chemical-vapour-deposition). The simplified decomposition reaction is given by:

$$SiHCl_3 + H_2 \leftrightarrow Si + 3\,HCl$$

During an additional heat treatment at 1500°C the temperature stable SiC is formed.

In the next processing step a Platinum/TiO2 –containing wash-coat is infiltrated into the DPF-material, to reduce the soot burn off temperature during the exhaust gas filtration. A detailed description of the coating procedure is given in [4].

The flow rate of a filter is dependent on the applied differential pressure and is one of the most important parameters to characterize a DPF.

The non linear relation between pressure drop and flow rate is described by the following equation:

$$\Delta p = \frac{\dot{V} \cdot s}{A} \cdot \left[\frac{\eta}{\alpha} + \frac{\rho \cdot \dot{V}}{\beta \cdot A} \right]$$

ρ = Viscosity coefficient $[m^2]$ β = Inertia coefficient $[m]$
s = Filter thickness $[m]$ $\dot{V}$ = Flow rate $[m^3/s]$
ρ = Fluid density $[kg/m3$ A = Filter surface $[m^2]$
ρ = Dynamic viscosity $[Pa{\cdot}s]$ $\Delta\rho$ = Pressure drop $[Pa]$

The equation can be divided into a laminar (linear) and a turbulent part.

For laminar flow, the simplified Darcy's equations can be used to show the relation between the variables:

$$\Delta p = \frac{\dot{V} \cdot s \cdot \eta}{A\alpha}$$

The flow late measurements have been performed up to pressure drops much higher than during operation of the filter material in the exhaust gas system of diesel engines.

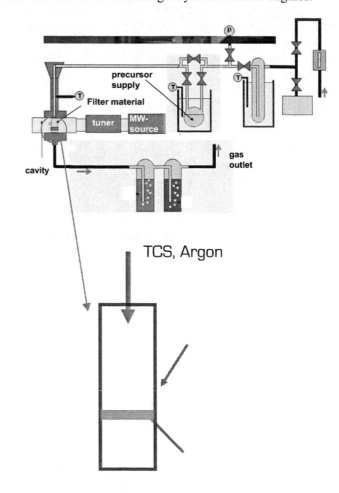

Figure 3: Experimental set-up for the microwave assisted trichlorosilane decomposition on the carbon pre-form.

 Cleantech 2007, ISBN 1-4200-6382-0

3 RESULTS

The development of the microstructure during the different process steps is shown in figure 4.

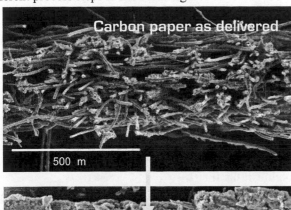

Carbon paper as delivered

500 m

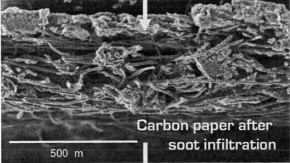

Carbon paper after soot infiltration

500 m

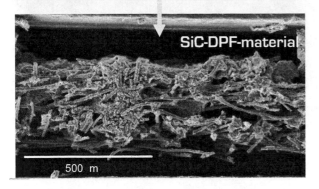

SiC-DPF-material

500 m

Figure 4: SEM-Cross section of the material after different processing steps

After dying, the carbon black infiltrated carbon paper material consists of graded porosity with a top layer of small pores and an increasing pore size down to large pores at the bottom.

By microwave decomposition of the trichlorosilane it is possible to cover the carbon black particles as well as the carbon fibres with fine grained silicon.

A SEM picture of the silicon coated carbon fibre is shown in figure 5.

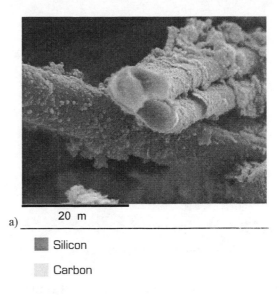

a) 20 m

■ Silicon

▨ Carbon

b)

Figure 5: SEM picture of a silicon coated carbon fibre and EDX-analysis of the carbon and silicon distribution

During the following heat treatment the carbon black is converted to SiC with maintaining the pore structure while the carbon fibres are partial converted to utilize the good mechanical properties of the carbon-fibre.

The XRD-analysis of the material after carbidization is shown in figure 6. The remaining carbon core enables an easy handling like bending, cutting and assembling of the filter material.

The resulting pore size distribution of the DPF compared to the pore size distribution of the carbon paper is shown in figure 7.

While the carbon paper meanly contains pores in the range of 30 to 300 m, the DPF has additional pores in the range of 100 nm. Even if these small pores are located in the small top layer of the filter, a strong effect on the permeability of the filter material can be measured (see figure 8).

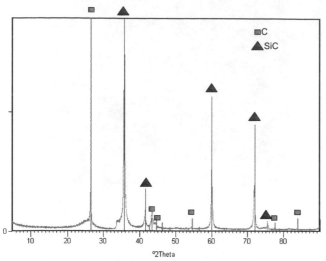

Figure 6: XRD-Phase analysis of the DPF after carbidization

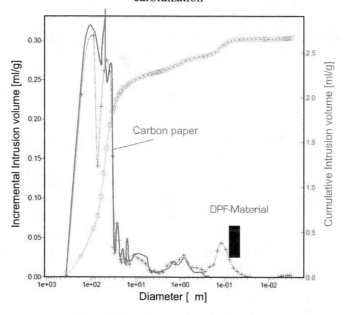

Figure 7: Pore size distribution of the DPF compared to the carbon paper as delivered.

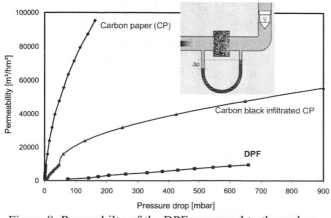

Figure 8: Permeabilty of the DPF compared to the carbon paper and the carbon black infiltrated carbon paper

4 CONCLUSION

A new processing route for a functionally graded deep bed DPF-system has been developed. Each step of the processing and also the selection of the material components (carbon paper, carbon black, standard catalysts trichlorosilane) enables an easy scale up.

Up to now the permeability of the DPF is in the same range as commercial filter systems, therefore especially the carbon black infiltration has to be further optimized to reduce the pressure drop in the top layer.

It could be shown that the microwave assisted pyrolysis of the silicon precursor (Trichlorosilane) offers an ideal method for homogeneous and controllable coating of the carbon material with silicon.

The carbon fiber core enables an easy handling and assembling of the filter.

Because of the graded pore size and porosity also the typical filter parameter like particle collection efficiency, storage volume and pressure drop are graded as shown in figure 8.

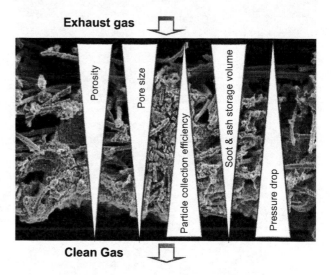

Figure 8: Functionally gradient concept for DPF

5 REFERENCES

[1] J. Adler, Ceramic Diesel Particulate Filters, Int. J. Appl. Ceram. Technol., 2 [6] 429–439 (2005)

[2] J. Oi-Uchisawa, A. Obuchi, S.Wang, T. Nanbaa, A. Ohi, Catalytic performance of Pt/MOx loaded over SiC-DPF for soot oxidation, Applied Catalysis B: Environmental 43, 117–129 (2003)

[3] T. Gerdes, M. Willert-Porada, H. Wolf, Development of Functionally Graded SiC-Based Hot Gas Filter, Multiscale and Functionally Graded Materials Conference 2006, Hawaii, USA

[4] T. Schubert, Synthese, Verarbeitung und Funktionsweise nanoskaliger anorganischer Materialien in galvanischen Energiewandlungssystemen, dissertation, University Bayreuth, 2005

Study on Non-Freon Air Cooling System Using Water Refrigerant

A. Sugawara[*], S. Nakamikawa[*], M. Kageyama[*], S. Yamazaki[*], N. Abe[*],
T. Itou[*], A. Tsurumaki[**] and K. Kawasaki[**]

[*]Department of Electrical and Electronic Engineering, Niigata University
8050 Ikarashi-2, Niigata 950-2181, Japan, akira@eng.niigata-u.ac.jp
[**]Department of Mechanical and Production Engineering, Niigata University

ABSTRACT

This paper aims for construction of a small, clean air cooling system using a water refrigerant and examines it experimentally. We offer a basic report for realization of the non-Freon air-cooling system using only water as a refrigerant. An experimental device consists of two groups, the first stage is composed a vacuum pump and a vacuum container of about 60 [liter] in volume, and the second stage has two heat exchangers (one is a heat load) and a water circulation pump. The variations of temperatures at each place are measured by thermocouples. An air-conditioner indoor unit or a heater as the heat load is used; the air conditioning measurement and the energy are measured, respectively. Consequently, in air-cooling system used only water refrigerant, the ability to cool a room air enough is provided. The development of a new vacuum pump which can exhaust a large amount of water vapor is introduced as an appendix.

Keywords: non-Freon, air-cooling system, water, vacuum pump, heat

1 INTRODUCTION

In recent years, global environment problems such as global warming and ozone depletion have been worried. The spread of air-conditioners contributes to an increase of work efficiency and improvement of comfortable living environment. On the other hand, alternatives, e.g. chlorofluorocarbon (HCFC, etc.), used for a refrigerant of air-conditioners don't destroy the ozone layer, but they are appointed as greenhouse gas, and the collection at the time of the disposal is obliged. However, it is difficult to collect all refrigerants.

Energy-saving product and non-Freon machinery have become active, and refrigerants using air-conditioners have zero or near to zero with ODP (an ozone destruction coefficient) and GWP (global warming potential). Interest in natural refrigerants like water, carbon dioxide, hydrocarbon, etc. has risen in the field of air conditioning. Water is superior for the cost and environment, and the handling is easy, too.

When water is located under vacuum, it boils till the pressure of saturation water vapor at the water temperature. The water temperature falls, since the boiling takes evaporative latent heat from water. In other words, we can get coldness by exhausting the vacuum container with water.

In food industry, there are some applications of water decompression cooling, e.g. moist lettuce or cut flowers, and decompression takes evaporation heat away from them rapidly [1], [2]. In resent years, Sanken Setsubi Kogyo Co. Ltd. (in Japan) and IDE technology Corporation (in Israeli) cooperated, and sell a water cooling system of the environment correspondence type that incorporated the steam turbo compressor [3]. This big cooling system which has freezing capacity of more than 350 [kW] is used in a gold mine in South Africa. In addition, there is absorption chiller of more than 100 [kW] using NH_3 or LiBr as the refrigerant. However, a small water decompression cooling system of several kW such as home use is not developed.

This paper aims for construction of a small, clean air cooling system using a water refrigerant and examines it experimentally. We offer a basic report for realization of the non-Freon air-cooling system using only water as a refrigerant.

2 WATER REFRIGERANT

There are some natural refrigerants, i.e. hydrocarbon (isobutane, propane), CO_2, ammonia, air, and water. Water exists in the natural world abundantly, has not the toxicity, burn ability, and is cheap. In addition, characteristics of water show the specific heat at constant pressure of 1.846 [kJ/(kg · K)], the latent heat of 2457 [kJ/kg] at 17.5 [℃], and water can remove large thermal energy by evaporation.

International Committee of Weights and Measures in 1990 was adopted an experimental formula which was reported about the saturated water vapor pressure P_s by D. Sonntag in 1986 [4]. The saturated water vapor pressure calculated by the experimental formula is shown in Fig. 1.

3 COEFFICIENT OF PERFORMANCE

Efficiency of the refrigerator is evaluated in COP (Coefficient of Performance). COP in the experiment that installed an air-conditioner indoor unit is calculated as follows.

First, the relative humidity Φ is measured. Using P_s, the relative humidity Φ is converted into the absolute humidity X.

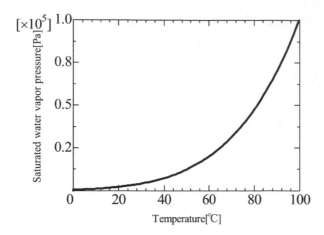

Fig. 1. A curve of the saturated water vapor pressure.

$$X = 0.622 \times \frac{\Phi P_s}{1.013 \times 10^5 - \Phi P_s} \quad (1)$$

where 0.622 is the ratio of molecular weight of steam/air.

Next, using Eq. (1), the ratio of enthalpy h is calculated by,

$$h[\text{kJ/kg}] = 1.005 \times T + (2447 + 1.846 \times T) \times X \quad (2)$$

where the specific heat at constant pressure of dehydration air is 1.005 [kJ/(kg(DA)・K)], T is the temperature [K], the evaporative latent heat of water is 2447 [kJ/kg] at 22.0 [℃], and the specific heat at constant pressure of water vapor is 1.846 [kJ/(kg・K)]. The specific volume V is calculated using the ideal gas equation by,

$$V[\text{m}^3/\text{kg}] = \frac{(287.13 + X \times 461.7) \times T}{(1 + X) \times 1.013 \times 10^5} \quad (3)$$

where the gas constants of dry air and water vapor are 287.13 [J/(kg・K)] and 461.7 [J/(kg・K)], respectively.

Therefore, the cooling capacity C_c at the quantity of wind 0.12 [m³/s] is,

$$C_c[\text{kW}] = \frac{h_{inlet} - h_{outlet}}{V} \times 0.12 \quad (4)$$

Then, COP is obtained by,

$$\text{COP} = \frac{\text{Cooling capacity[kW]}}{\text{Integrated power[kW]}} \quad (5)$$

4 EXPERIMENTAL SETUP AND METHOD

4.1 Cooling experiment using air-conditioner indoor unit

Experimental setup is shown in Fig. 2. The experimental device consists of two groups, the first stage is composed a vacuum pump and a vacuum container of about 60 [liter] in volume, and the second stage has two heat exchangers (one is a heat load) and a water circulation pump.

In the first stage, the primary cooling water of 5 [liter] is located in a vacuum chamber. A heat exchanger is soaked in a plastic container which is thermally isolated with the stainless vacuum chamber. When the vacuum chamber is evacuated by a water ring vacuum pump, the temperature of the primary cooling water falls down so that thermal energy is taken as latent heat. In second stage, the cooled water by the heat exchanger inside the vacuum chamber is flowing into another heat exchanger of an air-conditioner indoor unit (product by Corona Co., CSH-ES282-W type) with flow speed of 2.6 [litter/min.] by the circulation pump. The amount of secondary circulating water is 5 [liter]. The room air as heat load exchanges the heat and becomes a cold wind from an outlet of the air-conditioner indoor unit. The cold wind isolates with a room air by a plastic sheet.

The variations of temperature at each place, i.e. primary cooling water temperature TCvw, air-conditioner indoor unit entrance water temperature TCci, the exit temperature TCco, room air temperature TCai, and the cold air TCao, are measured by thermocouples. In addition, the humidity of the room air and the exit from the air-conditioner indoor unit are measured by each humidity sensor. The absolute humidity and enthalpy characteristics are calculated using Eqs. (1) and (2). The power integrated by the water ring vacuum pump and the circulation water pump are measured by a wattmeter. The integrated power equals to the denominator in Eq. (5).

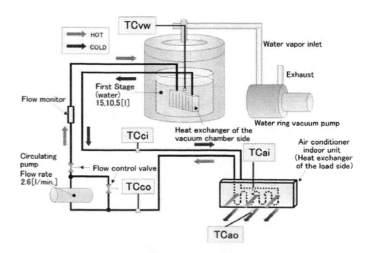

Fig. 2. Experimental setup.

Cleantech 2007, ISBN 1-4200-6382-0

4.2 Cooling experiment using heat load

As a heat load instead of the air-conditioner indoor unit, a heater of 0.24–1.33 [kW] is used. The quantity of the primary cooling water is 5 [liter], the secondary circulating water is 5 [litter] with flow speed of 3.0 [litter/min.]. Electric power consumption (a denominator of Eq. (5)) and each refrigerant water temperature in Fig. 2 are measured.

5 EXPERIMENTAL RESULTS AND DISCUSSIONS

5.1 Results for air-conditioner indoor unit

The variation of temperature at each place and the integrated power are shown in Fig. 3, respectively. During the experiment, TCai is kept at almost the constant temperature of 29 [℃]. Just after the experiment starts, TCvw decreases suddenly and the others also decrease gently. The experimental data shows a steady state in 20 minutes after the experiment starts. In the steady state during 30-45 [minute], the mean dry-bulb temperature difference between TCai and TCao is about 7.2 [℃].

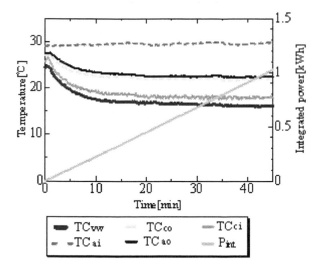

Fig. 3. Characteristics of temperature and integrated power.

The variations of relative humidity at the room air inlet and cold air outlet are shown in Fig. 4. The inlet relative humidity is approximately constant, but the outlet relative humidity increases, and the characteristic becomes approximately constant in 20 [minute] after the experiment starts. The reason seems that the relative humidity is defined as the quantity of water vapor for unit volume divided by the saturation steam density at the temperature. The saturation steam density depends on the temperature decreasing, but the quantity of water vapor is almost same.

The relative humidity is converted by Eq. (1) to the absolute humidity. Using Eq. (2), the enthalpy is calculated and the COP is depicted in Fig. 5. The COP shows low

value at the time of the vacuum pump starting, but it shows about 1 for the steady state.

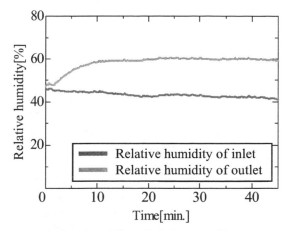

Fig. 4. Relative humidity (Primary cooling water of 5 [liter]).

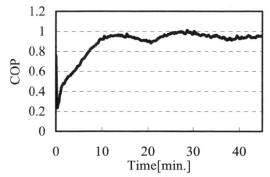

Fig. 5. COP （Primary cooling water of 5 [liter]）.

5.2 Results for heat load

The secondary circulating water is supplied a constant heat load (1.03 [kW]), and the characteristics are shown in Fig. 6. The experimental data shows a steady state in about 40 [minute] after the experiment starts. In the steady state during 50-70 [minute], the mean temperature difference ΔT between heater inlet and outlet is about 4.7 [℃]. The variation of the quantity of the primary cooling water is shown in Fig. 6 right axis. The primary cooling water of 15 [liter] at 25 [℃] decreases linearly and the rate of the quantity becomes 15 %.

The heat load supplied by the heater, i.e. the cooling capacity Cc, is given from the thermal balances Q_1 at the first stage and Q_2 at the second stage by

$$C_c = Q_1 = 2442 \times \Delta m \tag{6}$$
$$Q_2 = 4.2 \times \Delta T \times \Delta q \tag{7}$$

where the evaporative latent heat of water is 2442 [kJ/kg] at TCvw= 24 [℃], Δm [kg] is the quantity of evaporation, the specific heat of liquid water is 4.2 [kJ/(kg・K)], Δq

[kg/s] is the flow rate of the circulation water. The difference, i.e. $Q_1 - Q_2$, means the thermal leak from the vacuum chamber and depends on the temperature difference between the room temperature and the primary cooling water. COP is about 1.2 for the steady state as an average of Δm for 10 [s] substituting the electric power of about 1.3 [kW] and Eq. (6) to Eq. (5).

Consequently, in air-cooling system used only water refrigerant, the ability to cool a room air enough was provided.

However, the vacuum pump in this study is a water ring vacuum pump. Since this vacuum pump can't evacuate over the pressure of saturation water vapor of the water seal, the vacuum pump needs a lot of water seal with about 2 [℃]. We considered using oil-sealed rotary pump, oil free vacuum pump and etc., but there is no ability to exhaust a large amount of water vapor. Therefore, the development of a new vacuum pump which can exhaust a large amount of water vapor and can evacuate up to 1/100 of atmospheric pressure will be needed.

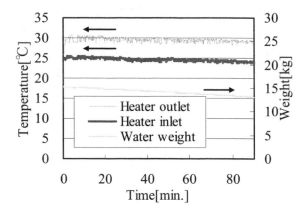

Fig. 6. Experimental results for heat load.

6 CONCLUSIONS

A small air-cooling system of around several kW using only water refrigerant as a non-Freon technology was examined. The vacuum-cooling could cool down the room air about 22 [℃] for the ambient temperature of 29 [℃]. The COP was about 1.0.

Consequently, cooled water was made by vacuum-cooling and the circulating exchanged cold energy to the room air. The non-Freon air-cooling system using only water refrigerant will be realized.

APPENDIX

The engine vacuum pump has the following characteristics. An engine of a car is used as it is, and the modified cylinder head which has some reverse-check valves made by FRP sheet of 0.3 mm in thickness can exhaust gas by the pressure difference between the internal and external automatically (see Fig. 7). The new vacuum is simple structure, cheap, and easy construction. The water temperature characteristics exhausting water vapor for 3 [liter] of this vacuum pump (400 [rpm]) were measured (see Fig. 8). Consequently, it is cleared that the new vacuum pump can work enough for the air cooling system using water refrigerant.

Fig. 7 Engine vacuum pump.

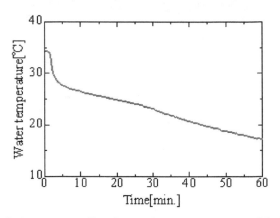

Fig. 8. Vacuum-cooling by engine vacuum pump (No heat load).

REFERENCES

[1] A. Tambunan, Y. Sagara, Y. Seo, and Y. Kawagoe. "Image Analysis on Temperature Distribution within Lettuce Undergoing Vacuum Cooling". Agricultural Engineering International: the CIGR Ejournal. Vol. VII. Manuscript FP 04 002. January, 2005.

[2] Da-Wen Sun（Corresponding author), Tadhg Brosnan. Extension of the vase life of cut daffodil flowers by rapid vacuum cooling. International Journal of Refrigeration 22 (1999) 472-478.

[3] http://www.jiakanto-koryu.org/sheet/sheet_F-1-20020829.pdf (in Japanese).

[4] D.Sonntag: "Important new Values of the Physical Constants of 1986, Vapor Pressure Formulations based on the ITS-90, and Psychrometer Formulae" Z.Meteorol.70 (1990)5,340-344.

Conjugates of Magnetic Nanoparticle-Enzyme for Bioremediation

You Qiang [1], Amit Sharma[1], Andrzej Paszczynski [2], and Daniel Meyer[1]

[1] Department of Physics, University of Idaho, Moscow, ID 83844, USA, youqiang@uidaho.edu
[2] Environmental Biotechnology Institute and Department of Microbiology Molecular Biology and Biochemistry, University of Idaho, Moscow, ID 83844, USA

ABSTRACT

Enzymes are proteins that function as biocatalysts in bioremediation. One of the major concerns in environmental applications of enzymes is their short lifetime. Enzymes lose their activity due to oxidation, which results in less stability and a shorter lifetime thereby rendering them less efficient. An effective way to increase the stability, longevity, and reusability of the enzymes is to attach them to magnetic iron nanoparticles. If enzymes are attached to the magnetic iron nanoparticles then we can easily separate the enzymes from reactants or products by applying a magnetic field. With this aim, two different catabolic enzymes, trypsin and peroxidase, were attached to uniform core-shell magnetic nanoparticles (MNP's), produced in our laboratory. Our study indicates that the lifetime and activity of enzymes increases dramatically from a few hours to weeks and that MNP-Enzyme conjugates are more stable, efficient, and economical. We predict that MNPs shield the enzymes preventing them from becoming oxidized. This results in an increased lifetime of the enzymes. Because of the high magnetization (~140 emu/g) of our MNPs, nanoparticle-enzyme conjugates can efficiently be magnetically separated, making enzymes more productive. We also found that the enzyme structure plays a major role in efficient attachment of MNPs

Keywords: enzyme, magnetic nanoparticle, conjugate, bioremediation, environmental application

1 INTRODUCTION

Small size, high surface area and low toxicity has made magnetic iron nanoparticles most promising element for various fields such as biomedical and environmental applications. Since the mid-1970s, MNPs have been widely studied for their applications in various areas of biomedical science from therapeutic agent targeting and magnetic resonance imaging (MRI) to cell separation and purification [1]. Drug targeting and delivering, gene therapy, and hyperthermia are active biomedical research areas with great promise for disease treatment. But much less effort has been applied to developing MNPs in the environmental research field. Recently, a field demonstration was performed in which nanoscale (100-200 nm in diameter) bimetallic (Fe/Pd) particles were gravity-fed into groundwater contaminated by chlorinated aliphatic hydrocarbons [2]. The nanoparticles were found to be uniquely suited for subsurface delivery and dispersion and rapidly degraded chlorinated contaminants such as trichloroethylene (TCE) by reductive dehalogenation during oxidation of the zero-valent metals.

Contaminations in soil and water are major concerns of environment. For example, 2, 4, 6-trinitrotoluene (TNT) is anthropogenic pollutants present in environment. It's found in solid form and it slowly dissolves, gradually polluting soil and groundwater. Pure iron Fe(0) nanoparticles can biotransform one or more nitro group of TNT into amino group detoxifying the area, which is contaminated because of TNT [7, 9]. Presence of enzymes in a biochemical reaction accelerates the rate of reaction. During reaction enzyme retain their property, thus they can be cost effective if we could reuse them [14]. Enzymes can be reused if we can immobilize them by attaching them on a solid surface, this will make it easier to separate enzymes from the solution. If enzymes are attached to the magnetic iron nanoparticles then we can easily separate the enzymes from reactants or products by applying a magnetic field. Short lifetime of enzymes outside the living cell also limits their applications [9]. Attempts have been made repeatedly to increase the stability of enzymes by encapsulating biomolecules in silica gels but repeatability and long term stability still remains a concern [16]. Lack of stability of enzymes during storage is also one of the issues with enzymes.

The magnetic nanoparticle-enzyme conjugates (MNP-Es) will have a major advantage over metal-only particles such as those described by Elliot and Zhang [2] that react stoichiometrically with substrates in equimolar reactions rather than catalytically; zero-valent metals are quickly consumed by water passivation and/or contaminant reduction. In contrast, particle-bound enzymes, when stabilized to prevent protein degradation, can act as true catalysts, turning over many moles of substrate molecules before ultimate enzyme inactivation. Moreover, immobilization of bioactive molecules on the surface of magnetic nanoparticles is of great interest because the magnetic properties of these bioconjugates promise to greatly improve the active delivery, recovery, and control of biomolecules in environmental and other applications.

Rossi et al. [3] covalently conjugated the enzyme glucose oxidase to 20-nm (diameter) magnetite (Fe_3O_4) nanoparticles in their development of glucose sensors. The iron oxide nanoparticles were derivatized with amino groups using 3-(aminopropyl)triethoxysilane, and glucose was attached to the amino linkers. The amount and activity of the immobilized enzyme was increased relative to that with only physical adsorption processes. Covalent immobilization also increased the stability of the enzyme. The nanoparticles maintained their bioactivity at 4°C for up to 3 months. The direct binding of an *Aspergillus niger* glucose oxidase via a carbodiimide linkage to magnetic nanoparticles was found to be very effective, resulting in bound enzyme efficiencies between 94-100% and increased enzyme resistance to thermal and pH-dependent denaturation [4]. The same reaction was used to examine cholesterol oxidase (CHO) properties after binding to Fe_3O_4. Stability and activity of CHO was enhanced after attachment to magnetic nanoparticles, improving the potential for use of this enzyme in various biological and clinical applications [5]. Ohobosheane et al. [6] demonstrated biochemical modification of silica-based nanoparticles whose surfaces were linked to glutamate dehydrogenase and lactate dehydrogenase allowing them to function as biosensors and biomarkers. The immobilized enzyme molecules were shown to retain excellent enzymatic activity in respective reactions.

Here we reported a new method to cross linking enzymes with bifunctional reagents which help in increasing the life time of enzyme; this process involves the crystallization of enzymes Proteins sometimes don't get efficiently crystallize or proteins are locked during crystallization making them inactive [17]. We have found an efficient way of binding enzymes. Attaching enzymes to magnetic iron nanoparticles extends their lifetime from few hours to weeks. In this paper, we described how surface area plays a major role in attaching enzymes to the nanoparticles and how iron nanoparticles help in immobilizing enzymes.

2 EXPERIMENT

Monodispersive core-shell iron nanoparticles were produced using third generation cluster deposition apparatus [10-12]. The size of the nanoparticles was controlled by varying the growth distance, power, and helium and argon gas ratio. Uniform ~20 nm size iron nanoparticles were deposited on a plastic substrate. The nanoparticles were then removed from the plastic surface and collected in the solution of pH 7. Magnetic moment of the iron oxide nanoparticles produced in our lab is ~140 emu/g (4). Fig. 1(a) shows a TEM image of ~20 nm iron nanoparticles before enzymes were attached, and Fig. 1(b) is the image of nanoparticles uniformly dissolved in the solution.

Two catabolic enzymes were attached to the nanoparticles namely trypsin and horseradish peroxide C (HRP). To prevent denaturation and leaching nanoparticles were coated with 3-aminopropyl triethoxy silane, thus prolonging the stability of the magnetic nanoparticle. The first reaction shown in Fig. 2 is an example of silanization of MNP passivated with ferric-oxyhydroxy-polymer with 3-

aminopropyltriethoxysilane. Commercially available cross-linking agents were used to attach activated enzymes to nanoparticles. Four different coupling reagents: **SANH** (succinimidyl 4-hydrazinonicotinate acetone hydrazone); **C6-SANH** (C6-succinimidyl 4-hydrazinonicotinate acetone hydrazone); **SFB** (succinimidyl 4-formylbenzoate) and **C6-SFB** (C6-succinimidyl 4-formylbenzoate) were used. These reagents prevent homopolymerization of MNP's and enzymes, and provide variability in spacer-arm length from 5.8 to 14.4 Å.

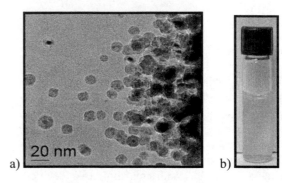

Fig. 1: TEM image of ~20 nm iron nanoparticles, (b) is the image of the nanoparticles uniformly dissolved in the solution of pH 7.

This length variability results in MNP-E conjugates with different rotational properties, which in turn influences enzyme active center accessibility and directly affect MNP-Es enzymatic activity. The SANH converts primary amines to hydrazinopyridine moieties and SFB converts primary amines to benzaldehyde moieties. These two moieties cross-link producing final MNP-E congeners. The random polymerization of multiple enzyme molecules or MNP particles is avoided. The C6-SANH and C6-SFB contain six-carbon aliphatic linkers between functional groups.

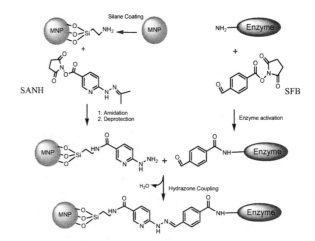

Fig. 2: Reaction scheme used for producing MNP-Es using amino-silane-coated MNP's.

The enzymes were covalently linked with nanoparticles by reacting them separately with amino-silane or peptide coated nanoparticles. After the modifications the MNP's were purified of excess reagent. The hydrazine/hydrazide-modified MNP's were reacted with the aldehyde-modified molecule to yield the desired MNP-E conjugates (Fig. 2). Both reaction mixtures (enzyme + SANH and MNP's + SFB) were incubated for 3 hours at room temperature with no shaking in buffer (pH=7.3). The concentration of SFB and SANH were in 10 molar excess of protein or MNP's. After activation the excess of the heterobifunctional coupling agent and the buffer replacement (to the conjugation buffer, pH=4.7) was performed in single chromatographic step using a Sephadex G25, PD10 column.

3 RESULTS AND DISCUSSION

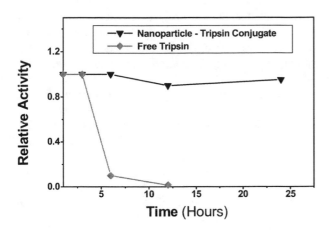

Fig. 3: Stability of MNP-trypsin at pH=7 at 5 °C.

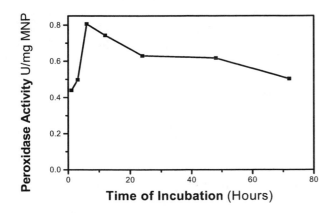

Fig. 4: Optimization MNP-Peroxidase coupling reaction time (Conjugate Peroxidase—SANH/ MNP's- SFB)

During coupling reaction activated MNP and enzyme solution were agitated using rocking shaker. The optimal molar ratio of enzyme to MNP was 10:1 and optimal pH was 4.7. We attached MNP's to trypsin, and found the stability of the trypsin is very good. Fig. 3 shows the relative activity of free trypsin and nanoparticle-trypsin

conjugate. It can be clearly seen that free trypsin looses its activity after about six hours (1) in comparison to the MNP-E conjugate which is active for more than twenty five hours. We also attached MNP to peroxidase and checked the activity of the peroxidase after every three hours. As seen in Fig. 4 peroxidase was active for more than seventy hours. Stability of enzymes was now tested by incubating them for 5 weeks.

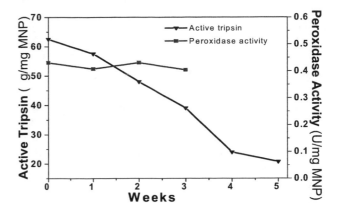

Fig. 5: Stability of MNP-Es in weeks.

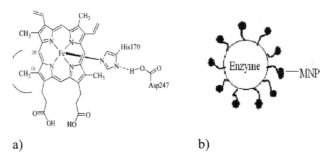

Fig. 6: a) Structure of horseradish peroxidase b) Enzyme with the envelope of MNP

Fig 5 shows the stability of enzyme with MNP. MNP's and enzymes were joined to each other by covalent bond using a heterobifunctional cross linkers. We used SANH and SFB as cross linkers to attach MNP's and Enzymes. These cross linkers enveloped the MNP's on enzymes. We estimated the density of enzymes conjugated to the MNP's using protein concentration measurement techniques. Knowing available surface area, protein amount bonded, and enzyme dimensions, we were able to calculate the fraction of surface covered by a given enzyme; trypsin covered 20% of the available surface of MNP and peroxidase covered 25% of the available surface area of MNP. Horseradish peroxidase C has the iron atom at the center Fig. 6a. Experiments are performed with iron immobilized on the surface of HRP for biosensors and biochip applications [15]. Theorell in 1942 published that

HRP and some of its derivative has the magnetic properties. Magnetic moment of our iron oxide nanoparticles produced is ~140 emu/g [10-12]. Presence of iron atom at the center of enzyme and high magnetic moment of nanoparticles resulted in increased attachment of peroxidase to nanoparticles in comparison to trypsin. Enzyme looses its activity due to the process of oxidation or other processes. The enzyme starts self digesting which results in the loss of the activity. We predict that magnetic nanoparticle envelope's the HRP and trypsin enzyme thus completely shielding it as seen in Fig 6b. This slows down the process of oxidation which results in the increase in lifetime as seen in Fig. 4 & Fig. 5. The productivity and cost efficiency of enzymes could be increased if we could reuse them. Iron nanoparticles being magnetic, we are able separate MNP-E conjugates after the reaction and immobilize enzymes making them more productive. We did this by applying a magnetic field of 45 gauss to MNP-E conjugates in vile [Fig. 1b]. MNP-E conjugates were effectively separated in solution in about 30 seconds.

4 CONCLUSION

Iron nanoparticles significantly increase the lifetime of enzymes thus making enzymes more productive. Immobilizing enzymes by attaching to iron nanoparticles makes it more stable this is because iron nanoparticles cage the enzymes by preserving or improving their activity. Size of the nanoparticle plays a very important role in attachment of enzymes. The higher rate of enzyme attachment to the nanoparticle was due to large surface area offered by iron nanoparticle of size ~20 nm. The fact that trypsin and peroxidase are active for more than five weeks implies that our iron nanoparticles are not toxic at this condition. MNP-E conjugates are also efficiently immobilized by using an external magnetic field.

5 ACKNOWLEGEMENT

This work is supported by NIH-INBRE (P20 RR016454), Murdock Charitable Trust and DOE-EPSCoR (DE-FG02-04ER46142).

REFERENCES

1. W. Andra, and H. Nowak. Magnetism in Medicine. 1998, Wiley-VCH, Berlin.
2. D. W. Elliott and W. X. Zhang. "Field assessment of nanoscale bimetallic particles for groundwater treatment". *Environ. Sci. Technol.* 2001, 35:4922-4926.
3. L. M. Rossi, A. D. Quach, and Z. Rosenzweig. Glucose oxidase-magnetite nanoparticle bioconjugate for glucose sensing. *Anal. Bioanal. Chem.* 2004, 380(4):606-613.
4. G. K. Kouassi, J. Irudayaraj, and G. McCarty. Activity of glucose oxidase functionalized onto magnetic nanoparticles. *Biomagn. Res. Technol.* 2005, 200, 3(1):1-10.
5. G. K. Kouassi, J. Irudayaraj, and G. McCarty. "Examination of cholesterol oxidase attachment to magnetic nanoparticles". *J. Nanobiotechnology* 2005, 3(1):1-9.
6. M. Qhobosheane, S. Santra, P. Zhang, and W. Tan. 2001. Biochemically functionalized silica nanoparticles. *Analyst* 126:1274-1278.
7. Jungbae Kim, Yuehe Lin, Jay W. Grate "Single-Enzyme Nanoparticles on Nanostructured Matrices" 2003 Biological Sciences PN03083/1746
8. S. K., Ahuja G. M. Ferreira, and A. R. Moreira. 2004. Utilization of enzymes for environmental applications. *Crit. Rev. Biotechnol.* 24:125-154.
9. R.W.S. Weber, D.C. Ridderbusch and H. Anke: 2,4,6-Trinitrotoluene (TNT) tolerance and biotransformation potential of microfungi isolated from TNT-contaminated soil. *Mycological Research* **106**: 336-344, 2002
10. You Qiang, Jiji Antony, Amit Sharma, Sweta Pendyala, Joseph Nutting, Daniel Sikes and Daniel Meyer, "Novel Magnetic Core-Shell Nanoclusters for Biomedical Applications", Journal of Nanoparticle Research, 8, 489, (2006).
11. J. Antony, Y. Qiang, Donald R. Baer and C. M.Wang, "Synthesis and Characterization of Stable Iron-Iron Oxide Core-Shell Nanoclusters for Environmental Applications", J. of Nanoscience and Nanotechnology, 6, 568-572 (2006).
12. Y. Qiang, J. Antony, M. G. Marino, and S. Pendyala, "Synthesis of Core-Shell nanoclusters with High Magnetic Moment for Biomedical Applications", IEEE Transactions on Magnetics, 40(2004) 6, 3538-3540.
13. Kevin O' Grandy "Biomedical application of magnetic nanoparticle" Journal of Physics D: Applied Physics: Editorial 36,131 (2002)
14. Dongfang Cao, Pingli He, Naifei Hu " electrochemical biosensors utilizing electron transfer in heme proteins immobilized on Fe_3O_4 nanoparticles Analyst,2003,128,1268-1274
15. D. L. Graham, H. Ferreira J. Bernardo P. P. Freitas J. M. S. Cabral "Single magnetic micro sphere placement and detection on-chip using current line designs with integrated spin valve sensors: Biotechnological applications" Journal of Applied Physics volume 91, 10, 2002.
16. Jacques Livage, Thibaud Coradin and C'ecile Roux "Encapsulation of biomolecules in silica gels" J. Phys.: Condens. Matter 13 (2001) R673–R691
17. Chandrika P Govardhan "Crosslinking of enzymes for improved stability and performance" Current Opinion in Biotechnology 1999, 10: 331-335

Recovery of Caustic Soda in Textile Mercerization by Combined Membrane Filtration

J. Yang[*], C. Park[**] D. Lee[*], S. Kim[*]

[*]Korea Institute of Industrial Technology, Chonan 330-825, South Korea, sykim@kitech.re.kr
[**]Kwangwoon University, Seoul 139-701, South Korea, chpark@kw.ac.kr

ABSTRACT

This study is intended to find the optimum operational conditions for the recovery of caustic (NaOH) solution from mercerization in textile process. For this, the silt density index (SDI) of ceramic membranes, the fouling property of NF membranes, the optimum conditions for the membrane regeneration through chemical cleaning, the optimum removal conditions of total organic carbon (TOC), turbidity, color and the permeate flux through the membranes were investigated. As a result, a combined membrane process using ceramic membrane (first step) and polymeric membrane (second step) was found to be suitable for the removal of total suspended solid (TSS), residual organics, turbidity including color and the recovery of caustic solution from wastewater caustic stream in mercerization process. The permeated caustic solution can be reused, which could offer economical benefits through the reduction of chemical use and the cost of wastewater treatment.

Keywords: recovery, caustic soda, mercerization, ceramic membrane, combined membrane filtration

1 INTRODUCTION

General industrial textile processes consist of pretreatment, desizing, scouring, bleaching, mercerizing, dyeing, and finishing. They are not only heavy consumers of energy and water but also producers of large amounts of chemical pollutants. Their impact on the environment is mainly related to the exhaustion of non-renewable water resources and the production of wastewater, which requires an appropriate pretreatment before discharge into recipient water bodies. The caustic wastewater from textile mercerization is hot and alkaline, containing one to five percent sodium hydroxide, but it contains few fiber impurities since most of them are removed in the upstream processes [1]. Therefore, the membrane technology is suitable for the treatment of various textile effluent streams and the recovery of valuable chemicals from them. The polymeric nanofiltration/reverse osmosis (NF/RO) membrane can selectively permeate relatively small organic molecules and ions from the textile mercerization. Using this type of membrane, the removal of fiber impurity and turbidity with color, the recovery of sodium hydroxide, and the recycle of process water from textile mercerization can be achieved. However, the NF/RO membrane filtration has problems of easy fouling, which often results in low flux and poor separation efficiency [2]. Therefore, a pretreatment step, such as the microfiltration/ultrafiltration (MF/UF) membrane process, is required.

2 MERCERIZATION

Mercerization is a process in which textiles (typically cotton) are treated with a caustic solution to improve properties such as fiber strength, shrinkage resistance, luster, and dye affinity. The caustic solution actually rearranges the cellulose molecules in the fiber to produce these changes. Higher-end fabrics may be double or triple mercerized for additional benefits. The fabric is first immersed in a caustic solution of about 18-25% strength and a relatively cool temperature of 16-32°C. A series of rollers (timing cans) are used to keep the fabric flat and smooth while controlling the time of caustic exposure. The fabric is then sprayed with rinse water and then washed with a neutralizing chemical before final drying (figure 1). In this study, a combined membrane process was applied, not only to improve the rejection efficiencies and flux recovery, but also to recycle the purified caustic solution back into the process.

3 EQUATIONS

The permeate flux of MF/UF can be expressed in terms of membrane resistances as in Eq. (1).

$$J \equiv \frac{1}{A}\frac{dV}{dt} = \frac{\Delta P}{\mu(R_m + R_c + R_p + \cdots)} = \frac{\Delta P}{\mu(R_t)} \qquad (1)$$

Where, μ is viscosity of the fluid, ΔP is transmembrane pressure, R_t is total resistance, R_m is intrinsic membrane pressure, R_c is cake layer resistance and R_p is pore plugging resistance.

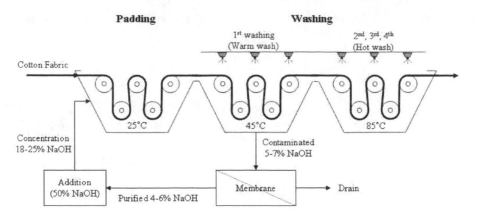

Figure 1: Caustic recovery process of mercerization effluent.

4 METERIALS AND METHODS

4.1 Membrane

Three ceramic membranes (MF with 0.5μm pore size, MF with 0.1μm pore size and UF with 0.05μm pore size) manufactured by pall Co., USA, and three polymeric membranes (UF with MWCO of 1k Da, NF with MWCO of 200 Da and RO with MWCO of 50 Da) manufactured by Osmonics Inc. (USA) were used in this study. The characteristics of the membranes are summarized in Table 1.

Item	Ceramic membrane (Pall Co., USA)		Polymeric membrane (Osmonics Inc., USA)		
	MF	UF	UF	NF	RO
Active layer	Zirconia	Titania	Polyamide thin film (TF)		
Support layer	α-Alumina		Polysulphone		
Module type	Tubular		Plate		
pH range	1-14		2-12		
pressure	1-4 bar		1-40 bar		
temperature	1-100℃		1-50℃		
SDI	-		< 5	< 5	< 3
Filtration area	0.005 m2		0.0136 m^2		

Table 1: Characteristics of the membranes used in this study

4.2 Combined membrane process

The dual-membrane process was composed of MF/UF and NF/RO for the recovery of caustic solution (Figure 2). A ceramic membrane unit of tubular type was used for the pretreatment process. It was designed for a product flow rate of 25.2 m^3/h and TMP of 1 to 4 bar in the lab scale. A polymeric membrane unit of plate type was used for main process. It was designed for a product flow rate of 7.5 m^3/h and TMP of 1 to 40 bar in the lab scale. The hybrid system was composed of a 6 L feed tank, where the caustic wastewater was circulated at a constant speed, circulating pump and crossflow flow membrane module.

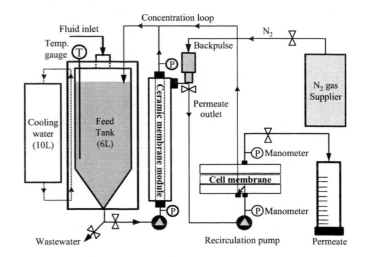

Figure 2: Schematic flow diagram of the experimental system.

4.3 Analysis

SDI, TSS, TOC, turbidity, conductivity, NaOH concentration and color were measured to analyze the sample. Representative samples were collected daily for the analysis. The permeate samples were taken from permeate valve located after membrane units (Figure 2). SDI value was measured following ASTM D4189-45. TSS value was measured using the Standard Methods 2540 D. The concentration of organic compounds was measured using a TOC analyzer (Multi N/C 3000, Analytikjena, Germany). Turbidity value was determined using a turbidity meter (Orion AQUAfast, Thermo Electron Co., USA). Conductivity value was measured using a conductivity meter (Multi 340i, WTW Inc., Germany).The concentration of NaOH was measured using a sodium electrode (Orion 4 star, Thermo Electron Co., USA). The decolorization rate was measured using a UV-Vis spectrophotometer (UNVIKON XS, BIO-TEC Ins., Italy).

5 RESULTS AND DISSCUTION

5.1 Membrane permeability

Permeate flux is an important factor to be considered during process design. In this study, two main parameters, membrane pore size and TMP, were examined for their effect on the permeate flux of pure water and caustic wastewater. The permeate flux values of pure water and caustic wastewater at different operation pressures were measured and plotted in Figure 3. When the same membrane was used, the average permeate flux of pure water was higher than that of caustic wastewater. This result is due to the decreased permeate flux by higher total resistance and viscosity of the caustic wastewater. As for the caustic wastewater, the permeate flux through the 0.5μm pore size membrane was higher than that through the 0.1 and 0.05μm pore size membranes. The difference of the permeate flux might be due to the different characteristics of the cakes formed on the membrane. When the cross-flow velocity is identical, the permeate flux is dependent on the size of deposited particles. The initial flux of 0.5μm pore size membranes was higher than those of 0.1 and 0.05μm pore size membranes, so the particle size of cake that formed on 0.5μm pore size membranes surface might be larger and the cake resistance was smaller than those of 0.1μm and 0.05μm pore size membrane [3].

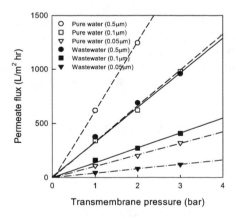

Figure 3: Dependence of the permeate fluxes of pure water and caustic wastewater on TMP (Temperature 30°C, velocity 4.4 m/s, initial volume 6 L).

5.2 Membrane selectivity

The NF/RO filter was required not only to remove large suspended solids but also to remove smaller suspended solids and colloidal substances. Membrane fouling index is an important parameter in the design of the integrated NF/RO membrane process. SDI values of 0.5μm, 0.1μm and 0.05μm pore size membranes were measured to be 11.8, 5.8 and 4.3, respectively. The colors (absorbances) of the permeates were 0.78, 0.63 and 0.43, respectively. In this study, UF ceramic membrane of 0.05μm pore size was selected for NF/RO process was, since, in generally, the SDI values of pretreated permeate prior to NF/RO are no more than 5.

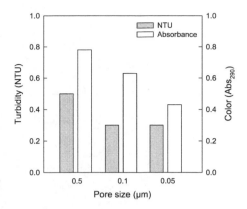

Figure 4: Pore size dependent properties (turbidity and color) of permeates obtained through ceramic membrane filtration (temperature 30°C, initial volume 6 L, TMP 2 bar).

5.3 Effect of temperature

Temperature is an important factor that affects the permeate fluxes of membranes. As shown in Figure 4, the steady flux increased when the temperature was higher than 10°C. Increase of the flux with temperature was consistent with the decrease in viscosity of the fluid in this temperature range. This means that the increase of flux with temperature was due to the decrease in viscosity of caustic wastewater [3]. Such relationship can be derived from Eq. (1). Permeability flux of the caustic wastewater through UF ceramic membrane increased by 2.4% when the temperature increased by 1°C.

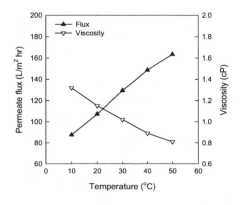

Figure 4: Effect of temperature on steady-state flux and permeate viscosity (velocity 4.4 m/s, initial volume 6 L, TMP 1 bar).

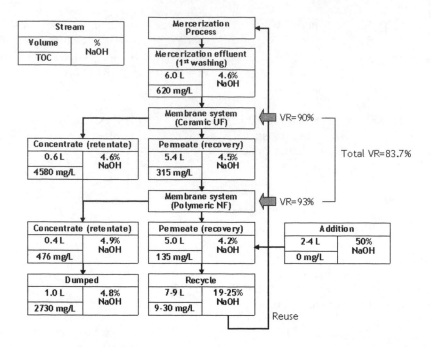

Figure 5: Mass balances for sodium hydroxide recovery and recycling by combined membrane system.

5.4. Combined membrane process

Table 2 shows the effects of combined process of ceramic UF and polymeric NF on rejection efficiencies of caustic wastewater from mercerization process. Only for the UF membrane, the rejection of TSS and turbidity was more than 99.0%, while the color and TOC rejection were about 74.7% and 49.2%, respectively. The combined membrane process of UF and NF membranes was found to treat the caustic wastewater effectively. This combined membrane process showed >99.9% TSS and turbidity, 87.7% color, 78.2% TOC removals.

Item	Caustic wastewater	UF Permeate water	NF permeate water
NaOH	4.6 w/w %	4.5 w/w %	4.2 w/w %
Conductivity	184 mS/cm	175 mS/cm	165 mS/cm
pH	13<	13<	13<
Absorbance (λmax_{290})	1.7	0.43	0.21
TSS	140 mg/L	0 mg/L	0 mg/L
TOC	620 mg/L	315 mg/L	135 mg/L
Turbidity	30 NTU	0.3 NTU	0.0 NTU

Table 2: Removal efficiencies of unit process (UF) and combined process (NF) by optimum conditions.

5.5 Mass balances

Figure 5 shows the caustic solution and total organic carbon mass balances for a single regeneration under the optimum conditions of the combined process. The combined membrane process showed 91.3% sodium hydroxide recovery with 83.7% volumetric recovery.

6   CONCLUSIONS

In order to recover caustic solution from caustic wastewater, a combined membrane process was used. As a result, clean caustic solution was obtained with high purity, which can be recycled back into the process. It is expected that this system also can be applied to other processes like polyester caustification.

ACKNOWLEDGMENTS

This research was supported by the Ministry of Science and Technology (National Research Laboratory Program).

REFERENCES

[1] E. K. Choe, E. J. Son, B. S. Lee, S. H. Jeong, H. C. Shin and J. S. Choi, Desalination 186, 29–37, 2005.
[2] T-K. Kim, C. Park and S. Kim, J. Clen. Prod. 13, 779-786, 2005.
[3] Y. Zhao, Y. Zhang, W. Xing and N. Xu, Chem. Eng. J. 111, 31-38, 2005.

Desorption studies of model contaminants from recycled PET in dry air and nitrogen atmosphere by thermogravimetric analysis

A. S. F. Santos[*], J. A. M. Agnelli[**]; Eliton S. Medeiros[***], Elisângela Corradini[***]; Sati Manrich[**]

[*]Laboratory of Chemical Processes and Particle Technology, Institute for Technological Research, São Paulo, SP, Brazil, ameliafs@ipt.br
[**]Department of Materials Engineer, Federal University of São Carlos, São Carlos, SP, Brazil, agnelli@power.ufscar.br, sati@power.ufscar.br
[***]Embrapa Instrumentação Agropecuária, São Carlos, SP, Brazil, eliton@cnpdia.embrapa.br

ABSTRACT

The closed-loop recycling of PET bottles has to prove its ability to decontaminate recycled plastics into a level that offers a negligible risk to public health and does not compromise the organoleptic properties of packed foods. Our previous work [1] resulted in a new bottle-to-bottle process that employs dry air atmosphere under ideal mass and heat transfer conditions to remove contaminants from polymer matrix. The experimental results showed that dry air was able to increase the productivity of PET super-clean technologies, based on solid-state polymerization processes (SSP). In this work we have evaluated the difference in desorption rate of model contaminants from recycled PET in dry air and nitrogen atmosphere by thermogravimetric analysis. Furthermore, it was investigated if sample exposed to synthetic air and nitrogen atmosphere during the thermogravimetric analysis presented any difference in their thermal properties. According to the results obtained in this work, the thermogravimetric analysis was not sensitive to detect difference in the rate of contaminant desorption from PET in synthetic air and in nitrogen atmosphere and any evidence of interaction between PET matrix and exposed atmosphere has not been observed by DSC analysis.

Keywords: desorption, surrogates, PET, atmosphere, thermogravimetry.

1 INTRODUCTION

The PET bottle-to-bottle recycling is a good example of a successful technology that allowed an increase of recycled plastic market as well as a reduced environmental risk through minimization of plastic waste volume. These technologies, also known as super-clean technologies, generally consist of conventional mechanical recycling followed by further steps that employ heat, vacuum, inert atmosphere, solid-state polymerization, solvent extraction, chemical surface treatments, vacuum degassing, supercritical fluid extraction (SFE) and steam distillation [2].

Among the super-clean technologies mentioned above, solid-state polymerization processes are the most disseminated ones [3-9]. These super-clean technologies are a kind of thermal extraction process, which main variables consist of type and composition of atmosphere, temperature, particle size, type/nature of equipment ad agitation system.

At the Residue Recycling Center in Federal University of São Carlos (UFSCar), comparative tests of the extraction rates of benzophenone from PET in dry air, vacuum or inert gas have demonstrated that the diffusion of this contaminant from plastic surface was the fastest in dry air [10]. Based on this result, a new super-clean process was proposed [1], which differs from others, once it uses only the conventional drying and crystallization recycling steps, at the upper temperature and dry-air flux limit. This process was developed to decontaminate the recycled plastic to the appropriate purity imposed by FDA and by the International Life Sciences Institute (ILSI) for recycled PET to be used in direct food contact applications. The new technology could be added to other ones with the possibility of increasing productivity rates and reducing both energy consumption and raw material costs, all advantages that go in hand with demands for clean technologies.

Aiming to confirm the previous results by a characterizing technique other than gas chromatography, this work evaluated the difference in desorption rate of model contaminants from recycled PET in dry air and nitrogen atmosphere by thermogravimetric analysis. Furthermore, any difference associated with the physical aging process of the sample treated in each atmosphere was also investigated through differential scanning calorimetry (DSC) analysis.

2 EXPERIMENTAL

2.1 Materials

All surrogates (toluene, trichloroethane, eicosan and benzophenone) and solvent (hexane) were analytical grade (purity higher than 99 %), and used without further purification.

Two-liter PET bottles donated by Plastipak Packaging do Brasil Ltda. (Campinas, Brazil) with intrinsic viscosity of 0.80 dL.g^{-1}were used. The bottles were cut, and the central parts of these bottles, due to their uniformity in thickness, crystallinity, stretch ratio and carboxyl content, were used. These parts were subsequently cut into symmetric squares of 8x8 mm to standardize the physical, morphological and structural properties of PET flakes and reduce the variation of surrogates' content in PET matrix [11].

2.2 Contamination Condition

Virgin square PET flakes were soaked in spiked solution containing 15 % (w/v) of toluene and trichloroethane, 10 % (w/v) of d-limonene, 5 % (w/v) of benzophenone and 3 % (w/v) of eicosan dissolved in hexane. The mixture was placed in a special cell with flakes completely immersed in the surrogate solution. Then the cell was placed in an incubator and maintained at 40° C for two weeks. Afterwards, the flakes were removed from the spiked solution, rinsed with hexane and dried with a paper towel.

2.3 Thermogravimetric Analysis (TGA)

The contaminated PET squares were heated to 60°C using heating rate of 50 °C/min and maintained isothermally at this temperature in a thermogravimetric analyzer (Shimadzu 50) to follow the mass loss for two hours under synthetic air and nitrogen atmosphere. The purge gas flow rate was set at 20 mL/min for both cases. All tests were carried out in duplicate.

2.4 Differential Scanning Calorimetry (DSC)

In order to verify any change in the thermal characteristics of PET squares exposed to synthetic air and nitrogen atmosphere during the thermogravimetric analysis, they were analyzed by differential scanning calorimetry (DSC).

The DSC curves were recorded on a DSC Shimadzu 50 apparatus, with purge inert gas (N_2) flow rate of 50 mL/min. The samples were heated from 30 to 280 °C using heating rate of 10 °C/min. The crystallinity degree was determined by the ratio between melt enthalpy of sample and melt enthalpy of a sample 100% crystalline (119.8 J/g) [12].

3 RESULTS

In order to model the desorption rate from thermogravimetric analysis, it was assumed that the extraction process is diffusion-controlled and, consequently, the solvation or dissolution/dispersion stages were negligible [13, 14]. The kinetic of desorption obeys Fick's law and thus presents a linear behavior with the square of time at the first stages of desorption [15]. Figure 1 shows the thermogravimetric results calculated according to these hypotheses.

As could be observed, the desorption rate of contaminants in synthetic air atmosphere was slightly higher than in nitrogen. Nevertheless, considering the mean deviation of these results, they seemed similar. Probably, the sensitivity of this analysis for the range of mass loss evaluated was not enough to detect any difference between the rate of desorption of contaminants from PET in these two atmospheres. Therefore, those results could not confirm our previous study, which was carried out using chromatographic analysis and observed an increase of about 60% in diffusion coefficient of benzophenone in dry air atmosphere when compared with that one in nitrogen atmosphere [9].

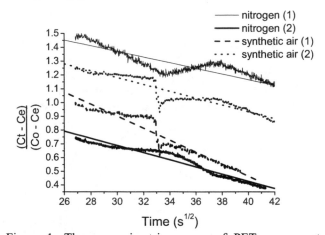

Figure 1: Thermogravimetric curves of PET squares at different atmospheres. Nitrogen (2) f=0.75, synthetic air (1), f=1, synthetic air (2), f = 1.25, nitrogen (1), f = 1.5. f, multiplying factor. 1 e 2, duplicates.

The DSC curves are depicted in Figure 2. These results did not show any significant difference in thermal properties of PET exposed to synthetic air or nitrogen atmosphere (Table 1).

Atmosphere	No.	Tg (°C)	Tm (°C)	Crystallinity (%)
synthetic air	1	95.9	250.3	28.4
	2	95.6	249.5	27.3
nitrogen	1	94.2	248.8	27.0
	2	93.3	249.3	26.1

Table 1: Tg, Tm and crystallinity values of PET samples exposed to different atmospheres at 60°C.

Once has not been observed any increase in Tg of PET samples exposed to dry air atmosphere in relation to that one observed in nitrogen atmosphere, no specific interaction between that atmosphere and PET matrix or change in PET matrix process of enthalpic relaxation due to the type of atmosphere employed could be identified. Furthermore, a slightly higher Tg has just been observed for higher crystalline samples, as expected due to the greater restrained amorphous phase.

Once thermal treatment has been carried out below the Tg temperature of this polymer, the differences in crystallinity degree of PET samples could be inherent to PET sample variability. Nevertheless, the DSC analysis was not the most sensitive technique to measure Tg, mainly for semi-crystalline samples.

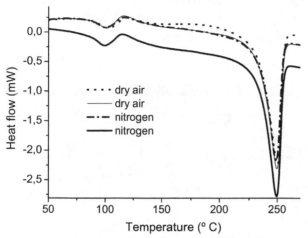

Figure 2: DSC curves of PET samples isothermally treated at 60° C in different atmospheres.

Other possible factors that probably contributed to improve the desorption rate of benzophenone from PET in dry air atmosphere were the reduced water content of dry air; the co-diffusion mechanism between contaminants and oxygen from dry air atmosphere through the polymer matrix, since nitrogen atmosphere is an inert gas of lower molar mass; and others. Nevertheless, none of them have been investigated in this article.

4 CONCLUSIONS

The results achieved in this work indicated that thermogravimetric analysis was not sensitive to detect difference in the rate of contaminant desorption from PET in synthetic air and in nitrogen atmosphere, considering the level of contaminant evaluated. In the same way, no difference in thermal properties, i.e. of interaction between PET matrix and exposed atmosphere, has been observed by DSC analysis.

5 ACKNOWLEDGMENTS

The authors are grateful to FAPESP for the financial support of this study.

REFERENCES

[1] FAPESP; UFSCar. S. Manrich, A. S. F. Santos; J. A. M. Agnelli. "Processo de descontaminação de poliéster reciclado". 14 jun 2004. N°. 002993. INPI-SP (in portuguese).

[2] S. Manrich, A. S. F. Santos. "An overview of recent advances and trends in plastic recycling". In: Conservation and Recycling of Resources. Ed. Nova Science Publishers, 2006.

[3] BÜHLER AG. A. Christel; C. Borer; T. Hersperger. "Method and device for decontaminating polycondensates". WO 01/34688, 17 Maio 2001.

[4] THE COCA-COLA COMPANY. M. Rule. "Process for removing contaminats from polyesters and controlling polymer molecular weight". U.S.P. 6.103.774, 15 Ago. 2000.

[5] OHL APPARATEBAU&VERFAHRENSTECHNIK GmbH. F. Rudiger. "Method for recycling PET flakes". U.S.P. 6.436.322, 20 Ago. 2002.

[6] VISY PLASTICS PTY LTD. E. Kosior. "Processo para preparar um poli(etileno tereftatalato) para contato com alimentos e poli(etileno terftalato) reciclado". PI 0.014.094-5, 21 Maio 2002.

[7] WELLMAN, Inc. C. S. Nichols; C. Moore. "Food quality polyester recycling". U.S.P. 5.876.644, 02 mar. 1999.

[8] KRONES AG. "Krones implements first bottle-to-bottle recycling facility". PETticker, Maio 2002. Access at: www.petnology.com

[9] "PET bottle to bottle". Recycling News, Linz, 1-2, 2000.

[10] A. S. F. Santos, Phd Thesis, Universidade Federal de São Carlos, 2004 (in portuguese).

[11] A. S. F. Santos, J. A. M. Agnelli, S. Manrich. In Proceedings of Brazilian Polymer Congress, 6., Gramado, 2001, CD-ROM.

[12] T. Toda, H. Yoshida, K. Fukunishi. "Structure and molecular motion changes in poly(ethylene terephthalate) induced by annealing under dry and wet conditions". Polymer, 36 (4), 699-706, 1995.

[13] X. Lou; H. G. Janssen; C. A. Cramers. Anal. Chem. 69, 1598, 1997.

[14] R. W. Limm, H. C. Hollifield. Food Addit. Contam. 12, 609, 1995.

[15] J. Comyn. "Polymer permeability". London: Elsevier Applied Science Publishers Ltd, 1988.

MOBILE ASBESTOS DECONTAMINATION BY MICROWAVE HYBRID HEATING

Andreas Rosin*,**, Thorsten Gerdes*,**, Monika Willert-Porada*
*Chair of Material Processing, University of Bayreuth, D-95440 Bayreuth
**InVerTec, Institut fuer Innovative Verfahrenstechnik e.V.

ABSTRACT

Within the European project "New Safe and Cost Effective Techniques Against Asbestos Risks" different methods for decontamination of asbestos in asbestos containing wastes, abbreviated as ACW from building industry were investigated. A mobile asbestos decontamination unit was developed, aiming at the complete disintegration of asbestos fibres on-site. The process is based on a chemo-thermal treatment. Disintegration of fibres is achieved by addition of chemical reactants at temperatures of 300 to 500 °C. Heating is performed by a combination of microwave and infra-red heating devices. Microscopic and phase analysis show a complete destruction of asbestos fibres.

1 INTRODUCTION

Asbestos is the generic name given to a class of natural, fibrous silicates. The international ban of asbestos started during the 1980ies, e.g. in Germany 1980, and in the United States 1983. The need for disposal of asbestos containing materials is growing, since the waste stream from the extensive use of such materials in the 1960s and 1970s in industrial and domestic infrastructures is significantly increased, because asbestos containing products are reaching the end of life time. A cost effective, safe and reliable disposal should include the complete disintegration of asbestos fibres. Usually, a wide range of concentrations and mixtures of different asbestos phases is present in ACW. Destruction of asbestos fibres difficult, because of their excellent thermal and chemical resistance [1,2], and the heterogeneous material mixture.

Traditional asbestos abatement is organized by onsite stripping work and subsequent transport and landfilling. A mobile process would decontaminate the hazardous waste onsite and significantly reduce the need for hazardous waste transport. As shown in figure 1, complete disintegration of asbestos fibres into non-hazardous phases can be achieved either by heat treatment, by chemical additives, or a combination of both.

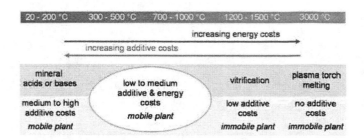

Figure 1: Concepts of a asbestos decontamination processes with different energy and additive demands.

High temperature processes like plasma arc treatment [3] or melting/ vitrification processes [4, 5] require high amounts of energy and substantial investment cost, therefore are not the preferred technologies for cost effectiveness. High temperature processing plants are immobile, therefore ACW have to be transported from the stripping site to the processing plant. Chemical treatment [6, 7] or mechanical destruction by grinding [8] are alternatives, however, full disintegration is not easy to achieve by such methods.

In ACW the fibres are usually embedded in an organic or inorganic matrix, a thermal or chemical impact on asbestos fibres is therefore limited. Supporting additives must first penetrate the matrix material to expose the asbestos fibres to chemical disintegration.

The present study therefore investigates the disintegration of various ACW upon combined chemo-thermal treatment, in contact with alkali melts under microwave radiation in a laboratory scale reactor. The results were used for the development of a pilot plant, utilised for testing of different ACW under on-site conditions within a special testing facility.

Most ACW originate from construction materials and consist of carbonates, sulphates and silicates. In alkali melts, like caustic soda, these substances are dissolved easily.

Hence, asbestos fibres are released from the surrounding matrix and become susceptible to the chemical attack of the melt, which by adjusting the process temperature leads to full disintegration of the siliceous asbestos fibres, according to the following reactions:

$$CaCO_3 \text{ (s)} + 2 \text{ NaOH (l)} \rightarrow Na_2CO_3 \text{ (s)} + Ca(OH)_2 \text{ (s)}$$

$$CaSO_4 \text{ (s)} + 2 \text{ NaOH (l)} \rightarrow Na_2SO_4 \text{ (s)} + Ca(OH)_2 \text{ (s)}$$

$$SiO_2 \text{ (s)} + 2 \text{ NaOH (l)} \rightarrow Na_2SiO_3 \text{ (s)} + H_2O \text{ (g)}$$

The melting point of caustic soda is 320 °C, enabling an operation of a mobile processing unit at moderate temperatures. The process generates low emissions due to absorption of gaseous compounds like CO_2 or SO_2 in the melt.

2 EXPERIMENTAL

A glove-box equipped with means to handle asbestos materials safely at lab-scale conditions was build.

The box provides a working compartment with a modified domestic microwave oven and two material locks with self-closing sliding doors, which separate the asbestos-contaminated working area from the laboratory environment.

The inner lock has a water shower to clean any material taken to the outside. Inside the working compartment a slightly reduced pressure is applied. An air cleaning unit prevents airborne fibres to be released. During experiments the composition of the exhaust gas is examined by a gas analyser (see Figure 2).

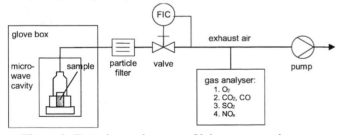

Figure 2: Experimental set-up of laboratory equipment.

Asbestos sample (typically 20 to 50 g) and caustic soda are filled into 100 ml quartz crucibles, which provide acceptable resistance to high temperatures and corrosion. The crucible is insulated by a microwave transparent material and covered by a glass hood. After microwave treatment the melt is cooled down and dissolved in water. After several washing steps the solid residues are further separated by sedimentation. The dried and sealed samples are analysed by SEM, EDX and XRD.

3 RESULTS

A minimum ratio of 3:1 of NaOH to ACM was found to be reasonable for treatment of ACW with different matrix materials like, e.g., gypsum, cement or lime. The process is easy to control. NaOH is melting under microwave radiation rapidly.

Because the NaOH-melt absorbs even more readily microwave energy than the solid NaOH, the microwave power level can be reduced to compensate heat losses and to keep the temperature constant as soon as a melt pool has formed. However, it is observed that microwave penetration into liquid caustic soda melt is limited to few centimetres.

After heat treatment the product is dissolved in water in order to separate the solid residues of the ACW from the caustic melt.

It is shown that caustic soda effectively destroys any asbestos fibres at either 450 °C/ 30 minutes or 400°C/ 45 minutes at lab-scale (Figure 3). The product consists chemically of the same elements like asbestos fibres, i.e. silicon, carbon, oxygen and different metals like Mg or Fe, but is free of sodium. As gaseous by-product mainly water vapour is released, with minor amounts of SO_2 or NO_x detected in the exhaust gas by IR-measurements. It is therefore assumed, that substantial amounts of CO_2 and other volatile oxides are directly absorbed or reacting in the caustic soda melt.

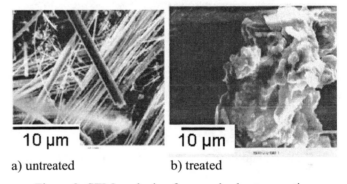

a) untreated b) treated

Figure 3: SEM analysis of sprayed asbestos coating a) untreated and b) treated in caustic soda melt, 45 min. at 400°C.

3.1 SCALE-UP

For scale-up of the process direct microwave heating of ACW collected on-site in steel drums is chosen. It suits well the demands of a robust on-site system, the steel drum serves as re-usable or one-way containment for collected ACW and as process vessel for the chemo-thermal treatment.

Microwave drum heating was tested with 70 and 200 litre drums on "artificial" ACW made from wet sand, rock wool, expanded perlite and a fibrous dummy material. As heat source a single 2.45 GHz, 2 kW magnetron was installed on top of the drum. The scale-up experiments demonstrate, that the microwave field is focussed along the axis of the drum, causing hot zones in the centre of the barrel. A more homogeneous microwave field distribution inside the drum was achieved by a triple arrangement of magnetrons (result of project partner Plazmatronika, Poland, see figure 4).

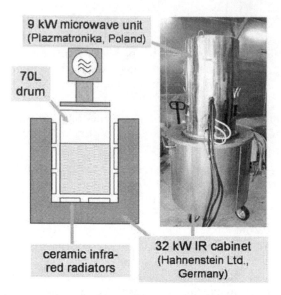

Figure 4: Design of mobile, hybrid-heated prototype unit.

The most important limitation when heating caustic soda by microwave absorption comes from the reduction of microwave penetration depth in the melt. Therefore, heat loss to the exterior has been compensated by application of infra-red radiators as additional heat source to provide hybrid heating. The pilot plant underwent tests for disintegration of ACW on a kilogram scale. The infra-red power is 32 kW, the microwave power is 9 kW.

3.2 PILOT PLANT TEST

The pilot plant test is performed at an experimental waste treatment area (see Figure 5).

Figure 5: Operation of prototype unit at experimental waste centre (CMS, France).

Several tests are carried out on sprayed asbestos coating (80-90% asbestos) and gypsum based plaster (10-15%

asbestos). Both ACW typically contain a lot of moisture due to wetting procedures during removal. The volume of the drums is 70 litres. Because of the low density of sprayed asbestos the amount of ACW is limited to 6 kg plus 25 kg caustic soda, and 10 kg of gypsum plaster with 40 kg of caustic soda, respectively. The treatment takes 2-4 hours, to reach 450 - 500°C followed by a dwell time of 45 minutes. Afterwards the drums are removed from the heating station and cooled down. Both ACW yield a homogeneous product. As concluded from cross-checked analysis a complete destruction of asbestos fibres is achieved.

A comparative test with conventional heating only reveals significantly different results. Two different layers are formed during treatment in the drum: a red-coloured, porous upper layer, and a greyish-white layer with a dense crystalline structure at the bottom.

4. DISCUSSION AND CONCLUSION

Alkali melts are able to dissolve asbestos minerals efficiently into non-fibrous compounds, independently of matrix materials. The caustic soda process requires moderate temperatures in the range of 400 to 500°C. Apparently microwave radiation causes convection streams that keep the melt in movement, even if microwave penetration is significantly reduced during melt formation. The average process time at pilot scale is 3 hours. Post-processing, i.e. dissolution in water, solid-liquid separation and drying, is necessary to obtain an inert and disposable product. Phase analysis and optical microscopy in designated liquid media of the raw product and the solid residue show a complete destruction of asbestos fibres. The solid residue is sodium free, i.e. all of the caustic soda is removed by the water treatment of the product. The recovery and re-use of caustic soda is feasible, and operation costs could be reduced by 70% to 0.75 €/kg ACW. The pilot plant tests demonstrated the efficiency of the hybrid heating system to achieve homogeneous treatment. Based on the presented results the development of an industrial process could be designed like illustrated in Figure 10.

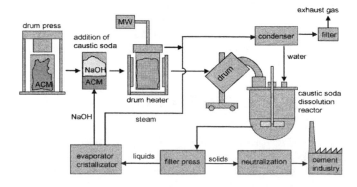

Figure 10: Development of mobile, industrial-scale process.

ACKNOWLEDGEMENTS

The project was funded by the European Commission under the 5th Framework Programme (project ref. G1RD-CT-2001-00498). The excellent collaboration with the project partners is gratefully acknowledged.

REFERENCES

[1]   A. Rosin, T. Gerdes, M. Willert-Porada, N. Kondratenko, MOBILE ASBESTOS DECONTAMINATION BY MICROWAVE HYBRID HEATING, 10th International Conference on Microwave and High Frequency Heating, AMPERE, Modena, Italy, 2005, pp 458-461

[2]   Team of authors, "Der Rohstoff Asbest und seine Verwendung", Freiburger Forschungshefte, VEB Deutscher Verlag für Grundstoffindustrie, Leipzig (1973)

[3]   R. Poiroux, M. Rollin, "High Temperature Treatment of Waste: From Laboratories to the Industrial Stage", Pure & Appl. Chem., Vol. 68, No. 5 (1996), pp. 1035-1040

[4]   A.U. Clausen, V.R. Christensen, S.L. Jensen, "Method of converting asbestos cement into a harmless product". U.S. Patent 5,614,452 (1997)

[5]   D. Roberts, J.H. Stuart, "Vitrification of asbestos waste", U.S. Patent 4,820,328 (1989)

[6] J.M. Blessing, "Chemical Decomposition of Asbestos", Proceedings of Davos Recycle'93 Conference (1993)

[7]   M. Porcu et al., "Self-Propagating Reactions for Environmental Protection: Treat-ment of Wastes Containing Asbestos", Ind. Eng. Chem. Res. Vol. 44 (2005), p. 85-91

[8]   P. Plescia et al. "Mechanochemical treatment to recycling asbestos-containing waste", Waste Management, Vol. 23 (2003), pp. 209–218

Serpentine Fluidic Structures for Particle Separation

A. Kole, M. H. Lean and J. Seo

Palo Alto Research Center Incorporated
Palo Alto, CA 94304, ashutosh.kole@parc.com

ABSTRACT

The abstract proposes size and mass based separation and concentration of particles including biological agents suspended in fluidic media in a serpentine channel structure. On the curved sections of the serpentine channel, the interplay between the outward directed centrifugal force and the inward directed transverse pressure field from fluid shear allows for separation of particles. Methods currently employed for particle separation include: mechanical sieving, sedimentation, hydrodynamic chromatography, and electrophoresis. These techniques are batch processes and require large investments in equipment and set-up time for each run. This present study details a filter-less continuous process which employs flow velocity and tailored channel geometry to achieve separation and segregation of particles over a large dynamic size range which can span micro-scale to macro-scale fluid capacities.

Keywords: serpentine channel, centrifugal force, pressure field, particle separation

Nomenclature:

V	=	Flow Velocity
P	=	Pressure
F_{cf}	=	Centrifugal Force on the Particle
F_{id}	=	Inward directed dynamic forces
F_{vd}	=	Force due to viscous drag
R	=	Radius of curvature of the channel
η	=	Dynamic Viscosity of the fluid
m	=	Mass of the particle
a	=	Radius of the particle assuming it to be spherical

1 BACKGROUND

Particle separation and sorting represents an important requirement especially in biological and chemical processes for both macro-scale and miniaturized lab-on-chip applications. The techniques used today for this purpose fall into two broad categories – mechanical sieving and external force field. Mechanical sieving involves the use of filters as a physical barrier and has its own disadvantages such as clogging, reduction in performance with time and high cost for filters designed for smaller size particles.

Sedimentation, chromatography, electrophoresis and Field Flow Fractionation (FFF) are techniques based on external force. More recent developments in microfluidics based filter-less particle separation system include work based on the *Zweifach–Fung* effect [1], Pinched Flow Fractionation (PFF) [2, 3], SPLITT Fractionation [4], Ultrasonic particle separation [5].

Microfluidics based centrifugal separation has been reported by Brenner [6] which is essentially a miniature centrifuge constructed on a rotating disk with polymer microstructures to carry the fluid. Centrifugal separation in a curved microchannel is also presented in [7], which is based on generation of secondary flow called Dean's vortices in the transverse plane, owing to the curvature. These vortices push the particles towards the outer side of the curved channel, which are collected through the bifurcation.

Though most of these techniques have seen exponential growth, it should be noted that all the above work in this field has a variety of shortcomings. For example, most of them require an additional external force. Moreover, many of these techniques are limited to batch processing and are scaled to handle only minute volumes of samples. Further, many of these processes are typically designed for only a centrifugal mode of operation. The current demand is for continuous flow, high throughput and low cost processes.

2 PROPOSED TECHNIQUE

Here we present a filter-less technique and a system for size and mass based separation of particles. The technique utilizes only channel geometry, radius of curvature and flow velocity within the channel to exert the required force on particles within the fluid to separate out either to the inside or to the outside channel walls. Strategically located collection chambers help in collection of particles from the flowing fluid. The technique is a continuous process which separates particles over a large dynamic size range and which can span from micro-scale to macro-scale fluid capacities.

2.1 Analytical consideration for flow in a curved channel

Particles within a flowing fluid experience a combination of forces acting on them. In the scenario of laminar fluid flow in a curved channel, particles have to face a competition between the outward directed centrifugal force and the inward directed hydro-dynamic forces, which result from pressure variation, channel geometry and flow conditions. Along with these forces there is also the viscous drag, which transports the particles forward in the flow direction. The resultant of these forces causes a net force in one direction, which affects particle motion with respect to the channel axis. Fig 1 shows the different forces acting on a particle of a spherical shape.

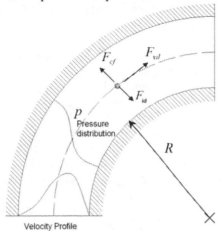

Figure 1: Segment of a curved channel showing various forces acting on the particle along with the velocity profile and the pressure distribution

The expressions for the centrifugal, transverse pressure driven, and viscous drag forces acting on the levitated particle can be expressed as follows:

$$F_{cf} = \frac{mV^2}{R(r)} = \frac{4}{3}\pi a^3 \frac{V^2}{R(r)} \tag{1}$$

$$F_{id} = p\pi a^2 \tag{2}$$

$$F_{vd} = 6\pi\eta aV \tag{3}$$

The particles will move outwards if $F_{cf} > F_{id}$, or

$$\frac{4}{3}\pi a^3 \frac{V^2}{R(r)} > p\pi a^2 \tag{4}$$

Or, $a > p\dfrac{R(r)}{V^2}\dfrac{3}{4}$ (5)

Equation (5) can be used to determine the lower bound for levitated particle size that will move outwards for any given geometry, pressure and velocity of flow. Particles smaller than this lower bound will move inwards, or

$$a < p\frac{R(r)}{V^2}\frac{3}{4} \tag{6}$$

The distance of travel before the particle migrates across the flow channel (transverse direction) is dependent on the relative magnitudes of F_{vd} and F_{id}

Also since $F_{id} \propto a^2$ and $F_{vd} \propto a$, larger particles will be more affected by the flow induced transverse pressure drop directed towards the inner surface. The transverse pressure may be derived by considering peripheral flow in a concentric cavity where the parabolic profile fits:

$$V_\theta = V_0(r - r_1)(r_2 - r) \tag{7}$$

And r_1 and r_2 are the inner and outer radii, respectively. The radial pressure drop, p, is given by (for R>>r):

$$p = \int_{r_1}^{r_2} \frac{\rho V_\theta^2}{R(r)} dr$$

$$= V_0^2 \frac{\rho}{R}[\frac{r^5}{5} - \frac{(r_1 + r_2)r^4}{2} + \frac{(r_1^2 + 4r_1r_2 + r_2^2)r^3}{3}$$

$$- r_1r_2(r_1 + r_2)r^2 + r_1^2r_2^2 r] \tag{8}$$

It can be seen from the equation (8) that the pressure is a function of the velocity of flow, density of particles and the radius of curvature of the curved channel.

At a higher velocity the centrifugal force, directed away from the centroid of the curvature, is dominant on the particles, pushing the particles towards the outside wall [7, 8]. At a lower velocity, the centrifugal force is not strong enough to push particles towards the outside and to trap them near the outside wall. The particles can settle at the base of the channel in the slow flow and migrate toward the inner wall by the hydrodynamic pressure induced by the secondary flow or Dean's vortex. The Dean's number can be calculated by equation (9),

$$De = \text{Re}\sqrt{\frac{D_{eq}}{2R}} \tag{9}$$

where, Re is the respective Reynolds number and D_{eq} is the hydraulic diameter. Placing a collection chamber strategically towards the outer wall will capture the particles from the flow with high efficiency. The collection chamber should be designed in such a way that the flow pattern within the collection chamber is decoupled from the flow within the channel adjacent to it. A strategically located collection chamber at the inner wall of the channel will capture these particles in a similar fashion as that in the higher velocity case.

3 EXPERIMENTS

3.1 Prototype

Experimental prototypes were built to prove the concept. The schematic of the experimental prototype is shown in Fig. 2. The width of the channel used is 5 mm. The radius of curvature of both the curved sections is 28 mm. The thickness of the whole structure is 500 μm (0.05 cm). The collection chambers are located strategically so that the particles will move into them as soon as they encounter a resultant directional force. Collection chambers A & D lay on the out side of the curved channel profile while chambers B & C lay on the inside of the curved channel profile. The fluid containing particles enters through the inlet, which is connected to a peristaltic pump.

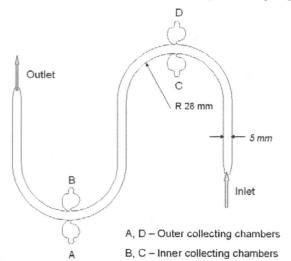

Figure 2: Schematic of the experimental prototype to prove the concept.

The fluid passes through the two curvatures of the serpentine channel and comes out through the outlet. Two different flow rates were used for the experimental purpose *viz.* 5 ml/min and 14 ml/min.

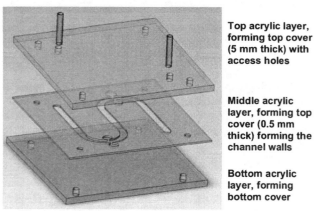

Top acrylic layer, forming top cover (5 mm thick) with access holes

Middle acrylic layer, forming top cover (0.5 mm thick) forming the channel walls

Bottom acrylic layer, forming bottom cover

Figure 3: Three component layers of a serpentine cell.

The channels were formed by cutting silicone sheets to the required dimensions using a laser cutter. The acrylic sheet formed top and bottom covers and also provided holes for the inlet and outlet. The different layers were bonded together with the help of screws. The channel was primed with DI water to remove bubbles and then a solution containing particles was flowed at two different flow rates forming the parts of two different experiments. Figure 3 shows a schematic of the flow cell that was used for the experiments.

3.2 Experimental results and discussion

Two different results were observed at the two different inlet flow rates. At a flow rate of 14 ml/min, the centrifugal

force on the particle dominates over the force due to the hydrodynamic pressure gradient experienced by the particles. This force is directed outwards away from the centroid of the radius of curvature. The Reynolds number for the flow is 85 and Dean's number is 10 in this case. The Dean's vortices keep the particles re-suspended within the channel and prevent them from settling down. Thus, due to the centrifugal force encountered by them, the particles move away from this center of the curvature when they come across the curved section. This outward force acting on the particle pushes them into the collecting chambers A and D which are situated outwards of the curvature. Fig. 4 shows particle laden fluid flowing through the channel from the inlet to the outlet along a serpentine path at an inlet flow rate of 14 ml/min. It can be seen that the particles get diverted into the outer collection chambers A and D and not in the chambers B and C which are situated inwards of the curvature just opposite of the outer collection chambers. Fig. 4 (a) and (b) shows enlarged views of the collection chambers which are also shaped to de-couple fluidics but allow particle capture.

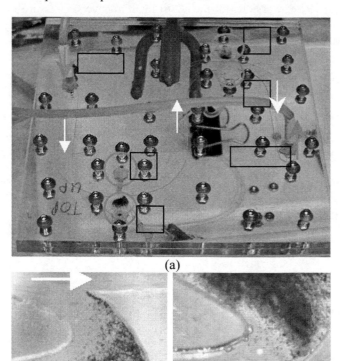

(a)

(b)   (c)

Figure 4: (a) Experimental set-up showing that the particles being collected in the outer collection chambers A and D at a flow rate of 14 ml/min and showing an expanded view of the collection chambers A (a) and D (b). The arrow indicates the direction of flow.

At a flow rate of 5 ml/min, the hydrodynamic forces on the particle due to the pressure difference across the channel were dominant compared to the centrifugal force experienced by the particles. These forces are directed inwards towards the centroid of the radius of curvature. The Reynolds number for the flow is 30 and Dean's number is 3 [7, 8]. Due to the slow velocity, the particles settle down and the velocities of the Dean's vortices are not sufficient to get them re-suspended. The centrifugal force is not large enough to push them towards the outer wall. Thus particles move towards the center of the curvature when they come across the curved section. This inward force acting on the particle pushes them to the collecting chambers B and C which are situated inwards of the curvature. Fig. 5 shows particle laden fluid flowing through the channel from the inlet to the outlet along a serpentine path at an inlet flow rate of 5 ml/min. It can be seen that the particles get diverted into the inner collection chambers B and C and not in the chambers A and D which are situated outwards of the curvature just opposite of the inner collection chambers. Fig. 5 (a) and (b) shows enlarged views of the collection chambers which are shaped to decouple fluidics but allow particle capture.

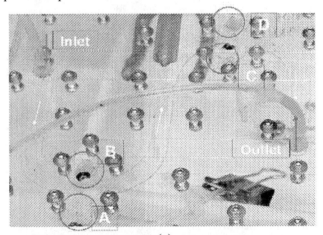

(a)

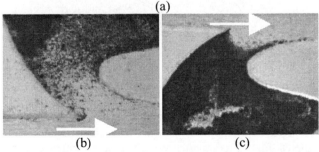

(b)   (c)

Figure 5: (a) Experimental set-up showing that the particles being collected in the inner collection chambers B and C at a flow rate of 5 ml/min and showing an expanded view of the collection chambers B (a) and C (b). The arrow indicates the direction of flow.

4   DISCUSSION AND CONCLUSION

This method solves the issue of an external field required for manipulating particles in the fluid. In the proposed design, geometric controls are only required on the channel width, height and the radius of curvatures of the curved sections. Another advantage of this method is that

continuous particle separation of a wide range of liquid volumes can be achieved when compared to techniques such as centrifugation or chromatography. As compared to other continuous particle separation processes, the advantage is the simplicity in geometric control of the device. The channel widths need not be comparable with the size of the particle under question. The magnitude and the direction of the force on the particle can be manipulated just by changing these geometric parameters and the flow rate.

By easily altering the channel widths and the radius of curvatures of the curved sections, particles of decreasing size can be collected at discrete collection chambers placed along the length of the serpentine channel. The correct selection of dimensions can limit the device for a particles size separation range below 10 μm. This is the typical range of biological cell size. The collection efficiency can be improved by, along with dimensional changes, the strategic placement of collection chambers. A microscale version of such a device can be easily fabricated with simple techniques and can be easily integrated inline with other components in a Lab-on-a-chip type environment. The simplicity comes with the fact that a use of external field is eliminated. This makes the whole micro-scale analysis device much simple and reliable.

REFERENCES

[1] Yang S., Zhan J., **"Particle Separation in Microfluidic channels using flow control"**, Proceedings of IMECE04'.

[2] Takagi J., Yamada M., Yasuda M., Seki M., **"Continuous particle separation in a microchannel having asymmetrically arranged multiple branches"**, Lab on a chip 2005.

[3] Zhang X., Cooper J., Monaghan P., Haswell S., **"Continuous flow separation of particle within an asymmetric microfluidic device"**, Lab on a chip 2006.

[4] Narayanan N., Saldanha A., Gale B., **"A microfabricated electrical SPLITT system"**, Lab on a chip 2005

[5] Kapishnikov S., Kantsler V., Steinberg V., **"Continuous particle size separation and size sorting using ultrasound in a microchannel"**, J. Stat. Mech. (2006)

[6] Brenner T., **"Polymer Fabrication and Microfluidic Unit Operations for Medical Diagnostics on a Rotating Disk"**, Dissertation at Institute of Microsystems, University of Frieburg, December 2005

[7] Ookawara, S., Higashi, R., Street, D., and Ogawa, K. **"Feasibility Study on Concentrator of Slurry and Classification of Contained Particles by Micro-Channel"**, Chem. Eng. J., v.101, 171-178 (2004)

[8] Sudarshan, A., Ugaz, V., **"Multivortex micromixing"**, PNAS 2006

Sorption of Ethylbenzene, Toluene and Xylene onto Crumb Rubber from Aqueous Solutions

L. Alamo-Nole[*], F. Roman[**] and O. Perales-Perez[***]

[*]Department of Chemistry, University of Puerto Rico, Mayagüez, lan20547@uprm.edu
[**]Department of Chemistry, University of Puerto Rico, Mayagüez, froman@uprm.edu
[***]Department of Engineering Science & Materials, University of Puerto Rico, Mayagüez, ojuan@uprm.edu

ABSTRACT

Samples of waste tires crumb rubber mesh 14-20, produced by REMA Inc., were contacted with 30ppm aqueous solutions of ethylbenzene (E), toluene (T) and xylene (X) to evaluate the corresponding sorption capability. The concentration of the sorbent varied from 0.1 to 10 g crumb rubber/L. Solution aliquots were withdrawn at different times and analyzed by GC-MS to monitor the progress of the sorption process. Obtained results confirmed the capability of crumb rubber to remove ETX compounds from aqueous solutions. This removal efficiency was dependent on solution pH and crumb rubber concentration. The ethylbenzene concentration dropped from 30 ppm down to 1.4 ppm in the first 30 minutes of contact when 10 g/L of crumb rubber were used. The maximum removal of xylene, ethylbenzene and toluene were 99, 95 and 77%, respectively, at pH 6. The corresponding uptake capacities were 55, 48 and 24 mg/g crumb rubber. The sorption efficiency of crumb rubber was xylene > ethylbenzene > toluene.

Keywords: waste tire, crumb rubber, sorption, recycling, water treatment

1 INTRODUCTION

To assure the quality of life and health of living beings is of vital importance. Despite of existent environmental protection policies, pollution events are still frequent not only in Puerto Rico but also in the rest of the world. One of the most common pollution sources is represented by oil derivatives generated by refining, transport, uses and residues treatment activities. Among oil derivatives, high concentrations of aromatic compounds such as toluene (T), ethylbenzene (E) and xylene (X) have been detected in oil and gasoline [1]. These compounds can also get mobilized into the aqueous phase, which would make the contamination problem even worse. The maximum contaminant levels (MCL) established by US-EPA in drinking water for ethylbenzene, toluene and xylene are 0.7, 1.0 and 10 mg/L, respectively [1]. Exposure to ETX solvents can cause disturbances in the central nervous system, and damage to kidney and liver [2].

Several approaches to remove ETX compounds from water have been reported in the technical literature. Granular activated charcoal (GAC) is the most common adsorbent. Other approaches consider zeolites and surfactant-modified zeolites [3]; but the prohibitive costs involved with the sorbent fabrication limit their applicability to treat large volumes of polluted effluents. Evidently, the ideal sorbent should exhibit removal capacities comparable to commercial products under cost-effective conditions. Moreover, the sorption of organic pollutants will require a polymeric matrix with non polar components.

On the other hand, there are at least 275 million waste tires in stockpiles in the U.S. It has been estimated that the amount of discarded tires reaches 10 billion every year worldwide. Although, markets now exist for 76% of these waste tires, up from 17% in 1990, the remaining is still stockpiled, or land filled [4]. Crumb rubber is composed of a complex mixture of elastomers like polyisoprene, polybutadiene and styrene-butadiene. On a rubber-composition point of view the major components of tires are rubber vulcanized with sulfur (1.1%), stearic acid (1.2%), ZnO (1.9%), extender oil (1.9%) and carbon black (31.0%) [5]. Carbon black is used to strengthen the rubber and improve its abrasion resistance. This component should exhibit similar adsorbing characteristics as activated charcoal, a well known agent used to remove organic and inorganic compounds from aqueous and gaseous effluents, a fact that makes viable the removal of target species through sorption/adsorption mechanisms [6-10]. Stearic acid would also behave as an ionic exchanger. Moreover, non-polar organic pollutants are expected to interact with the rubber matrix via van der Waals interactions [11].

Under these premises, the present work addresses the evaluation of waste tires crumb rubber as potential sorbent for ETX pollutants from aqueous solutions.

2 MATERIALS AND METHODS

2.1 Materials

Waste tires crumb rubber was provided by REMA Inc., a tire rubber recycling company located in Caguas, Puerto Rico. The crumb rubber mesh 14 -20, (Figures 1and 2), was washed with deionized water for 24 hours and dried at

Cleantech 2007, ISBN 1-4200-6382-0

room temperature. The average rubber particle size was estimated at 2.45 mm. Ethylbenzene, toluene and o-xylene were ACS certified grade and were used without further purification. Low concentrations of these compounds assure complete solubility in water. Accordingly, the adsorbates concentration was kept constant at 30 ppm in all tests.

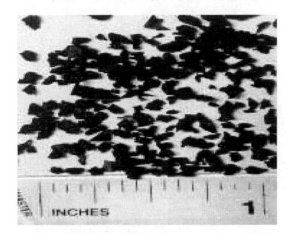

Figure 1: Crumb rubber mesh 14–20, as provided by REMA, Inc.

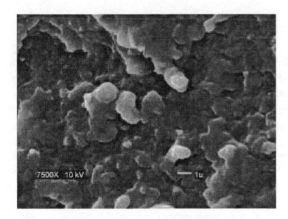

Figure 2: Typical SEM image of crumb rubber. Although the surface exhibits significant roughness, no mesopores were observed.

2.2 Crumb Rubber Chemical Stability Test

In order to evaluate the possibility to release metals from crumb rubber, 1 gram of the sorbent was contacted with 100 mL of distilled water at pH 1.5, 3.0, 6.0 and 9.0 for 24 hours. HNO_3 or NaOH 10% w/w solutions were used to adjust the pH values. Solution samples were withdrawn at the end of the contact period and submitted for copper, cadmium, arsenic, zinc, lead and chromium analyses by Inductively Coupled Plasma Optical Emission Spectroscopy (ICP-OES). EPA 200.7 rev 5.0 protocol was followed. All analyses were run by triplicate.

2.3 Sorption Experiments

Crumb rubber, at concentrations varying from 0.10 g/L to 10 g/L, were contacted with 30ppm ETX aqueous solutions. ETX solutions were directly prepared in 250-mL amber bottles leaving an optimum 10-mL head space. The solution head-space was sampled at 0.5, 1, 2 and 6 hours intervals using a Solid Phase Micro Extraction Method (SPME) and submitted for quantitative analyses by GC-MS.

2.4 Solution Analyses

After selecting the most suitable sampling technique, the SPME fiber made of polydimethylsiloxane was placed in the crumb rubber-ETX solution head-space for 15 seconds and then into the GC-MS injector. There, it was treated thermally for 2 minutes to desorb the analytes. The concentrations of ETX were determined using GC-MS Q Plus with ion tramp, electron impact ionization of 70 eV and an ion trap mass spectrometer in the Single Ion Monitoring mode at 65, 91 and 92 m/z. Quality controls (QC) samples of 30 ppm for each solvent were run in all experiments and new calibration curves were conducted when the lack of fitting error was higher than 15%.

3 RESULTS AND DISCUSSION

3.1 Crumb Rubber Chemical Stability Test

As evidenced by the data in table 1, the release of metals from crumb rubber is below EPA regulations even for pH values as low as 1.5. In all cases, the terminal concentrations were below EPA regulations for drinking water. The zinc detected in solutions can be attributed to the dissolution of ZnO, which is a constituent of tire rubber. Since the sorption of ETX compounds in our work took place at extremely short contact times (30 minutes, or so) and at pH 6, there are no concerns about the potential release of toxic metal ions during the sorption process.

	pH values in solutions				EPA regulation mg/L
	1.5	3.0	6.0	9.0	
Cu	0.0828	0.043	ND	0.0001	1.3
Cd	0.0023	ND	0.001	ND	0.005
As	0.04	ND	ND	ND	0.050
Zn	2.38	1.11	0.41	0.29	5.0
Pb	ND	ND	ND	ND	0
Cr	0.05	0.09	ND	ND	0.1

Table 1. Metals release, in mg/L, from crumb rubber at different pH values. ND: no detected by ICP-OES

3.2 Sorption of ETX

The removal of ethylbenzene, toluene and xylene by mesh 14-20 crumb rubber at pH 6.0 was highly efficient and rapid. Above 80% of the pollutants was removed in the first 30 minutes of contact, which is in good agreement with previous works [12, 13]. Terminal ETX concentrations exhibited a rising trend when the concentration of crumb rubber was less than 10 g/l. This behavior was attributed to the expected decrease in the amount of sorption sites for fewer amounts of the sorbent. Figure 3 shows the variation in the sorption of ethylbenzene with time and concentrations of crumb rubber. Almost 95% of ethylbenzene was removed by using 10 g/L of crumb rubber at pH 6. The removal percentage went down to 17% for a crumb rubber concentration as low as 0.1 g/L, i. e. one-hundred times less crumb rubber than at 10 g/L. The sorption efficiency of crumb rubber was pretty high for xylene followed by ethylbenzene and, lastly, toluene (Figure 4-a). In turn, a 5% decrease in the removal efficiency of ethylbenzene and xylene was observed for pH 1.5 (Figure 4-b). The sorption of organic compounds should not be dependent on the solution pH value; however, the release of zinc ions during the contact time could be conducive to some structural modification of the polymeric structure. The composition of crumb rubber is the key for its sorption capability. Isoprene and butadiene are hydrocarbon chains present in crumb rubber that can interact with the alkyl groups in the organic pollutants [14]. For instance, the presence of methyl groups in the structure of o-xylene can account for its rapid and efficient removal. In turn, the length of the ethyl group –that is larger than the methyl group- can be related to the comparatively less adsorption of ethylbenzene when compared to xylene. Toluene has the largest solubility in water (515 mg/L at 25°C), thus its partition between crumb rubber and water is expected to be significantly different from the observed for xylene and ethylbenzene (with solubility of 200 mg/L and 152 mg/L, respectively). Furthermore, the remarkable hydrophobicity of xylene and ethylbenzene can also explain their fast removal by crumb rubber.

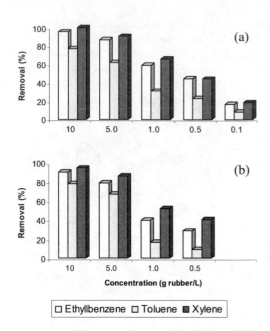

Figure 4: Comparative sorption behaviour of ETX by crumb rubber at pH 6.0 (a), and 1.5 (b). The Initial ETX and crumb rubber concentrations were 30ppm, and 10 g/L, respectively.

3.3 Freundlich Isotherms

Freundlich's equation permitted a very good fitting of the sorption data for ethylbenzene, toluene and xylene. Values of $1/n \leq 1$ indicate a rapid removal of the adsorbate at high concentrations. The other isotherm parameter is the K_f constant that is related to the loading factor of the sorbent; a large K_f value will mean high removal capacity. These two parameters have been calculated from the experimental data and are shown in Table 2. This table includes the data from other groups that also worked with scrap tires, although with different particle sizes. As observed, our results are in good agreement with reported information and confirms the capability of waste tires crumb rubber as a cheap and efficient sorbent for ETX compounds

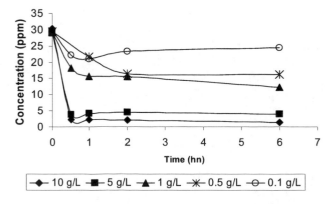

Figure 3: Removal of ethylbenzene as a function of contact time and concentrations of crumb rubber, in g/L, at pH 6. The ethylbenzene concentration was 30 ppm.

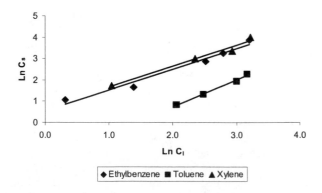

Figure 5: Freundlich isotherms for ethylbenzene, toluene and xylene at room temperature and pH 6.0.

Cleantech 2007, ISBN 1-4200-6382-0

Studied Analytes	Kim et al. [12]		Guanasekara et al.[13]		This study	
	1/n	K_f	1/n	K_f	1/n	K_f
Ethylbezene	0.998	1014	--	--	0.9728	1750
Toluene	0.991	297	0.999	264	1.2683	166
O-Xylene	0.990	1011	--	--	0.9913	1949

Table 2: Freundlich's parameters for ETX sorption by crumb rubber

4 CONCLUSIONS

The capability of waste tires crumb rubber mesh 14-20 to remove ETX pollutants from aqueous solutions has been verified. The best removal efficiency was observed for xylene followed by ethylbenzene and toluene. The corresponding removal efficiency was 99, 95 and 77% at pH 6.0. No significant variation was observed at other pH values.

5 ACKNOWLEDGEMENTS

We are grateful to Rubber Recycling and Manufacturing Company (REMA), the Puerto Rico Water and Environmental Research Institute (PRWRERI), the TOYOTA Foundation and the solid Waste Management Authority of Puerto Rico (ADS) for their support to this research

REFERENCES

[1] EPA. Puerto Rico National Priority Site Fact Sheets. 2005. www.epa.gov

[2] Wilbur, S. and Bosch, S. "Interaction Profile for: Benzene, Toluene, Ethylbenzene, and Xylenes (BTEX)". Agency for Toxic Substances & Disease Registry (ATSDR). 2004. Accessed online at www.atsdr.cdc.gov/interactionprofiles/ip05.html

[3] Ranck, M., Bowman, R., Weeber, J., Katz, L. and Sullivan, E. "BTEX Removal from Produced Water Using Surfactant-Modified Zeolite". *J. Environ. Eng.* 131(3) 434-442. 2005.

[4] ISRI. Institute of Scrap Recycling Industries, Inc. "Rubber Recycling Rolls Along". 2001. http://www.isri.org/industryinfo/rubber.htm

[5] Amari, T.; Themelis, N. and Wernick, I. "Resource Recovery from Used Rubber Tires". *Resources Policy*, 25, 170-188. 1999.

[6] Arocha, M.; Jackman, A. and McCoy, B. "Numerical Analysis of Sorption and Diffusion in Soil with Microspores, Macrospores, and Organic Matter". *Comp. Chem. Eng.* 21, 489. 1997.

[7] Knocke, W. and Hemphill, L. 1981. "Mercury (II) Sorption by Waste Rubber". *Water Res.*, 15(2), 275.

[8] Rowley A.; Husband, F. and Cunningham, A. "Mechanism of Metal Adsorption from Aqueous Solutions by Waste Tire Rubber". *Water Res.* 18(8), 981. 1984.

[9] Sameer, A. and Fawzi, B. "Adsorption of Cooper Ions on to Tire rubber". *Adsorption. Sci.& Tech.* 18(8), 685. 2000.

[10] Zarraa M. "Adsorption Equilibria of Single-component and Multi-component Metal Ions on to Scrap Rubber". *Adsorption Sci. &Tech.* 16(6), 493. 1998.

[11] Stom, W. and Morgan, J. 1996 "Aquatic Chemistry. Chemical Equilibria and Rates in Natural Waters", Third Edition. Wiley-Interscience Publication. USA.

[12] Kim, J.; Park, J. and Edil, T. "Sorption of Organic Compounds in the Aqueous Phase onto Tire Rubber". *J.Environ. Engng.* 123(9), 827. 1997.

[13] Guanasekara, A.; Donovan, J. and Xing, B, A. "Ground Discarded Tire Remove Naphthalene, Toluene, and Mercury from Water". *Chemosphere,* 41, 1155. 2000.

[14] Unnikrishnan, G.; Thomas, S. and Varghese, S. "Sorption and Diffusion of Aromatic Hydrocarbons through Filled Natural Rubber". *Polymer*, 37, 2687. 1996.

Growth and Some Enzymatic Responses of *E. coli* to Photocatalytic TiO$_2$

A. Erdem, D. Metzler, H.W. Chou, H.Y. Lin and C. P. Huang[*]

Department of Civil and Environmental Engineering, University of Delaware, DE
* Corresponding author: huang@ce.udel.edu

ABSTRACT

The effects of photocatalytic nano-TiO$_2$ on the survival or die-off of *E coli* (e.g. TB1) were investigated under ambient conditions. Experimentally, 18-h *E. coli* culture was exposed to photocatalytic nano-TiO$_2$ at various concentrations, e.g., 0 to 1,000 mg/L and particle sizes, e.g., 3 to 55 nm both in darkness and the presence of several light sources including a simulated solar light. Preliminary results indicated that there was bacteria die-off in the presence of nanoscale TiO$_2$ in dark. Generally it appears that the growth rate decreases as the particle size decreases. The presence of light irradiation significantly enhanced the killing of *E. coli* due to additional photocatalytic activity. Upon exposure of *E. coli* to nano-TiO$_2$ the photocatalytic activity that was generated has markedly increased the production of MDA, TTC and GST. SEM observations vividly indicate cell wall damages.

Keywords: E.coli, TiO$_2$, toxicity, die-off, photocatalysis

1 INTRODUCTION

The photocatalytic destruction of organic compounds in polluted air and water has been extensively studied. Following the work of Matsunaga, et al [1], an interest has grown in using this process for water disinfection [2 - 8]. Different mechanisms involved in the bactericidal action of TiO$_2$ photocatalysis have been proposed [1, 9 - 12]. The above studies suggest that the cell membrane is the primary target of reactive photogenerated oxygen species attack. Oxidative attack of the cell membrane leads to lipid peroxidation. The combination of cell membrane damage and oxidative attack of internal cellular components, results in cell death.

TiO$_2$ in the anatase crystal form is the most used semiconductor with a band gap of 3.2 eV or more. Upon excitation by light whose wavelength is less than 385 nm, the photon energy generates an electron hole pair on the TiO$_2$ surface. The hole in the valence band can react with H$_2$O or hydroxide ions on the TiO$_2$ surface to produce hydroxyl radicals (OH·), and the electron in the conduction band can reduce O$_2$ to produce superoxide ions (O$_2^{·-}$). Both holes and OH· are extremely reactive with organic compounds including microbial tissues. Matsunaga, et al. [1,2] reported that microbial cells in water could be killed by contact with a TiO$_2$-Pt catalyst upon illumination with near-UV light for 60 to 120 min. The findings of Matsunaga, et al [1,2] have prompted various researchers to use TiO$_2$ as a sterilization agent to disinfect drinking water. Most notably, Cai, et al. have attempted to kill cancer cells with the TiO$_2$ photocatalyst [13]. Blake et al. (1999) have reviewed the medical application of photocatalysts [14].

Several researchers have proposed the mechanisms of the killing of microorganisms using photocatalytic TiO$_2$. Matsunaga, et al. were the first to believe that direct photochemical oxidation of intracellular coenzyme A to its dimeric form was the cause of decreases in respiratory activities that led to cell death [1,2]. Saito, et al. proposed that the photocatalytic TiO$_2$ was disrupted the cell membrane and the cell wall of *Streptococcus sobrinus* AHT, as evident of leakage of intracellular K$^+$ ions that caused cell death [9]. Sakai, et al. reported the leakage of intracellular Ca^{2+} ions from cancer cells [15,16]. Sunada, et al. reported direct damage to outer membrane of *E. coli* as evident of the destruction of the endotoxin, an outer membrane component [6].

Although studies have demonstrated the importance of the bactericidal affects of the TiO$_2$ photocatalyst, the main mechanism causing the photocatalytic killing process has not been well-established yet. In this paper we compared the toxicity of TiO$_2$ toward bacteria exemplified by *E. coli* under dark and light conditions. Additionally we also studied the enzymatic responses of E. coli in terms of MDA formation and TTC reduction analysis.

2 MATERIALS AND METHODS

2.1. Culture of microorganisms:

E. coli TB1 strain was grown aerobically in 100 mL of Luria-Bertani (LB) broth at 37°C on a rotary shaker (200 rpm) for 18 h. The cells were harvested by centrifugation at 7,800g for 10 min, and resuspended in 10% glycerol and LB Broth mixture. The final optical density at 660 nm of the suspension was determined by measuring the turbidity with a Hach DR/2000 spectrophotometer to calculate the growth rate of the cells.

2.2. Experimental setups:
Five experimental runs were conducted in the laboratory (Table 1).

2.2. Photocatalytic reaction

Four different sizes of TiO_2 were used in experiments; P25 Degussa (75% anatase, 25% rutile, 30 nm), R5 Reade (99% anatase, 5 nm), R10 Reade (99% anatase, 3 nm), U100 (75% anatase, 25% rutile, 100 nm) particles. A stock suspension (10 g L^{-1}) in four different particle sizes was prepared with LB Broth media and kept refrigerated in the dark until use. *E.coli* was added to an aliquot of the stock suspension immediately prior to toxicity run. The final concentrations ranged from 0.01 to 1 g L^{-1}. All experiments were conducted in continuously stirred aqueous slurry solutions to ensure maximal mixing and to prevent possible settling of the TiO_2 particles.

Table 1: Experiments conducted

Run #	Culture / Media	Initial Conc.	Light Source	Instrument	Incubation
I	2 mL / 1 L	10^9 CFU mL^{-1}	No Light	Jacketed beaker, 200 rpm	37 °C for 24 h
II			Halogen (100 W)		
III	0.1 mL / 1 L		UV (100 W)		
IV			Solux (70 W)	Shaker, 200 rpm	
V			Agrosun (40 W)		

2.3. Cell viability:

The numbers of viable cells in cell suspensions that were subjected to the light and dark treatments were determined by plating serially diluted suspensions onto LB agar plates. The plates were incubated at 37° C for 24 h, and then the numbers of colonies on the plates were counted by using Fisher Acculite 133-8002 model colony counter.

2.4. Determination of lipid peroxidation:

Lipidperoxidation was determined by monitoring the formation of Malondialdehyde (MDA). Quantification of MDA was done following the methods described by Esterbauer, et al. [17] and Maness, et al. [10]. The method is based on the formation of pink MDA- Thiobarbituric Acid (TBA) adducts which has an absorption maximum in acidic solution at 532 nm. The concentrations of the MDA formed were calculated based on a standard curve for the MDA (Sigma Chemical Co.) complex with TBA. The extent of lipid peroxidation was expressed in nanomoles of MDA per milligram (dry weight) of cells.

2.5. Determination of cellular respiration:

The reduction of 2,3,5-triphenyltetrazolium chloride (TTC) to its reduced product, 2,3,5-triphenyltetrazolium formazan (TTF), was measured as described by Maness et al (1999), with minor modifications. The concentrations of the TTF formed were determined based on a standard curve for freshly prepared TTF (Sigma Chemical Co.) in methanol.

The rate of O_2 or TTC reduction was expressed in nanomoles of O_2 or TTF per minute per milligram (dry weight) of cells.

3 RESULTS

3.1. Effects of TiO_2 concentrations on cell viability

In order to study the killing mechanism, a high concentration (10^9 CFU ml^{-1}) of *E. coli* cells and TiO_2 concentrations ranging from 0.1 to 1 g L^{-1} were used to examine any change in cellular processes resulting from TiO_2 biocidal action. (Fig. 1, 2 and 3). The results indicated that *E. coli* underwent a two-stage response to TiO_2 particles; a rapid decrease in population within the first 15 min followed by a slow decrease upon extended treatment time, e.g., 30-45 min. No significant difference in terms of bacterial die-off was observed in the size range of 3 to 55 nm. An optimal particle concentration of 0.01g L^{-1} for the survival of *E. coli* was observed.

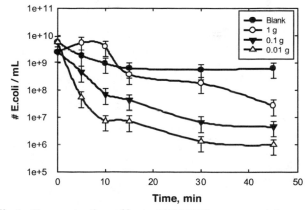

Fig 1: Concentration effect of R10 TiO_2 nanoparticles on the die-off of *E.coli* under dark condition (Run I).

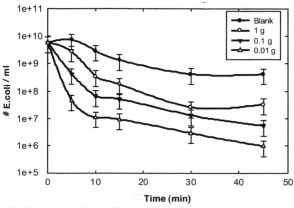

Fig 2: Concentration effect of P25 TiO_2 nanoparticles on the die-off of *E. coli* under dark condition (Run I).

3.2. Effect of irradiated TiO_2 on lipid peroxidation

To estimate membrane damage, we examined the production of MDA, a product of lipid peroxidation, by *E. coli* cells. The effects of irradiated TiO$_2$ on MDA formation in *E. coli* cells under dark and light conditions were also determined (Figure 6). *E. coli* cells were incubated with 0.001 to 0.3 g L^{-1} of TiO2 and were irradiated with Solux70 (8 W m^{-2}). Figure 6 shows that cells in the dark produced comparably low levels of MDA than in the light.

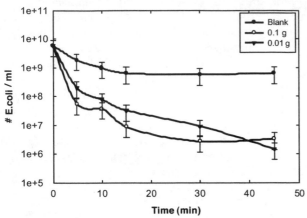

Fig 3: Concentration effect of U100 TiO$_2$ nanoparticles on the die-off of *E.coli* under dark condition (Run I).

Table 2: The percent kill of *E. coil*

Time, Min	Particle size				
	0	R10 3 nm	R5 5 nm	P25 25nm	U100 55 nm
5	68.15	91.87	90.13	99.84	92.86
30	90.12	99.89	99.64	99.91	99.77

Experimental conditions: 0.1g L^{-1}; particles sizes 5 nm; reaction time: 30 min; under dark (Run I)

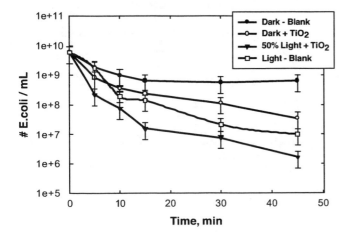

Fig 4: Die-off of E. coli as affected by Concentration. Experimental conditions: particle concentration: 0.5 g L^{-1} Particle: R5 TiO$_2$; light: Halogen (Run II)

3.3. Effect of TiO$_2$ on cellular respiratory activity

Since the bacterial cell membrane contains essential components of the respiratory chain, it was reasonable to investigate the effect of TiO$_2$ photocatalysis on cellular respiratory activities. Respiration was monitored by studying the reduction of TTC to TTF. Succinate was used as the electron donor in both assays. Figure 7 shows the reduction of TTC to TTF in 2 h under dark and light conditions.

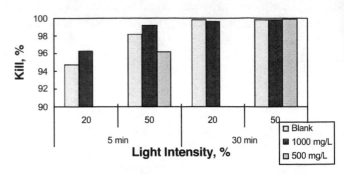

Fig 5: Stage-II Die-off of E. coli. Experimental conditions: Particle: R5 TiO$_2$; light: Halogen (Run II).

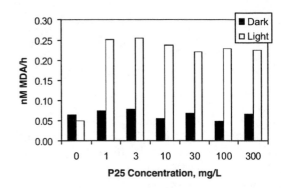

Fig 6: The effects of irradiated TiO$_2$ on MDA formation in *E. coli* cells under dark and light conditions (Run IV).

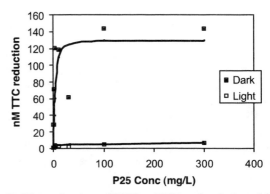

Fig 7: The reduction of TTC to TTF under dark and light conditions (Run IV)

Cryo-SEM images were taken in order to see the physical damage of the cells under dark and light conditions. Figure 8 shows that the effect of irradiated TiO$_2$ on the cell membrane is higher than the effect of TiO$_2$ under dark condition.

4 CONCLUSIONS

Results show that in darkness, nanoparticles still exhibited adverse effect on the growth of bacteria. The die-off of bacteria takes in two stages; fast die-off followed by a second slow killing. The smaller particle concentrations and larger primary particle size appear to be more damaging to the bacteria under dark conditions.

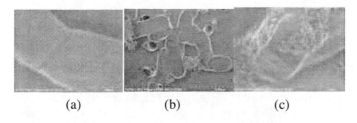

(a)   (b)   (c)

Fig 8: The SEM images of *E.coli* with no treatment (a), in the presence of P25 TiO$_2$ at 0.1 g L^{-1} under dark (b) and in the presence of P25 TiO$_2$ at 0.1 g L^{-1} P25 TiO$_2$ under light (c).

In the presence of light, results show that the light source plays an important role on the toxicity of TiO$_2$ to bacteria. Among four different light sources used, the presence of halogen light was adequate to kill bacteria without involving photo-catalyst, TiO$_2$. Larger concentrations of TiO$_2$ appear to be beneficial to the bacteria, since the concentrated solution would shield the bacteria from long wavelength UV and other activated TiO$_2$ complex.

Results of MDA analysis show that the lipid peroxidation process was dependent on the presence of both light and TiO$_2$. Under dark, the effects of irradiated TiO$_2$ on MDA formation in *E. coli* cells were lower than the MDA production in the light. The results also showed that small concentrations of TiO$_2$ caused a high lipidperoxidation.

Cellular respiratory activity was examined by TTC reduction to TTF. The results indicated that in 2 h under light condition, the bacteria became unhealthy and TTC reduction in light was lower than in dark.

Overall, results demonstrated that TiO$_2$ was able to deactivate common infectious water-borne microorganisms such as *E. coli*.

5 ACKNOWLEDGEMENTS

This research was funded by U.S.-EPA Science to Achieve Results (STAR) Program (Grant # R-831 72101). One of us, Ayca Erdem, wishes to thank the Akdeniz University, Antalya, Turkey, for the award of a fellowship.

6 REFERENCES

[1] T. Matsunaga, R. Tomoda, T. Nakajima, and H. Wake, FEMS Microbiol. Lett. 29 (1985) 211.

[2] T. Matsunaga, R. Tomoda, T. Nakajima, N. Nakamura, and T. Komine, Appl. Environ. Microbiol. 54 (1988) 1330.

[3] C. Wei, W. Y. Lin, Z. Zainal, N. E. Williams, K. Zhu, A. P. Kruzic, R. L. Smith, and K. Rajeshwar, Environ. Sci. Technol. 28 (1994) 934.

[4] M. Bekbolet, and C. V. Araz, Chemosphere 32:5 (1996) 959.

[5] Bekbolet, M., Wat. Sci. Tech.35 (1997) 95.

[6] K. Sunada, Y. Kikuchi, K. Hashimoto, and A. Fujishima, Environ. Sci. Technol. 32 (1998) 726.

[7] P. A. Christensen, T. P. Curtis, T. A. Egerton, S. A. M. Kosa and J. R. Tinlin, Appl. Cat.B 41 (2003) 371.

[8] A. G. Rincon and C. Pulgarin, Appl.Cat. B 44 (2003) 263.

[9] T. Saito, J. M. T. Iwase, and J. Horie, Photochem. Photobiol. B: Biol. 14 (1992) 369.

[10] P. C. Maness, S. Smolinski, D. M. Blake, Z. H. E. J. Wolfrum, and W. A. Jacoby, Appl. Environ. Microbiol. 65 (1999) 4094.

[11] Z. P. C. Huang, D. M. Maness, E. J. Blake, S. L. Wolfrum, W. A. Smolinski, and J. Jacoby, Photochem. Photobiol. A: Chem. 130 (2000) 163.

[12] K. T. Sunada, K. Watanabe, and J. Hashimoto, Photochem. Photobiol. A: Chem. 156 (2003) 227.

[13] R. K. Cai, K. Hashimoto, Y. Itoh, Y. Kubota, and A. Fujishima, Bull. Chem. Soc. Jpn. 64 (1991) 1268.

[14] D. M. Blake, P. C. Maness, Z. Huang, E. J. Wolfrum, W. A. Jacoby, and J. Huang, Sep. Purif. Methods 28 (1999) 1.

[15] R. X. Cai, K. Hashimoto, T. Kato, K. Hashimoto, A. Fujishima, Y. Kubota, E. Ito, and T. Yoshioka, Photomed. Photobiol. 12 (1990) 135.

[16] H. Sakai, E. Ito, R. X. Cai, T. Yoshioka, K. Hashimoto, and A. Fujishima, Biochim. Biophys. Acta 1201 (1994) 259.

[18] H. Esterbauer, and K. H. Cheeseman, Methods Enzymol. 186B (1990) 407.

Novel FD SOI Devices Structure for Low Standby Power Applications

Ming-Wen Ma[a], Tien-Sheng Chao[b], Kuo-Hsing Kao[b], Jyun-Siang Huang[b] and Tan-Fu Lei[a]

[a] Inst. and Dept. of Electronics Eng., National Chiao Tung Univ., Hsinchu, Taiwan, R.O.C
1001 Ta-Hsueh Road Hsinchu, Taiwan 30050, R.O.C, tflei@faculty.nctu.edu.tw
[b] Inst. and Dept. of Electrophysics, National Chiao Tung Univ., Hsinchu, Taiwan, R.O.C,
tschao@mail.nctu.edu.tw

ABSTRACT

In this paper, full-depleted SOI devices with source/drain extension shift and high-κ offset spacer were investigated in detail. The calculated results show that the source/drain extension shift can decrease off-state leakage current I_{off} significantly by utilizing the extra electron barrier height in source/drain extension shift region to reduce standby power dissipation. However, the on-state driving current I_{on} is also sacrificing simultaneously. In order to overcome this drawback, the high-κ offset spacer is used to increase the on-state driving current I_{on} effectively due to the enhanced vertical fringing electric field to elevate the channel voltage drop and reduce series resistance. Consequently, a nanoscale FD SOI device with 8-nm S/D extension shift and TiO_2 offset spacer can possess high driving current I_{on} and ultra-low leakage current I_{off} about 0.003 times lower than conventional SOI structure.

Keywords: Silicon-on-insulator (SOI), S/D extension shift, high-κ offset spacer dielectric, fringing electric field.

1 INTRODUCTION

High-performance and low-power transistors with high on-state driving current I_{on} and low off-state leakage current I_{off} are required for state-of-the-art CMOS technology [1]. Full-depleted silicon-on-insulator (FD SOI) devices have been anticipated to play a significant role for next generation technology. According to International Technology Roadmap for Semiconductors (ITRS) [1], the 65-nm node SOI processes with 32-nm channel length and 12-Å oxide thickness could be the mainstream CMOS technology. Until now, lots of efforts have been devoted to it in recent years [2]-[6]. For nanoscale MOSFETs, standby leakages current I_{off} has become serious challenges to the reliable circuit design. However, high on-state driving current I_{on} and low off-state leakage current I_{off} would not be obtained simultaneously. Therefore, a novel technology for achieving this target is required urgently. Here, we find that the source/drain (S/D) extension shift away from the gate edge can provide an extra electron barrier height to decrease the off-state leakage current I_{off}. However, this method will result in an ultra-high series resistance when device operates in on-state to degrade the on-state driving current I_{on}. In order to improve the on-state driving current

I_{on}, many high-κ materials are employed widely as the gate dielectric to increase the gate capacitance [7]-[9]. Recently, high-κ material has been used as the offset sidewall spacer dielectrics to improve the on-state driving current I_{on} for the thin-film transistors (TFTs) [10][11]. In the same way, we can utilize S/D extension shift to reduce the off-state leakage current I_{off} and to improve the on-state driving current I_{on} by using high-κ offset sidewall spacer.

In this paper, the 65-nm node SOI devices with different S/D extension shifts and offset sidewall spacer dielectrics are studied by two-dimensional (2-D) device simulator MEDICI [12]. Finally, the optimized structure is derived to obtain the high on-state driving current I_{on} and ultra-low off-state leakage current I_{off} simultaneously.

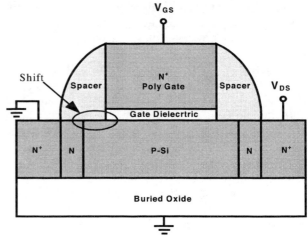

Figure 1. The FD SOI device structure with shifted source/drain extension.

2 SIMULATION PROCEDURE

Nanoscale SOI device structure used in MEDICI simulation is with a channel length of 32-nm, gate oxide thickness of 12-Å, offset sidewall spacer width of 30-nm, and body thickness of 15-nm [1][6], respectively. Five shifts in S/D extension from the gate edge, -5, 0, 5, 10 and 15-nm are used to investigate the off-state leakage current I_{off}. In order to utilize fringing electric field to improve the on-state driving current I_{on}, different dielectric constant (κ) of the offset spacer on the fringing electric field of the device, including air (ε_r=1), SiO_2 (ε_r=3.9), Si_3N_4 (ε_r=7.5), HfO_2 (ε_r=25) and TiO_2 (ε_r=80), were used. Note that the width of spacer was fixed at 30-nm in the simulation [6].

3 RESULTS AND DISCUSSION

Figure 1 shows the FD SOI device structure with a shifted S/D extension. This shifted S/D extension results in a significant decrease of the off-state leakage current I_{off}, i.e. V_{GS}=0V and V_{DS}=1.0V, as shown in Fig. 2. Figure 2 indicates a great reduction in the off-state leakage current I_{off} by several orders with widened shift- width. It is because the shifted S/D extension region yields a extra p-type region outside the gate edge to provide a much lower surface channel potential, as shown in Fig. 3, to increase the electron barrier height, resulting in reducing the off-state leakage current I_{off}. This electron barrier height is significantly increased with the widened S/D extension shift, resulting in a progressive reduction of the off-state leakage current I_{off} as shown as Fig. 2.

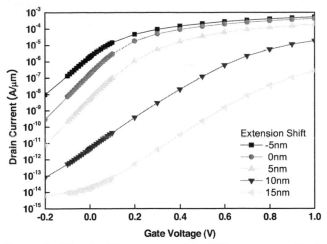

Figure 2. The I_{DS}-V_{GS} curve with different source/drain extension shift.

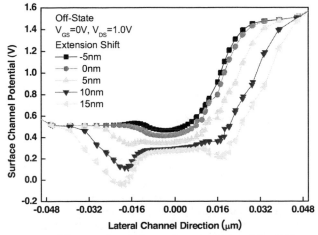

Figure 3. The surface channel potential with different source/drain extension shift in off-state.

However, Fig. 2 also indicates that the degradation in the on-state driving current I_{on}, i.e. V_{GS}=1.0V and V_{DS}=1.0V, happens simultaneously with the widened S/D extension shift. This degradation of the on-state driving current I_{on} has many accounts. The dominant account is that the widened S/D extension shift introduces a series resistance $R_{S/D}$ because of the extra p-type region. This extra p-type surface region could be inverted by the assistant of the vertical fringing electric field via the offset sidewall spacer.

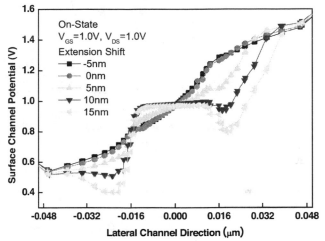

Figure 4. The surface channel potential with different source/drain extension shift in on-state.

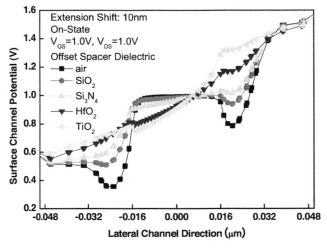

Figure 5. The surface channel potential with 10nm extension shift and different high-κ offset spacer in on-state.

The inversion layer in shifted S/D extension region maybe not formed as the extra p-type region is too wide or the vertical fringing electric field is too weak to introduce enough minority carriers. Therefore, the devices would turn on difficultly when the S/D extension shift is over 10-nm as shown in Fig. 2. Therefore, we can expect that this introduced electron barrier height could not be surmounted as the devices are in on-state as shown in Fig. 4. Figure 4 apparently indicates that two electron barrier heights exist in S/D extension shift region as the shift is over 10-nm.

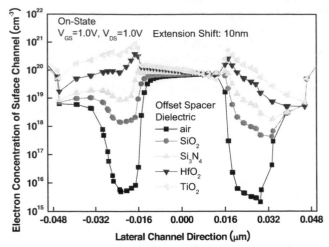

Figure 6. Electron concentration of surface channel with different spacer dielectrics in on-state region.

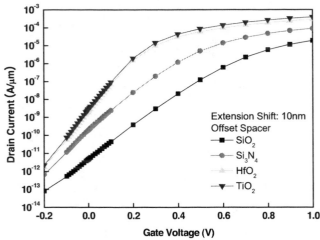

Figure 7. The I_{DS}-V_{GS} characteristics of four dielectric offset spacer SOI devices.

Although the off-state leakage current I_{off} and on-state driving current I_{on} decrease at the same time, the reduction magnitude of on-state driving current I_{on} is not as much as off-state leakage current I_{off} when the shifted width is below 10-nm. To optimize devices' performance, one can try to improve the slight degradation in on-state driving current I_{on} and maintain much lower off-state leakage current I_{off} by adopting suitable shift-width. As mentioned above, high-κ materials are used as the offset sidewall spacer dielectrics to enhance this vertical fringing electric field. Consequently, using the high-κ spacer to enhance the vertical fringing electric field for an improved on-state driving current becomes the most feasible approach. Figure 5 shows the surface channel potential with fixed 10-nm S/D extension shift and different high-κ offset sidewall spacers in the on-state. With the offset sidewall spacer dielectric constant

increasing, the vertical fringing electric field was significantly enhanced to elevate the surface channel potential, resulting in reducing the electron barrier height in S/D extension shift region efficiently as shown in Fig. 5. Moreover, this enhanced vertical fringing electric field can increase the voltage drop across the surface channel and induce more minority carriers in the S/D extension shift region, including S/D extension region, to ensure the complete inversion of the extra p-type region and reduce the series resistance significantly as shown in Fig. 6.

Figure 7 shows the I_{DS}-V_{GS} characteristics of different offset sidewall spacer dielectrics. It indicates a great improvement of on-state driving current I_{on} and subthreshold swing by using high-κ dielectric as offset sidewall spacer. Similarity, the off-state leakage current I_{off} is also degrading with the increasing of on-state driving current. However, it is still useful as both technologies of high-κ offset sidewall spacer and shifted S/D extension are collocating. For an ultra-low off-state leakage current I_{off} with constant high on-state driving current I_{on} design requirement, a 8-nm shift of S/D extension shift region with TiO_2 high-κ offset sidewall spacer device can perform. Consequently, a nanoscale FD SOI device with 8nm S/D extension shift and TiO_2 offset spacer can possess high on-state driving current I_{on} and ultra-low off-state leakage current I_{off} about 0.003 times lower than conventional SOI structure is derived as shown in Fig. 8.

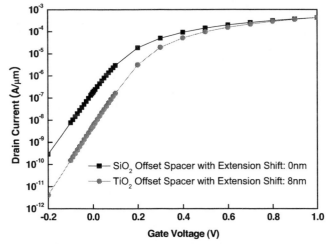

Figure 8. The I_{DS}-V_{GS} characteristics of SiO_2 spacer with zero extension shift and TiO_2 offset spacer with 8nm extension shift, respectively.

4 CONCLUSIONS

A novel nanoscale FD SOI structure possesses ultra-low off-state leakage current I_{off} and keeps high on-state driving current I_{on} is proposed for the first time. It utilizes the shifted S/D extension region to provide an extra electron barrier height to reduce off-state leakage current I_{off}. In addition, high-κoffset sidewall spacer is used to elevate the

Cleantech 2007, ISBN 1-4200-6382-0

voltage drop across surface channel and reduce series resistance to improve on-state driving current I_{on} significantly. This method provides a feasible approach to achieve both low standby current I_{off}, while maintaining high driving current I_{on} simultaneously. Consequently, a 65-nm node FD SOI device with 8-nm S/D extension shift and TiO_2 offset spacer can possess high on-state driving current I_{on} and ultra-low off-state leakage current I_{off} about 0.003 times lower than conventional SOI structure. This structure is one of the promising devices to be considered for next generation devices or beyond.

5   ACKNOWLEDGEMENT

This work is supported by the National Science Council, Taiwan, under contract No: NSC-94-2215-E-009-071.

REFERENCES

[1]   *International Technology Roadmap for Semiconductors 2003*.

[2]   B. Doris, M. Ieong, H. Zhu, Y. Zhang, M. Steen, W. Natzle, S. Callegari, V. Narayanan, J. Cai, S. H. Ku, P. Jamison, Y. Li, Z. Ren, V. Ku, D. Boyd, T. Kanarsky, C. D'Emic, M. Newport, D. Dobuzinsky, S. Deshpande, J. Petrus, R. Jammy, and W. Haensch, "Device design considerations for ultrathin SOI MOSFETs," in *IEDM Tech. Dig.*, 2003, pp. 631–634.

[3]   A. Vandooren, A. Barr, L. Mathew, T. R. White, S. Egley, D. Pham, M. Zavala, S. Samavedam, J. Schaeffer, J. Conner, B.-Y. Nguyen, B. E. White Jr., M. K. Orlowski, and J. Mogab, "Fully-depleted SOI devices with TaSiN gate, HfO_2 gate dielectric, and elevated source/drain extensions," *IEEE Electron Device Lett.*, vol. 24, no. 5, pp. 342–344, May 2003.

[4]   H. Y. Chen, C. Y. Chang, C. C. Huang, T. X. Chung, S. D. Liu, J.-R H.Y.-H Liu, Y. J. Chou, H. J. Wu, K. C. Shu, C. K. Huang, J. W. You, J. J. Shin, C. K. Chen, C. H. Lin, J. W. Hsu, B. C. Perng, P. Y. Tsai, C. C. Chen, J. H. Shieh, H. J. Tao, S. C. Chen, T. S. Gau and F. L. Yang, "Novel 20nm hybrid SOI/bulk CMOS technology with $0.183\mu m^2$ 6T-SRAM cell by immersion lithography," in *VLSI Tech. Dig.*, 2005, pp. 16-17.

[5]   F. L. Yang, C. C. Huang, C. C. Huang, T. X. Chung, H. Y. Chen, C. Y. Chang, H. W. Chen, D. H. Lee, S. D. Liu, K. H. Chen, C. K. Wen, S. M. Cheng, C. T. Yang, L. W. Kung, C. L. Lee, Y. J. Chou, F. Y. Liang, L. H. Shiu, J. W. You, K. C. Shu, B. C. Chang, J. J. Shin, C. K. Chen, T. S. Gau, P. W. Wang, B. W. Chan, P. F. Hsu, J. H. Shieh, S. K.-H. Fung, C. H. Diza, C.-M. M. Wu, Y. C. See, B. J. Lin, M.-S. Liang and J. Y.-C. Sun, C. Hu, "45nm node planar-SOI technology with $0.296\mu m^2$ 6T-SRAM cell," in *VLSI Tech. Dig.*, 2004, pp. 8-9.

[6]   F. L. Yang, C. C. Huang, H. Y. Chen, J. J. Liaw, T. X. Chung, H. W. Chen, C. Y. Chang, C. C. Huang, K. H. Chen, D. H. Lee, H. C. Tsao, C. K. Wen, S. M. Cheng, Y. M. Sheu, K.W. Su, C. C. Chen, T. L. Lee, S. C. Chen, C. J. Chen, C. H. Chang, J. C. Lu, W. Chang, C. P. Hou, Y. H. Chen, K. S. Chen, M. Lu, L. W. Kung, Y. J. Chou, F. J. Liang, J. W. You, K. C. Shu, B. C. Chang, J. J. Shin, C. K. Chen, T. S. Gau, B. W. Chan, Y. C. Huang, H. J. Tao, J. H. Chen, Y. S. Chen, Y. C. Yeo, S. K.-H. Fung, C. H. Diaz, C.-M. M. Wu, B. J. Lin, M.-S. Liang, J. Y.-C. Sun and C. Hu, "A 65nm node strained SOI technology with slim spacer," in *IEDM Tech. Dig.*, 2003, pp. 27.2.1–27.2.4.

[7]   N. Lu, H.-L. Li, M. Gatdner, D.-L. Kwong, "Improved device performance and reliability in high κ HfTaTiO gate dielectric with TaN gate electrode," *IEEE Electron Device Lett.*, vol. 26, no. 11, pp. 790–795, Nov. 2005.

[8]   K. Torii, T. Kawahara, K. Shiraishi, "Improvement of interfacial layer reliability by incorporation of deuterium into HfAlOx formed by D_2O-ALD," *IEEE Electron Device Lett.*, vol. 26, no. 10, pp. 722–724, Oct. 2005.

[9]   B. Doris, Y. H. Kim, B. P. Linder, M. Steen, V. Narayanan, D. Boyd, J. Rubino, L. Chang, J. Sleight, A. Topol, E. Sikorski, L. Shi, L. Wong, K. Babich, Y. Zhang, P. Kirsch, J. Newbury, J. F. Walker, R. Carruthers, C. D'Emic, P. Kozlowski, R. Jammy, K. W. Guarini, M. Leong, "MOS technology with metal gate and high-k," in *VLSI Tech. Dig.*, 2005, pp. 214-215.

[10]   Z. Xiong, H. Liu, C. Zhu, J. K. O. Sin, "Characteristics of high-κ spacer offset-gated polysilicon TFTs," *IEEE Trans. Electron Devices*, vol. 51, pp. 1304-1308, Aug, 2004.

[11]   Z. Xiong, H. Liu, C. Zhu and J. K. O. Sin, "A novel self-aligned offset-gated polysilicon TFT using high-κ dielectric spacers," *IEEE Electron Device Lett.* vol. 25, no. 4, pp. 194-195, 2004.

[12]   *User's Manual for MEDICI Two-Dimensional Device Simulation*, Synopsys Co., 2003.

Solar-Blind Dual-Band UV/IR Photodetectors Integrated on a Single Chip

D. Starikov [a,b,*], C. Boney [a,b], R. Pillai[b] and A. Bensaoula [b]

[a] Integrated Micro Sensors, Inc., 10814 Atwell Dr, Houston Tx 77096, USA, dstarikov@imsensors.com
[b] Texas Center for Advanced Materials, University of Houston, 724 S&R Building 1, Houston, Tx 77004, USA, bens@uh.edu

ABSTRACT

Employment of layered structures made of semiconductor materials with different optical absorption bands, is a new way of realizing either a broad spectrum photodetector or selective multiple band photodetectors. Such a concept based on structures fabricated using stacked semiconducting layers to obtain a multi spectral photoresponse is investigated in this paper. Based on the selected approach, fabrication of a dual-band UV/IR photodetector with a reasonable responsivity at room temperature has been demonstrated. The integrated device is capable of detecting optical emissions separately in the UV and IR parts of the spectrum. The responsivities of this device are ~0.01A/W, at a peak wavelength of 300 nm and ~0.08 A/W, at a peak wavelength of 1000 nm, respectively. The described dual-band photodetectors can be employed for false alarm-free fire/flame detection and advanced hazardous object or target detection and recognition in several industrial, military, and space applications.

Keywords: ultraviolet, infrared, photodetector, silicon, III nitrides

1. INTRODUCTION

Solid-state optical detectors, based on semiconductor materials, have replaced photoemissive devices in a wide variety of both commercial and military applications due to their broad spectral responsivity, excellent linearity, high quantum efficiency, large dynamic range of operation, and high potential for integration into large-format image arrays [1].

The spectral range of most semiconductor-based optical detectors is determined by optical absorption in the active semiconductor material layer at energies above the semiconductor band gap. As a result, narrow-band gap semiconductors, such as II-VI compounds in particular HgCdTe, are suitable for infrared detection. Si and some III-V compounds are perfect for detection in the visible (VIS) and near infrared (IR) range, and wide band gap semiconductor materials, such as diamond, SiC, and III nitrides, are superior for applications in the ultraviolet (UV) range.

Several military and industrial applications require simultaneous (or at least spatially registered/synchronized) detection of optical emissions in different spectral regions. Many such applications involve detection of flames, fires, and explosions that produce emissions in a wide range of the optical spectrum. Such emissions have distinct sharp peaks in both UV and IR that can be differentiated over the wide spectral range ambient light background by high-resolution, fast optical detectors, allowing time-resolved measurements in multiple spectral bands.

Multi-band detection capability featured in a single chip device became possible due to recent developments in the growth and processing of new semiconductor materials used for various spectral bands. Significant progress has been made lately in the development of UV detectors based on wide band gap materials. Several attempts to develop UV detector structures on diamond [2] were made by 1996, but due to the lack of high quality layers and insufficient doping levels, they did not result in practical devices. Visible-blind UV photodetectors have been fabricated on Silicon Carbide (SiC) substrates [3, 4], but the technology is relatively immature due to the lack of high quality large area substrates until few years ago (and no large area substrates prior to 1989 [1]).

Group III nitride materials are superior for advanced UV detector fabrication due to their wide direct band gap and high thermal, chemical, mechanical, and radiation tolerance. A large amount of research has been dedicated lately to the development of UV detectors based on GaN [5-7], GaN/AlGaN [8-10], and AlGaN [11]. Currently, attracting the most interest are AlGaN-based structures since they can allow detection in the very important UV range of 240-280 nm, which corresponds to the absorption range of solar radiation by the ozone layer [1].

In the area of IR detection, the conventional HgCdTe- and InSb-based detectors display high quantum efficiencies but are difficult to integrate into large arrays [1]. Detectors based on heterointernal photoemission (HIP) in Ge_xSi_{1-x}/Si heterojunctions have demonstrated additional opportunities for integration on Si wafers at sufficient sensitivities in the infrared range of 1-12 μm [12-15]. Large area SiGe-based HIP photodetector arrays of 400x400 pixels have been available for close to ten years [16].

Recent work on photodetector structures based on metal silicides [17, 18] has positively shown possibilities to extend the silicon-based detector's spectral range further into IR. However, such devices are limited to operation under cooled conditions, such as cryogenic temperatures or even lower.

The opportunity to grow III nitrides on Si wafers can be considered as the main key to the development of integrated multi-color detectors ranging from the UV to IR. Device-quality GaN layers have been demonstrated lately by several groups [19-24]. The new challenge is the growth of high quality InGaN and AlGaN layers on Si wafers in order to fabricate optoelectronic devices working in the range from UV to IR. Such an approach, based on stacked UV and IR pixels, removes the important fundamental problem of spatial alignment in multi wavelength array detectors.

In this work, we will demonstrate fabrication and characterization of a UV/IR visible-blind photodetector based on stacked semiconducting layers with desired properties, integrated on a single chip. The fabrication of the device is based on the growth of III-Nitride compounds on commercial Si wafers by Radio-Frequency Molecular Beam Epitaxy (RF MBE), which allows for precise control over the layer quality and composition at relatively high (up to 2 μm/hr) growth rates. In addition, employment of MBE allows for simple integration of molecular sources that can be used to grow III nitride compounds such as InN, GaN, InGaN, AlGaN, and AlN. Depending on the incorporation of In and Al in these compounds, the band gap of the epitaxial layer can be theoretically varied from about 0.8 eV (band gap of InN) to about 6.2 eV (band gap of AlN) [25]. MBE is also currently the method of choice for the fabrication of Si and SiGe based device heterostructures.

2. DEVICE LAYOUT

There are many approaches to employing stacked semiconducting layers in order to build multi-band photodetectors. In this work we focus on dual-band visible or solar-blind structures with sensitivities separately in the UV and IR parts of the spectrum.

Our approach is based on employment of III nitride layers grown on Si for fabrication of the UV-sensitive photodiode structures on the front side, and fabrication of the Si- or silicide-based IR-sensitive photodiode structures on the backside of commercial Si wafers. In this case the Si wafer also serves as a filter to visible light (Figure 1). The advantages of this approach include the relatively low cost and the availability of double side-polished Si substrates optimized for photodiode structure fabrication, while the drawbacks are the high lattice mismatch with the III nitride layers resulting in less than ideal crystalline quality, as well as

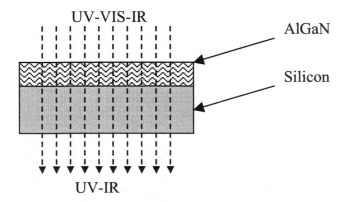

Figure 1: Approach based on a III nitride layer stacked on Si, and using the absorption properties of Si to filter out visible radiation.

the difficulties related to handling and processing of thin or even ultra-thin Si wafers that might be needed to optimize the IR radiation transmission and visible light blockage device characteristics.

3. EXPERIMENTAL REULTS

The UV/IR Photodiode structure, fabricated on a single chip, is shown in Figure 2. All III-nitride layers used for the photodiode fabrication were grown in a custom-made MBE chamber equipped with standard effusion cells for the group III components, such as: Ga, Al, In; and dopants, such as: Si, and Mg. Active nitrogen species are generated by an EPI Uni-Bulb radio-frequency (RF) plasma source. The details related to the III nitride growth and Schottky barrier formation are described in our previous publications [26-28].

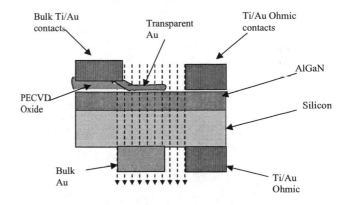

Figure 2: Schematic showing the processed dual structure chip with aligned Schottky and ohmic contacts on the front side and the backside.

In the present work the film growth experiments were carried out on commercial-grade 2" diameter thin (175 μm) Si <111> wafers. After the AlN buffer layer growth, the substrate temperature was lowered for the deposition of GaN. The GaN film was deposited at 700 °C, in order to strike a balance between film quality and background n-type doping, arising from Si diffusion out of the substrate. Interspersed in the 1.55μm of GaN were 3 thin (~10 nm) AlN layers deposited at a lower temperature and used to reduce stress in the film and hopefully reduce cracking of the final layers. We nevertheless observed a slight curvature of the resulting wafer as a result of the stress built up during the growth process and the use of a relatively thin silicon wafer.

Formation of Schottky barrier structures on n-GaN layers was started by a Plasma Enhanced CVD (PECVD) deposition of a 2μm thick tapered silicon oxi-nitride layer in order to provide continuity and isolation for a thin (100-200 Å) semi-transparent 1 mm diameter Au contacts. These contacts were deposited by thermal evaporation through a stencil mask partially on GaN surface and partially on the tapered oxide layer as shown in Figure 2. A thicker (2000 Å) Au contact deposited by e-beam evaporation was used also to form a Schottky barrier on the back side of the Si wafer. The ohmic contacts for both Schottky barrier structures and the semi-transparent Au layer sitting on the tapered oxide were formed by e-beam evaporation of Ti/Au (1000Å/1000Å) layers.

After fabrication of the Schottky barrier and ohmic contacts on both sides of the substrate, the sample was diced into smaller chips each having a single dual-band photodetector structure composed of 2 Schottky and 2 ohmic contacts on each side. The chips were then mounted onto insulating AlN ceramic plates that provide comparatively high thermal conductivity. Prior to chip attachment, the AlN plates were diced and a 5000 Å thick Au layer was deposited on one of the AlN ceramic plate sides by electron beam evaporation. The Au layer was then patterned using photolithography to match Schottky and ohmic contacts on the backside of the photodetector chip. The chips were then bonded to the patterned Au contacts on the AlN plate using a silver based high temperature electrically conductive resin and gel-type glue for improving the bonding strength and providing isolation between the Schottky and ohmic contacts. The metallization pads were then bonded to the TO-8 housing pins. Figure 3 shows a fully functional dual-band photodetector device packaged into a standard TO-8 housing.

Figure 3: Packaged integrated UV/IR photodetector prototype.

Spectral response measurements from the packaged photodetector were performed using an automated Jobin Yvon Triax 320 spectrometer. A calibrated xenon lamp was used for measurements in the UV and visible ranges; and a tungsten filament calibration lamp was used for measurements in the near and mid IR range.

The spectral response measured in two separate bands (Figure 4) indicates reasonable spectral selectivity of the device in the UV and the near IR bands. The peak responsivities at wavelengths of 265 nm in the UV and 1000 nm in the IR measured using an Oriel optical power meter at fixed values of the incident light power densities, were ~0.01 A/W (AlGaN-based structure) and ~0.08 A/W (Si-based structure), respectively.

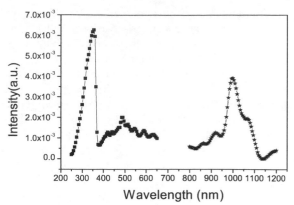

Figure 4: Spectral response from the dual band integrated photodetector measured separately in the UV and IR bands.

4. CONCLUSIONS

A viable approach to fabricating a dual-band UV/IR photodetector integrated on a single chip has been selected among various possible concepts that can be considered in the future. This concept based on employment of III nitride layers grown on commercial Si wafers was realized in an experimental visible-blind UV/IR photodetector operating at room temperature. The device was packaged into standard TO-8 housing and exhibited reasonable peak responsivities separately in the UV and IR parts of the spectrum.

Further improvements in the device performance can be achieved by optimization of the diode junctions, used in the photodetector fabrication. For example, replacement of Schottky barriers with p-i-n junctions might result in higher efficiencies.

The spectral response of such devices can be also improved by better separation of the UV and IR bands resulting in solar-blind devices and extension of the sensitivity further into the UV and IR bands. This can be achieved by optimization of the Si layer thickness and introduction of AlGaN layers with higher (more than 50%) Al content and replacement of Si-based junctions with silicide-based ones.

ACKNOWLEDGEMENTS

We acknowledge funds from an AF SBIR project to Integrated Micro Sensors Inc. (contract No. FA8103-04-C-0136, contact manager Mr. Joe Starzenski), as well as funds from the Institute of Space Systems on Operations (ISSO, University of Houston) and the NASA core funding to the Texas Center of Advanced Materials at the University of Houston.

REFERENCES

[1] M. Razeghi and A. Rogalski. Semiconductor ultraviolet detectors. J. Appl. Phys. 79 (10), 7433-7473, 1996.

[2] Michael D. Whitfield, Simon SM Chan, and Richard B. Jackman. Thin film diamond photodiode for ultraviolet light detection. Appl. Phys. Lett. 68 (3), 290-292, 1996.

[3] G. de Cesare, F. Irrera, F. Palma, and M. Tucci, E. Jannitti, G. Naletto and P. Nicolosi. Amorphous silicon/silicon carbide photodiodes with excellent sensitivity and selectivity in the vacuum ultraviolet spectrum. Appl. Phys. Lett. 67(3), 335-337, 1995.

[4] P. Mandracci, F. Giorgis, C. F. Pirri, and M. L. Rastello. Large area and high sensitivity a-Si:H/a-SiC:H based detectors for visible and ultraviolet light. Rev. Sci. Inst. 70(5), 2235-2237, 1999.

[5] J. M. Van Hove, R. Hickman, J. J. Klaassen, P. P. Chow, and P. P. Ruden. Ultraviolet-sensitive, visible-blind GaN photodiodes fabricated by molecular beam epitaxy. Appl. Phys. Lett. 70 (17), 2282-2284, 1997.

[6] Q. Chen, J. W. Yang, A. Osinsky, S. Gangopadhyay, B. Lim, M. Z. Anwar, M. Asif Khan, D. Kuksenkov and H. Temkin , Schottky barrier detectors on GaN for visible–blind ultraviolet detection. Appl. Phys. Lett. 70 (17), 2277-2279, 1997.

[7] Eva Monroy, Fernando Calle, Carlos Angulo, Pablo Vila, Angel Sanz, Jose Antonio Garrido, Enrique Calleja, Elias Muñoz, Soufien Haffouz, Bernard Beaumont, Frank Omnes, and Pierre Gibart. GaN-based solar-ultraviolet detection instrument Appl. Opt. 37 (22), 5058-5062, 1998.

[8] Wei Yang, Thomas Nohova, Subash Krishnankutty, Robert Torreano, Scott McPherson, and Holly Marsh. Back-illuminated GaN/AlGaN heterojunction photodiodes with high quantum efficiency and low noise. Appl. Phys. Lett. 73 (8), 1086-1088, 1998.

[9] E. Muñoz, E. Monroy, and F. Calle, F. Omnès and P. Gibart. AlGaN photodiodes for monitoring solar UV radiation. J. Geoph. Res. 105 (D4), 4865-4871, 2000.

[10] Cyril Pernot, Akira Hirano, Motoaki Iwaya, Theeradetch Detchprohm, Hiroshi Amano, and Isamu Akasaki. Solar-Blind UV Photodetectors Based on GaN/AlGaN p-i-n Photodiodes. Jap. J. Appl. Phys., Part 2, 39 (5A), L387-L389, 2000.

[11] F. Omnès, N. Marenco, B. Beaumont, Ph. de Mierry , E. Monroy, F. Calle, and E. Muñoz Metalorganic vapor-phase epitaxy-grown AlGaN materials for visible-blind ultraviolet photodetector applications. J. Appl. Phys. 86 (9), 5286-5292, 1999.

[12] H. Presting , M. Hepp, H. Kibbel, K. Thonke, R. Sauer, M. Mahlein, W. Cabanski, and M. Jaros. Midinfrared silicon/germanium based photodetection. J. Vac. Sci. Tech. B, 16 (3), 1520-1524, 1998.

[13] R. Strong, R. Misra, D. W. Greve, and P.C. Zalm. Ge_xSi_{1-x} infrared detectors I. Absorption in multiple quantum well and heterojunction internal photoemission structures. J. Appl. Phys. 82 (10), 5191-5198, 1997.

[14] J. R. Jimenez, X. Xiao J. C. Sturm, P. W. Pellegrini and M. M. Weeks. Schottky barrier heights of Pt and Ir silicides formed on Si/SiGe measured by internal photoemission. J. Appl. Phys. 75(10), 5160-5164, 1994.

[15] D. Krapf, B. Adoram, J. Shappir, A. Sa'ar, S. G. Thomas, J. L. Liu, and K. L. Wang. Infrared multispectral detection using Si/SixGe1-x quantum well infrared photodetectors. Appl. Phys. Lett., 78 (4), 495-497, 2001.

[16] H. Kibbel and E. Kasper. Vacuum 41, 929. 1990.

[17] C. Schwarz and H. von Ka¨nel, Tunable Infrared Detector with epitaxial Silicide/Silicon Heterostructures, J. Appl. Phys. 79 (11), 1996.

[18] T. L. Lin, J. S. Park, T. George, E. W. Jones, R. W. Fathauer, and J. Maserjian, Long-wavelength PtSi infrared detectors fabricated by incorporating a p+ doping spike grown by molecular beam epitaxy, Appl. Phys. Lett. 62, 254, 1993.

[19] Yasutoshi Kawaguchi,Yoshio Honda, Hidetada Matsushima, Masahito Yamaguchi, Kazumasa Hiramatsu, and Nobuhiko Sawaki. Selective area growth of GaN on Si substrate using SiO2 mask by metalorganic vapor phase epitaxy, Jap. J. Appl. Phys. Part 2, 37 (8B), L966-L969, 1998.

[20] Haoxiang Zhang, Zhizhen Ye, and Binghui Zhao. Epitaxial growth of wurtzite GaN on Si(111) by a vacuum reactive evaporation. J. Appl. Phys. 87 (6), 2830-2834, 2000.

[21] Shigeyasu Tanaka, Yasutoshi Kawaguchi, Nobuhiko Sawaki, Michio Hibino, and Kazumasa Hiramatsu. Defect structure in selective area growth GaN pyramid on (111)Si substrate. Appl. Phys. Lett. 76 (19), 2701-2703, 2000.

[22] D. Starikov, E. Kim, C. Boney, I. Hernandez, J.-W. Um, and A. Bensaoula. RF-MBE Growth of III-Nitrides for Micro Sensor Applications. 19th North American Conference on Molecular Beam Epitaxy, Tempe AZ,. Conf. Proc. 67, 2000.

[23] M.D. Craven, et. al., Appl. Phys. Lett. 84, 496, 2004.

[24] H.M. Ng, Non-polar GaN/AlGaN MQWs on r-plane sapphire, Appl. Phys. Lett. 80, 4369, 2002.

[25] H.X Jiang and J.Y. Lin , AlGaN and InAlGaN Alloys-Epitaxial Growth, Optical and Electrical Properties, and applications, Optoelectron. Rev. 10(4), 271–286, 2002.

[26] D. Starikov, N. Badi, I. Berishev, N. Medelci, O. Kameli, M. Sayhi, V. Zomorrodian, and A. Bensaoula. "Metal-insulator-semiconductor Schottky barrier structures fabricated using interfacial BN layers grown on GaN and SiC for optoelectronic device applications"; J. Vac. Sci. Technol. A: 17 (4),1235-1238, 1999.

[27] D. Starikov, I. Berishev, J.-W. Um, N. Badi, N. Medelci, A. Tempez, and A. Bensaoula. "Diode Structures Based on p-GaN for Optoelectronic Applications in the Near-Ultraviolet Range of the Spectrum". J. Vac. Sci. Technol B: 18(6), 2620-2623, 2000.

[28] D. Starikov, C. Boney, I. Berishev, I.C. Hernandez, and A. Bensaoula. "Radio-frequency molecular beam epitaxy growth of III nitrides for microsensor applications". J. Vac. Sci.Tech., B: 19(4), 1404-1408, 2001.

Characterization on Dioxin Emission of TiO$_2$ Nanoparticle-Encapsulating Poly(vinyl chloride) (TEPVC) compared to conventional PVC

H. Yoo[*], S. H. Kim[**] A. R. Lee[***], and S.-Y. Kwak[****]

School of Materials Science and Engineering, Seoul National University, San 56-1,
Sillim-dong, Gwanak-gu, Seoul 151-744, Korea
[*]friends1@snu.ac.kr, [**]hapful@dreamwiz.com, [***]aroasarang@hotmail.com, [****]sykwak@snu.ac.kr

ABSTRACT

TiO$_2$ nanoparticle-encapsulating poly(vinyl chloride) (TEPVC) was combusted in a well-controlled laboratory-scale incinerator at a temperature of 700 °C, and then dioxins (PCDDs, PCDFs and coplanar-PCBs) formed in the exhaust gases were analyzed by high resolution gas chromatography/high resolution mass spectrometry (HRGC/HRMS). TEPVC sample was prepared by the suspension polymerization of vinyl chloride monomer (VCM) with the 1.0wt% of surface modified amorphous TiO$_2$ nanoparticles based on the VCM weight reported in our previous study [1]. TEPVC shows valuable dioxin inhibition property due to the enhanced dispersibility of TiO$_2$ nanoparticles by encapsulation. Dioxin emission from TEPVC combustion was suppressed at the efficiencies of 39~82% by 0.926 wt% of encapsulated amorphous TiO$_2$ in TEPVC compared to conventional PVC.

Keywords: poly(vinyl chloride) (PVC), incineration, titanium dioxide (TiO$_2$), polychlorinated dibenzo-*p*-dioxins (PCDDs), polychlorinated dibenzofurans (PCDFs)

1 INTRODUCTION

Poly(vinyl chloride) (PVC) is widely used in extensive applications and more than 30 million tons of PVC are consumed annually in various commodities, so there is significant public concern about how the resulting wastes should be managed [2]. Incineration has competitive advantage in waste management and waste management depends mainly on incineration in many countries, but the incineration of PVC wastes can cause serious environmental problems because of the resulting formation of toxic chlorinated organic by-products such as polychlorinated dibenzo-*p*-dioxins (PCDDs) and polychlorinated dibenzofurans (PCDFs), generally known as dioxin.

Titanium dioxide (TiO$_2$) is a material of current interest due to their wide range of applications. It has been used in catalytic combustion processes as a catalyst for suppressing the emission of dioxins and their precursors through the absorption of chlorinated aromatic compounds followed by catalytic decomposition. In our previous study, we first prepared TiO$_2$ nanoparticle-encapsulating poly(vinyl chloride) (TEPVC) aiming at breakthrough to enhance the dispersibility of TiO$_2$ nanoparticles in PVC matrix to maximize catalytic activities of TiO$_2$ nanoparticles [1].

The present study is aimed to measure and compare the amount of dioxin emitted from both TEPVC and PVC incineration, estimating the dioxin inhibition efficiency of TEPVC. Amorphous TiO$_2$ was used in the preparation of the TEPVC because amorphous TiO$_2$ has negligible photocatalytic activity [3], thereby TEPVC being expected to undergo no photodegradation. But during the incineration of this polymeric material, the presence of amorphous TiO$_2$ is expected to suppress the emission of dioxin through the absorption of toxic organic compounds followed by catalytic decomposition.

2 EXPREMENTAL

2.1 Sample Materials

Amorphous TiO$_2$ nanoparticles were prepared through hydrolysis of titanium tetraisopropoxide (TTIP) according to the work of Inagaki et al [2]. 20 mL of TTIP dissolved in 40 mL of absolute ethanol by injection was gradually added in distilled water of 200mL under vigorous stirring and aged for 24 h at room temperature. The colloidal suspension was then filtered out and dried in a vacuum oven to give TiO$_2$ nanoparticles as a white solid.

The TEPVC sample was prepared using a previously reported method [1]. The encapsulation of amorphous TiO$_2$ in PVC resin was achieved through the suspension polymerization of vinyl chloride monomer (VCM) with the 1.0wt% of surface modified amorphous TiO$_2$ nanoparticles based on the VCM weight. Conventional PVC sample was prepared through suspension polymerization, too.

2.2 Characterization of TEPVC

Wavelength dispersive X-ray fluorescence (WD-XRF) spectrometry was used to determine the amorphous TiO$_2$ content of the TEPVC. The WD-XRF measurements were performed on a Ti element with a Shimadzu XRF-1700 sequential X-ray fluorescence spectrometer using lithium fluoride (LiF) as an analyzing crystal with a $2d$ value of 0.4028 nm. For the calibration of the quantitative analyses, WD-XRF analysis was also carried out for several standard samples, which were prepared by the mechanical mixing of PVC with various concentrations of TiO$_2$ from 0 to 2.0 wt%.

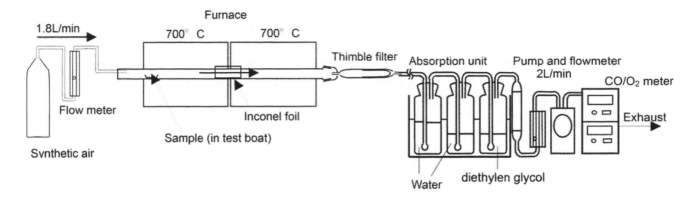

Figure 1: Schematic of the incineration test apparatus

The dispersity of amorphous TiO_2 nanoparticles in the TEPVC was investigated with transmission electron microscopy (TEM). For the TEM observations, the TEPVC grains were embedded in Gartan G-1 epoxy and cured at 60 °C for 90 min and ultra-thin cross-sections of the specimen was prepared by using a Leica Ultracut UCT ultracryomicrotome at room temperature. The TEM analyses were performed with a Jeol JEM-2000EXII at 200 kV of electron accelerating voltage.

2.3 Sample Combustion

Sample combustion, trapping of emission gas, and dioxin analysis were performed at Shimadzu Techno-Research, an official institute for dioxin measurement and analysis in Japan. TEPVC and PVC samples were first combusted in a well-controlled, laboratory-scale incinerator (Figure 1) at 700 °C. The small sample (approximately 25 mg) was placed onto a quartz boat and slid to the central position of tube furnace, then combusted. This operation was repeated for a total of 30 times, corresponding to 0.75 g of sample and 75 min of experimental period. Dioxins in exhaust gases were entrapped in an absorption unit composed of two water impingers, one diethylene glycol impinger, and the XAD-2 resin.

2.4 Dioxin Analysis

Dioxins entrapped in the absorption unit were determined by HRGC/HRMS in accordance with the Japanese standard method (Ministry of Health and Welfare of Japan, 1997). After solvent extraction with toluene for solid samples and with dichloromethane for liquid samples, a portion of the extract was spiked with $^{13}C_{12}$ labeled internal standard mixture containing one isomer each for tetra- to octa-chlorinated dibezo-p-dioxins and dibenzofurans and subjected to a column chromatographic clean-up procedure. This consisted of a multi-layer silica column (silica, 10% $AgNO_3$/silica, H22%- and 44% H_2SO_4/silica, silica, 2% KOH/silica, silica) and an aluminum oxide column.

HRGC/HRMS was preformed on a Waters Micromass Autospec Ultima mass spectrometer fitted with an Agilent HP6890 GC. SP-2331 (SUPELCO 60 m (length), 0.32 mm (internal diameter), 0.20 μm (film thickness)) coupled to DB-17HT (J & W 30 m, 0.32 mm, 0.15 μm) capillary column was applied to the determination of PCDD/PCDF congeners and HT8-PCB (SGE 60 m, 0.25 mm) capillary column was applied to determination of coplanar PCBs. The mass spectrometer was operated in selected ion monitoring mode (SIM) at a resolution > 10,000 and two ions were monitored for each congener group.

3 RESULTS AND DISCUSSION

3.1 Characterization of TEPVC

Figure 2 shows the dispersity of amorphous TiO_2 nanoparticles in the TEPVC. TiO_2 nanoparticles were well dispersed in PVC matrix, and the size of dispersed TiO_2 nanoparticles is in range of tens of nm.

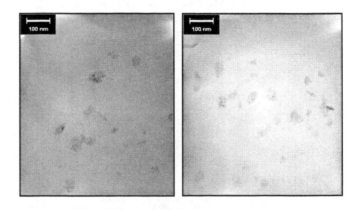

Figure 2: TEM images of the TEPVC.

WD-XRF spectrometry result indicates that amorphous TiO_2 nanoparticles were successfully incorporated within TEPVC during the encapsulation process. Amorphous TiO_2 content of TEPVC was 0.926 wt%.

	Name	TEQ factor	PVC		TEPVC	
			Concentration (pg/g)	Toxic equivalence (pg-TEQ/g)	Concentration (pg/g)	Toxic equivalence (pg-TEQ/g)
Polychlorinated dibenzo-*p*-dioxin (PCDD)	2,3,7,8-TeCDD	1	62	62	36	36
	1,2,3,7,8-PeCDD	1	130	130	70	70
	1,2,3,4,7,8-HxCDD	0.1	61	6.1	20	2.0
	1,2,3,6,7,8-HxCDD	0.1	77	7.7	22	2.2
	1,2,3,7,8,9-HxCDD	0.1	81	8.1	26	2.6
	1,2,3,4,6,7,8-HpCDD	0.01	300	3.0	66	0.66
	OCDD	0.0001	260	0.026	53	0.0053
	Total PCDDs	-	-	216.926	-	113.4653
Polychlorinated dibenzofuran (PCDF)	2,3,7,8-TeCDF	0.1	14000	1400	8100	810
	1,2,3,7,8-PeCDF	0.05	20000	1000	9400	470
	2,3,4,7,8-PeCDF	0.5	9600	4800	5000	2500
	1,2,3,4,7,8-HxCDF	0.1	18000	1800	7000	700
	1,2,3,6,7,8-HxCDF	0.1	15000	1500	6900	690
	1,2,3,7,8,9-HxCDF	0.1	4600	460	1900	190
	2,3,4,6,7,8-HxCDF	0.1	5200	520	2500	250
	1,2,3,4,6,7,8-HpCDF	0.01	37000	370	13000	130
	1,2,3,4,7,8,9-HpCDF	0.01	16000	160	5000	50
	OCDF	0.0001	49000	4.9	8600	0.86
	Total PCDFs	-	-	12014.9	-	5790.86
	Total (PCDDs+PCDFs)			12231.826		5904.3253
Coplanar polychlorinated biphenyl (Coplanar PCB)	3,4,4',5-TeCB (#81)	0.0001	1200	0.12	690	0.069
	3,3',4,4'-TeCB (#77)	0.0001	2200	0.22	1200	0.12
	3,3',4,4',5-PeCB (#126)	0.1	2000	200	1100	110
	3,3',4,4',5,5'-HxCB (#169)	0.01	710	7.1	380	3.8
	Non-ortho co-PCB		-	207.44	-	113.989
	2',3,4,4',5-PeCB (#123)	0.0001	630	0.063	360	0.036
	2,3',4,4',5-PeCB (#118)	0.0001	3700	0.37	1900	0.19
	2,3,3',4,4'-PeCB (#105)	0.0001	2200	0.22	1100	0.11
	2,3,4,4',5-PeCB (#114)	0.0005	2400	1.2	1100	0.55
	2,3',4,4',5,5'-HxCB (#167)	0.00001	1800	0.018	1100	0.011
	2,3,3',4,4',5-HxCB (#156)	0.0005	4500	2.25	2200	1.1
	2,3,3',4,4',5'-HxCB (#157)	0.0005	710	0.355	380	0.19
	2,3,3',4,4',5,5'-HpCB (#189)	0.0001	4500	0.45	2300	0.23
	Mono-ortho co-PCB	-	-	4.926	-	2.417
	Total coplanar PCBs			212.366		116.406
	Total dioxins			12444.192		6070.7313

Table 1: dioxins in the exhaust gases from conventional PVC and TEPVC combustion

3.2 Dioxin Formation from Incineration

Table 1 presents dioxins found in the exhaust gases from conventional PVC and TEPVC combustion in laboratory-scale incinerator. These results are shown on a pg/g basis, using the mass of samples consumed in the analysis. Toxic equivalence (WHO-TEQ) is calculated according to WHO-TEF (WHO/IPCS, 1998). In the case of PVC sample, PCDF congeners have been obtained in higher amounts than PCDD congeners; this is the usual trend [4]. The ratio of PCDFs to PCDDs of the combustion gas was 194, similar to the ratio of PCDFs to PCDDs of 230 in the case of TEPVC.

In the case of a blank sample, the amount of total dioxins (PCDDs, PCDFs, and co-planar PCBs) formed in the exhaust gas was 314 pg/g, which was 1/3589 of that formed from a conventional PVC sample and the TEQ was 1/100,000 of that from a conventional PVC sample, thus the dioxins from blank sample are negligible. The result of blank gas is shown on a pg/g basis assuming 1 g of sample used.

3.3 Suppression Efficiency of Dioxin Emission

The suppression efficiency of dioxin formation was calculated according to the equation

$$\text{suppression efficiency of dioxin formation (\%)} = \frac{C_{PVC} - C_{TEPVC}}{C_{PVC}} \times 100 \quad (1)$$

where C_{PVC} and C_{TEPVC} are the dioxin congener concentration of exhaust gases from PVC and TEPVC combustion, respectively.

Figure 4 shows the suppression efficiencies of PCDD/PCDF congeners' emission. The PCDD/PCDF congeners from TEPVC incineration are suppressed at the efficiencies of 42~82% compared to the PCDD/PCDF congeners from conventional PVC incineration. This result provides the generation of PCDDs/PCDFs from chlorinated plastics like PVC can be effectively suppressed by encapsulated amorphous TiO$_2$ nanoparticles. The higher chlorinated isomers of both PCDDs and PCDFs were more suppressed by amorphous TiO$_2$ nanoparticles encapsulated in the TEPVC.

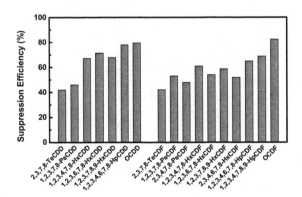

Figure 4: TEPVC's suppression efficiencies of PCDD/PCDF congeners' emission compared to the conventional PVC

Figure 5 shows the suppression efficiencies of coplanar PCB congeners' emission. The coplanar PCB congeners from TEPVC incineration are suppressed at the efficiencies of 39~54% compared to the coplanar PCB congeners from conventional PVC incineration.

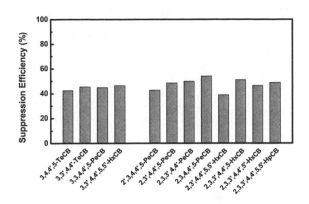

Figure 5: TEPVC's suppression efficiencies of coplanar-PCB congeners' emission compared to the conventional PVC.

4 CONCLUSIONS

TiO$_2$ nanoparticle-encapsulating poly(vinyl chloride) (TEPVC) shows valuable dioxin inhibition property due to the enhanced dispersibility of TiO$_2$ nanoparticles by encapsulation. Dioxin emission from TEPVC combustion was suppressed at the efficiencies of 39~82% by 0.926 wt% of encapsulated amorphous TiO$_2$ in TEPVC resin compared to conventional PVC. The results of this study show that the generation of the dioxins during combustion of PVC can be effectively suppressed by encapsulation of amorphous TiO$_2$ nanoparticles in the PVC resin.

REFERENCES

[1] H. Yoo, S. H. Kim, S.-Y. Kwak, IUPAC-PSK30, PO_B4_665, 2006.
[2] D. Braun, Prog. Polym. Sci., 27, 2171, 2002.
[3] B. Ohtani, Y. Ogawa, S.-i. Nishimoto, J. Phys. Chem. B., 101, 3746 (1997)
[4] I. Aracil, R. Font, J. A. Conesa, J. Anal. Appl. Pyrolysis 74, 465, 2005

Theoretical Study of the Reaction between OH Radicals and Formaldehyde adsorbed on Small Silica Clusters

Cristina Iuga* and Annik Vivier-Bunge*

*Departamento de Química, Universidad Autónoma Metropolitana-Iztapalapa,
Av. San Rafael Atlixco 186, Col. Vicentina, Iztapalapa 09340 D.F. Mexico,
email: annik@xanum.uam.mx

ABSTRACT

Heterogeneous reactions of atmospheric gases on aerosol particles may play an important role in atmospheric chemistry. However, the kinetics and mechanisms of adsorption and reaction of atmospheric gases on aerosol surfaces are not well understood. Clay particles are present in mineral dust in atmospheric aerosols, and radical reactions are thought to be heterogeneously catalyzed on them. In this work, quantum chemical methods are used to study the reaction of OH radicals with formaldehyde adsorbed on small $(SiO_4)n$ cluster models. We show that surface adsorbed formaldehyde can react in the presence of gas phase OH radicals to yield surface-bound formyl radicals and water. With the models employed, the reaction appears to be more favored on the silicate surfaces than in the gas phase. The effect of the model surface on the reaction mechanism is analyzed.

Keywords: atmospheric chemistry, mineral dust, pollutants, troposphere, OH radicals, formaldehyde, phyllosilicates, aerosols, reaction mechanism, small clusters.

1. INTRODUCTION

A major natural component of atmospheric aerosol is mineral dust, which enters the atmosphere from dust storms in arid and semiarid regions. About 33% of the earth's land surface is arid and a potential source region for this atmospheric mineral aerosol.

Mineral aerosol is a general expression for fine particles of crustal origin that are generated by wind erosion, and which consist mostly of silica and silicate minerals. Particles smaller than 10 μm have week-long atmospheric lifetimes [1] and they may be transported over thousands of kilometers. The transatlantic transport of Saharan dust to North America is a well-studied phenomenon (see reviews [1], [2] and recent articles [3], [4]).

The chemical and mineralogical composition of mineral aerosol is complex [5]. Globally, the most important minerals of the clay fraction (< 2 μm) transported in dust storms are illite, kaolinite, chlorite, and montmorillonite/smectite [6] whereas coarser particles mainly consist of quartz, feldspars, and carbonates [7]. Dust aerosols in the lower stratosphere consist almost entirely of clay particles [8].

Clay minerals, or phyllosilicates, are formed by sheets of SiO_4 tetrahedrons joined to a sheet of Al oxide octahedrons. The great diversity of these layered silicates is due to their capability for isomorphous substitution of various cations in the octahedral and tetrahedral sheets. Phyllosilicates have large specific surfaces and catalytic properties. Therefore, their presence in aerosols can be expected to play an important role in the heterogeneous chemistry of the troposphere.

The potentially reactive surface of mineral aerosols may be a significant sink for many volatile organic compounds in the atmosphere and consequently it could influence the global photooxidant budget. Laboratory studies, together with field observations and modeling calculations, have clearly demonstrated the importance of heterogeneous processes in the atmosphere. Some works have tried to quantify the effect of dust on tropospheric chemistry: Dentener et al. [1996] concentrated on heterogeneous uptake. He calculated that ozone concentration would decrease both because O_3 production decreased (N_2O_5 and HO_2 were taken up on dust) and because the O_3 molecules were themselves taken up on dust. Bian and Zender [2003] quantified the effect of dust on tropospheric chemistry due to both photolysis and heterogeneous update. They found that on a global average, O_3 decreases by 0.7%, OH decreases by 11.1%, and HO_2 decreases by 3.5 % when dust is added to the atmosphere. As discussed by Ravishankara, the ability to accurately predict the composition of the troposphere will depend on advances in understanding the role of particulate matter in the atmosphere and the extent to which heterogeneous reactions on solids and multiphase reactions in liquid droplets contribute to the chemistry.

The primordial role of OH radicals in the oxidative transformation of volatile organic compounds and other pollutants in the troposphere has stimulated interest in the study of their atmospheric reactions. Both experimental and theoretical studies of atmospheric reactions with OH radicals in the gas phase have been reported for a large number of reactions [9], [10], [11], [12], [13]. In most cases, these reactions are very fast, and accordingly, OH is the main oxidant of volatile organic species in the troposphere. However, the catalytic loss processes of atmospheric pollutants in the presence of OH radicals are not clear. Aerosols may promote the chemical reactions of OH radicals with adsorbed pollutants [14], [15].

Formaldehyde is an important component of the polluted troposphere. Concentrations of CH$_2$O measured in the remote troposphere are often lower and show considerably more variability than values computed from standard photochemical models [16], [17], [18], [19]. Some model studies have proposed that heterogeneous chemistry in aerosols and clouds could provide a fast sink for CH$_2$O [20], [21] though field observations offer no support for this hypothesis. Measurements of CH$_2$O gas-aerosol partitioning in surface air [22] indicate that only a small fraction of CH$_2$O is present in the aerosol. Observations in fogs and clouds show no evidence of CH$_2$O depletion relative to clear sky [23], [24], [25]. As formaldehyde is an important precursor to HOx in the troposphere, any heterogeneous interactions it may have with aerosol could potentially affect HOx levels, especially if it is removed from the troposphere. In a recent study by Carlos-Cuellar *et al.* examining the reaction of formaldehyde on SiO$_2$, [26] they found that the reaction of formaldehyde on SiO$_2$ are reversible.

Formaldehyde may, in principle, react with an OH radical according to two reaction paths: the abstraction of a hydrogen atom and the subsequent formation of a water molecule and a formyl radical; or, addition of the OH radical to the double bond, with the formation of the H$_2$C(OH)O• alkoxy radical. In the gas phase at room temperature, the abstraction reaction is favoured. The experimentally determined Arrhenius parameters [27], [28] indicate that the activation energy barrier is negative and very small. The gas phase formaldehyde + OH reaction has been studied by Galano *et al.* [12] who showed that a complex mechanism, involving the formation of a pre-reactive complex, explains adequately the observed negative activation energy of the reaction and the preference for the abstraction path. However, if the formaldehyde molecule were anchored to a clay surface, the branching ratio between the two reaction paths could be significantly altered.

Theoretical studies on clusters or molecular models of silicates have been performed by several authors, many of them concerning models for zeolites [29], [30], [32]. Sauer and co-workers have made an exhaustive revision of quantum mechanical models used to study molecule-solid interactions [31]. These methods have been shown to provide useful results for local properties of solids. Molecular Van der Waals complexes between adsorbed molecules and surfaces have been studied [32] and adsorption energies have been reproduced for a large number of compounds. The adsorption of methanol, formaldehyde and formic acid on hydrohylated silica surface has been studied [33] using DFT methods and ONIOM calculations on small clusters. To the best of our knowledge, there has been no report of a radical-molecule reaction mechanism study being performed for a molecule adsorbed on a surface.

The cluster model has been widely used to represent the active sites of solid catalysts, because it allows the performance of full geometry optimizations of minima and transition states as well as frequency calculations with high quality quantum chemical methods. However, it has some disadvantages, such as the appearance of boundary effects where the cluster is separated from the crystal framework, or the neglect of the long-range electrostatic effects caused by the Madelung potential of the solid.

The rigid tetrahedron SiO$_4$ is the building block of all siliceous materials, from zeolites to quartz and amorphous silica. At the surface, the structures terminate either in a siloxane group with oxygen on the surface, or in isolated, vicinal or geminal silanols [34], [35], [36].

The easiest and chemically best defined procedure for making a cluster neutral is to saturate the dangling bonds resulting from a homolytic cut with monovalent atoms, normally hydrogen [37]. Different small clusters have been used in the literature to model silicate surfaces [38], [39].

In this work, the mechanism of the formaldehyde + OH hydrogen abstraction reaction will be studied, with the formaldehyde attached to a model silicate cluster. It is clear that the results will depend significantly on the choice of the cluster model. As a first tentative approach, simple monomer, dimer and trimer clusters will be used. Our aim is to identify a computational method and a cluster size and geometry that is adequate to yield reliable results for this type of reactions.

2. METHODOLOGY

In order to study the reaction of an OH radical with a volatile organic compound adsorbed on the surface of a mineral dust aerosol model, we have chosen the reaction of a formaldehyde molecule with an OH radical. On the one hand, formaldehyde is a very reactive polar molecule that is easily adsorbed on surfaces, and on the other hand, many theoretical and experimental data are available for this gas phase reaction.

At first, different cluster surfaces have been modeled using several quantum chemistry methods. Then, formaldehyde is adsorbed on the surface, and an OH radical is allowed to approach the adsorbed formaldehyde. In this work, the BH&HLYP/6-311G** method was chosen. The gas phase reaction of formaldehyde with an OH radical was calculated at this level. Reactants, the pre-reactive complex, the transition state and the final intermediate product were optimized at this level of theory. The corresponding energy profile and optimized structures are shown in Figure 1.

In this mechanism, the reaction occurs in two steps: at first, a fast pre-equilibrium between the reactants and the pre-reactive complex is established, followed by an internal rearrangement leading to the elimination of a water molecule:

Step 1: RHC=O + OH• $\rightleftharpoons$ [RHC=O ---- HO•]   (1)

Step 2: [RHC=O ---- HO•] $\rightarrow$ RC=O• +H$_2$O   (2)

The calculated structures are similar to the ones previously found theoretically by Alvarez Idaboy et al [12]. The calculated activation energy barrier al BHandHLYP/6-311g** level is 0.23 kcal/mol, very close to the experimental results, which vary between +0.4 and -0.4 kcal/mol [40].

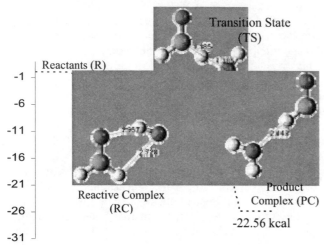

Figure 1: Gas phase reaction energy profile.

The surface models have to represent the structural features of the system simulated as closely as possible, but they have to be limited in size. In this work, three surface models have been considered, corresponding to the monomer, the dimer and the trimer.

Standard ab initio quantum chemical calculations were performed using the standard density functional BHandHLYP method as implemented in the Gaussian 03 program package. All geometries were fully optimized at this level using the Berny analytical gradient method. The unrestricted approximation was used for radicals. Normal mode analyses were carried out at the same level to confirm the nature of the various stationary points, only positive eigenvalues for minima and one negative eigenvalue (imaginary frequency) for transition states. The motion along the reaction coordinate corresponds to the expected transition vector. Corrections for zero-point energy (ZPE) (vibrational energy at 0°K) were taken from the force constant analysis and added to the total energies.

3. RESULTS AND DISCUSSION

The optimized cluster geometries are shown in Figure 2 In all cases, full geometry optimization was achieved. It can be observed that for all surface models, the Si-O distance varies between 1.61Å and 1.64Å, and the O-H distance is about 0.95 Å, both in very good agreement with the experimental data for clay minerals. The O-Si-O angle varies between 102.9 and 115.7.

The initial step starts with the adsorption of a formaldehyde molecule. In all cases, the formaldehyde molecule is adsorbed on the model surface in a plane that is perpendicular to the surface. A weak adsorption complex is formed. When OH dangling bonds are available, one of the formaldehyde H atoms is oriented towards the surface O atoms while its oxygen atom is attracted to a surface terminal hydrogen. Both non-bonding H...O distances correspond to weak hydrogen bonds. In our calculations, the formaldehyde molecule is allowed to move freely until it reaches the optimum adsorption site. All surface-adsorbate complexes have been fully optimized at the BHandHLYP/6-311G** level. Figures 4.(a), 5.(a) and 6.(a) show the optimized geometry of formaldehyde adsorbed on the surface models considered in this work. Relevant bond lengths are indicated on the figures.

The adsorption energy is defined as the difference between the total electronic energy of the surface-adsorbate complex and the isolated molecule and cluster, including ZPE corrections:

$$E_{adsorption} = E_{surface-adsorbate\ complex} - (E_{molecule} + E_{cluster}) + \Delta ZPE \qquad (3)$$

Adsorption energy results are summarized in Table 1. They are similar for all surface models, and they are negative, indicating that formation of the adsorption complex is exothermic.

The OH radical attack on adsorbed formaldehyde occurs in the following way. At first, the positively charged hydrogen atom of the OH radical approaches a lone pair of the oxygen atom to form a stable pre-reactive complex. In this complex, the OH radical lies in the plane of the CHO group (see figures 4.(b), 5.(b) and 6.(b)).

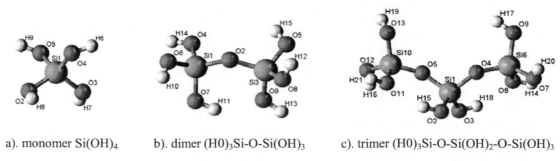

a). monomer Si(OH)$_4$ b). dimer (HO)$_3$Si-O-Si(OH)$_3$ c). trimer (HO)$_3$Si-O-Si(OH)$_2$-O-Si(OH)$_3$

Figure 2: BHandHLYP/6-311G** cluster models.

Cleantech 2007, ISBN 1-4200-6382-0

Prereactive complexes seem to be common in all radical-molecule reactions, and they are due mainly to the long-range Coulombic interactions between the reactant molecules. The geometries of the pre-reactive complexes obtained for the reaction between OH radical and formaldehyde adsorbed on the surfaces are very similar in all cases and very close to those obtained in the gas phase. Concerning the stabilization energy of the pre-reactive complexes (which is equal to E_{-1}), the results obtained with the three different surface models employed range from 3.18 kcal/mol to 5.19 kcal/mol.

From this structure, the oxygen of OH may flip, in the plane, toward the hydrogen to be abstracted as the energy increases to a maximum at the transition state. The transition state for this reaction step has been identified, and the vibrational motion associated with the imaginary frequency shows that the reaction involves an H atom migration. Calculated transition state energies are lower (for the dimer and trimer models) or slightly higher (in the case of the monomer) than the energy of the reactants, thus yielding an effective negative activation energy for the reaction. Thus the larger models predict that reaction is considerably faster on the surface than in the gas phase.

However, concerning the heats of reaction (Table 1), the employed method predicts the reaction path to be less exothermic than in the gas phase.

The energy profile of the reaction is shown in Figure 3. The stabilization energy of the prereactive complexes (E_{-1}), the effective activation energies $E_a^{eff} = E_2 - E_{-1}$ at 0 K, and the heat of reaction $\Delta H = E_P - E_R$ at 298 K, of the adsorbed formaldehyde + OH reaction are given in Table 1. The ZPE corrections have been included in all the energy differences.

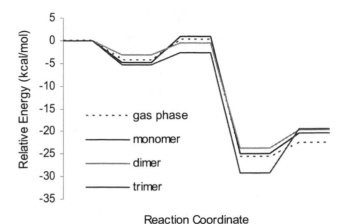

Figure 3: Reaction energy profile.

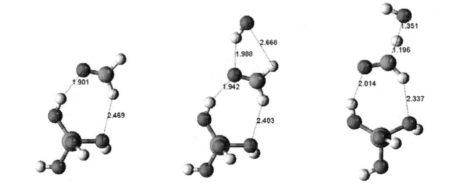

a). Surface-adsorbate complex   b). Pre-Reactive Complex (RC)   c). Transition State (TS)

Figure 4: Reaction on monomer surface.

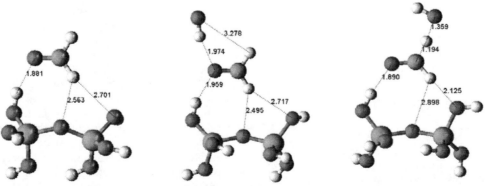

a). Surface-adsorbate complex   b). Pre-Reactive Complex (RC)   c). Transition State (TS)

Figure 5 : Reaction on dimer surface

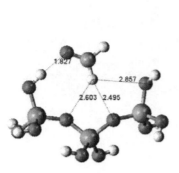

a). Surface-adsorbate complex.

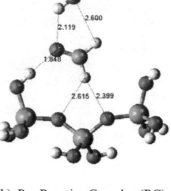

b). Pre-Reactive Complex (RC).

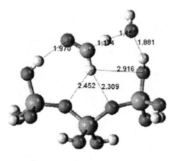

c). Transition State (TS).

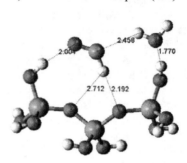

c). Product Complex (PC).

Figure 6: Reaction on trimer surface.

Table 1.- BHandHLYP/6-311g** relative barrier energies (including the ZPE corrections) and Heat of Reaction Energies (ΔH), in kcal/mol, of the OH radical reaction with formaldehyde adsorbed on different surface models (in kcal/mol).

Surface model	$E_{adsorption}$	E_{-1}	E_2	$E_a{}^{eff}$	ΔH	Imaginary frequency (cm^{-1})
gas phase	-	-5.19	5.42	0.23	-22.56	-932
monomer	-6.38	-4.92	5.83	0.92	-20.35	-1,204
dimer	-7.77	-3.18	2.59	-0.59	-19.68	-1,147
trimer	**-9.14**	**-4.99**	**2.23**	**-2.76**	**-29.19**	**-659**

$E_{-1} = E_{stabilization} = E_{RC} - E_R$

$E_2 = E_{TS} - E_{RC}$

$E_a{}^{eff} = E_{TS} - E_R$

$\Delta H = E_P - E_R$

3. CONCLUSION

In this work, the mechanism of the formaldehyde + OH hydrogen abstraction reaction have been studied, with the formaldehyde attached to a model silicate cluster. We have shown that, if small cluster models are used to represent SiO_2 surfaces, the reaction of adsorbed formaldehyde with OH radicals yields surface-bound formyl radicals and water. The geometries of the equilibrium structures (surface-adsorbate complexes, pre-reactive complexes and transition states) are very similar in all cases and very close to those obtained in the gas phase.

The obtained results are compared with those available in the literature for the formaldehyde + OH radical reaction in the gas phase. The convergence of the results with increasing size of the surface model seems to indicate that the presence of the surface favors the reaction rate.

Extension of the present work in several directions (larger surface molecular models, periodic models, inclusion of thermodynamic data, consideration of the cavity structure) is both possible and necessary in the future, and it is being performed.

Cleantech 2007, ISBN 1-4200-6382-0

REFERENCES

[1] J. M. Prospero, Proc. Natl. Acad. Sci. U.S.A., 96, 3396, 1999.

[2] J. M. Prospero, S. Guerzoni, R. Chester, Eds. Kluwer: Dordrecht, The Netherlands, 1996.

[3] J. M. Prospero, I. Olmez, M. Ames, Water, Air, Soil Pollut. 125, 291, 2001.

[4] K. D. Perry, T. A. Cahill, R. A. Eldred, D. D. Dutcher, T. E. Gill, J. Geophys. Res. 102, 11225, 1997.

[5] G. Coude-Gaussen, P. Rogon, G. Bergametti, L. Gomes, B. Strauss, J. M. Gros, M. N. Coustumer, J. Geophys. Res. D8, 9753, 1987.

[6] R. Chester, Sci. Geol. Mem. 88, 23, 1990.

[7] L. Gomes, D. A. Gillette, Atmos. Environ. 27A, 2539, 1993.

[8] R. L. Miller, I. Tegen and J. Perlwitz, J. Geophys. Res. 109, D04203, 2004.

[9] R. Atkinson, Atmos. Environ. 34, 2063, 2000.

[10] R. Atkinson, D. L. Baulch, R. A. Cox, R. F. Hampson, J. A. Kerr, M. J. Rossi, J. Troe, J. Phys. Chem. Ref. Data 28, 191, 1999.

[11] V. Hugo Uc, A. Grand and A. Vivier-Bunge, Journal of Molecular Structure: THEOCHEM 684, 171-179, 2004.

[12] J. R. Alvarez-Idaboy, N. Mora-Diez, R. Boyd and A. Vivier-Bunge, J. Am. Chem. Soc. 123, 2018 2001.

[13] J. R. Alvarez-Idaboy, N. Mora-Díez, A. Vivier-Bunge. J. Am. Chem. Soc. 122, 3715, 2000.

[14] S. Oh and J.M. Andino, Atmos. Environ. 36, 149–156, 2002.

[15] M. Sørensen, M. D. Hurley, T. J. Wallington, T. S. Dibble and O. J. Nielsen, Atmos. Environ. 36, 5947-5952, 2002.

[16] S. C. Liu, M. Trainer, M. A. Carroll, G. Hubler, D. D. Montzka, R. B. Norton, B. A. Ridley, J. G. Walega, E. L. Atlas, B. G. Heikes, B. J. Huebert and W. Warren, J. Geophys. Res., 97, 10 463–10 471, 1992.

[17] X. Zhou, Y-N. Lee, L. Newman, X. Chen and K. Mopper, J. Geophys. Res. 101, 14 711–14 719, 1996.

[18] D. J. Jacob, B. G. Heikes, S-M. Fan, J. A. Logan, D. L. Mauzerall, J. D. Bradshaw, H. B. Singh, G. L. Gregory, R. W. Talbot, D. R. Blake, G. W. Sachse, J. Geophys. Res. 101, 24 235–24 250, 1996.

[19] G. P. Ayers, R. W. Gillett, H. Granek, C. de Serves and R. A. Cox, Geophys. Res. Lett. 24, 401–404, 1997.

[20] R. B. Chatfield, Geophys. Res. Lett. 21, 2705-2708, 1994.

[21] J. Lelieveld and P. J. Crutzen, Nature, 343, 227-233, 1990.

[22] W. Klippel and P. Warneck, Atmos. Environ. 14, 809-818, 1980.

[23] M. C. Facchini, S. Fuzzi, J. A. Lind, M. Kessel, H. Fierlinger-Oberlinninger, M. Kalina, H. Puxbaum, W. Winiwarter, B. G. Arends, W. Wobrock, W. Jaeschke, A. Berner and C. Kruisz, Tellus 44B, 533–544, 1992.

[24] W. C. Keene, B. W. Mosher, D. J. Jacob, J. W. Munger, R. W. Talbot, R. S. Artz, J. R. Maben, B. C. Daube and J. N. Galloway, J. Geophys. Res. 100(D5), 9345–9357, 1995.

[25] J. W. Munger, D. J. Jacob, B. C. Daube, L. W. Horowitz, W. C. Keene and B. G. Heikes, J. Geophys. Res. 100D, 9325–9333, 1995.

[26] S. Carlos-Cuellar, P. Li, A. P. Christensen, B. J. Krueger, C. Burrichter, V. H. Grassian, J. Phys. Chem. A, 107, 4250, 2003.

[27] R. Atkinson, D. L. Baulch, R. A. Cox, R. F. Jr. Hampson, J. A. Kerr, M. J. Rossi, J. Troe, J. Phys. Chem. Ref. Data 26, 521, 1997.

[28] W. B. DeMore, S. P. Sander, D. M. Golden, R. F. Hampson, M. J. Kurylo, C. J. Howard, A. R. Ravishankara, C. E. Kolb and M. Molina, J. JPL Publication 97-4, 1997.

[29] A. C. Lasaga, Reviews in Mineralogy 31, 23-86, 1995.

[30] C. I. Sainz-Díaz, V. Timón, V. Botella and A. Hernández-Laguna, American Mineralogist 85, 1038-1045, 2000.

[31] J. Sauer, P. Ugliengo, E. Garrone, V. Saunders, Chem. Rev. 94, 2095-2160, 1994.

[32] J. Sauer, P. Ugliengo, E. Garrone, V. Saunders, Chem. Rev. 94, 2095-2160, 1994.

[33] Xin Lu, Qianer Zhanga and M. C. Lin, Phys. Chem. Chem. Phys. 3, 2156E2161, 2001.

[34] A. Burneau, O. Barrés, J. P. Gallas and J. C. Lavalley, Langmuir 6, 1364, 1990.

[35] C. E. Bronnimann, R. C. Zeigler and G. E. Maciel, J. Am. Chem. Soc. 110, 2023, 1988.

[36] S. Leonardelli, L. Facchini, C. Fretigny, P. Tougne and A. P. Legrand, J. Am. Chem.Soc. 114, 6412,. 1992.

[37] J. Sauer, P. Ugliengo, E. Garrone, and V. R. Saunders, Chem. Rev. 94, 2095-2160, 1994.

[38] J. Sauer, J. Chem. Rev. 89, 199, 1989.

[39] I. Papai, A. Goursot, and F. Fajula, J. Phys. Chem. 98, 4654, 1994.

[40] The NIST Chemical Kinetics Data Base, NIST Standard Reference Database; U.S. Dept. of Commerce, Technology Administration, National Institute of Standards and Technology: Gaithersburg, MD, 17-2Q98.

An investigation of natural nano-particles for cleaning

Mike Nero, Bao Tran

Naturell Clean, Inc 2953 Bunker Hill Lane Ste 400 Santa Clara, CA 95054
mike@naturellclean.com , bao@tranassoc.com

ABSTRACT

Traditionally cleaning products have been produced using man made chemicals such as synthetic surfactants made of petroleum distillates, vegetable oils, or large synthetic alcohols. The chemicals used in most cleaning products can often have a negative effect on the environment. Our focus is on replacing these man made compounds with naturally occurring nano-particles which act as catalysts.

The catalysts are produced using an organic medium (sea kelp), a variety of microbes, and water. The transformation of these simple ingredients into catalysts such as enzymes and bio-surfactants is accomplished in a process that resembles the brewing of wine. This makes production relitivelly simple and inexpensive, simply put the ingredients togather, provide the right conditions, and let nature take its course.

Keywords: Catalysts, enzymes, bio-surfactants.

1 Catalysts

A catalyst decreases the activation energy of a chemical reaction. Catalysts participate in reactions but are neither reactants nor products of the reaction they catalyze. Catalysts work by providing an (alternative) mechanism involving a different transition state and lower activation energy. The effect of this is that more molecular collisions have the energy needed to reach the transition state. Hence, catalysts can perform reactions that, albeit thermodynamically feasible, would not run without the presence of a catalyst, or perform them much faster, more specific, or at lower temperatures. Catalysts *cannot* make energetically unfavorable reactions possible — they have *no* effect on the chemical equilibrium of a reaction because the rate of both the forward and the reverse reaction are equally affected. The net free energy change of a reaction is the same whether a catalyst is used or not; the catalyst just makes it easier to activate.eaction helps to accelerate the same reaction. They work by providing an alternative pathway for the reaction to occur, thus reducing the activation energy and increasing the reaction rate. Catalysts generally react with one or more reactants to form a chemical intermediate that subsequently

2 Enzymes

Enzymes are proteins that catalyze (*i.e.* accelerate) chemical reactions. In these reactions, the molecules at the beginning of the process are called substrates, and the enzyme converts them into different molecules, the products. Almost all processes in the cell need enzymes in order to occur at significant rates. Since enzymes are extremely selective for their substrates and speed up only a few reactions from among many possibilities, the set of enzymes made in a cell determines which metabolic pathways occur in that cell. Like all catalysts, enzymes work by lowering the activation energy ($\Delta G^{\ddagger}$) for a reaction, thus dramatically accelerating the rate of the reaction. Most enzyme reaction rates are millions of times faster than those of comparable uncatalyzed reactions. As with all catalysts, enzymes are not consumed by the reactions they catalyze, nor do they alter the equilibrium of these reactions. However, enzymes do differ from most other catalysts by being much more specific.

Cleantech 2007, ISBN 1-4200-6382-0

3 Bio-Surfactants

Bio-Surfactants are a product of microbial action on an organic medium. Unlike other surfactants, bio-surfactants are effective at either end of the pH scale, and at either hot or cold temperatures. They affect the surface tension of liquids in which they are dissolved. They can lower the water's surface tension from 72 mN/m to 27 mN/m at a concentration as low as 20 µM. Bio-Surfactants accomplish this effect as they occupy the intermolecular space between water molecules, decreasing the attractive forces between adjacent water molecules, mainly hydrogen bonds, creating a more fluid solution that can go into tighter regions of space increasing water's wetting ability. Bio-Surfactants reduce the surface tension of water by adsorbing at the liquid-gas interface. They also reduce the interfacial tension between oil and water by adsorbing at the liquid-liquid interface. Many bio-surfactants can also assemble in the bulk solution into aggregates. Some of these aggregates are known as micelles. The concentration at which bio-surfactants begin to form micelles is known as the critical micelle concentration or CMC. When micelles form in water, their tails form a core that is like an oil droplet, and their (ionic/polar) heads form an outer shell that maintains favorable contact with water. When bio-surfactants assemble in oil, the aggregate is referred to as a reverse micelle. In a reverse micelle, the heads are in the core and the tails maintain favorable contact with oil.

4 Sea Kelp

We chose kelp as the basic ingredient for this product because it is a rich source of amino acids, minerals, and enzymes such as Amylase, Diastase, Phosphatase, Catalase, Cytochrome, Lactic, Oxidoreductases, Transferases, Hydrolases, Lyases, Isomerases, Pepsin, Trypsin, Thioredoxin, Peroxidase, Bromoperoxidase, Mannuronan-C5-epimerase, D-Glucanase, b-Lactamase, Penicillinase.

Analyses of kelp shows that it's mineral properties are quite different from other vegetation. Here is a breakdown:

Moisture	5.4
Protein	8.3
Dietary Fiber	37.0
Crude Fiber	3.5
Total Ash (minerals)	31.0
Ash free of salt	16.0
Fat	0.6
Salt (NaCl)	15.2

Amino Acids (% of total Amino Acids)			
Alanine	10.3	Leucine	4.9
Arginine	2.8	Lysine	4.3
Aspartic Acid	11.5	Phenylalanine	2.7
Cystine	3.0	Serine	5.0
Glycine	5.0	Threonine	6.0
Glutamic Acid	12.4	Tyrosine	3.4
Histidine	1.4	Valine	3.3
Isoleucine	2.5		

Trace minerals and elements	%
Ag Silver	0.000004
Al Aluminum	0.193000
Au Gold	0.000006

B Boron	0.019400		Mo Molybdenum	0.001592
Ba Barium	0.001276		N Nitrogen	1.467000
Be Beryllium	Trace		Na Sodium	4.180000
Bi Bismuth	Trace		Ni Nickel	0.003500
Br Bromine	Trace		O Oxygen	Undeclared
C Carbon	Undeclared		Os Osmium	Trace
Ca Calcium	1.904000		P Phosphorous	0.211000
Cb Niobium	Trace		Pb Lead	0.000014
Cd Cadmium	Trace		Pd Palladium	Trace
Ce Cerium	Trace		Pl Platinum	Trace
Cl Chlorine	3.680000		Ra Radium	Trace
Co Cobalt	0.001227		Rb Rubidium	0.000005
Cr Chromium	Trace		Rh Rhodium	Trace
Cs Cesium	Trace		S Sulphur	1.564200
Cu Copper	0.000635		Se Selenium	0.000043
F Florin	0.032650		Sb Antimony	0.000142
Fe Iron	0.089560		Si Silicon	0.164200
Ga Gallium	Trace		Sn Tin	0.000006
Ge Germanium	0.000005		Sr Strontium	0.074876
H Hydrogen	Undeclared		Te Tellurium	Trace
Hg Mercury	0.000190		Th Thorium	Trace
I Iodine	0.062400		Ti Titanium	0.000012
Id Indium	Trace		Tl Thallium	0.000293
Ir Iridium	Trace		U Uranium	0.000004
K Potassium	1.280000		V Vanadium	0.000531
La Lanthanum	0.000019		W Tungsten	0.000033
Li Lithium	0.000007		Zn Zinc	0.003516
Mg Magnesium	0.213000		Zr Zirconium	Trace
Mn Manganese	0.123500			

Cleantech 2007, ISBN 1-4200-6382-0

REFERENCES

[1] Recognizing the Best in Innovation: Breakthrough Catalyst". *R&D Magazine*, September 2005, pg 20.

[2] Bairoch A. (2000). "The ENZYME database in 2000". *Nucleic Acids Res* **28**: 304-305. PMID 10592255.

Aided Transport of Nano-Iron in Clay Soils Using Direct Electric Field

L. Hannum[*] and S. Pamukcu[**]

[*]Virginia Tech, 200 Patton Hall, Blacksburg, VA, USA, hannum@vt.edu
[**]Lehigh University, Bethlehem, PA, USA, sp01@lehigh.edu

ABSTRACT

The influence of direct current electric fields on the possible electrokinetic delivery of nano-iron slurry in clay for the purpose of fast and effective remediation of soil contaminants has been tested. Nano-iron can be introduced to soil hydraulically in slurry form, but in tight clay soil, delivery of a uniform distribution of the slurry may be difficult to achieve for effective remediation. Additionally, the limited life of the nano-iron particles, previously shown in field studies of being on the order of 4-8 weeks, further emphasizes the need for the nano-iron particles to reach the contaminated site efficiently before the particles oxidize and become ineffective. This study demonstrates that by integrating electrokinetics with nanotechnology, the transport of nano-particles can be electrokinetically enhanced for subsurface remediation of tight clay soils where transport time and process efficiency may be an issue.

Keywords: electrokinetic remediation, nano-iron, redox, electro-osmosis, electro-migration

1 INTRODUCTION

In most field situations, the contaminants are found adsorbed onto soil surfaces, iron-oxide coatings, soil colloids and natural organic matter. Most contaminants are retained in clay interstices as hydroxycarbonate complexes, or present in the form of immobile precipitates and products in soil pore throats and pore pockets that "lace" the vadose zone. This exacerbates the situation as the available technologies, such as in-situ bioremediation, chemical treatment or the traditional pump-and-treat method may not be able to treat the entire site effectively in low permeability soils.

Electrokinetics is the sustainable process of applying a low current electrical field across a porous medium, such as soil, to induce ionic migration and pore fluid movement. The process is a proven, sustainable technology that can transport liquids and slurries in clay soil at a significantly higher rate than hydraulic methods [1, 2].

In previous investigations [3], the application of direct electric current in soils was observed to contribute to the success of the desired transformation reactions by not only providing the "driving force" necessary for the delivery of active reagents, but also by lowering the energy for the redox reactions to occur. This increase was attributed to the possible over-potential created by double-layer polarization of the clay surfaces leading to spontaneous Faradaic processes under the applied field.

2 BACKGROUND

2.1 Recent Research on Electrokinetics for Soil Decontamination

Remediation of heavy metal contaminants from cohesive, fine-grained soils has presented numerous challenges to geoenvironmental engineers. First, the inherent low permeability of the medium makes traditional pump and treat methods of contaminant removal infeasible. The hydraulic head required to force the flow of water through the soil voids at a predetermined flow rate, would be impractical beyond small-scale laboratory tests. Second, in-situ treatment methods such as bioremediation are generally costly and ineffective in the removal of metallic wastes [1]. Third, heavy metal contaminants have a tendency to sorb onto clay soil particle surfaces, thereby reducing ionic mobility and making conventional treatment methods even more difficult to apply successfully [2].

Recent work on reducing subsurface contamination has been conducted on the effects of changing the oxidation state of heavy metal precontaminated kaolinite clay. It has been shown that ferrous iron Fe (II) can be transported electrokinetically through a Cr (VI) pre-contaminated kaolinite soil bed and result in significantly more Cr (III) present at low to slightly acidic pH regions in the saturated clay [3].

2.2 Nanotechnology in Soil Decontamination

Previous studies have demonstrated that nano scale particles can be used effectively to transform or remove chlorinated hydrocarbons (CHCs) and heavy metals to less toxic or non-toxic hydrocarbons or oxidation states [3,4,5,6,7].

Laboratory tests have proven successful, with greater than 99% removal of TCE using nanoscale iron particles. Pilot field tests of the effectiveness of nano-iron as an environmental catalyst conducted at an industrial research facility waste disposal area showed that nano-iron particles

Cleantech 2007, ISBN 1-4200-6382-0

can remain reactive for 4 to 8 weeks. These tests also determined that nanoparticles had the greatest influence approximately 6 to 10 m around the injection well and could flow with groundwater for over a distance of 20 m. The injection of nano-iron particles in both the field and lab tests showed a drop in the redox potential [7].

2.3 Integration of Electrokinetics and Nanotechnology

Nano-iron can be introduced to soil hydraulically in slurry form and delivered via gravitational flow to the contaminated site. This method has proven successful in remediating a site contaminated by chlorinated solvents [7]. However, in tight clay soil, delivery of a uniform distribution of the slurry may be difficult to achieve for effective remediation due to the limited life of the nano-iron particles before deterioration. Electrokinetics (EK) is a proven, sustainable technology that can transport liquids and slurries in clay soil at a significantly higher rate than hydraulic methods [1,2,3]. The combination of these two technologies has not been studied before and may complement each other in the geoenvironmental field.

3 EXPERIMENTAL PROGRAM

3.1 Nano-iron

Polymer coated dispersed nano-iron, developed at Lehigh University, was used as the iron input into the system. Due to the polymer coating (polyvinyl alcohol-co-vinyl acetate-co-itaconic acid) the iron particles tend to repel each other, and as a result, remain suspended in solution rather than settling over time as occurs with bare nano-iron. The nano-iron particles are zero-valent (Fe^0) and have a diameter less than 100 nm for 80% of slurry batch sample. Zero-valent iron is an effective electron donor regardless of its particle size [7,8].

3.2 Electrokinetic Equipment

A modified electrophoretic (EP) cell (Econo-Submarine Gel Unit, model SGE-020) was used for the experiment and is composed of a rectangular, acrylic, translucent box with a rectangular (20 cm x 18.5 cm) sample tray placed inside it, as shown in Figure 1. To each side of the tray are two liquid chambers, which were used to hold electrolyte salt solution during the experiments.

The tray was modified by the installation of seven platinum wires of 0.25 mm diameter stretched across the tray transversely. The wires were glued to the acrylic tray in three locations using electrically conductive cement adhesive. The wires were labeled E1 to E7 from the anode side, as shown in Figure 1.

The voltage was supplied using an HP E3620A Dual Output DC Power Supply. The reference electrode used to take redox potential readings was an Accumet glass body

Calomel Reference Electrode 13-620-51 (Fisher Chemicals SP 138-500). For pH and temperature readings, the Extech Instruments ExStik II Refillable pH meter PH110, was chosen due to its ability to read the pH of both liquids and wet solids. Finally, the Perkin Elmer AAnalyst 200 Atomic Absorption Spectrometer was used to measure the iron concentration of liquid and soil samples.

3.3 Procedure

Saturated Georgia kaolinite clay and 2.0M NaCl solution was used to make a soil paste with 60% water content. A 2 mm thick uniform layer of the paste was spread onto the EP cell tray over the platinum wires.

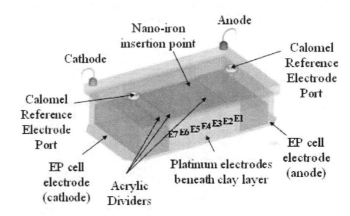

Figure 1: Modified Electrophoretic Cell Schematic

The anode and cathode chambers were filled with 675 mL each of the 2.0 M NaCl solution. The water level was 1.0 mm above the soil bed to ensure saturation throughout the experiment, but to not overtop the acrylic dividers, to ensure that all nano-iron transport in the system occurs through the soil layer. The nano-iron was inserted in the anode side groove, as shown in Figure 1. A constant potential of 5.0 V was applied across the electrodes to remain in the linear range of the power supply and to prevent excessive gas generation caused by electrochemical reactions at the electrodes [3].

Measurements were taken over 46 hours, with more frequent readings earlier on. The redox potentials were measured at each auxiliary electrode (E1-E7) with reference to the Calomel electrode. At the end of 46 hours, three soil samples were extracted above each auxiliary electrode for a total of 21 samples. The samples were diluted and analyzed using U.S. EPA method 7000A for atomic absorption spectrometry [9].

4 RESULTS

4.1 Redox Potential

It was determined that the redox potential reduced across the soil bed from anode to cathode, indicative of nano-iron transport from anode to cathode, as shown in Figure 2. The presence of nano-iron pushed the redox potential to higher positive values at low pH (anode side), while lowering it to higher negative values at high pH (cathode side) than the electrokinetic effects alone. The diffusion of nano-iron without the electrical field showed no activation of the iron, as indicated by little or no change in the redox potential for the diffusion sample in Figure 2. These results showed that nano-iron was both transported and activated by the applied electrical field.

There were marked spatial and temporal oscillations of redox potential in the electrokinetic tests with nano-iron, as shown in Figure 3. These were more pronounced than those in the control or diffusion control tests, indicating enhanced nano-iron movement in the system. Redox potential fluctuated in time early on and then settled to a constant profile after about 17.5 hrs.

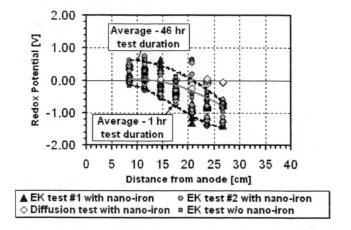

Figure 2: Redox Potential vs. Distance - Test Comparison

The post-test stoichiometric analysis of the measured quantity of Fe^{+2} at the sampling locations versus the corresponding redox potentials was compared to the Nernst equation predictions for half reaction. The data agreed well with the Nernst prediction at higher pH regions and displayed positive shift from the Nernst relation at low pH regions, possibly due to clay surface polarization. A comparison of the nano-iron movement between the electrokinetically enhanced movement and the diffusion test in Figure 4 shows the effect of 5.0 V potential applied to the system at 46.0 hrs. As observed, the electrokinetically enhanced transport activated the nano-iron corrosion at the cathode side, while the diffusive system did not show any sign of corrosion.

4.2 pH

The iso-electric point (IEP) of the nano-iron, which is the critical value at which the net surface charge is zero, occurs at a pH of 8.3 and was found to be independent of iron concentration [8]. The iron nanoparticles will be positively charged when the pH is less than 8.3 and

negatively charged for pH values greater than 8.3. The zeta potential of kaolinite is also a strong function of the pH of the system. Typically, the zeta potential will vary from -50 mV in alkaline conditions to 0 at approximately a pH of 2.0. Therefore, for pH values between 2.0 and 8.3, the nanoparticles will be attracted to the negatively charged clay surfaces and will stay on the surface, contributing to surface reactions. However, for pH values less than 2.0 or greater than 8.3, the iron will be released into the pore fluid and transported through the porous media.

Comparing the pH distributions shown in Figure 5, pH greater than 8.3 typically occurs at electrodes E6 and E7. pH of less than 2.0 was not experienced in the system. Therefore, it is expected that increased nano-iron transport will occur around E6 and E7, which may explain the increased negative redox potentials of E6 and E7 in Figures 2 and 3.

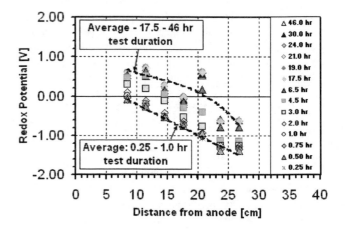

Figure 3: Redox Potential vs. Distance – Time Effects

The electrokinetically enhanced transport and activation of nano-iron were evidenced by the higher negative potentials achieved in nano-iron specimens on the cathode side, at the same pH for the same test durations of the control specimens. Figure 6 shows the variation of redox potential (E_h) versus pH. The variation of E_h vs. pH in the electrokinetics only specimen follows closely the lower bound for electrolysis of water, while when nano-iron is transported using electrokinetics the system displays higher oxidation potential for iron.

5 CONCLUSIONS

Analyte, containing polymer coated nano-iron particles, was electrokinetically injected into the soil to demonstrate the enhanced transport of nano-iron in clay soils using an applied electric field.

The significance of the results obtained in this study is two fold: 1) electrokinetically enhanced transport of highly reactive nano-particles designed for subsurface remediation is possible in tight clay soils where transport time and process efficiency are critical; 2) direct redox measurements of the host medium in an integrated application of two technologies as described above can provide adequate evidence of their performance in the field.

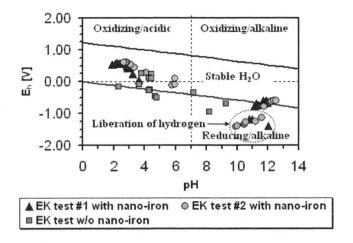

Figure 6: Redox Potential vs. pH – Test Comparison

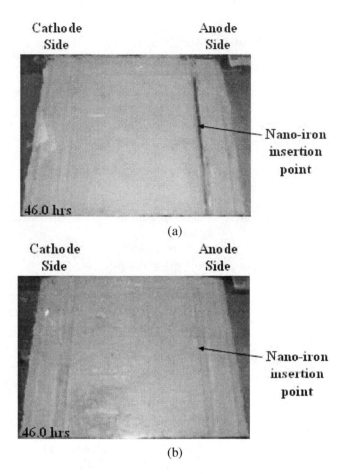

(a)

(b)

Figure 4: Soil Bed at 46.0 hrs
(a) Diffusion Only (b) Electrokinetically Enhanced

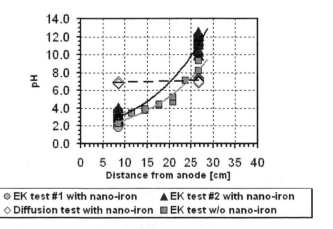

Figure 5: Spatial pH Distribution – Test Comparison

REFERENCES

[1] Reddy, Krishna R. and Parupudi, Usha S. (1997). "Removal of Chromium, Nickel, and Cadmium from Clays by In-Situ Electrokinetic Remediation." *J. Soil Contamination*, 6(4):391-407.

[2] Yeung, Albert T. and Datla, Subbaraju. (1995). "Fundamental Formulation of Electrokinetic Extraction of Contaminants from Soil." *Can. Geotech. J.*, 32:569-583.

[3] Pamukcu, Sibel, Weeks, Antoinette, and Wittle, Kenneth J. (2004). "Enhanced Reduction of Cr (VI) by Direct Electric Field in a Contaminated Clay." *Environ. Sci. Technol.*, 38, 1236-1241.

[4] Elliott, D.W. and Zhang, W. (2001)."Field assessment of nanoscale bimetallic particles for groundwater treatment." *Eviron. Sci. Technol.*, 35, 4922-4926.

[5] Engelmann, M.D., Doyle, C.J., and Cheng, I.F. (2001)."The complete dechlorination of DDT by magnesium/palladium bimetallic particles." *Chemosphere*, 43, 195-198.

[6] Li, F. and Vipulanandan, C. (2003). "Microemulsion Approach to Nanoiron Production and Degradation of Trichloroethylene." *Proceedings of CIGMAT-2003 Conference*, Houston, TX, II-1-3.

[7] Zhang, Wei-xian. (2003). "Nanoscale Iron Particles for Environmental Remediation: An Overview." *J. Nanoparticle Research.* 5:323-332.

[8] Sun, Y.P., Li, X.Q., Cao, J., Zhang, W.X , and Wang, H.P. (2006). "Characterization of Zero-valent Iron Nanoparticles." *Advances in Colloid and Interface Science*, 120:47-56.

[9] U.S. Environmental Protection Agency. "Test Methods for Evaluating Solid Waste, Physical/ Chemical Methods," 3.3, 2004.

Nano-Catalysts and Colloidal Suspensions of Carbo-Iron for Environmental Application

K. Mackenzie, H. Hildebrand, F.-D. Kopinke

Helmholtz Centre for Environmental Research – UFZ, Department of Environmental Technology,
D-04318 Leipzig, Germany

ABSTRACT

The present paper shows two different examples for *ex-situ* and *in-situ* water treatment using nano-sized materials. Two novel colloidal particles have been developed and tested in initial studies for i) the *in-situ* generation of sorption/reaction barriers based upon zero-valent iron on sorption-active carbon carriers (carbo-iron) for application in subsurface water treatment and ii) the selective catalytic elimination of halogenated hydrocarbons using Pd on magnetic carriers as agent for *ex-situ* waste water treatment.

The common ground for both applications is the utilization of nano-particles for dehalogenation reactions in the aqueous phase in order to minimize mass transport limitation and therefore permit high decontamination rates. These water treatment applications have been selected for presentation in order to show the wide applicability of nano-sized materials in environmental technology.

Keywords: zero-valent iron, carbo-iron, dechlorination, Pd catalyst, sorption-assisted reaction, water treatment

1 INTRODUCTION

During the last ten years, intensive research has been carried out on the development of suitable materials for reactive barriers in subsurface application. Permeable reactive barriers (PRBs) on the basis of zero-valent iron have already been constructed at more than 120 sites worldwide and count as an approved technology [1]. The halogenated organic compounds (HOCs) are removed from the groundwater according to the equation: $R\text{-}X + Fe^0 + H_2O \rightarrow Fe^{2+} + R\text{-}H + OH^- + Cl^-$.

Recently, the application of nano-sized particles has been advanced, because colloidal solutions of metal particles can easily be injected into the aquifer without the necessity of extensive underground work. This fact and the potential of nano-sized zero-valent iron (ZVI) to migrate in groundwater to form a reactive zone have stimulated the research on nano-particles [2].

Literature studies on nano-sized ZVI reveal on the one hand desirable properties such as high reactivity and injectability. On the other hand, undesirable properties are also described, such as a tendency to agglomeration and untimely sedimentation and hence a limited mobility under aquifer conditions [3]. The hydrophilic nature of the iron surface makes it not well suited for source remediation.

The utilization of hydrophilized carriers or poly(acrylic acid) helped to mobilize colloidal suspensions of ZVI [3].

The addition of emulsifying substrates forms emulsion droplets which contain the ZVI particles in water surrounded by an oil-liquid membrane. Emulsified zero valent iron (EZVI) has been successfully tested in a field-scale demonstration [4]. EZVI proved miscible with the organic phase; therefore, contact between the reaction partners was realized. Nevertheless, the addition of chemicals to the reactive agent is necessary in order to achieve the close contact between the reactants. In addition, the emulsified particles show even less mobility.

The objective of our work is to develop materials with tailored properties for *in-situ* generation of sorption/reaction barriers for subsurface water treatment at low cost and without the need for additional chemical supply. The surface properties should be suitable for both plume and source treatment.

ZVI in every form and particle size has its limitations concerning the pollutant spectrum which can be treated; e.g. iron completely fails for the dehalogenation of aromatic substances, such as PCBs, halogenated benzenes and phenols. However, the utilization of catalytic hydrodehalogenation with Pd catalysts can solve this problem. Palladium catalysts have proved to be well suited for promoting hydrodehalogenation reactions in the aqueous phase according to the equation $C_lH_mX_n + n\,H_2 \rightarrow C_lH_{m+n} + n\,HX$ [5]. The present paper aims at a treatment technique designed for special industrial wastewater contaminated with only small amounts of halogenated hydrocarbons – amounts which are nevertheless large enough to make a discharge into municipal sewage works impossible. The consequence is the necessity of expensive and energy-intensive incineration of aqueous waste. Therefore, especially for medium-sized enterprises, a decentralized selective wastewater dehalogenation treatment brings not only ecological credibility but also an important economic advantage.

Our research in the field of *ex-situ* water treatment follows the aim of detoxifying the water by a *selective* destruction of the HOCs in reductive hydrodehalogenation reactions on nano-catalysts containing palladium. By detoxification we mean that the persistent HOCs are converted into dehalogenated organic compounds which can easily be removed by biodegradation in a wastewater treatment plant.

2 EXPERIMENTAL

Carbo-iron. The support material activated carbon (AC, e.g. SA Super from Norit), was ground in the presence of deionized water (horizontal mill 200 AHM, Alpine

Hosokawa). 500 g of this AC (D_{50} = 0.8 m, D_{90} = 1.6 m) and 1080 g Fe(NO_3)$_3$ · 9H_2O (= 150 g Fe^{3+}) were mixed in 1 L of water and shaken. The water was removed and the re-ground residue was reduced at 550 to 580°C under H_2 atmosphere for 8…10 h. Before application as a dehalogenation agent, the freshly produced carbo-iron was subject to a mild deactivation step. In water, carbo-iron can easily be extracted from unwanted by-products by magneto-separation. The Fe^0 content of the composite material was 20 wt-% (measuring the H_2 evolved by GC/WLD after addition of a stoichiometric excess of HCl).

Pd/Fe₃O₄. The magnetite particles (Aldrich, 20-30 nm) were spiked in aqueous solution with Pd in the form of Pd(ac)$_2$. After decolorization of the orange solution (visible after magnetoseparation), H_2 treatment ensured complete reduction to Pd^0. The catalysts contained 0.1 wt-% Pd.

Dehalogenation studies. The dehalogenation activities of both types of particles were studied in batch and column experiments using probe HOCs such as carbon tetrachloride, trichloroethene, 1,2-dibromoethene and, for the catalytic reaction, also chlorobenzene. Kinetic studies were carried out by following the educt disappearance and the product evolution (HOCs by total extraction and GC/MS analysis of the extracts; methane, ethene and ethane by analysis of headspace concentrations, and halogenide by IC analysis of the reaction solution).

3 RESULTS AND DISCUSSION

Carbo-iron. Based on approved methods in abiotic water treatment by PRBs (pollutant sorption and pollutant destruction by chemical reaction) this study focused on AC as sorbent and ZVI as reactant. Both materials were considered as basis for a new composite material to be provided in such a form that stable colloidal solutions can be introduced into aquifers via injection wells.

Because pure nano-iron is neither sufficiently mobile nor well-suited for source remediation, we decided to follow an alternative approach. This approach is based on finely-ground AC with a D_{50} particle size of 0.8 m which is quasi-soluble, forming stable colloidal solutions in water over a wide concentration range. We gave this material additional reactivity by impregnating it with iron salts then reducing it at elevated temperatures with hydrogen. The procedure results in AC with ZVI nano-clusters. We call this new composite material carbo-iron. With this reagent a new remediation strategy can be followed – the *in situ* generation of a permeable sorption/reaction barrier in contaminated aquifers.

Figure 1 shows that both components (AC and Fe^0) are in close contact. Iron forms predominantly clusters in the size range of 20-50 nm. However, larger crystallites (> 100 nm) can also be found. The TEM bright-field image and the selected area diffraction (SAD) pattern study of the Fe particles (e.g. the larger black areas on the right-hand side of the picture) prove the crystallinity of those Fe^0 particles.

Our concern was that a close contact of AC and Fe^0 does not automatically mean that hydrophobic substances which are sorbed to the AC carrier are easily available for reduction at the Fe centers. However, dehalogenation experiments using carbon tetrachloride, trichloroethene, 1,1,1,2-tetrachloroethane and 1,2-dibromoethene proved the suitability of carbo-iron as dehalogenation agent in the aqueous phase.

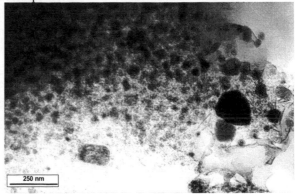

Figure 1: Transmission electron microscopy (TEM) bright-field image of carbo-iron (20 wt-% Fe^0)

Pure nano-sized ZVI can be produced by reduction of iron precursors using $NaBH_4$ in aqueous solution (Fe^B) or H_2 in the gas phase at elevated temperatures (RNIP). The differences between these two types of nano-particles concerning structural properties and reactivity in dehalogenation reactions of HOCs have been intensively studied [6,7]. The chemical nature of the iron surface was found to be very important for the particle reactivity (B-doped surface versus predominantly Fe oxide shells). Magnetite shells are believed to hinder the electron transfer from the iron to the HOC [7]. The synthesis of carbo-iron is similar to the RNIP production. Therefore, particular attention was turned to a mild deactivation step avoiding the formation of a retarding magnetite shell around the Fe nano-structures. According to our experience the deactivation step is very important for the reactivity and longevity of the ZVI.

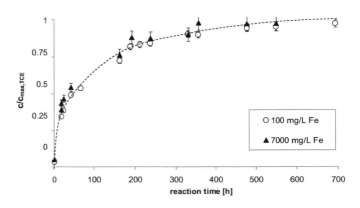

Figure 2: Observed reaction kinetics for the dechlorination of trichloroethene to ethene/ethane using various concentrations of carbo-iron. ($c_{0,TCE}$ = 30 mg L^{-1}, 0.1 M NH_3/NH_4^- buffer, yields: $c_{max,chloride}$ = 100%, $c_{max,C2\ hydrocarbons}$ = 40...45%)

In the batch experiments all of the mentioned HOCs were dehalogenated in a manner comparable to using pure nano-ZVI. The observed reaction rate using 7g L^{-1} Fe^0 was in the same order of magnitude as known from the suspensions of conventional nano-iron with the same Fe concentration. However, the reductive dehalogenation takes place with a comparable reaction rate apparently independent of the carbo-iron concentration. Figure 2 shows that widely varying carbo-iron concentrations seem to have no drastic effect on the observed reaction rate.

Equations (1) to (4) show, that under the plausible assumptions that almost all TCE is adsorbed on the AC ($m_{HOC,adsorbed} >> m_{HOC,dissolved}$) and that the reaction takes place on the surface, the reaction rate is determined by the Fe^0 content of the carbo-iron and not by the total Fe^0 concentration, with $x_{Fe} = c_{Fe}/c_{AC}$ as the Fe^0 content of AC carrier and $c_{HOC,adsorbed} \approx c_{HOC,total}/c_{AC}$ in [mg$_{HOC}$ g$_{AC}$$^{-1}$].

for ZVI:

$$\frac{dc_{HOC}}{dt} = k_{Fe} \cdot c_{Fe} \cdot c_{HOC,dissolved} \qquad (1)$$

$$\ln(\frac{c_{HOC,0}}{c_{HOC,t}})_{total} = k \cdot c_{Fe} \cdot t \qquad (2)$$

for carbo-iron:

$$\frac{dc_{HOC}}{dt} = k'_{Fe} \cdot c_{Fe} \cdot c_{HOC,adsorbed} \qquad (3)$$

$$\ln(\frac{c_{HOC,0}}{c_{HOC,t}})_{total} = k'' \cdot x_{Fe} \cdot t \qquad (4)$$

Therefore, it is plausible that the reaction rate in the carbo-iron is independent of the Fe concentration applied as long as enough reductant is present. In terms of longevity in plume control this does not bring any advantage over pure nano-iron, but in terms of material reduction for less extensive plumes it certainly can. In addition, the AC carrier has a collecting function for hydrophobic pollutants.

The mobility of carbo-iron has been examined in column tests (column length l = 75 cm, c$_{carbo-iron}$ = 100 mg L^{-1}). Particles which passed through the column were regarded as mobile. In order to describe the mobility of the particles in the columns we defined the mobility as an operational parameter:

$$m_{mobile}[\%] = \frac{m_{carbo-iron,out}}{m_{carbo-iron,in}} \cdot 100\,\% \qquad (5)$$

Comparison of carbo-iron mobility to that of pure nano-iron already showed a marked increase in the transport length of the particles. However, addition of polyanionic stabilizers, such as humic acids (see Fig. 3, right column), results in longer transport lengths and a more homogeneous formation of the reaction zone. Experiments with carboxymethyl cellulose are currently in progress. First results show that carboxymethyl cellulose is able to

stabilize carbo-iron particles even better. Both stabilizing agents are environmentally benign. Humic acids are natural materials; carboxymethyl cellulose is very inexpensive and commercially available in food grade purity.

Figure 3:
Mobility of carbo-iron
Left: unstabilized particles with m$_{mobil, (l = 75 cm)}$ = 85%
Right: stabilized with 5 wt-% humic acid resulting in m$_{mobil, (l = 75 cm)}$ = 90%.

The penetration of the reducing agent into DNAP phases is regarded as pre-condition for source remediation. Carbo-iron has with activated carbon *a priori* a hydrophobic carrier. Figure 4 shows the different wetting behaviour for the more hydrophilic nano-iron particles (middle) in comparison to carbo-iron particles (right). The samples were slightly shaken. The figure shows the easy penetration of the hydrophobic carbon particles into the organic phase.

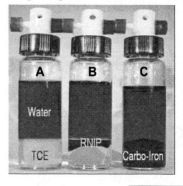

Figure 4:
Comparison of wetting properties of ZVI reagents in a TCE DNAPL phase under dyed water (A)
(B): Reactive nano iron particles (RNIP, from Toda Kogyo Corp.)
(C): Carbo-iron (ZVI on activated carbon)

Pd/Fe₃O₄ nano-particles. Let us come back to the other example of the application of nano-particles for environmental technology: the *ex-situ* water treatment using Pd catalysts. The specific catalyst activity can be expressed as follows

$$A_{Pd} = \frac{V_{water}}{m_{Pd} \cdot \tau_{1/2}} = \frac{\ln(c_{t1}/c_{t2})}{\ln 2 \cdot c_{Pd} \cdot (t_2 - t_1)} \qquad (6)$$

(in [L g^{-1} min^{-1}])

with V_{water} as the water volume applied, m_{Pd} and c_{Pd} as the Pd mass and concentration and $\tau_{1/2}$ as the HOC's half-life, obtained from its disappearance kinetics. t_1 and t_2 are two arbitrarily chosen sampling times; c_{t1} and c_{t2} the corresponding HOC concentrations [8]. The value of $A_{Pd,i}$ is equivalent to a first-order rate coefficient according to $k_{obs} = \ln 2 \cdot A_{Pd,i} \cdot c_{Pd}$ [s^{-1}]. Equation (6) helps to evaluate the effectiveness of Pd catalysts. Pd activities of 0.1 L g^{-1} min^{-1} were found in field tests for groundwater treatment using granular Pd catalysts such as Pd on γ-alumina [9].

However, the inherent Pd activity is higher by several orders of magnitude. The catalytic activity of the described particles for model HOCs such as chlorobenzene or TCE is extremely high: up to 3000 L/(g$_{Pd}$ min). This is several orders of magnitude higher than can be achieved in fixed-bed arrangements of granular catalysts for groundwater treatment. If in environmental applications only a fraction of this reactivity can be reached, Pd catalysts will eventually make their way into environmental technology as an approved tool.

We regard magnetite particles as attractive carriers when adding hydrogen or other hydrogen sources because they are quite resistant against various milieu parameters. Hydrogen has a water solubility of $S_{H2}^{15°C} = 0.84$ mM at $p_{H2} = 100$ kPa. This may be insufficient for the treatment of highly contaminated wastewaters. For example, from the stoichiometry of the hydrodechlorination reaction it follows that only up to 28.6 mg L^{-1} of trichloroethene can be reduced with dissolved hydrogen ($C_2HCl_3 + 4 H_2 \rightarrow C_2H_6 + 3 HCl$). In practice, however, much higher HOC concentrations may occur. Formic acid as an alternative H-source was found to be as reactive as H_2 under acidic conditions, but less reactive under alkaline conditions. In the framework of our studies in the field of wastewater treatment using nano-sized catalysts for clean-up of HOC-contaminated wastewater, formic acid has already been tested and found to be suitable as a H-donor. Nano-sized Pd-magnetite catalysts can even be stored as a suspension in formiate solution (pH > 8) without loss of activity. This mixture can be seen as part of a wastewater-treatment toolkit. Studies involving other H-donors are in progress.

The catalytic material was tested with various technological options, including batch and continuous flow reactors, depending on the type and amount of wastewater to be treated. The performance of a laboratory flow-through reactor revealed a Pd activity of > 3000 L/(g min).

The extraction of the magnetite particles from the reaction suspension by means of magnetoseparation was highly effective. The results of the laboratory tests are altogether very promising.

Due to the fact that catalysts in the nano-scale are still novel tools in water treatment, it is necessary to characterize and evaluate both their potential and any possible risks they may pose. The particle size not only determines the catalytic activity; it may also be possible for such tiny particles to pass through cell membranes, which bears the threat of their possible chemical reactions with cell constituents. Therefore, investigations concerning size-dependent cell toxicity of various types of nano-catalysts are necessary. The chemical potential of such reactive systems is enormous, but it can only be exploited responsibly on the basis of a thorough ecotoxicological analysis of possible risks in the framework of an interdisciplinary approach. The evaluation of the risks of such nano-particles for living cells plays an important role within the presented studies; of course we hope that the benefits will be found to outweigh the potential risks.

4 CONCLUSIONS

With carbo-iron (20 wt-% ZVI), a novel material is provided which forms stable colloidal solutions up to high concentrations and Ca^{2+} concentrations. Stabilizers such as polyanionic substances can help to keep the particles mobile under aquifer conditions. Humic acid as a natural product can act as stabilizer; however, carboxymethyl cellulose is able to stabilize much higher particle concentrations of carbo-iron and facilitates long transport distances.

Carbo-iron has proved its dehalogenation activity for chlorinated and brominated C_1- and C_2-hydrocarbons in aqueous solution. Furthermore, the hydrophobic nature of carbo-iron permits its distribution in NAPL phases where dehalogenation can be performed without the application of additives. This is a great advantage over pure nano-sized ZVI, which requires additives for source treatment.

Extremely active catalysts for cyclic batch applications can be produced on the basis of ferromagnetic nano-sized carrier colloids containing only traces of Pd (0.1 wt-%). Nano-sized Pd catalysts have been successfully tested for waste water problems at the laboratory scale. Dehalogenation using pure Pd colloids or Pd on colloidal supports shows the true inherent activity of Pd clusters which is several orders of magnitude higher than reached in fixed bed arrangements due to minimized mass transfer limitations. Magnetite particles are suitable catalyst carriers because of their high resistance against milieu parameters. Their magnetism allows a complete extraction of the nano-particles from the treated water by magneto-separation. Beside the study of the opportunities and the high potentials of such catalyst systems, another objective of our work is an evaluation of their environmental risks.

REFERENCES

[1] Simon, F.-G., Meggyes, T.: *Land Contamination & Reclamation*, 8, 103-116, 2000.

[2] Zhang, W.-x. and Elliot, D. W.: *Remediation* 16(2), 7-21, 2006.

[3] Schrick, B., Hydutsky, B.W., Blough, J.L. and Mallouk, T.E.: *Chem. Mater.*, 16, 2187-2193, 2004.

[4] Quinn, J., Geiger, C., Clausen, C., Brooks, K., Coon, C., O'Hara, S., Krug, T., Major, D., Yoon, W.-S., Gavaskar, A., Holdsworth, T.: *Environ. Sci. Technol.*, 39, 1309-1318, 2005.

[5] Matatov-Meytal, Y., Sheintuch, M.: *Ind. Eng. Chem. Res.* 37, 309-326, 1998.

[6] Nurmi, J., Tratnyek, P., Sarathy, V., Baer, D., Amonette, J., Pecher, K., Wang, C., Linhan, J., Matson, D., Penn, R., Driessen, M.: *Environ. Sci. Technol.*, 39, 1221-1230, 2005.

[7] Liu, Y., Majetich, S., Tilton, R., Sholl, D., Lowry, G.: *Environ. Sci. Technol.* 39, 1338-1345, 2005.

[8] Mackenzie, K., Frenzel, H., Kopinke, F.-D.: *Appl. Catal. B: Environmental*, 63, 161-167, 2005.

[9] McNab, W. W., Ruiz, R., Reinhard, M.: *Environ. Sci. Technol.* 34, 149-153, 2000.

An integrated NMR/nanosensor system for sensitive detection of environmental toxins and harmful microbes

J. Manuel Perez and Charalambos Kaittanis

Nanoscience Technology Center, Department of Chemistry and Biomolecular Science Center
12424 Research Parkway, Suite 400, Orlando, FL 32826, jmperez@mail.ucf.edu

ABSTRACT

The use of magnetic nanoparticles in conjunction with NMR detection technologies has lead to significant improvements in the detection of various molecular targets with high sensitivity and selectivity in complex media. Recently, superparamagnetic iron oxide nanosensors have been designed to quantify various biomolecular targets, demonstrating high sensitivity and specificity [1]. Using this technique various targets such as nucleic acids (DNA and mRNA), proteins and even viruses have been detected, with a sensitivity in the low femtomole range (0.5 – 30 fmol) for DNA. The observed changes in T2 are directly proportional to the concentration of the target in solution and can be easily detected by existing magnetic resonance (NMR/MRI) techniques. In this report, we present recent work geared towards the detection of pathogens and toxins.

Keywords: pathogens, toxins, sensing, iron oxide, magnetic nanoparticles

1 INTRODUCTION

The principle underlying the detection mechanism of these magnetic relaxation nanosensors is based on their ability to switch between a dispersed and clustered (or assembled) state upon target interaction, with a concomitant change in the spin-spin relaxation time (T2) of the solution's water protons (Figure 1). This NMR-based detection approach requires no separation; it is robust to interferences and has been performed in whole blood, lipid emulsion, and tissue culture media. These magnetic nanosensors are composed of a 4-8 nm core of superparamagnetic iron oxide core surrounded by a crosslinked dextran coating, resulting in a nanoparticle of 30-40 nm in size. In order to enhance their stability and functionality, the dextran coating is crosslinked with epichlorohydrin and followed by ammonia treatment to incorporate accessible functional amino groups. These nanoparticles exhibits great stability, even under extremely harsh conditions, such as a 30-minute incubation at 120 °C, with neither size nor coating alterations. Conjugation of biomolecules such as peptides, proteins, antibodies and oligonucleotides to the nanoparticles, using various crosslinking agents and conjugation chemistries, creates target specific nanosensors that have been utilized in the detection of various targets.

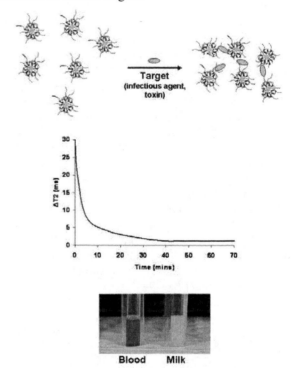

Figure 1. Magnetic nanosensor principle. The detection of the target via magnetic relaxation is fast and can be done in opaque media.

In particular, pathogen- and toxin-specific nanosensors can be designed by conjugating the corresponding antibody on the nanoparticle via Protein G. Additionally, fluorescent dyes have been conjugated to the magnetic nanosensors allowing for multimodal (magnetic and optical) sensing.

2 VIRAL SENSING

One of the first applications of the use of these magnetic nanosensors for the identification of pathogens, was the sensing and quantification of viral particles in serum [2]. It was hypothesized that the presence of multiple copies of a viral protein on the viral coat facilitates multivalent interactions between the multi-epitope virions and the magnetic nanosensors. This feature would enhance nanoassembly formation; therefore, promoting high sensitivity for viral detection. Having this in mind, magnetic nanoparticles were designed to sense a specific virus, such as herpes simplex virus-1 (HSV-1) and adenovirus-5 (ADV-5). Using these viral-sensing magnetic nanosensors, low levels of virus (5 virions per 10 μL) were detected in serum. The identification of viral particles with magnetic nanosensors outperforms the detection of viruses with contemporary PCR techniques, providing quick and easy-to-read results with minimal artifacts, and without the need for protein removal or sample amplification. In the future, acknowledging the potential of this assay, viral-specific nanoparticles can be designed for the detection and determination of the localization of viruses *in vivo*, serving as viral-specific MRI agents.

3 BACTERIAL SENSING

Most recently, we have been able to develop bacteria-specific magnetic nanosensors [3]. For these studies, we used *Mycobacterum avium* spp. *paratuberculosis* (MAP), as our model organism. We selected this bacterium because its growth in culture is difficult, slow and its identification with current methods is not easy. Furthermore, this bacteria is known to be present in the blood and milk of cattle and it is known to be responsible for Johne's disease in cattle and presumably Crohn's disease in humans. The development of bacterial-specific nanosensors has allowed the sensing of this specific bacterial target (MAP) in complex media (whole milk and blood) with high specificity and sensitivity. The MAP nanosensors used in this study were prepared by conjugating anti-MAP antibodies to superparamagnetic iron oxide nanoparticles via Protein G. Upon addition of the bacteria, formation of the bacterial-induced nanoassembly was detected via magnetic relaxation measurements almost immediately in both phosphate buffer and milk. The specificity of our nanosensors towards the bacteria (MAP) was tested by comparing the sensors response to various other types of bacteria. Figure 2 shows that the sample that had MAP alone (115 Colony Forming Units [CFUs] in 10 μL) had the highest change in T2, while the samples that had other bacteria (~10^6 CFUs in 10 μL) demonstrated minimal T2 changes. More importantly, the sample containing a mixture of bacteria, including MAP (77.5 CFUs in 10 μL), was identified as MAP-positive, despite the presence of interference, underlying the specificity of our nanosensors. Detection and quantification of MAP in milk was done by incubating MAP-spiked whole milk with MAP nanosensors (2.1 μg Fe/μL). We found that the change in T2 was indirectly proportional to the MAP concentration, supporting our proposed detection model. Reliable quantification of MAP from 15.5 to 775 CFUs (R^2=0.93) was achieved after a 30-minute incubation at room temperature (Figure 3). In control experiments using

nanoparticles with no MAP antibody minimal changes in T2 were observed. This approach, apart from sensitive and fast, is independent of the sample's optical properties, requires minimum sample preparation and can be used at the points-of-care. This method provides a novel approach for microbial detection that can potentially expedite decision making in a broad range of fields including the clinical, environmental and agricultural sectors.

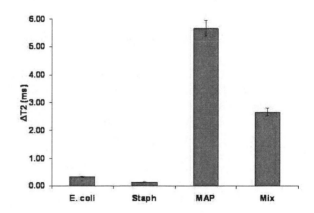

Figure 2. Specificity of the MAP nanosensors

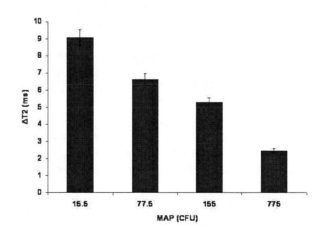

Figure 3. Concentration dependent behavior of MAP magnetic nanosensors.

4 POTENTIAL FOR SENSING TOXINS

The fact that the developed magnetic nanosensors can facilitate rapid detection of a molecular target without extensive sample preparation or target amplification in turbid samples and cam work in the presence of interferences makes them attractive for developing sensing technologies to monitor the presence of toxins and pathogens in environmental and clinical samples. Complex and opaque media such as blood, cell suspensions, culture media, lipid emulsions and even whole tissue can be used. Additionally, there is no need for sample immobilization onto a flat surface, such as in the case of microarrays, facilitating faster hybridization and monitoring of binding kinetics. Current work is geared toward developing technologies for the detection of ricin toxin, anthrax, and other highly pathogenic bacteria, among others applications.

5 MAGNETIC RELAXATION DETECTORS

Using high-throughput NMR, hundreds of environmental samples can be screened, drastically speeding up the screening of a library of nanoparticles for detecting a particular target (toxin or pathogen). In addition, high-throughput NMR can be used to screen multiple samples collected at various locations for the presence of a particular target in a matter of hours. It would be difficult to implement a portable and deployable system for environmental sensing and point-of-care diagnostics that required a bulky detection system (NMR or MRI). Since spectroscopic or 3D information is not required to measure the target-induced changes in magnetic relaxation, a simple magnetic relaxometer (0.47 T Bruker MiniSpec) will be sufficient. In fact, most of the work published in the literature has been done using a 0.47 T Bruker MiniSpec. However, a briefcase-sized NMR relaxometer would be ideal for this application. To advance this work, we are collaborating with a company to develop such a device and initial testing reveals that it can sense for various toxins with high sensitivity and selectivity in a matter of minutes.

REFERENCES

[1] (a) Perez, J. M.; Josephson, L.; O'Loughlin, T.; Högemann, D.; Weissleder, R. *Nat Biotechnol.* 2002, 20(8): p. 816-20. (b) Perez, J.M., Josephson, L., Weissleder, R *ChemBioChem* 2004; 5: 261-264.

[2] Perez, J. M.; Simeone, F. J.; Saeki, Y.; Josephson, L.; Weissleder, R. *J. Am. Chem. Soc.* 2003, 125(34), 10192-10193.

[3] Kaittanis, C., Saleh, A. N., Perez J.M., *NanoLetters* 2007, 7(2), 380-383

Acknowledgments: We gratefully acknowledge support through the National Cancer Institute by a Career Award to JMP (CA101781).

Nanotechnology and the Global Poor:
United States Policy and International Collaborations

R. Rodrigues*, T. Lodwick**, R. Sandler***, and W.D. Kay****

*Santa Clara University, School of Law, Santa Clara, CA, USA, rubenr@alum.mit.edu
**Nanotechnology and Society Research Group (NSRG),
Northeastern University, Boston, MA, USA, lodwick.t@neu.edu
***NSRG, Northeastern University, Boston, MA, USA, r.sandler@neu.edu
****NSRG, Northeastern University, Boston, MA, USA, w.kay@neu.edu

ABSTRACT

Nanotechnology and nanomanufacturing have tremendous potential for benefiting the global poor—the approximately 2.77 billion people in the world that live on less that 2 dollars per day (purchasing power parity). For example, nanotechnologies may help provide reliable local energy production and potable water availability, increased agricultural efficiency, inexpensive medical diagnostics and treatments, and greater access to technology and information more generally [1]. This paper examines existing and potential pathways for promoting nanotechnology and nanomanufacturing that benefit the global poor either by directly meeting their needs or supporting nascent industries in developing countries. Informal international collaborations as well as formal international research partnerships are discussed, as is the role of international organizations. However, special attention is given to U.S. policy. Recommendations regarding intellectual property licensing, incentivizing research on pro-poor nanotechnologies, and promoting collaborations between U.S. and developing world researchers are made. In the long run, a nanotechnology research and development strategy conducive to realizing the possibilities for nanotechnology to benefit the global poor might constitute an effective form of foreign aid that would also benefit the U.S. by promoting stability and security in developing nations and creating new markets for U.S. companies.

Keywords: global poor, United States, international collaborations, foreign aid, nanotechnology

1 BARRIERS TO NANOTECHNOLOGY IN THE DEVELOPING WORLD

Perhaps the most basic barrier to conducting nanotechnology research is equipment costs. One way for a researcher in a developing nation to reduce these costs is by collaborating with other researchers, either from another developing nation (South-South collaboration) or from a developed nation (North-South collaboration). Each type of partnership has benefits and limitations. While South-South research is more likely to focus on developing world problems, resources may still be constrained; and while North-South collaboration enables access to high-tech facilities, little incentive exists for developed world researchers to partake in such collaborations.

The lack of incentives for researchers in the developed world to aid the developing world is a critical barrier to diffusing nanotechnology. There is little or no financial incentive for developed world researchers to make the required effort to work with developing world researchers. Similarly, there are very few funding sources that exist to provide incentives for developed world researchers to independently address the social problems facing the developing world (pro-poor research).

Relevant to the issue of incentives is patents. Patents are being aggressively sought with respect to all aspects of nanotechnology research, including equipment to conduct research. Such layered patenting drives costs prohibitively high for developing world researchers [2]. As with pharmaceuticals, the exclusive rights associated with patents also are likely to be a major barrier to bringing nanotechnology to the developing world.

Among the non-technical barriers to nanotechnology research and applications in the developing world are bad governance, anti-science attitudes, lack of health care, and lack of resources in all levels of education. In cases where these problems are severe, applying pro-poor technologies (as part of broader development strategies) will probably be more feasible than attempting to build scientific capacity.

An additional, substantial non-technical barrier is the political realities of foreign aid. There is often conflict over how and where foreign aid should be spent, and in the past aid to develop scientific capacity has been criticized when more fundamental needs go unmet [3]. Moreover, nanotechnology remains unproven and the risks associated with nanomaterials largely unknown, which is particularly problematic for developing nations with meager regulatory capacities. Furthermore, the U.S. is a global competitor in nanotechnology, so there is a focus within U.S. nanotechnology policy on assuring status as a global leader, rather than helping others.

2 PATHWAYS FOR NANOTECHNOLOGY

A United Nations Millennium Project report suggested that, "Nanotechnology is likely to be particularly important in the developing world, because it involves little labor, land, or maintenance; it is highly productive and inexpensive; and it requires only modest amounts of materials and energy." [4]

Nanotechnology is already a global phenomenon. Developing and poorer countries are contributing to worldwide nanotechnology research, and some are even moving toward commercialization in certain areas. Many developing nations, particularly Brazil, Mexico, Argentina, and South Africa, have official nanotechnology research initiatives. Some poorer nations, such as Nigeria and Colombia have seen individuals contribute research on nanotechnology. Countries of all economic levels all over the world are participating in nanotechnology research.

2.1 Nanotechnology Research in the Developing World

The South African Nanotechnology Strategy (SANS) is an example of developing world nanotechnology research that engages both the social and industrial potential for nanotechnology [5]. South Africa's pro-poor research seeks to not only apply nanotechnology towards the problems of poor South Africans, but also to the greater problems faced across all of southern Africa. Technical projects SANS currently works on include those involving nano-membranes for delivering clean water, improved solar cells, and the development of fuel cells. For example, South Africa funded a pilot nanofiltration water treatment plant which proved to be a successful and sound alternative for water filtration in rural South Africa [6].

South Africa's industrial research seeks to apply nanotechnology in areas such as nanocomposites, nanomaterials, and other areas that can enhance Africa's resource-based industries. South Africa already has a commercialized nanotechnology innovation in the area of nanocatalysis, which it hopes to use to add value and productivity to its active mining industry. By focusing industrial nanotechnology research on domestic industries, South Africa will be able to add value and productivity to its industries, both nascent and already established, and export technologies abroad. This approach can serve as a model for other developing nations.

In South America, Argentina and Brazil have created a partnership known as the Argentinean-Brazilian Nanotechnology Center, which focuses on industrial and social applications of nanotechnology, as well as pro-poor technologies. Brazil, similar to SANS, is applying industrial nanotechnology research to one of the nation's major industries, agriculture. Brazil's agricultural research organization, EMBRAPA, has already developed a "food-taste" sensor capable of rapidly detecting various parameters of locally grown agricultural products.

2.2 Partnerships and Collaborations

SANS and the Argentinean-Brazilian Center constitute formal South-South partnerships, since they have a regional emphasis and a desire to work with other developing nations. Although these exist, along with more informal South-South collaborations, it is more common to find North-South partnerships and collaborations.

Although informal North-South collaborations exist, as discussed above, there are few incentives to promote them. Moreover, it is difficult to track and quantify informal collaborations, which in turn makes it difficult to assess their robustness or effectiveness. Here the focus is on more formalized North-South partnerships.

In 2005, Argentina, in addition to developing its own nano research labs, established a partnership with Lucent Technologies and its Bell Labs facility in New Jersey. When Argentinean scientists need to conduct research with equipment more sophisticated than is currently available in-country, they can work with Bell Labs researchers and facilities. This partnership gives Argentineans access to more advanced technology, reducing equipment costs, and gives Lucent Technologies access to new talent and marketable research, thus providing incentive for the partnership. Also, as with South Africa, Argentina's nanotechnology research focuses on a preexisting domestic industry, semiconductor fabrication.

Academic institutions in Mexico have partnered with the University of Texas at Austin (UT) to form the International Center for Nanotechnology and Advanced Materials (ICNAM) [7]. Again, this partnership provides Mexican researchers access to more advanced equipment and provides UT with exposure to new talent. Also, Albany Nanotech and the University of Albany College of Nanoscale Science and Engineering have forged a formal partnership with the Centro de Investigación en Materials Avanzados (CIMAV) in Mexico. Both sides in this partnership are capable of advanced nanotechnology research, so the goals are to leverage the resources of each institution and provide opportunities for researchers to expand their knowledge through international collaboration.

The European Union has encouraged North-South collaborations through its Framework Programmes (FP). Under FP, a developing nation can receive research funding if it applies in collaboration with an EU nation. South African researchers have already collaborated with EU researchers under FP on nanotechnology research [8]. Another product of FP is NanoforumEULA, which funds exchange visits for 20 Latin American researchers at research organizations in the EU each year [9]. The goal is to create lasting relationships and provide developing world researchers with access to high-tech nanotechnology facilities.

Cleantech 2007, ISBN 1-4200-6382-0

3 U.S. FOREIGN AID POLICY

The U.S. currently spends more than any other nation on foreign aid in real terms, yet it spends among the least as a percentage of gross national income (GNI) when compared to other major donor governments [10]. Currently, there is no national effort to bring nanotechnology to the developing world, but there are many potential avenues to do so.

3.1 Potential Pathways

The Bill and Melinda Gates Foundation, which funds pro-poor research, provides a model the U.S. could follow. Fabio Salamanca-Buentello and Peter Singer have suggested creating a similar international fund to support development-focused nanotechnology research [11]. The U.S. could attempt to achieve this through multilateral organizations such as the U.N., or it could set up its own fund through nanotechnology grant providers such as the National Science Foundation (NSF) and the National Institutes of Health (NIH). This approach is not likely to build research or industrial capacity in developing nations, but might be effective in promoting development of pro-poor technologies.

Although the developing world would benefit from funds for pro-poor research, it would benefit more from assistance that helps to build research, industrial, or economic capacity domestically. The U.S. could offer funding for North-South collaborations related to nanotechnology and development in several ways. The NIH and NSF could make such collaborations a criterion for certain types of grants. Also, the Department of State's Science, Technology, and Engineering Mentorship Initiative, which currently seeks to enhance Iraq's scientific capacity by forming relationships between Iraqi and U.S. scientists, could be expanded to other developing nations.

The U.S. could also use multilateral avenues, such as the U.N. or the World Bank to build scientific capacity in developing nations. The U.N. has established international shared-use research facilities, and the Third World Academy of Sciences (TWAS) has proposed creating a shared-use nanotechnology research lab in sub-Saharan Africa [12]. Additionally, the U.N. and TWAS have established South-South researcher exchanges which allow knowledge expansion without encouraging a 'brain-drain' from developing nations [13]. The U.S. could encourage and fund the expansion of these multilateral routes to promote nanotechnology in developing nations.

However, the greatest potential for a broad initiative rests with the main foreign aid organizations, the U.S. Agency for International Development (USAID) and the Millennium Challenge Corporation (MCC), which have experience funding development related research. Although USAID currently lacks any programs linking nanotechnology and development, its efforts to bring biotechnology to developing nations serve as a promising framework for nanotechnology. USAID has funded partnerships between U.S. research organizations and developing world scientists to tackle specific agricultural issues. For example, with USAID funding, researchers at Purdue University have worked closely with African scientists to develop a strain of sorghum resistant to the parasitic weed striga. After many years, a successful strain was developed which has helped prevent famine and ensure food security [14]. In addition to establishing and supporting partnerships, USAID's biotechnology efforts include sponsoring developing world students for U.S. graduate degrees and supporting agricultural education in participating countries. USAID also helped develop India's Department of Biotechnology. In addition to assisting with scientific capacity, USAID works with nations to build regulatory capacity to ensure safe biotechnology practices. Each of these types of efforts--building partnerships and collaborations, supporting education in the U.S. and in-country, building institutional capacity, and researcher exchanges--could be extended to nanotechnology.

Another possible approach for USAID is to foster public-private alliances between developing nations and U.S. companies. USAID's Global Development Alliance has successfully established hundreds of such alliances [15]. For example, to help reduce the digital divide, USAID allied with Cisco Systems to establish Networking Academies in over 40 developing nations, resulting in thousands of graduates [16].

In recent history, USAID has done a poor job at building scientific capacity in developing nations, primarily because of a decline in technical expertise in the agency [17]. Unless USAID reorganizes itself to include more technical experts, it is better for the agency to provide the incentives needed to overcome the existing barriers to nanotechnology research and development in the developing world. Although USAID is not the most technologically sophisticated U.S. agency, it is the most capable agency with respect to development, which makes it the best positioned agency to oversee the infusion of nanotechnology into development.

MCC is a new aid agency which offers unique opportunities for nanotechnology. Unlike USAID, MCC has a screening process for aid candidates, and currently relatively few nations are eligible for aid [18]. MCC seeks to reduce poverty through sustainable economic growth, and it does so through multi-year investment plans known as Compacts, which are managed primarily by the developing nation. Compacts typically involve hundreds of millions of dollars and therefore can make a significant impact on a developing nation. Nanotechnology has not been incorporated into any Compacts, but the reason MCC is enticing as a nanotechnology infusion vehicle is because an eligible nation could develop a Compact with the goal of building an independent nanotechnology research or manufacturing sector. Moreover, Compacts can take multifaceted approaches, so a nano-oriented Compact could combine education, research, domestic industry, and regulatory capacity all at once.

As discussed above, the current patent system is likely to complicate many of these and other efforts to assist the global poor through nanotechnology. Patent exemptions for innovations with beneficial humanitarian applications or innovative classes of patents that reward innovations according to their contribution to assisting the global poor would help to overcome some of these barriers [19]. Patents have been a major obstacle in developing nations with regards to biotechnology. To ensure that nanotechnology can be fostered in developing nations it is essential that intellectual property barriers are minimized.

4 RECOMMENDATIONS

The most effective and beneficial way for U.S. policy to encourage nanotechnology's contribution to resolving the problems of global poverty and development is through targeted research partnerships and collaborations fostered and supported by USAID, as well as through MCC research Compacts focused on building sustainable research or manufacturing capacity.

REFERENCES

[1] Salamanca-Buentello et al, Nanotechnology and the Developing World," PLoS Medicine, 2(5), e97, 2005; Hillie, Munasinghe, Hlope, Deraniyagala, "Nanotechnology, Water, and Development," Meridian Institute Global Dialogue on Nanotechnology and the Poor. Available online at: http://www.merid.org/nano/waterpaper; Meridian Institute, "Nanotechnology and the Poor: Opportunities and Risks." Available online at: http://www.meridian-nano.org/gdnp/paper.php

[2] Lemley, "Patenting Nanotechnology," Stanford Law Review, 58 (2), 601, 2005.

[3] National Research Council, "The Fundamental Role of Science and Technology in International Development," National Academies Press, 36, 2006. Available online at: http://www.nap.edu/catalog/11583.html#toc

[4] Juma, Yee-Cheong, "Innovation: Applying Knowledge in Development," United Nations Millennium Project, Task Force on Science, Technology, and Innovation, 70, 2005.

[5] Republic of South Africa, Department of Science and Technology, "The National Nanotechnology Strategy." Available online at: http://www.dst.gov.za/publications/reports/Nanotech.pdf

[6] Hillie et al, "Nanotechnology, Water, and Development," 35-38.

[7] International Center for Nanotechnology and Advanced Materials, The University of Texas at Austin, http://www.engr.utexas.edu/icnam/about/index.htm

[8] European – South African Science and Technology Advancement Programme, http://www.esastap.org.za/

[9] NanoforumEULA, MESA+, http://www.mesaplus.utwente.nl/Links/nanoforumeula

[10] Statistical Annex of the 2006 Development Co-operation Report, Organization for Economic Co-operation and Development, http://www.oecd.org/dac/stats/dac/dcrannex

[11] Salamanca-Buentello et al, "Nanotechnology and the Developing World," e97.

[12] International Centre for Science and High Technology, United Nations Industrial Development Organization, http://www.ics.trieste.it/; Third World Academy of Sciences, http://www.twas.org

[13] Third World Academy of Sciences, "Building Scientific Capacity: A TWAS Perspective," 25, 2004. Available online at: http://www.twas.org

[14] USAID, "Improving Lives through Agricultural Science and Technology," 3-4, 2003. Available online at: http://www.usaid.gov/our_work/agriculture/improving_lives7-03.pdf

[15] National Research Council, "The Fundamental Role of Science and Technology in International Development," 28.

[16] USAID, "The Global Development Alliance: The Private Revolution in Global Development," 92-97, 2006. Available online at: http://www.usaid.gov/our_work/global_partnerships/gda/report2006.html

[17] National Research Council, "The Fundamental Role of Science and Technology in International Development, 5.

[18] For more detail on MCC screening and eligibility, as well as current MCC Compacts, please visit: http://www.mcc.gov

[19] Pogge, "A New Approach to Pharmaceutical Innovations," 21 June 2005, On Line Opinion. Available online at: http://www.onlineopinion.com.au/view.asp?article=3559

Nanotechnology and the Water Market: Applications and Health Effects

F.S. Mowat[*] and J.S. Tsuji[**]

[*]Exponent, Inc., Menlo Park, CA, USA, fmowat@exponent.com
[**]Exponent, Inc., Bellevue, WA, USA, tsujij@exponent.com

ABSTRACT

Nanotechnology has enormous potential for creating cost-effective, simple, and efficient tools for addressing water-supply challenges while preventing the creation of potentially toxic byproducts. The success of these technologies using nanomaterials, however, will depend on whether the nanoparticles and fibers can be confined and isolated from human and environmental receptors and on assessments of the potential health and environmental risks if exposures do occur. To truly qualify as "green" technology, applications of nanotechnology will need to demonstrate that such exposures and potential concerns for health and environmental risk can be managed adequately. This paper describes some of these technologies, focusing on nanofiltration and disinfection, desalination, and environmental remediation, as well as implications to human health and the environment.

Keywords: filtration, desalination, disinfection, remediation, health effects

1 MOTIVATION

Safe and adequate supplies of water are vital for agriculture, industry, recreation, and human consumption. The World Health Organization (WHO) estimates that 2.4 billion people worldwide lack basic sanitation, and 3.4 million people, mainly children, die annually from water-related diseases [1]. Increasing challenges exist in maintaining sufficient clean water supplies due to extended drought, more stringent health-based standards (e.g., new 10-ppb arsenic standard in the U.S.), increasing populations and water demands worldwide, chemical and biological contamination threats, and potentially toxic byproducts from conventional treatments (e.g., from chlorination). Nanotechnology holds the promise of cost-effective and efficient solutions to these challenges, and will likely play a critical role in water-related technologies in the future.

2 NANOTECHNOLOGY APPLICATIONS IN THE WATER MARKET

The world market for water and wastewater is anticipated to increase from $287 billion in 2004 to $412 billion by 2010 [2]. Applications of nanotechnology in the water market include disinfection and filtration, real-time and remote monitoring, groundwater and subsurface treatment or remediation, and wastewater treatment and recycling (Table 1).

Nanotechnology	Example Application	Advantages
Alumina fibers	Disinfection, nanofiltration	Improved clogging resistance; high adsorption
Metals, semiconductors	Disinfection, nanofiltration, desalination	Generally recognized as safe
Zero-valent iron (ZVI)	Groundwater remediation	No toxic by-products; long-lasting; highly effective
Beads, resins	Remediation	Rapid adsorption; little solid waste; may be re-usable
Membranes, clays/zeolites	Wastewater recycling, desalination, purification	Can be selective for chemical; high surface area and adsorption
Capsules	Wastewater recycling, remediation	Re-usable; can be selectively engineered
Dendrimers	Wastewater recycling, remediation, filtration	High capacity for metals; able to self-assemble
Nanotubes	Purification, treatment	Can be selectively engineered
Composites	Treatment	Large surface area and adsorption capacity

Table 1: Examples of nanotechnology applied to the water market (details and citations provided in main text).

2.1 Nanofiltration and Disinfection

Nanofiltration is becoming more important in wastewater treatment as a pressure-driven membrane separation process [3,4], wherein disinfection and removal of solids, bacteria, and other materials is required. Current limitations in filtration technologies occur when properties of the membrane are exceeded and either allow the medium being filtered to pass through or prevent solution from passing due to fouling and clogging [3]. These limitations likely occur most frequently at the industrial level, where filtration needs involve high throughput of different solutions.

Nano-sized or nano-loaded membranes and filters working under principles of low-pressure reverse osmosis (RO) provide several advantages to filtration, including superior properties (e.g., high permeability and retention of organics), high surface area, and ability to accommodate high flow rates [3]. In addition, "smart" nanomembranes could contain embedded sensors to automatically change performance and sensitivity based on sensing of contaminant or solution differential across the membrane or could be engineered for selected effluents [3,5]. In particular, ceramic and nanoclay/zeolite membranes have the combined advantage of high chemical, mechanical, and thermal resistance and large surface area and absorption capacity through cation exchange; they also are relatively inert and can be engineered to be selective for a particular effluent [5–8]. Although generally expensive, their capacity for multiple use and regeneration would greatly reduce associated costs. Alumina nanofibers improve clogging resistance and fouling rates due to their highly electropositive surface, which attracts submicron and nano-sized particles [9]. In some applications, alumina nanofibers are equipped with charged polymer brushes or are pleated to increase surface area and adsorption capacity [9].

Nanometals and other semiconductor materials (e.g., titanium dioxide [TiO_2], palladium-coated gold or iron, and zinc oxide [ZnO]) also play a role in chemical degradation and disinfection, wherein reaction of the metals with ultraviolet light can oxidize harmful microorganisms [10–13]. This reaction generally results in oxidation of organic pollutants to carbon dioxide [10], and has been found to degrade phenol [10], toluene [5], *E. coli* [11,14], trichloroethylene (TCE) [13,15,16], polychlorinated biphenyls [16], carbon tetrachloride [17], and other volatile organic compounds [3,5]. These metals can also be doped to include various antibacterial metals (Ag, TiO_2, ZnO), to target certain agents [18]. An added benefit is that many metals used in these applications are generally recognized as safe and occur naturally in the environment.

2.2 Monitoring

Nanoscale solid-state sensors are currently being developed by the University of California at Davis and the U.S. Environmental Protection Agency to provide real-time remote detection and to aid in facilitating the process of monitoring and treating pollutants [19]. These sensors can be engineered to be selective for certain chemicals—such as arsenic and chromium—allowing them to be more efficient than "broad range" monitoring devices. Other researchers, at Arizona State University, are developing high-performance, low-cost sensors to provide early warning and prevention of metal ion contamination using quantum effects exhibited by many nanoparticles for use in nanoelectrodes [20].

2.3 Desalination

Fresh water is important for sustaining all forms of life and it is becoming less available due to effects of global warming, environmental destruction, and loss of freshwater sources [2]. Up to 80% of the salt in surface waters results from natural erosion, although anthropogenic sources such as use of water softeners and land-use practices also contribute to concentrated salts [3]. Current desalination processes are expensive and time consuming, and are sensitive to fouling [3]. RO and multi-effects distillation are two types of desalination processes that could benefit from nanotechnology. These processes could include running salinated water through a series of membranes, distillation, and condensation to rapidly and effectively remove salts from water [3]. Nanoengineered membranes could be used for efficient and relatively cheap desalination of water supplies, and would offer the added benefit of being portable and easy to clean [8].

2.4 Groundwater Remediation

Nano-sized particles have been found to be efficient and expedient in groundwater remediation, without producing large volumes of waste or forming toxic by-products [17,21,22]. In addition, nanobeads and nanoresins have produced rapid absorption of different contaminants, with little waste generated they allow for multiple uses, making them highly cost-effective [23]. In some cases, nano-sized metals are incorporated into the resin beads to increase specificity, absorption, and stability [23]. Similarly, dendrimers, which are three-dimensional branched structures with a high degree of surface functionality and diversity, have a large capacity to trap metal ions, which can then be filtered out [24]. Dendrimers also have the ability to self-assemble, resulting in branching and increased surface area for adsorption. Nanocapsules (nanoparticle with a "hole" in it) or nanoemulsions could be engineered to contain a particular substance to be released under controlled conditions to treat a particular chemical or contaminant [25].

Zero-valent nano-iron (ZVI or Fe^0) has been particularly successful in groundwater remediation trials [17,21,22]. ZVI has been shown to be effective in dechlorination of organic solvents, transformation of fertilizers, detoxification of certain pesticides, and immobilization of contaminant metals [17,22,26,27]. In a pilot-scale study at Hunter's Point Shipyard in northern California, injection of ZVI resulted in removal of 99.1% of total chlorinated solvents and a 99.2% reduction in TCE concentrations over a period of three weeks (Figure 1) [26]. The formation of by-products often seen during traditional TCE destruction was not observed. In addition, metals present in the soil, such as manganese and arsenic, were not mobilized during treatment.

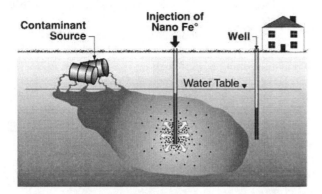

Figure 1. Example of a pilot-scale study of groundwater and subsurface treatment using ZVI (adapted from [27]).

2.5 Wastewater Treatment and Recycling

Many of the technologies described above can be used to treat and recycle wastewater. For example, systems that use nanoscale forms of TiO_2 and magnetic nanoparticles to decompose organic pollutants and remove salts and heavy metals from liquids enable the use of contaminated wastewater for irrigation and drinking [8,28]. Nanoelectrocatalytic systems could also be used to purify contaminated, salinated water for purposes of drinking and irrigation [8]. Nanomembranes could be functionalized to remove a particular component in the wastewater during treatment, prior to recycling. In addition, carbon nanotubes could be used for wastewater treatment. These nanotubes have been successful in filtering petroleum and treating contaminated drinking water [29].

3 IMPLICATIONS FOR HEALTH AND THE ENVIRONMENT

Although currently proposed treatments using nanotechnology are not expected to result in toxic by-products, as in chlorination, exposure to nanoparticles may be a concern because their small size allows them to remain in suspension in water. In addition, nanoparticles, again by virtue of their small size, may more readily cross biological membranes and barriers, where they can exert potentially toxic effects [30]. Other concerns are that even relatively inert substances may become more reactive, and thus toxic, by virtue of their small size and higher surface area. These issues must be considered from both a human health and environmental standpoint.

3.1 Environmental Issues

Nanoparticles may enter the environment either intentionally (e.g., ZVI) or accidentally, and the environment is the ultimate sink for these materials. In the environment, nanoparticles show enhanced dispersal and mobility due to their small size, and are thought to be more reactive. Nanoparticles could be liberated into water supplies, where exposure of humans and both aquatic and terrestrial organisms could occur. The presence of coatings, a surface charge, or other surface properties may result in enhanced interactions with bacteria, algae, and other microorganisms in the environment, and may result in bioaccumulation and possibly biomagnification up the food chain [31].

To minimize potential for exposure, nanotechnologies used in the water market should be evaluated for containment within a durable matrix (e.g., resin), such that any accidental release of the treatment product would not result in release of nanoparticles into the environment. In addition, it might be beneficial if the nanoparticles used were able to aggregate upon release, allowing for either rapid removal from the system (e.g., through filtration) once the desired treatment was achieved, or increasing the likelihood that the larger particle would fall out of solution. Several authors have demonstrated that metal-ethylenediaminetetraacetic acid complexes may result in chelation and subsequent extraction of nanometals used in treatment [32–34].

3.2 Human Health Issues

When evaluating human health, a four-step risk assessment process is generally used, which encompasses hazard identification, dose-response assessment (toxicity), exposure assessment, and risk characterization (Figure 2).

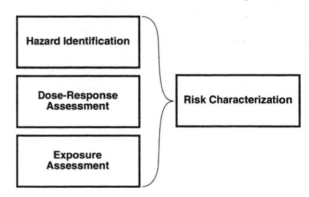

Figure 2. The risk assessment process (adapted from National Research Council, 1983).

Hazard identification considers issues related to the composition, size, structure, and presence of surface charges or coatings. In general, the hazards of novel materials will be less predictable than small-scale versions of known substances [35]. Relatively little is known about the toxicity or dose-response of nanomaterials, and information to date is from *in vivo* studies, primarily as short-term studies in rodents or bench-scale studies using isolated cells or tissues, using selected materials. The primary routes of exposure for humans are oral, dermal, and inhalation routes, with workers and academic researchers likely experiencing the highest exposures. Potential exposure to nanomaterials must be considered for the entire life cycle (e.g., from synthesis to disposal) of the material or product. For nanomaterials within products, assessment of the intrinsic material properties of the product will be necessary to evaluate the degree of containment over time. Finally, in risk characterization, uncertainties must be dealt with to fully understand health risks.

The available research indicates that toxicity cannot be predicted simply as a function of size. Rather, many factors are important for determining relative toxicity, including chemical composition and particle shape and structure [35,36]. Much of the scientific research indicates that toxicity may be related more to surface area than to mass [30], although smaller particles are not necessarily more toxic (Tsuji et al. 2006). Additional determinants of exposure and toxicity for nanomaterials in water applications include aggregation potential, surface charges, and, degree of containment [35–38]. The effect of these factors is likely material specific, and generalizations may be misleading. For example, aggregation may increase the size of particles beyond the nano range, but for at least one type of nanoparticle (C_{60} carbon, or fullerenes), aggregation increased solubility and toxicity to bacteria [39]. These factors will need to be considered in risk assessments and safety evaluations of these new technologies, and specific issues regarding the environment and human health are described further below.

4   CONCLUSIONS

Nanotechnology holds great promise for improving the efficacy and efficiency of water treatment. Most scientific reviews of nanomaterials have concluded that the risks associated with these substances can be managed, but due to the paucity of information regarding toxicity, more health and environmental effects research is needed. Some studies show that smaller particles are not necessarily more toxic, but the ability of these smaller particles to disperse and become mobile in both the environment and the human body must be evaluated. If nanomaterials are used in the water market for filtration, disinfection, and other treatment, product-specific implications should be considered, particularly for products that are designed to stay or otherwise degrade *in situ*.

REFERENCES

[1] WHO, Action against infection: A newsletter for WHO and its partners, 2001.

[2] H. Kaiser Consultancy, "Water markets worldwide briefing conference," London, 2006.

[3] U.S. Bureau of Reclamation/Sandia National Laboratories, "Desalination and water purification technology roadmap," Report #95, 2003.

[4] D. Ollis, Ann NY Acad Sci, 984, 65, 2003.

[5] W. Song et al., Environ Sci Technol, 39, 1214, 2005.

[6] D. Yaron-Marcovich et al., Environ Sci Technol, 39, 1231, 2005.

[7] E. Pitoniak et al., Environ Sci Technol, 39, 1269, 2005.

[8] E. Court, et al., Nanotoday, 14, 2005.

[9] F. Tepper, L. Kalendin, and C. Hartmann, Water Conditioning & Purification, February, 55, 2005.

[10] J. Byrne et al., Appl Catal B, Environ, 17, 25, 1998.

[11] P. Dunlop et al., J Photochem Photobiol A, Chem, 148, 355, 2002.

[12] R. Serpone and F. Khairutdinov, Stud Surface Sci Catal, 103, 41, 1996.

[13] M. Nutt, Environ Sci Technol, 39, 1346, 2005.

[14] P.-C. Maness et al., Appl Environ Microbiol, 65(9), 4094, 1999.

[15] D. Ollis, Chemistry, 3, 405, 2000.

[16] Water Resources Research Institute, Annual Technical Report, 2003.

[17] J. Nurmi, Environ Sci Technol, 39, 1221, 2005.

[18] www.nanophase.com.

[19] EPA, "Innovation and research for a clean environment," National Center for Environmental Research, 2002.

[20] N. Tao, "A nanocontact sensor for heavy metal ion detection," Proceedings EPA Nanotechnology and the Environment, 28, 2002.

[21] S. Joo, Environ Sci Technol, 39, 1263, 2005.

[22] S. Kanel, Environ Sci Technol, 39, 1291, 2005.

[23] SolmeteX, Inc., www.solmetex.com.

[24] M. Daillo, Environ Sci Technol, 39, 1366, 2005.

[25] E. Acosta, Environ Sci Technol, 39, 1275, 2005.

[26] R. Mach, "Micro-scale ZVI treatment of groundwater," Federal Remediation Technologies Roundtable, 2004.

[27] J. Quinn, Environ Sci Technol, 39, 1309, 2005.

[28] F. Salamanca-Buentello et al., PLoS Med, 2(4):0300.

[29] A. Goho, Science News, August, 2004.

[30] G. Oberdörster et al., Environ Health Perspect, 113, 823, 2005

[31] E. Oberdörster, Environ Health Perspect, 112, 1058, 2004.

[32] Prairie et al., Environ Sci Technol, 27, 1776, 1993.

[33] Borrell-Damian and D. Ollis, J Adv Oxid Technol, 4, 125, 1999.

[34] Chen and D. Ollis, Colloids & Surfaces, 151, 339, 1999.

[35] J. Tsuji et al., Tox Sci, 89(1), 42, 2006.

[36] Royal Society, "Nanoscience and nanotechnologies: Opportunities and uncertainties," 2004.

[37] A. Maynard et al., J Toxicol Environ Health, 67, 87 2004

[38] D. Warheit et al., Exp Lung Res, 29, 593, 2003.

[39] J.D. Fortner et al.,Environ Sci Technol, 39, 4307, 2005.

Nano-structures materials for Energy Direct Conversion and Fuel Breeding

L. Popa-Simil, I.L.Popa-Simil

LAVM LLC, Los Alamos, NM 87544, USA

ABSTRACT

Almost all the modern applications (e.g. terrestrial and space electric power production, naval, underwater and railroad propulsion and auxiliary power for isolated regions) require a compact-high-power electricity source. One solution to reduce the greenhouse emissions and delay the catastrophic events occurrences may be the development of massive nuclear power. More, there is a concern that the modern civilization may exhaust the oil based energy resource within few decades. Thus, it is better to find other sources of energy that can replace the Carbon based energy resources. The actual basic conceptions in nuclear reactors are at the base of bottleneck in enhancements. The actual nuclear reactors look like high security prisons applied to fission products. It is not about release of the fission products, but to give them the possibility to acquire stabile conditions outside the hot zones, in exchange for advantages – possibility of enhancing the nuclear technology in power production. Three main developments are possible by accommodating the materials and structures with the phenomenon of interest like the high temperature fission products free and direct conversion nuclear reactors, cleaner nuclear fuel breeding and the fusion energy harvesting.

Keywords: nano structure, nanograin, breeding, nuclear fuel

1 INTRODUCTION

The actual planet thermal power is about 14 Tw, which does not include the transportation. The total irradiation power received from sun is about 200 Pw, which shows that soon the technological power is becoming 0.1% from the average power received from the sun and the thermal pollution started to count in the planetary energetic balance.

Global warming and global dimming, the two facets of the climate change put the civilization in front of capital choices. The message from the nature says now:"Stop burning carbon!" because the chemical mechanism influence over planetary climate is orders of magnitude more efficient than the direct thermal pollution. That translated means: "It is still OK not having high efficiency but burning carbon is prohibited", and even if everything will be set by tomorrow there will be a hard price to be paid next hundred years for what have been achieved up to date. In front of this decision the world realized that there is NO good solution, no really clean and reliable energy, most of the solutions having distorted images by the politico economic speculations based on partial knowledge of the population. That turns into the following brief evaluation based on what means to have a power unit and how reliable that power source is.

To produce 1 Gwday of thermal energy is required 1 kg of ^{239}Pu or ^{235}U, less than ¼ cup of material generating about same amount of primary waste but less dense, and 3 times more secondary waste from the neutron capture. The initial investment scales about 2-4 \$/w, while the price of fuel is another 5,000 \$/day, with lifetime >30 y.

To get the same amount with oil it takes about 2.6 kt, and more than 4 kt with coal, which means about 2,500 m^3/day, releasing twice as much CO_2. The initial investment is about 0.5-2 \$/w and the cost of fuel is about 2.5 mil. \$/day, with a lifetime >30 years.

To get this amount from the wind may take about 20-100 windmills, with about 1 windmill/ha takes about 1 km^2, but not for the whole day. For only ⅓ of day with solar power in the best deserts it may take 4 km^2 and about 100 Mt equipment, but does not work in cloudy days. The investment costs are 4-8 \$/w and the other costs and lifetime are undefined.

Adding a such unreliable source in a national grid, obliges all the baseline producer to waste an equivalent power between ½ and ¾ to dumb in the river just to be in stand by to deliver in the moments the renewable energy device is not receiving from the nature.

The power reliability national security and territorial safety are supplementary conditions that aggravate the problem. In order to harvest efficiently some renewable power there is necessary to have a reliable fast response backup which to take the fluctuations from the nature and keep the right balance between the supply and demand.

The nuclear alternative remains the only good compact solution, but for backup the response time is not good enough, and that requires hydro or gas turbines.

2 THE NUCLEAR OPTION

The nuclear option seems to be among the few alternatives that comply to the most of the requirements and allows safe efficient renewable energy harvesting. In fact, it has its own drawbacks, making it unlikely for most of the population.

The most acute problem is that of the nuclear waste and the cost impact of the hazard exposure.

More the nuclear industry is mature, with more than 60 years of experience, and most of the technologies are near their physical limits, any small improvement being expensive and difficult to achieve.

The main drawbacks in nuclear power production come from the fundamental concepts applied to nuclear fuel.

The main problems of the nuclear fuels are:
- Low thermal conductivity of the fuel which drives to low power density (about 200-500 w/cm^3)
- Inappropriate temperature distribution which induces stress and mechanical failures
- Aggravation of the mechanical and thermal properties with the burnup

that drives to the direct consequences as:
- short fuel lifetime inside the nuclear reactor of about 10-30 months
- highly radioactive nuclear fuel, and important amount of low radioactive highly toxic waste
- high toxicity chemical processing as Purex and Urex to extract Plutonium and Uranium
- bottleneck of disposal process in the geological storage of the nuclear fuels
- anticipated nuclear fuel peaking in about 50 years if the actual slow development trend is maintained

All these drawbacks, which make the nuclear technology less attractive, and less clean and environmentally friendly by abusing environment resources, have the spring in obsolete concepts and technologies applied by highly conservative people.

The most important is related to inappropriate treatment of the fission products in opposition with the nature's laws, of equally distributing the mass and energy on all the available freedom degrees (space and time). That is why the imprisonment concepts based on up to nine confinement layers with the sole purpose of immobilizing the fission products on spot drives to high costs for any enhancement.

In fact, this paper do not state that these fission products are less dangerous as is scientifically established and have to be released, it only states that a better nature understanding with better physics may bring a relief.

3 THE MICRO-HETERO-STRUCTURE

The fission products induced damage was observed and addressed since 1950 in the dispersion fuel research[1, 2], attempting to create a new fuel material called "cermet".

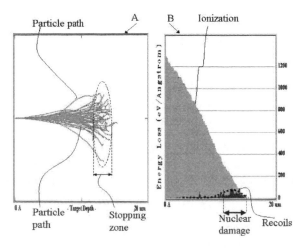

The cermet studies showed that the smaller the fuel particle is, so are the fission products damage, but the damage is transferred to another solid that has its limitations. The MC simulation [3] of the fission process in Fig. 1 clearly showed that up to 80% of the path the fission products are inducing less than few percents of damage; all the damage takes place towards the end of the range.

This result driven to the concept of the tri-layer micro hetero structure [4], where the fuel is made shorter than the effective fission products range. This development offered the possibility of releasing the fission products out of the nuclear fuel with the possibility of being collected outside the reactor's hot zone and separated accordingly.

The behavior of the poison accumulation in the nuclear fuel in the hypothesis of using a micro-beads fuel, coated in a stabilizer material and washed by a liquid metal drain fluid is presented in Fig.2.

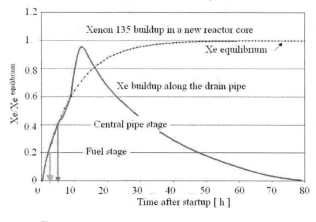

Fig. 2 shows that in a PWR nuclear reactor the ^{135}Xe is building up to an equilibrium value [5], being continuously burned out due to its high n absorption cross-section of about several thousands higher than the fissile material. If the reactor is stopped the Xe peaks and then decays as shown in Eq. 1.

$$^{135}Te \xrightarrow{<.5\,min} {}^{135}I \xrightarrow{6.7h} {}^{135}Xe \xrightarrow{9.2h}$$
$$\xrightarrow{9.2h} {}^{135}Cs \xrightarrow{2\times10^6\,y} {}^{135}Ba \qquad (1)$$

The other isotopes simply reach the power dependent equilibrium value, and remain there following the power.

The advantage of the drain system is that after generation in short time the fission products are removed from the high neutron flux zone and the poisoning effect is dimmed simultaneously with a simplification in the decay schemes where the n absorption branches are missing.

The relation in Eq 2. gives the total amount of radioactive waste.

$$\frac{1}{\langle n_{fission}-1 \rangle} = \frac{\sum_{fissile}\sigma_i N^i}{\sum_{absorbtion}\sigma^j N_j + S_L} = \frac{F}{L} \qquad (2)$$

where S is the fission term, generating the fission products while L is the loss term containing the leakage through the surfaces S_L and the neutron absorption term.

To minimize the nuclear secondary waste given by the so called "neutron activation" process is needed to design the nuclear reactor in such a manner that most of the absorption to be made in fertile materials as depleted uranium or thorium, or other desired isotopes instead in structural and shielding materials.

Here comes in place the concept of using super-grade isotopic enriched materials that often comes into conflict with the so-called "non-proliferation" [6-9] forethoughts and international and states own capabilities distrust.

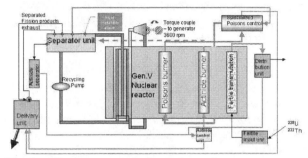

Fig. 3 – The structure of a isotopic nuclear reactor

In such a nuclear reactor the fuel is made from isotopic enriched material as ^{235}U, or ^{239}Pu, surrounded by fertile transmutation unit, specialized poison and residual actinides burner systems.

This type of reactor requires a more complex operation, similar to a "living being" not to a "hot rock", and manifesting life like cycles. The advantages are related to ecologic compatibility releasing less volume of waste and less toxic.

4 THE NANO-BREEDING STRUCTURE

To obtain super-grade material with minimal chemistry a new process based on nuclear selectivity of recoils have been developed. Fig. 4 shows the principal operation diagram. A fast neutron n n is heating a ^{238}U or ^{232}Th nucleus located into a nano grain of cluster size. Due to the

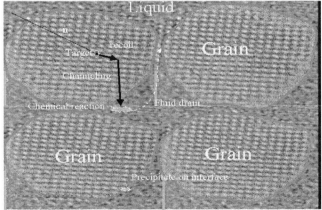

Fig. 4 – The recoil selective isotopic separation

impulse transfer and further beta decay as in Eq. 3 the nucleus becomes a Frenkel defect being driven out by a process of diffusion enhanced by cluster rejection.

$$^{238}_{92}U + ^{1}_{0}n + \ddot{v} = ^{239}_{92}U + \beta + \ddot{v} = ^{239}_{93}Np + \beta + \ddot{v} = ^{239}_{94}Pu \quad (3)$$

The cluster-enhanced rejection is applied to the transmuted nucleus as ^{239}Np and is acting as cluster-attractive force for ^{238}U recoiled by neutron scattering.

The combined effect of cluster enhanced selective diffusion with selective recoil reaction drives to a high concentration of ^{239}Np in the drain fluid, which takes it out

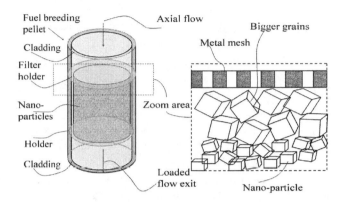

Fig.5 – The nano-breeding pellet

from the hot zone. Fig. 5 shows a nano-structure stabilized under a gradient of particle magnitude based on nano-flow of a reactive fluid selective to the new product and inert for the fertile material, having in the same time small n absorption cross-section.

The collateral results in nuclear medicine interest isotopes showed a difference between the low purities predicted by the recoil model, versus the high collection yield and purities obtained experimentally [10].

5 HIGH BURNUP STRUCTURES

The micro-hetero structure fuel made of a refractory micro mesh material used as mechanical support to deposit nuclear fuel beads made of oxides or carbides incorporating fissile material These fuel beads are coated in a refractory material, making a delta layer with role in faceting and adhesion forces matching between the fuel bead and the drain liquid metal. The fuel made with beaded mesh structure is looking like an elastic, compressible felt. When squeezed, the material reactivity is increasing because some of the liquid metal is eliminated. During the operation, the fissile material is burned and eliminated into the drain liquid as fission products. In this way the fissile material mass in the bead is reducing, while the drain fluid mass is increasing if the volume is maintained constant. By squeezing the fuel the ratio is corrected and the reactivity of the fuel is brought back to criticality. As shown in Fig. 6 the new fuel is introduced at the large end and while it advances in burnup is pushed towards the narrow end, being squeezed.

In this manner more than 99% of the initial fissile material load may be burned, the limitation being brought by the mechanical safety of the highly irradiated structure.

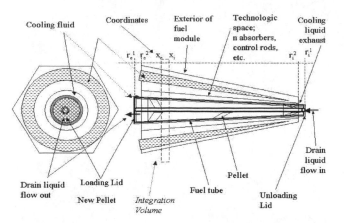

Fig. 6 – Variable geometry nuclear reactor fuel channel

A near "perfect burning reactor" it is achieved, which delivers as nuclear waste the mass of the burned nuclear structure and the residual structure activated mass, which represents another 10-20% from the fuel mass. The rest of the leakage neutrons are used to generate new fuel almost twice as much than burned. By this procedure the entire reserves of nuclear fuel, Uranium, Thorium may be converted in energy, and is expanding the nuclear fuel resources more than 200 times.

6 DIRECT NUCLEAR ENERGY CONVERSION INTO ELECTRICITY

Developing a repetitive structure of nano-layers there is

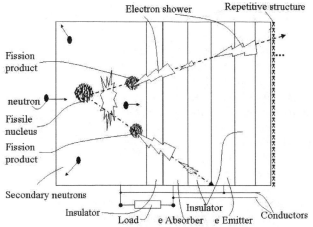

Fig. 7 – The direct conversion principle

possible to collect the knock-on electrons and to obtain a voltage between the layers and electric current. The device looks like a super-capacitor, loaded by the fission or any other radiation energy, as shown in Fig. 7 and discharged on an electric Load. This structure may contain or not

nuclear fuel and may even operate in cryogenic conditions, giving by quantum effects efficiencies higher than 90%.

The device may be used with fusion nuclear reactors too, where the final product might be He, with no radio toxicity.

7 CONCLUSIONS

The new developments brought by nano-technologies transforms the actual nuclear industry into the cleanest environmentally friendly power production.

The revolutionary enhancements possible to be brought to nuclear power production improve almost all parameters by more than two orders of magnitude, the progress being similar to that obtained by the transition from the massive stone wheel to the actual metallic mesh based tires.

REFERENCES

1. White D.W., B.A.P., Willis A.H.,, *Irradiation Behavior of Dispersion Fuels.* Fuels Elements Conference. KAPL-P, 1957. **Proceedings**.
2. Weber C.E., *Progress on Dispersion Elements.* Progress in Nuclear Energy,, 1959. **2**(Series 5. Metallurgy and Fuels).
3. Zigler James, *PARTICLE INTERACTIONS WITH MATTER, SRIM - The Stopping and Range of Ions in Matter.* The Stopping and Range of Ions in Matter, 2006. **IBA - Ion Beam Analysis,**(ibm.web).
4. POPA-SIMIL L., *Long life single load reactor fuel.* Proceedings, 2006. **ICAAP 2006**(1): p. 140-148.
5. HEEMOON K., P.K., et al., *Xenon Diffusivity in Thoria-Urania Fuel.* Nuclear Technology, 2004. **147**(1): p. 149-156.
6. POUCHON M. A., N.M., HELLWIG C., INGOLD F., DEGUELDRE C.,, *Cermet sphere-pac concept for inert matrix fuel.* Journal of Nuclear Materials, 2003. **319**: p. 37-43.
7. Highlights, U., *Toward a National Energy Strategy".* Energy Inside, 2001.
8. BURAKOV B., A.E., *Immobilization of Excess Weapons Plutonium in Russia: A Review of LLNL Contract Work.* Proc. Meet. For Coordination and Review of Work,, 2001. **UCRL-ID-143846**: p. 229-234.
9. DZIADOSZ D., A.T.N., SAGLAM M., SAPYATA J.J.,, *Weapons-Grade Plutonium-Thorium PWR Assembly Design and Core Safety Analysis.* Nuclear Technology, 2004. **147**(1): p. 69-83.
10. POPA-SIMIL L., *New Compact Targets for Radio-Pharmaceuticals Production.* WTTC-7, 1997. **Procedings**(News From The Labs.): p. 17-18.

Building Green with Nanotechnology

G. Elvin

Green Technology Forum, Indianapolis, IN, USA, elvin@greentechforum.net[1]
Ball State University, Muncie, IN, USA

ABSTRACT

Nanotechnology, the manipulation of matter at the molecular scale, is opening new possibilities in green building through products like solar energy collecting paints, high-insulating translucent panels, and heat-absorbing windows. Even more dramatic breakthroughs are now in development such as spray-on solar collecting paint, windows that shift from transparent to opaque with the flip of a switch, and environmentally friendly biocides for preserving wood. These breakthrough materials are opening new frontiers in green building, offering unprecedented performance in energy efficiency, durability, economy and sustainability. This paper provides an overview of nanotechnology applications for green building, with an emphasis on the energy conservation capabilities of architectural nanomaterials in green building.

Keywords: green building, green technology, clean technology, nanotechnology, architecture

1 THE GREEN BUILDING IMPERATIVE

Green building is one of the most urgent environmental issues of our time. Buildings are responsible for 40 percent of the emissions responsible for global climate change and 42 percent of the electricity consumed in the U.S. Waste from building construction accounts for 40 percent of all landfill material, and sick building syndrome costs an estimated $60 billion annually. Clearly, buildings play a large part in our current environmental dilemma.

But they also offer an opportunity to improve environmental quality and occupant health. Green building is a catch-all phrase encompassing efforts to reduce waste, toxicity, and energy and resource consumption in buildings. The green building movement has grown to the point that major cities like Chicago and Seattle now require new buildings to comply with strict environmental standards. As public and private clients alike call for more sustainable buildings, and architects become increasingly adept at designing them, a dramatic shift is emerging, from buildings that harm the environment to ones that heal it.

Architects, owners and builders committed to green building, however, are often frustrated by limited material choices. Buildings need light, for example, but current windows are extremely poor insulators, leading to increased energy consumption. Similarly, alternatives to polyvinyl chloride (PVC) pipe for plumbing are healthier than this known carcinogen but can be costly. Now, however, a new frontier is opening in building materials as new products and possibilities are introduced by nanotechnology.

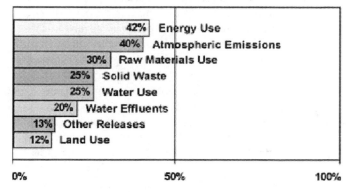

Figure 1: Environmental impacts of buildings (Levin, 1997).

2 NANOTECHNOLOGY FOR GREEN BUILDING

Nanotechnology, the control of matter at dimensions of roughly one to one hundred billionths of a meter, is bringing dramatic changes to the building industry. Products incorporating nanotechnology were valued at $13 billion last year, with sales expected to top $1 trillion by 2015. Already, nanotechnology has brought us self-cleaning windows, smog-eating concrete, and wi-fi blocking paint. And in the world's nanotech labs work is underway on illuminating walls that change color with the flip of a switch, nanocomposites as thin as glass yet capable of supporting entire buildings, and photosynthetic building facades.

Many nanotechnology-enabled products and processes now on the market can help create more sustainable buildings. Those in development today that can be brought to market offer even more promise for dramatically improving the environmental performance of buildings. Nanotechnology-enabled advances for green building include new materials such as carbon nanotubes and insulating nanocoatings, as well as new processes including photocatalysis and nanoporous filtration. Nanomaterials can contribute to green building by improving the strength, durability, and versatility of structural and non-structural materials, reducing their toxicity, and improving insulation.

2.1 Improved Material Strength

Material strength is critical in a building, defining its structure, longevity, and resistance to gravity, wind, earthquake, and other loads. A load-bearing structural material's strength/weight ratio is particularly important because stronger, lighter materials can carry greater loads per unit of material. Greater strength/weight ratio therefore means fewer materials, which in turn means fewer resources and energy consumed in production.

Nanotechnology promises significant improvements in structural materials both through nano-reinforcement of existing materials like concrete and steel, and through the introduction of altogether new materials like carbon nanotubes. Reinforcement of concrete is a particularly rich field, with experimentation taking place in dozens of labs around the world. Concrete is the most common human-made material by volume, and its key ingredient, portland cement, is by itself responsible for almost ten percent of the emissions responsible for global climate change.

"Development of nano-binders can lead to more than 50 percent reduction of the cement consumption, capable to offset the demands for future development and, at the same time, combat global warming," report Sobolev and Ferrada-Gutiérrez [1]. Nanofiber reinforcement has been shown to improve the strength of concrete significantly. Even simply grinding portland cement into nanoparticles has been shown to increase compressive strength four-fold [2].

Steel's high strength/weight ratio and combination of compressive and tensile strength make it the material of choice for tall buildings. But it can be weakened by corrosion and is extremely heavy. Corrosion can be inhibited, however, using a proprietary steel manufacturing technology by MMFX which creates a laminated atomic-scale structure similar to plywood. [3].

Sandvik now offers Nanoflex, a high-modulus steel of extreme strength to create thinner and even lighter components than those made from aluminium and titanium. Sandvik Nanoflex is currently being used in medical equipment, such as surgical needles and dental tools. [4].

In addition to reinforcing existing materials, nanomaterials like nanotubes and buckypaper may lead to new building materials. Carbon nanotubes, sheets of graphite one atom thick formed into a cylinder, can be more than 50 times stronger than high-carbon steel [5]. Nanotubes offer the promise of structures many times lighter yet stronger than steel. Also known as buckytubes, a reference to Buckminsterfullerenes (carbon molecules twice as hard as a diamond,) they may form the building blocks of tomorrow's architecture, along with buckypaper, a planar material composed of buckytubes.

Buckypaper may have non-structural applications as well. Researchers at the University of Texas at Dallas have produced transparent carbon nanotube sheets stronger than steel and so thin that a square kilometer would weigh only 30 kilograms. The nanotube sheets combine high transparency with high electrical conductivity and could be used as electrodes for bright organic light emitting diodes in energy-efficient displays and solar cells [6].

2.2 Improved Building Insulation

Nanotechnology may also play a role in conserving energy through improved building insulation. Because of their high surface/volume ratio, nanofibers can trap large amounts of air, increasing a product's insulating ability. This is the principle behind one of the most popular nanomaterials for building now on the market—Nanogel. Nanogel insulation is a form of aerogel, the lightest weight solid in the world. It has a content of 5 percent solid and 95 percent air. The high air content means that a 3.5-inch thick Nanogel panel can offer an insulating value of R-28 in a 75 percent translucent panel [7].

Because it traps air at the molecular level, an insulating nanocoating even a few thousands of an inch can have a dramatic effect. Nansulate HomeProtect ClearCoat employs Hydro-NM-Oxide, which the manufacturers say is by far the world's best thermal insulation medium—R-10 to R-13 per inch, as compared to polyurethane foam (R-6.64 per inch) and fiberglass batts (R-3.2 per inch) [8].

Other products integrate insulating nanoparticles with conventional materials. Masa Shade Curtains are coated with metal nanofilm to block ultraviolet rays and improve insulation. The stainless steel film absorbs infrared rays, blocking out sunlight and lowering room temperatures by 2-3° C more than conventional products [9].

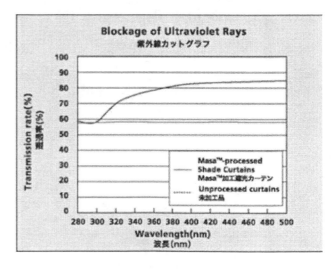

Figure 2: Ultraviolet ray transmission (Suzutora, 2007).

2.3 Nanocoatings

Some nanocoatings are used to insulate materials, while others protect wood, metal and masonry. Scratch-resistant nano-coatings are becoming a viable alternative to polyurethane coatings that can produce harmful volatile organic compounds (VOCs). Nano-engineered ultraviolet curable coatings by Ecology Coatings, for example, not only contain no toxic solvents, they contain no water, eliminating the need for heat curing, cutting manufacturing energy consumption by 75 percent [10].

DuPont is also working on nanoparticle paint for autos. The paint, licensed from Ecology Coatings, is cured using ultraviolet (UV) light at room temperature, rather than in the 400° F ovens required for conventional auto paint.

"We are in the early stages of a profound industry change," said Bob Matheson, technical manager for strategic technology production at DuPont. Ecology Coatings is also working on coatings that would make a wide range of reprocessed organic and waste materials durable enough to be used as building materials.

Nanocoatings can break down dirt as well, and PPG and Pilkington offer self-cleaning window glass. The Jubilee Church in Rome by Richard Meier & Partners Architects features self-cleaning concrete panels. Photocatalytic titanium dioxide nanoparticles are built into the precast panels, making them shed dirt. Depolluting nanocoatings that trap airborne pollutants in a nanoparticle matrix and decompose them can be applied to almost any surface cleansing surrounding air.

"Among other things, we want to construct concrete walls that break down vehicle exhausts in road tunnels," said Karin Pettersson, a spokeswoman for Swedish construction giant Skanska. "It is also possible to make pavings that clean the air in cities."

2.4 Air Filtration

The EPA rates indoor air quality among the top five threats to human health, and estimates medical expenses due to poor indoor air quality at $60 billion per year. Nanotechnology can improve indoor air quality through detection and filtration of unwanted airborne particles, and by reducing or eliminating offgassing and VOCs in finishes and cleansers. A new nano device, capable of detecting fungus that attacks wood, for instance, has been developed by a team of Polish scientists. Aspergillus versicolor is very a common mold fungus known to produce carcinogenic toxins. The fungus is detected using nanofibers whose properties change in the presence of the fungus and spark an electrical signal [11].

The high surface/volume ratio of many nanofibers makes them excellent materials for filters. Their small size enables them to trap fine unwanted particles while still providing sufficient airflow. Samsung Electronics' Nano e-HEPA (High Efficiency Particulate Arrest) is one of several nano-enabled air filtration systems now on the market. A metal dust filter coated with silver nanoparticles eliminates 99.7 precent of influenza viruses, and another nano-filter eliminates all noxious VOC fumes from paint, varnishes and adhesives [12].

2.5 Energy Conservation and Conversion

Buildings consume nearly 50 percent of all energy used in the U.S., and lighting and other appliances use one third of the energy used in buildings. The amount of energy used by lighting could be reduced greatly by nanotechnology.

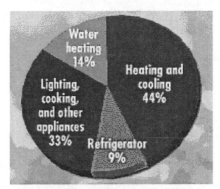

How We Use Energy In Our Homes

Figure 3: Residential energy consumption (US Dept of Energy, 2004).

"This will completely change the way we use lighting," said Professor Ravi Silva of The Advanced Technology Institute at the University of Surrey, developers of one such project. "Ultra Low Energy High Brightness Light (ULEHB) will produce the same quality light as the best 100 watt light bulb, but using only a fraction of the energy and last many times longer."

These new ultra low energy lighting devices will be fabricated using carbon nanotube-organic composites. ULEHB lighting may offer a cost efficient and clean replacement solution for mercury based fluorescent lamps and many other low efficiency heat-producing light sources. The technology can also be used for low cost solar cell production [13].

Similar hybrid organic-inorganic materials, organic light-emitting devices (OLEDs), are highly efficient and long-lived natural light sources in which ultra-thin layers of organic molecules are deposited on glass or transparent plastic. Since OLEDs are transparent when turned off, they could be installed as windows to mimic the feel of natural light after dark. Almost any surface in a home, whether flat or curved, could become a light source, including walls, curtains, ceilings, cabinets and tables.

The sun offers us a continuous source of clean, free energy. Solar represents less than .5 percent of today's energy market, but is growing 30 percent annually. Solar collection devices currently rely on silicon technology, but new solar nanotechnologies based on thin film materials, nanocrystalline materials, and conducting polymeric films offer the prospects of cheaper materials, higher efficiency, and flexible features.

Nanosolar produces thin-film solar cells that can be seamlessly integrated into building facades. The company is purchasing sites to build the world's largest factory for making solar power cells, which will ultimately produce enough cells to power 325,000 homes, tripling the U.S. production of solar cells [14].

Konarka has developed light-activated "power plastic" that is flexible, lightweight, lower in cost and more versatile in application than traditional silicon-based solar cells. It is made from conducting polymers and nano-engineered materials that can be coated or printed onto a surface, making it possible to incorporate a range of colors and patterns. Power plastic is bringing power-generating capabilities to building components including awnings, roofs, windows, and window coverings [15].

Work is even underway to create spray-on polymer-based solar collecting paint. "You just paint it on," said Wake Forest University Professor David Carroll of the nano-phase material with an efficiency of 6 percent, double that of similar cells, but still well shy of silicon's 12 per cent efficiency. "I strongly believe we can get there within the next year," Carroll said of the 12 percent efficiency goal [16].

2.6 Environmental Sensing

In just five years, the global market for nanosensors is expected to top $17 billion. Within a decade, nanosensors will be collecting and transmitting vast amounts of information about our environment and its users. Sensors smaller than a penny are already available that detect airborne toxins like carbon monoxide in and around a building. These nanosensors will conserve energy and resources by collecting and transmitting data on changes in temperature, humidity, and other indoor air quality factors.

Buildings will incorporate a rich network of interacting, intelligent objects, from light-sensitive photochromic windows to user-aware appliances. Tomorrow's green buildings will be constantly changing, as their components continuously interact with their users, their environment, and each other. These dynamic environments will be almost organic in their ability to learn and respond to changes, and architects will need to learn to design for change.

REFERENCES

[1] Sobolev and Ferrada-Gutiérrez, "How Nanotechnology Can Change the Concrete World: Part 2," American Ceramic Society Bulletin, No. 11, 16-19, 2005.

[2] Garcia-Luna and Bernal, "High Strength Micro/Nano Fine Cement," Proceedings of the 2nd International Symposium on Nanotechnology in Construction, 285-292, 2005.

[3] "A $276 Billion Problem," MMFX Technologies Corporation, http://www.mmfxsteel.com/index.shtml, 2007.

[4] "Ultra High Strength Stainless Steel Using Nanotechnology," http://www.azonano.com/details.asp?ArticleID=338, 2003.

[5] Min-Feng Yu et. Al., "Strength and Breaking Mechanism of Multiwalled Carbon Nanotubes Under Tensile Load," Science 287, 637-640, 2000.

[6] "U.T. Dallas-Led Research Team Produces Strong, Transparent Carbon Nanotube Sheets," http://www.utdallas.edu/news/archive/2005/carbon-nanotube-sheets.html, 2005.

[7] "Daylighting Systems with Nanogel," http://www.cabot-corp.com/cws/businesses.nsf/CWSID/cwsBUS200509130810AM2399?OpenDocument&bc=Products+%26+Markets/Aerogel/Overview&bcn=23/4294967102/1000&entry=product, 2007.

[8] "Ultra Thin High Performance Protective Insulation & Mold Prevention Coating," http://www.industrial-nanotech.com/nansulate_home_protect.htm, 2007.

[9] "Blockage of Ultraviolet Rays and Heat Insulation," http://www.suzutora.co.jp/MASA/MASA_ENG/ex03.html, 2007.

[10] Elvin, "Nanocoatings Transforming Automotive, Solar Cell and Wireless Industries," http://www.nanotechbuzz.com/50226711/nanocoatings_transforming_automotive_solar_cell_and_wireless_industries.php, 2006.

[11] "Nanotechnology Comes to the Rescue of Mouldy Wood," http://cordis.europa.eu/fetch?CALLER=EN_NEWS&ACTION=D&SESSION=&RCN=27181, 2006

[12] "Samsung Launches Nano e-HEPA Air Purifier System," http://www.azonano.com/details.asp?ArticleID=560, 2004.

[13] Elvin, "Variable Mood Lighting for Walls and Ceilings with Nanotubes," http://www.nanotechbuzz.com/50226711/variable_mood_lighting_for_walls_and_ceilings_with_nanotubes.php, 2006.

[14] Elvin, "Nanosolar Facility to Triple US Solar Production," http://www.nanotechbuzz.com/50226711/nanosolar_facility_to_triple_us_solar_production.php, 2006.

[15] "Konarka Builds Power Plastic that Converts Light to Energy – Anywhere," http://www.konarka.com/, 2007.

[16] Elvin, "Nanocoatings Transforming Automotive, Solar Cell and Wireless Industries," http://www.nanotechbuzz.com/50226711/nanocoatings_transforming_automotive_solar_cell_and_wireless_industries.php, 2006.

[1] Green Technology Forum, 9801 Fall Creek Rd. #402, Indianapolis, IN, 46256, www.greentechforum.net, elvin@greentechforum.net

Cleantech 2007, ISBN 1-4200-6382-0

Heat is the Enemy of Energy Efficiency: a Novel Material for Thermal Management

David Forder[*], Paul A. Fox[**]

[*]TAG Technology, 11-19 Bank Place, Melbourne, VIC, 3000 Australia, mail@tagtechnology.com
[**]CT Capital, 1702-L Meridian Ave #138, San Jose, CA, 95125, paul.fox@ctcapitalgroup.com

ABSTRACT

Control over the flow of heat has been a major challenge to engineers for centuries. A modern day example is a data center, where cooling accounts for 50% of the total energy costs. At the structural scale, a dark colored roof will increase the air conditioning load. Inside the building, the metal enclosures of the servers will trap and re-radiate heat internally. At a semiconductor scale, the processing power of the chips themselves is limited by their ability to dissipate heat.

A new nano-chemical, developed by TAG Technologies, can impede the flow of heat through materials in one direction only. This new material can be added to almost any product to control the flow of heat, save energy, and reduce greenhouse gas emissions.

Examples are presented in this paper including data centers, cool roofs, consumer goods packaging, refrigerated shipping, automotive, high voltage electrical conductors, and semi-conductors.

Keywords: thermal management, energy efficiency, data center, cool roof, semiconductor, greenhouse gas

1 INTRODUCTION

Engineers facing thermal issues should now consider the use of TAG technology, a novel material technology which can impede the flow of heat in one direction only.

This new material can be added to almost any product to control the flow of heat, save energy, and reduce greenhouse gas emissions. This paper describes:

- How TAG technology increases the infrared rejection and emissivity of a surface, thereby impeding the flow of heat in one direction while increasing its flow in the other.
- The example of a data center which shows the many applications where TAG could be used to increase the overall efficiency of the operation.
- How a TAG coated "cool roof" would pay back in 3 years compared to 4 years for the best competing technology.

- How TAG technology can improve thermal management in a range of other applications including consumer goods packaging, electronics, logistics, industrial processes, electrical transmission, automotive, and semiconductors.
- How, when assessing the applicability of TAG technology and tuning its characteristics, a number of factors must be considered.
- The future focus of research into TAG technology.

2 DESCRIPTION OF TAG TECHNOLOGY

TAG® is a unique additive to paints, coatings, membranes and films that can slash the cost of heating or cooling anything.

When added to a polymer matrix, TAG creates an interference effect, similar to a diffraction grating, blocking inbound radiant energy at the desired wavelengths while increasing the outbound emissivity in desired wavelengths.

Through this mechanism, TAG increases the effective infrared rejection and emissivity of a surface, thereby impeding the flow of heat in one direction while increasing its flow in the other (Figure 1).

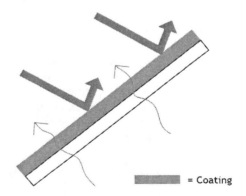

Figure 1: TAG Mechanism

The electromagnetic mechanism of TAG is still under investigation. Based on work carried out at Australia's leading energy technology research organization, the University of New South Wales, the increase in infrared rejection compared to a bright white surface is approximately 15% [1]. Our testing has also shown the emissivity compared to a black surface to be increased by around 7% for the selected wavelengths.

More importantly for commercialization, empirical testing by third parties has shown a significant beneficial impact in a number of industrial applications, several examples of which will be described in this document.

The mechanism of TAG is not influenced by color. Therefore, designers can get thermal performance as well as aesthetic appeal. Figure 2 shows the impact of TAG on the heating and cooling curve of a black painted sample under a standard sun. The untreated sample is the topmost curve. The underside of the TAG treated sample was significantly cooler as shown in the curve below. A bright white sample is shown on the bottom curve for reference. TAG has even been shown to have a similar beneficial effect in clear films for windows.

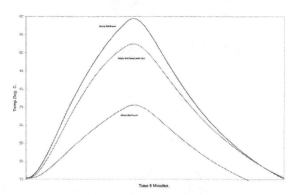

Figure 2: Impact of TAG on a Black Sample

A key benefit of the one-way effect of TAG, when compared to the use of insulation in thermal control, is that insulation slows the flow of heat in both directions. For example, the insulation of a building may slow the rate of heating during the day, but once the sun goes down it also slows the rate at which the interior can cool.

The TAG material is an organic compound which is non-toxic and indeed edible. The exact nature of the compound will be revealed in future papers subject to progress in IP protection. Considerable intellectual property and know-how is associated with the tuning of TAG to specific parts of the electromagnetic spectrum.

3 DATA CENTER EXAMPLE

According to the Lawrence Berkeley Lab, about 17 percent of US server farms are located in the San Francisco Bay Area and Silicon Valley, requiring about 80 megawatts of power to run. Aggregate electricity use for servers doubled over the period 2000 to 2005 both in the U.S. and worldwide. Of that increase in power usage, 90% is due to an increase in the volume of servers in data centers [2]. Blade servers are an important part of this growth and tend

to increase the power usage and density, while methods such as server virtualization decrease usage but increase power density. These trends will exacerbate both equipment and data center cooling issues for designers.

In the data center of the future, TAG could be used on the roof, cladding, ducting, plant, racks, enclosures of routers and servers, heat sinks, even the chips themselves to improve the dissipation of heat. In an integrated design, TAG could be used to optimize the flow of heat from the processor and electrical equipment and out into the environment.

Cooling accounts for 50% of energy costs in a typical data center [3], while thermal issues also have a significant impact on the energy efficiency of the servers, routers and power conversion equipment.

The building envelope is a simple entry point for the use of TAG in data centers. Paint enhanced with TAG could be applied to the roof to reduce the solar load on the air-conditioning. In California, such a project would be partially subsidized by the local utility as a demand management measure.

We have recently modeled the impact of a TAG coating compared to a dark colored roof and the best available current "cool roof" technology. The model was based on the DOE Cool Roof Calculator developed by the Oak Ridge National Laboratory (www.ornl.gov), the product data base at the Cool Roof Rating Council (www.coolroofs.org), the paint manufacturers' data sheets, and discussions with roofing companies.

We assumed a 100,000 square foot building located in Sacramento, California with an existing R-5 roof and an "average" existing cooling and heating system (i.e. this example does not have the high internal heat loads of a dedicated data center).

As shown in Table 1 below, TAG reduces the cooling load by increasing the effective solar reflectance and infrared emmittance of the roof. Moreover, unlike other cool roof technologies, TAG does not have a deleterious impact on heating costs during winter. This is because the TAG can be tuned so that infrared emmittance at those wave lengths are reduced. This peculiar effect is also extremely valuable when significant amounts of heat are being produced inside the building, e.g. in a data center.

A TAG coated roof would pay back in 3 years compared to 4 years for the best competing technology, while the first cost is only 12% higher. Importantly, the investment payback is now much less than the critical 4 year (~25%) hurdle rate.

Longevity is a concern for building owners. A typical "best in class" cool roof coating has a solar reflectance of 88% when new, but this will fall to 84% after 3 years due to weathering. Moreover accumulated dirt can reduce solar reflectance down to as little as 60-65%. The unique mechanism of TAG means that the effect of weathering is significantly reduced while dust has no impact on TAG.

Item	Black Roof	Light Roof	Best Current Cool Roof	Cool Roof with TAG
Cooling Load [Btu/ft²/ year]	15,398	9,377	2,980	1,372
Heating Load [Btu/ft²/year]	16,955	16,210	20,560	16,210
Cooling Saved ($)		8,800	18,200	20,562
Heating Saved ($)		1,500	(7,000)	n/a
Demand Charge Saved		6,600	13,300	14,906
Total Energy Cost Saving		16,900	24,500	35,468
Savings c.f. Light Roof			45%	110%
Total Cost of Painting ($)		99,000	101,280	111,792
Total Cost/sf		$0.99	$1.01	$1.12
Pay Back (years)		5.9	4.1	3.2

Table 1: Comparison of Cool Roof Technologies

While savings are important to building owners, internal temperatures are interesting from both the point of view of actual temperatures, and the perception of occupants. Independent testing [4] showed that TAG could keep the interior of a building 3°C to 7°C (5.4~12.6°F) cooler. As a rule of thumb, up to 50% more energy is required to decrease temperature by each additional 1 degree Celsius.

Much of this reduction is radiant heat from the interior surfaces of the building. A further rule of thumb is that where mean radiant temperature (MRT) differs from air temperature by more than 2°C, every 1°C rise in MRT requires a compensating drop in DBT of approximately 2°C [4]. This would therefore imply that the use of TAG to reduce the MRT might allow an increase in the air temperature, i.e. under cooling conditions the thermostat could be set higher to achieve the same level of perceived comfort at an even lower cost.

4 OTHER APPLICATIONS

This new material can be added to almost any product to control the flow of heat, save energy, and reduce greenhouse gas emissions.

For consumer packaged goods, independent testing of aluminum beverage cans coated with TAG showed 50% faster cooling. This type of application can lead to coatings that keep beverage cans colder longer, or cook food in cans much faster, to speed production filling lines or enable faster turnover at retail locations. Similar tests with food packaging demonstrated that food could last 60% longer in retail stores, resulting in substantial reductions in waste.

Another important part of the supply chain where TAG finds application is in shipping containers. Independent testing by the Kuwait Institute of Scientific Research found that the interior of containers painted with a TAG coating closely tracked the ambient temperature in the shade, and remained 9 to 14°C cooler than an un-insulated container and a surprising 7-11°C cooler than an insulated container. These differences would represent significant savings on logistic costs and wastage.

In automotive applications, TAG could help boost driver comfort and fuel efficiency. Independent testing of armored vehicles showed that TAG paint could keep the interiors 23% cooler at the peak of the day. TAG can also be used on engine components to increase the efficiency of heat dissipation, cool the engine, and increase the overall fuel efficiency.

Another application for TAG is reducing the temperature of high voltage conductors (Figure 3). Independent testing by Downer Industries showed that a clear coating containing TAG was able to reduce conductor temperature by 23°C, resulting in a sag reduction of 235% compared to the control. Using TAG would allow a T&D company to reduce power line sag for a given current (thereby increasing life or reducing maintenance) or increase the current capacity for given sag (thereby increasing the carrying capacity of the transmission line).

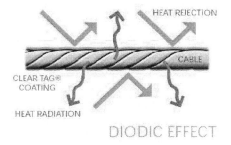

Figure 3: Impact on High Voltage Conductor

Thermal management is a particular concern for the semiconductor industry. The trend for higher power in smaller chips means that the heat flux of today's chips is similar to that experienced by the tiles of the space shuttle upon re-entry into the Earth's atmosphere. Unless this heat flux is addressed, chip performance can be reduced or the chip's lifespan could by shortened. Further research is required to determine the application of TAG in semiconductor packaging.

5 ASSESSING TAG APPLICATIONS

Assessing the applicability of TAG technology requires an understanding of the specific thermal issues to be solved, the chemical composition of the host polymer, the impact of substrates, and the design of measurement techniques used to tune the additive to a particular application. These issues will be the subject of future papers.

6 FUTURE RESEARCH

Future research will focus on the unique mechanism by which TAG technology influences the emissivity and infrared rejection of surfaces, as well as further empirical application testing.

REFERENCES

[1] Largent and Bardos, "Reflection Characterization of Coated Samples" Solarch, UNSW, 2001.

[2] J.G. Koomey, "Estimating the Total Power Consumption by Servers in the US and the World", Lawrence Berkeley national Laboratory, February 15, 2007

[3] Cisco, APC & Emerson Network Power data. 2006

[4] King, S. "Summary Expert Opinion", Solarach, University of New South Wales, march 2005

Self-Assembled Soft Nanomaterials from Renewable Resources

George John

Department of Chemistry
City College of the City University of New York
Convent Avenue at 138th Street, NY 10031
john@sci.ccny.cuny.edu

ABSTRACT

A set of amphiphilic glycolipids were synthesized from cardanol (a by-product of cashew industry) and diaminopyridine (DAP). These amphiphiles encompass self-assembling units such as long hydrophobic saturated or unsaturated chain, open or closed sugar as headgroup and aromatic (phenyl or DAP) as linker. Amphiphiles from both series (cardanyl and DAP) exhibited excellent self-assembling properties to produce various lipid based materials ranging from structurally unordered fibers to highly uniform nanotubes. Their self-assembling properties were investigated by various techniques including EF-TEM, SEM, XRD and DSC. The nanotubes are comprised of bilayer structure with interdigitated alkyl chains associated through hydrophobic interactions, hydrogen bonding and $\pi-\pi$ stacking. The self-assembling behavior of cardanyl glucosides and the synthetic analogues from diaminopyridine were compared. The tubes derived from DAP amphiphiles contain accessible 2,6-diaminopyridine linker that can interact with thymidine and related nucleosides through multipoint hydrogen bonding, thereby quenching the intrinsic fluorescence of the aromatic linker. These results clearly showed that efficient molecular design, and synthesis of novel amphiphiles from renewable resources will lead to supramolecular nanostructures and nanomaterials, otherwise under-utilised.

Keywords: organic soft materials, amphiphiles, self-assembly, lipid nanotube, renewable resources.

1 INTRODUCTION

The self-assembly of low molecular weight building blocks into nanoscale molecular objects has recently attracted considerable interest in terms of the bottom-up fabrication of nanomaterials [1]. Soft nanotubular structures represent a potentially powerful architecture generated through self-assembly of amphiphilic molecules [2-6]. Several classes of amphiphiles are known to provide these materials including lipid-modified peptides [3], bolaamphiphiles [4], and sugar-lipid conjugates [5]. The building blocks currently used in supramolecular chemistry are synthesized mainly from petroleum-based starting materials. However, bio-based organic synthesis presents distinct advantages for the generation of new building blocks since they are obtainable from renewable resources. An example of the latter is the sugar-derivatized cardanols [6], which consist of a carbohydrate head group and an aliphatic alkyl chain connected through a phenyl moiety [6a]. Under optimal solution conditions, these alkylphenylgluco-pyranosides form fibrous aggregates and nanotubes upon dispersion in water, particularly when the alkyl chain is unsaturated leading to a bent structure that induces supramolecular chirality [7]. Despite the ability of a wide range of compounds to assemble into coiled fibers [6, 7], no clear design rules have been formulated. Such information is critical to advance applications in medicine [8], chemical and biological sensing [9], and sub-micro-Total Analysis System (sub-μ-TAS) designs [10]. Here we report the facile nanotube preparation from cardanol based glycolipids, easily available from plant crop-derived resources, and further the design, synthesis and utility of a fully synthetic analogue from diaminopyridine (DAP) derivatives. Specifically, by combining simple monosaccharides, fatty acids, and diamino-aromatic linkers, we generated an array of morphologies ranging from fibers that lack structural regularity to highly uniform nanotubes.

2 EXPERIMENTS

2.1 Cardanyl Glucoside (1 and 2)

Cardanol (a mixture of long chain phenol differing in the degree of unsaturation in the side chain) was obtainable by double vacuum distillation of cashew nut shell liquid (CNSL) [11]. Cardanyl glucosides **1** and **2** were synthesized in two steps from penta-O-acetyl β-D-glucose and the corresponding phenol as reported earlier [6]. The glycosidic bond was formed in a Lewis acid-catalysed (borontrifluoride etherate) reaction at room temperature to give the β-product exclusively.

1: a (5%)+b (50%)+c (16%)+d (29%) **2**: a (100%)

Cardanyl glucosides

The acetylated β-glucopyranosides were purified by recrystallization from ethanol and deprotected quantitatively using trimethylamine in aqueous methanol. The crude products were purified by silica-gel column chromatography and recrystallization in methanol afforded a cardanyl glucoside **1** and its saturated homologue **2**. The self-assembled fibrous structure was prepared by dissolving 1-5 mg in 50-100 mL of water at boiling temperature. The clear solution obtained was cooled to room temperature at ambient conditions. The helical fibers obtained from compound **1** were aged to several days at ambient conditions yielded tubular structures.

2.2 Diaminopyridine (DAP) Amphiphiles (3 and 4)

1,3-diaminopyridine was coupled with β-D-glucose by reductive amination and further purification procedures afforded monomer components used in this study [12]. Of particular interest were compounds **3** and **4**, which formed fibrous assemblies upon dispersing in water following vortexing at 100°C for 30 min, slow cooling to room temperature, and incubation for 12 h.

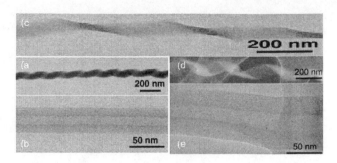

DAP amphiphiles

3 RESULTS AND DISCUSSION

The cardanyl glucoside mixture **1** self-assembled into characteristic helically-coiled ribbons after 12-24 h incubation [Fig. 1(a)], whereas the saturated analogue **2** gave a helically-twisted morphology [Fig. 1(c)]. Thus, we found that the difference in degree of unsaturation on the hydrophobic long chain phenols could generate variety of morphologies on self-assembly. This might be one of the first examples of morphological control of helical nanofibers by unsaturation of the hydrophobic segment in a single tail amphiphile. Interestingly, coiled ribbons formed from **1** were gradually converted into a tubular structure over several days, but the twisted ribbon remains intact even after many months. High resolution transmission microscopy (TEM) revealed that the obtained lipid nanotubes have uniform inner diameters of 10-15 nm, outer diameters of 50-60 nm, and extended lengths of 10-1000μm [Fig 1(b)] and the aspect ratio is ~1000. Nevertheless the cardanyl glucosides produced intriguing morphologies from self-assembly in water; those materials lack the ability of efficient binding of target molecules. This prompted us to design and synthesis of DAP amphiphiles capable of generating functional nanostructures, utilizing the chemistry of cardanyl glucosides.

Figure 1. TEM images of (a) an individual helically coiled nanofiber from cardanyl glucoside mixtures, **1** (b) an individual nanotube from cardanyl glucoside mixtures, **1**. (c) a helically twisted nanofiber from saturated glucoside, **2**, and TEM images of self-assembled morphologies of DAP lipids, **3** in water, (d) helical coiled fibers and (e) nanotubes. The dimensions of the nanotubes were uniform and reproducible.

The synthesis of DAP derivatives was accomplished by combining simple monosaccharides, fatty acids, and diamino-aromatic linkers, which on self-assembly in water generated an array of morphologies ranging from fibers that lack structural regularity to highly uniform nanotubes. Visible and fluorescence microscopy showed bundles of self-assembled structures, with the latter showing identical fluorescence properties as free DAP (λ_{ex}=263 nm; λ_{em}= 535 nm). TEM images after 4 h showed that the precipitates from **3** formed helical ribbon morphologies, which upon aging for an additional 12 h yielded nanotubes with an outer diameter of 60-80 nm and an inner diameter of ca. 20 nm (Fig. 1d, e). DSC analysis of the nanotubes from **3** showed a gel-to-liquid crystalline phase transition temperature (T_m) of 70°C, which was far higher than the T_m of 42°C for the unsaturated cardanyl glucoside system [6a,b]. Despite the relatively high T_m; the nanostructures from **3** were not highly crystalline. The lack of the double bond in the alkyl chain to give **4** had a significant effect on the morphology of the resulting nanostructure. Instead of nanotubes, the precipitates from **4** formed fibrous structures with a thickness of ca. 80-150 nm. These fibers had a relatively high T_m of 90°C, indicating more significant crystalline packing compared to the unsaturated systems from **3**. Hence, the unsaturated oleic acid moiety appears to be critical in nanotube formation.

To gain insight into the molecular orientation and packing profile within the assembled morphologies from **3** and **4**, and to understand why the two similar monomers gave strikingly different self-assembled morphologies, we examined the wet and dry forms of the self-assembled nanofibers *via* small angle X-ray scattering. The molecular length of the amphiphiles was calculated by CPK modelling on the basis of single crystalline data of oleic acid [13], and then X-ray diffraction patterns were obtained. The small-angle diffraction patterns of the nanotubes from **3** revealed ordered reflection peaks with a long period of 3.5 nm, which is substantially smaller than twice the extended molecular length of **3** (*d*-spacing of 3.03 nm by the CPK molecular modelling).

These results strongly suggest that the nanotubes from **3** (nanotube from **1** also adopts similar morphology) form a bilayer structure with interdigitated alkyl chains associated through hydrophobic interactions (Fig. 2). Moreover, according to powder X-ray diffraction analysis, the

Cleantech 2007, ISBN 1-4200-6382-0

glucopyranoside moieties of the bilayer participate in strong intermolecular hydrogen bonding, which results in a highly ordered chiral packing structure.

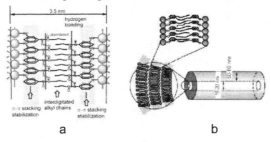

a   b

Figure 2. Shematic representation of (a) DAP, **3** solid bilayers (b) molecular packing within the lipid nanotube.

This combination of hydrophobic and hydrogen bonding interactions appears to favour the formation of the nanotubular structure. Conversely, the diffraction pattern of the nanofibers from **4** (molecular length of 3.2 nm) indicated a shorter *d*-spacing of 3.3 nm, which translates into a greater degree of interdigitation of the bilayer structure. The "kink" in self-assembled structures from **3** appears to reduce the crystallinity of the nanostructure enabling more facile formation of a nanotube.

Aminopyridine derivatives are known to function as artificial receptors that can bind various ligands through complementary multipoint H-bonding [14]. Hence, we reasoned that the DAP residue could serve as a functional recognition element, and in the process, its intrinsic fluorescence would be affected by selective interaction with external ligands. To test this hypothesis, we added water-soluble compounds that can undergo H-bonding to the nanotube from **3**. Addition of up to 10 mM thymidine caused the nearly immediate quenching of fluorescence (Fig. 3a), with an apparent binding constant of $\sim 2.5 \times 10^3$ M^{-1}. In addition, thymidine analogs such as uracil and the anticancer compounds 5-fluorouracil also quenched fluorescence.

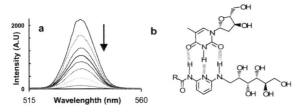

Figure 3. (a) Fluorescence quenching of DAP nanotubes by thymidine. The thymidine concentrations ranged from (top to bottom) 0 to 10 μM, emission (λmax = 535 nm), (b) schematic representation of the possible interaction between nanotubes of **3** and thymidine.

The fluorescence quenching was selective for nucleosides; β-D-glucose and urea, while capable of undergoing extensive H-bonding, did not quench the fluorescence of the nanotubes from **3**, even at concentrations as high as 16 mM (100 fold excess the nanotube concentration, based on the concentration of **3**). These results suggest that the interaction between the nanotubes and thymidine might occur through a three-point hydrogen bonded network [14] (Fig. 3b).

These and above results showed that the structural requirements for self-assembly of amphiphilic monomers

into highly organized and functional nanotubes have begun to be elucidated. These include the combination of strongly hydrophobic and hydrophilic moieties, a linker with suitable planarity, and hydrogen bonding interactions of hydrophilic groups that are favoured in sugars. In addition, substantial bending of the monomers is required, which arises from the *meta* orientation of the linker, along with unsaturation of the alkyl chain. This information can be used to design single chain amphiphiles that form high-axial-ratio nanostructures starting from simple molecules, which also contains molecular recognition groups that can be used to monitor the chemical selectivity of supramolecular aggregates towards guest binding, and for bionanocomposites. Taking together, present study clearly demonstrates the utility of plant/crop-based resources and their industrial by-products as an alternate feedstock for existing and new chemicals, detergents and functional soft materials.

4 ACKNOWLEDGEMENTS

Thanks are due to Dr. T. Shimizu, NARC, AIST, Japan and Prof. J. S. Dordick, Dept. of Chemical Engineering, RPI, Troy, NY for their support and encouragements.

REFERENCES

[1] G. M. Whitesides, J. P. Mathias and C. T. Seto, *Science* (254), 1312, 1991.

[2] (a) N. Nakashima, S. Asakuma and T. Kunitake, *J. Am. Chem. Soc.* (107), 509, 1985. (b) P. Yager and P. E. Schoen, *Mol. Cryst. Liq. Cryst.* (106), 371, 1984. (c) B. N. Thomas, C. R. Safinya, R. J. Plano and N. A. Clark, *Science* (267), 1635, 1995.

[3] (a) J. M. Schnur, *Science* (262), 1669, 1993. (b) T. Kunitake, *Angew.Chem. Int. Ed. Engl.* (31), 709, 1992.

[4] (a) J. -H. Fuhrhop and W. Helfrich, *Chem. Rev.* (93), 1565, 1993. (b) T. Shimizu, M. Kogiso and M. Masuda, *Nature* (383), 487, 1996.

[5] (a) D. F. O'Brien, *J. Am. Chem. Soc.* (116), 10057, 1994. (b) H. Engelkamp, S. Middlebeck and R. J. M. Nolte, *Science* (284), 785, 1999.

[6] (a) G. John, M. Masuda, Y. Okada, K. Yase and T. Shimizu, *Adv. Mater.* (13), 715, 2001. (b) G. John, J. H. Jung, H. Minamikawa, K. Yoshida and T. Shimizu, *Chem. Eur. J.* (8), 5494, 2002.

[7] (a) J. V. Selinger, M. S. Spector and J. M. Schnur, *J. Phys. Chem. B,* (105), 7157, 2001. (b) D. Berthier, T. Buffeteau, J.-M. Leger, R.Oda and I. Huc, *J. Am. Chem. Soc.* (124), 13486, 2002.

[8] X. Guo and F. C. Szoka Jr., *Acc. Chem. Res.* (36), 335, 2003.

[9] B. E. Rothenberg, B. K. Hayes, D. Toomre, A. E. Manzi and A. Varki, *Proc. Natl. Acad. Sci. USA,* (90), 11939, 1993.

[10] T. Vilkner, D. Janasek and A. Manz, *Anal. Chem.*(76),3373, 2004.

[11] J. H. P. Tyman, *Chem. Soc. Rev.* (8), 499, 1979.

[12] G. John, M. Mason, J. S. Dordick and P. M. Ajayan, *J. Am. Chem. Soc.*(126), 15012, 2004.

[13] J. Ernst, W. S. Sheldrick and J. –H. Fuhrhop, *Naturforsch.* (34b), 706, 1979.

[14] R. Shenhar and V.M. Rotello, *Acc. Chem. Res.* (36), 549, 2003.

Designing Today's CleanTech Research Facility:
Sustainable, Highly Technical Architectural and Engineering Design Applied to the New York State Alternative Fuel Vehicle Research Laboratory (AFVRL)

Joseph Ostafi, AIA, LEED® AP* and David F. Sereno, PE, Principal**

*Flad & Associates, 644 Science Drive, Madison, WI, jostafi@flad.com
**Affiliated Engineers, Inc., 5802 Research Park Boulevard, Madison, WI, dsereno@aeieng.com

ABSTRACT

The greatest challenges facing our world today are the need to reduce the pollutants entering our atmosphere AND the need to create fuels that can support our expanding global economies. The opportunity to support cutting-edge research on both challenges in one facility is truly inspiring. That is the basic premise behind the New York State Department of Environmental Conservation's *Alternative Fuel Vehicle Research Laboratory* (AFVRL).

The creation of working environments to support complex science while mitigating the impact of the facility on the natural environment has never been as important as it is today. The challenge of designing a laboratory for the science of fuels optimization and emissions reduction, while reducing overall energy consumption has yielded results surpassing the goals of the State of New York, setting them on a path to a ground-breaking new science facility.

Keywords: sustainable, LEED, architecture, engineering, environmental, emissions, laboratory, engine testing, facility

1 FACILITY MISSION

It is only fitting that a facility designed to study the crucial issues of fuel utilization and a cleaner environment respect the environment as much as possible. Functional performance is essential, but so is the need to "walk the walk" and develop the site sensitively, minimize non-renewable energy consumption, and create a stimulating place where people meet, collaborate, and solve complex problems.

The facility will house the State's Department of Environmental Conservation's (DEC) Bureau of Mobile Resources and Technology Division (BMRTD). Here scientists will analyze the transportation sector's emissions standards, translating accurate and repeatable data into policy. This once sequestered science becomes the focus of a state-of-the-art center that encourages academic interaction and public engagement.

The BMRTD has also facilitated numerous collaborative efforts involving research and emissions testing with the New York City Department of Environmental Protection (DEP), the United States Environmental Protection Agency (EPA), New York City Metropolitan Transit Authority (MTA), the New York City Taxi and Limousine Commission, New York State Energy Research and Development Authority (NYSERDA), and private and academic researchers.

2 BUILDING PERFORMANCE

The research and testing capabilities are required to accommodate a wide range of vehicular configurations and to accurately measure emissions derived from conventional and alternative propulsion fuels. The facility is also required to be technically and operationally accessible to other government agencies, authorities and industry groups in support of collaborative projects with shared objectives. The unique technological capabilities of the facility will be exploited to accommodate links with educational institutions in support of engineering science in graduate and undergraduate programs.

The facility will not only house an 800-horse-power chassis dynamometer, engine dynamometers, vehicular test/preparation area, chemical analysis, particulate matter, and clean weigh laboratory. The following are the core functional testing support functions within the building:

- Heavy-Duty Chassis testing cell
- Light/Medium Duty Chassis testing cell
- Engine testing cell
- Particulate Matter (PM) Lab
- Chemistry Analysis Lab
- Portable Emissions Measurement Systems Lab (PEMS)
- Filter weigh Particle Clean Lab Building Design

Set within a clean technology research and development campus (the first of its kind in the nation), the *AFVRL* project is a culmination of clean technology practice coupled with an appropriately contextual high-tech, Adirondack-style architecture. This has been achieved by utilization of local materials, primarily granite stone, and use of wood veneer products as accent ceilings and exterior treatment melded with clean lines of glass curtain wall systems and photovoltaic cells.

The building is organized by programmatic function. Half of the main floor is dedicated to vehicle testing and preparation, the other half dedicated to more people focused spaces like labs and offices. What happens in between and throughout the remaining building support areas are opportunities for employee interaction and collaboration. Circulation paths cross frequently, punctuated by amenities and natural places for pauses and impromptu conversation and engagement. Transparency is also an architectural feature that is celebrated wherever possible, further creating opportunities for individuals to witness one another's findings, thought process, and collaboration.

The form of the building roof was greatly inspired by the "Mobius Strip," a mathematical model used to explain a number of scientific mechanic scenarios as well as genetic replication sequences phenomena. This model is also known to be the design impetus of the recycling symbol which captures the mission of the facility as well as its people. The shape of the Mobius is designed into the building as a metaphor of its purpose, while serving as a mechanism tying very complex programmatic forms of the building together into one cohesive architectural solution.

Finally, form and function meld seamlessly in the new Alternative Fuel Vehicle Research Laboratory where the concepts behind the facility are driven by such disparate design criteria as:

- Accommodating the turning radii of a New York City articulating transit bus internal to the building
- Smart building organization - separation of noise, vibration, harshness and nano-scale measurement cleanroom scientific labs
- Successful integration of environmental building systems

3 SUSTAINABLE DESIGN INTEGRATION

Being part of a department that supports state environmental policy legislation, it is imperative that the bureau reflects its mission through the architectural and engineering design of this new laboratory. This facility exemplifies the spirit of the mission by employing renewable and innovative alternative energy sources as the ways and means to condition, power, and integrate the building into its natural setting.

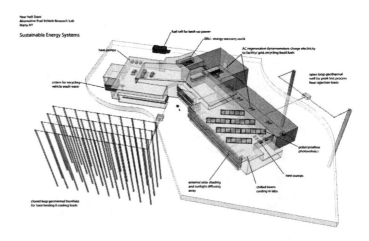

This is accomplished by employing innovative architectural and engineering sustainable building concepts into a performance based structure. More specifically, it is accomplished via the following systems:

- Geothermal (closed-loop) base load heating and cooling
- Geothermal (open well) for process peak load shaving
- Energy recovery wheels to offset high outside air requirements
- Electrical regeneration via AC dynamometers
- Passive and active day lighting strategies
- Photovoltaic system
- Water conservation, both on domestic and process loads

4 AUTHORS' BIOGRAPHIES

4.1 David Sereno, PE, Principal, Affiliated Engineers, Inc.

Mr. Sereno leads AEI's industrial practice and serves as project manager on many projects. Dave's mechanical engineering expertise includes designing transportation sector test / emissions facilities, alternative fuels test facilities, diesel engine assembly plants and aerospace test facilities. His experience spans a wide range of technical facilities, including alternative fuel test and emissions (gas, diesel, hydrogen, bio-diesel and fuel cells) test laboratories. Dave is currently the project manager for the New York State Alternate Fuel Vehicle Research Laboratory in Malta, New York.

4.2 Joseph Ostafi, AIA, LEED AP Project Architect/Designer, Flad & Associates

Mr. Ostafi has over nine years of architectural experience in research facilities for academic, corporate and government clients. His focus has been in the design of science and technology, laboratory planning, clean technology, and site master planning. An innovative designer, he thrives on projects that present unique and complex sets of challenges. Joseph is currently project architect and designer for the New York State Alternate Fuel Vehicle Research Laboratory in Malta, New York and Stony Brook University Advanced Energy Research and Technology Center.

5 COMPANY OVERVIEW

Over the past 25 years, Flad & Associates and Affiliated Engineers have achieved a national reputation for the design of some of the most important and complex research facilities in the country. Completed planning and design work totals more than 40 million square feet of laboratory and research space for academic, corporate and government clients including the Department of Energy. A significant focus of our staff has been on projects related to energy research, emissions, and fuels optimization, including projects that ensure reliable, economical, and alternative sources of energy to support sustained economic development without impairing the natural environment.

Good Policy Makes Good Business

Barry Cinnamon

California Solar Energy Industry Association
605 University Ave, Los Gatos, California, barry@akeena.net

ABSTRACT

In rapidly emerging industries like Clean Technology, lawmakers are challenged to come up with good public policy that will accomplish society's goals. Two notable new, well-funded policies such as the $3.2 billion California Solar Initiative (CSI) and the proposed "Securing America's Energy Independence Act (HR550)" -- which would provide a $3,000 per kilowatt federal tax credit for solar installations, are great for business.

For example, California's CSI will cost ratepayers $3.2 billion, but will provide over $6 billion in incremental savings by avoiding electrical infrastructure costs. The Governor's desire to achieve the results of this program within ten years - a short time frame for such an ambitious project is wise. Delay simply reduces the benefits available to Californians and dilutes the impact the state can have on cost-effectively improving its energy infrastructure, the environment, and the economy. As the timeframe for program implementation stretches out, the lost opportunity costs, in the form of infrastructure costs incurred, pollution not abated, and jobs not established, increase substantially.

Moreover, designing a long-term program provides certainty for the necessary manufacturing and installation infrastructure to meet these goals. If the experience of Japan is any indication, at the end of this ten-year period there will be a sustainable solar industry in the state that will no longer require incentives and will continue to provide the benefits outlined in this paper for future years and will be unlike any other in the U.S.

In summary, there is a set of guiding principles which will ensure the development and implementation of an effective, cost- and time-efficient solar systems program that will deliver the benefits envisioned by the Governor's Office. These principles are:

- Include and quantify all of the benefits and costs of solar power when evaluating the program

- Create dynamic, market-based incentives that drive end-customer behavior

- Base analyses on actual market data to ensure effectiveness

- Retain the 10-year timeframe to provide assurances for long-term industry investment

The true drivers of the solar market are end-customers making purchase decisions in each segment. Although there is a great deal of political support for solar power among various segments of the population, at the end of the day these purchase decisions are economic decisions in which the customer roughly calculates his/her own net benefit from adopting a solar technology.

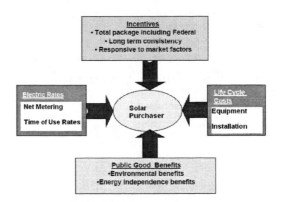

As shown in the above chart, the customer calculus includes all of the factors driving the cost of the solar system and comparing those to the potential savings the system can provide plus any "public good" factors (e.g., contributing to improving the environment). The current lack of a sufficiently adequate economic payback for many potential customers is a plausible explanation of why adoption rates are still relatively low, even thought very high levels of support has been expressed towards reusable energy.

For consumers, while the public good benefits of solar are often what may motivate a customer to investigate solar in the first place, it is important to note that the purely economic equation must work for a purchase decision to be made. A recent survey of photovoltaic dealers regarding their customers' characteristics indicates that those consumers who make decisions purely or even mostly on environmental grounds are a small percentage of total purchasers, and in fact, an increasing proportion of new customers were motivated primarily or exclusively by economic factors.

Commercial customers are entirely motivated by the economic benefits achievable through solar adoption, which include near-term savings, insulation from future electric rate increases, and promoting their own businesses through solar use.

Government customers are often similarly motivated, with some exceptions where the government entity wants to set an example by adopting solar.

Obviously, programs like CSI and HR 550 are great for business since they help to jump-start an entire industry. In addition, they are terrific for the environment. As a result, it's going to be a green light for investments in the solar industry with these policies in effect.

CSI and HR550 are examples of successful policies, but Cleantech companies should not stand on the sidelines waiting for the policy pendulum to swing their way. If a company or industry has solutions that, when fully implemented, are good public policies, then by all means these companies could actively advocate for favorable policies. Companies should indeed consider where their technologies are most beneficial, and then advocate for public policy that would accelerate their market adoption. Policies can be created by businesses.

REFERENCES

[1] "Solar Electric Dealers 2004", Coast Hills Partners.
[2] "The Economics of Solar Systems in California', Bernadette Del Chiaro, Environment California Research and Policy Center".
[3] California Solar Initiative

CLEAN TECH: NEW PARADIGMS AND COMMUNICATIONS CHALLENGES

Principle Author - Marianne Allison, Chief Innovation Officer, Waggener Edstrom Worldwide

Co-Authors – Andrew Pray, Haley Drage, Phil Missimore - Waggener Edstrom Worldwide

ABSTRACT

As clean technologies move into even greater prominence in the minds of consumers and on the screens and pages of the media world, the noise level surrounding the clean tech environment is rising as well. How will companies find their authentic voice? How will they reach their true target customers in ways that are impactful and with messages that resonate and encourage action?

This paper focuses on applying proven communications techniques—telling stories in ways that create memorable and differentiated communications--as a way to boost and enhance the profile of innovative clean-tech and nanotech companies.

Keywords: Marketing, PR, Communication, Strategy

INTRODUCTION

In recent years there has been a rising call from businesses, governments and consumers for the development of a diversified global energy future. The dynamics driving this are well known: a combination of the rising cost of fossil fuels, the finite nature of high quality hydrocarbons, environmental issues stemming from greenhouse gas emission and the geopolitical need to diversify energy resources.

With this in mind, investors, governments and environmentalists are eager to explore all avenues of energy and its creation. The search for the most cost-effective, most efficient, and most inexhaustible methods is on — will your communications help you lead or follow?

ENERGY – A SHIFTING PARADIGM

Energy as we know it — how we define it, where we get it and what we expect from it — is undergoing seismic changes. The energy market is facing new challenges as they work to market, distribute and protect their products. Indeed, energy companies, whether new or old, alternative or traditional, are experiencing a significant transformation. We need only look as far as global news headlines and policy developments to understand how the energy marketplace faces a daily churn of interest and exposure.

In a phrase, the energy industry is in the midst of a paradigm shift: a period characterized by upheaval, change, innovation, uncertainty, competing visions and diverse priorities. As clean technologies move into even greater prominence in the minds of consumers and on the screens and pages of the media world, the noise level surrounding the clean tech environment is rising as well. How will companies find their authentic voice? How will they reach their true target customers in ways that are impactful and with messages that resonate and encourage action?

Communications challenges confront clean and nanotech companies—from start-ups to industry leaders—at every corner. Considering the high level of interest and media coverage on cleantech and nanotech, and its impact on investors, consumers and partners, the importance of effective communications has become even more critical.

So, what are you going to do about it?

AN OPPORTUNITY TO LEAD

Some companies are driving innovation and change, others are reacting to it and some are caught between these two extremes. Whatever your position, enterprises that exist in this tornado of innovation (we call it the Innovation Context™) are faced with market uncertainty, industry shifts, rapid evolution and breakthrough opportunities.

Through over 20 years of work with innovators across different markets, Waggener Edstrom Worldwide (WE) has observed that while critical innovations themselves are necessary, they're not sufficient for crossing the chasm. Thriving in the Innovation Context is also a <u>communications opportunity</u>: it spurs dialogue and debate and requires explanation and persuasion. Therefore, it also necessitates a special understanding of communications predicated on how audiences are influenced and the value of a story well told.

Innovative companies generally fall into one of the following scenarios within the Innovation Context:

> Established Leaders — "Big oil" companies tend to fall here, with all of the attendant opportunities (big resources/big stakes/big respect) and pitfalls (assumptions about complacency and defensiveness, big surface area for the challenger to attack.) Leaders have to aggressively reinvent themselves: For example, Shell Oil's "Renewables" division is forging ahead with strategic alliances in new cellulose ethanol technologies.

> Insurgent — These are the challengers—they can be edgier (everyone loves the David and Goliath story) but they have to prove their staying power in order to be more than a provocative irritant. Community Energy Inc. built America's first commercial wind farm in Atlantic City, N.J. Ocean Power Technologies (OPT) and the United States Navy are jointly testing OPT's wave energy technology.

Competitive — The challenge here is avoiding the feature-wars and building a sustained differentiation. Jockeying back and forth for market favor, Toyota Prius and the Accord Hybrid find themselves in competitive landscape. However, in a nod to grassroots inventors and enthusiasts, Toyota will begin to manufacture a "plug in" model of their flagship Prius hybrid.

Paradigm Shift — Paradigm shifters are the disruptors. Their role is almost entirely "alternative." Their communications imperative is articulating a compelling vision of a big destination, of course, but they can't look like mad scientists; they have to demonstrate an achievable path if they are to build an ecosystem, and persuade existing players to come along with them. In April of 2006, UBS AG and Diapason Commodities launched the first Global Biofuel Index, a way to introduce alternative fuel to investors in an entirely new way, validating both the category and their position as thought leaders in the category.

Regardless of innovation scenario or status as a large enterprise, small company or trade group, your success in the cleantech sector hinges on effective communication of your approach. Those who set the energy agenda and own the global conversation will do so by facilitating and driving the adoption of new ideas.

What we've learned is that when paradigms shift, both confusion and opportunity result. Eventually, a leader steps forward to fill the vacuum and becomes a leading voice in the new paradigm, are you ready to speak up and guide your customers, partners, governments, suppliers and employees to a new future?

SHIFTING SANDS: INTEGRATED COMMUNICATIONS & STAKEHOLDER INVOLVEMENT

Meanwhile, the media landscape that is so integral in your communications is changing – and quickly. The age of information means the democratization of information; it is more plentiful and accessible than ever before. Connected by the Internet, individuals can discover, process and share information across thousands of miles in a matter of seconds without having to "recheck" assumptions by what is published in mass media outlets. Now, a more unstable model of influence has emerged, a model that reflects how disparate people — from your partners to your employees to your legislators and regulators — interconnect to tell your story.

Although formal media institutions continue to wield considerable power , they are no longer the *only* sources of information and influence. The profound changes in the energy marketplace are catalysts for an explosion of information and participation by consumers in the information process. Recognizing the shift from a predictable communications model with sequential process and phased timing to a real-time, networked and relationship-driven model is the first step toward controlling the new influence model by understanding commonalities among each of these new influencers.

This new, fluid model of influence necessitates a holistic approach to communications: one that considers employees alongside customers, one that brings academics and public officials to the table early and often and one that employs real-time communications to offer transparency and engage stakeholders. Audiences now expect to participate and then experience and decide for themselves.

Tomorrow's leaders will involve myriad stakeholders (e.g. customers, partners, governments, independent organizations, etc.) in the energy paradigm shift, hear and include their voices and never underestimate the power of their involvement, particularly when introducing new ideas.

Thus a strategic approach to influence has to play out against this evolving backdrop. It requires a broad influence approach—we call it Integrated Influence—including media relations encompassing traditional media outreach and factoring in how traditional media themselves are morphing as well as new social media. And it also includes public affairs, analyst relations, investor relations and employee communications, and a deep conversation about business and social responsibility at the front end.

The ensuing conversation will influence of policy and perception. This process will be challenging; the influence model for the energy sector is truly unique, with interrelated influencers, from members of the mainstream media and government elites to academics and grassroots bloggers.

However, the potential rewards are rich for the company or organization which starts and extends the conversation globally, enabling connections with the influencers that matter most, changing perceptions and building trust and acceptance for new ideas and innovations — *their* ideas and innovations.

THE POWER OF Innovation Communications™

Launching innovations goes way beyond publicizing, beyond thought leadership, beyond ecosystem building—it is all of those things and more. A comprehensive approach to driving acceptance and adoption of an idea or concept, involves helping diverse audiences:

1. Discover, to build awareness

2. Learn, which helps individuals and organizations understand the basics—i.e., some construct for "how it works"

3. Envision, during which the future relevance of the innovator should be ensured (What does this mean to me? How does it work? Is it safe? Does this new thing work with my old thing? Where do I get supplies? Who will support it? etc...) and finally:

4. Inspire, building trust while encouraging action and the adoption of the new idea.

The sheer breadth of technology and competing forces in the clean tech sector can make this four-step process seem very daunting. Clearly some of the core elements of ANY good communications story are at play here—authenticity of voice, including key partners and customers in your communications outreach so that influentials hear voices of validation around your direction—they are communications constants. But in an environment where scientific awareness amongst broad consumer audiences is not high—and detailed information can be focusing, how can companies stand out?

ACTION

Ironically, perhaps, in the face of all this complexity, simplicity in the form of a great story is absolutely essential. Stories reveal a company's characters, their values, and at the core of Innovation Communications™ is a commitment to Storytelling. So what are some ways you can start to tell your story better?

1. **Stage your communications**- Who are you trying to reach and at which stage? Think about your audience plan as concentric circles: each audience requires a different influence plan and timetable and approach.

2 **Who cares?** - It is vitally important to not only have a clear sense of which media you need to have mindshare with, but who's driving the dialogue that influences them. In a time of uncertainty and disagreement about who's got the best vision, media will turn to a few critical voices to help them sort it out. For each audience segment you need to engage with, there are key opinion leaders, and they're not necessarily elites: they may be simply "the voice of" the little guy. Know who they are and make your communications all about influencing the influencers.

3 **Join the Dialogue** - Bringing your own voice into the discussion builds trust and allows you to socialize your ideas, explain your process, discuss your tradeoffs (innovation is always hard: people need to see the sincerity of your effort).

4 **Framing Your Story:** Georges Polti wrote that there are only 36 plots represented in every story ever told. How do you apply the universal concepts of storytelling to your communications? Do you have a typical "quest" plot, triumph of good over evil, sin/judgment and redemption?

5 **What Does Green Mean?** – More and more people are asking what does "good business" mean these days and most are responding by integrating Corporate Social Responsibility programs into their overall business model. Even when being green IS your business, know the burden and opportunity of being expected to walk the talk. In fact, the greener your proposition, the more opportunity for transparency (for better or for worse).

CONCLUSION

The lessons we've learned from our years of service with a broad range of innovators ring as true for clean technology as they do in deep technology – stories of innovation aren't enough. They need to be emboldened with powerful, thoughtful and strategic story-telling methodology in order to cut through the fray, separating your company from your competitors while establishing your voice as an industry and category champion.

Photovoltaics Innovation and Commercialization (PVIC) in Ohio

Dean M. Giolando[*], Robert Collins[**], Robert J. Davis[***]

[*]PVIC at The University of Toledo, Department of Chemistry, Toledo, OH, USA, dean.giolando@utoledo.edu
[**]PVIC at The University of Toledo, Department of Physics and Astronomy, Toledo, OH, USA, rcollin8@utnet.utoledo.edu
[***]PVIC at The Ohio State University, NanoTech West, Columbus, OH, USA, davis.2316@osu.edu

ABSTRACT

Photovoltaics Innovation and Commercialization (PVIC) was funded to strengthen the photovoltaics research and manufacturing base. Activities aim to eliminate market barriers faced by companies in the photovoltaics sector. Companies active in the photovoltaics industry, from those researching advanced materials development to those installing energy producing devices, advise and coordinate PVIC members. For maximum impact on increasing production efficiency and lowering costs, PVIC takes a vertically integrated approach from research in advanced materials to the fabrication of production-scale modules, to issues related to installation, and finally to aspects of customer acceptance. Collaborators in PVIC possess knowledge of how to overcome real-life problems arising in connecting a module to the electric grid and how to obtain customer support for building integrated PV designs. Companies along the entire value chain have been brought into PVIC.

Keywords: industry-academics, photovoltaics, materials, small business, manufacturing

1 INTENT OF PVIC

The Center for Photovoltaics Innovation and Commercialization (PVIC) supports activities centered on eliminating commercialization barriers currently facing companies in the photovoltaics (PV) sector. Ultimately, PVIC consists of a development center with an infrastructure attractive to companies already successfully marketing PV and to researchers of the future generations of PV devices. These activities bring together established companies and researchers seeking to be at the forefront of developments in the PV industry to seed the formation of new startup companies. To fulfill its mandate, PVIC concentrates on photovoltaics and on the generation of hydrogen fuel from PV-driven electrolysis of water. For maximum impact, PVIC takes a vertically integrated approach from research in advanced materials to the fabrication of production-scale modules, to issues related to installation, and finally to aspects of customer acceptance. Collaborators in PVIC possess knowledge of how to overcome production problems that arise in connecting a PV module to the electric grid and how to obtain customer support for integrated PV designs.

Companies along the entire value chain contribute to PVIC, and each has identified its current market status relative to shipping products or providing services to customers. Each company identified the commercialization barriers existing between their current positions and where they need to be in order to satisfy customer demands. Equipment, instruments, and techniques were identified to eliminate commercialization barriers faced by each of the companies involved in the Center. Budget items are selected because they overcome barriers faced by more than one company. Because such infrastructure is required for companies poised for near-term markets and for market entry in the future this ensures maximum leverage for an extended period of time.

Five thematic areas, reflecting the strengths of the PVIC partners and the current needs of the industry, emerge in this effort. They are:

(1) Acceleration of the commercialization of the next generation device technologies, in particular, those based on quantum dot structures incorporating both organic and inorganic components and those based on dye-sensitized metal oxides;

(2) Improvement in the current generation of thin film PV for lower production costs, higher deposition rates, and higher performance due to improvements in cell fabrication and characterization;

(3) Reduction of barriers involving fabrication of thin film (especially Groups II-VI CdTe and amorphous silicon) photovoltaic modules, including the use of ultralight weight substrate materials, improved high-volume deposition methods, advanced inline process monitoring, and improved understanding of material properties, as well as the packaging and environmental testing of these devices;

(4) Development in balance-of-systems (BOS) that tie these modules to actual power delivered in the home, business, or generating station;

(5) Promotion of the use of PV, including the training and organization of installation contractors, promotion of uniform PV-related building codes across the State, coordination of activities with major utility companies, and outreach efforts informing the public about the benefits of PV.

Photovoltaics is the focus of PVIC in part because Ohio has a long history of producing successful companies along the entire value chain of PV technology (PV module fabrication, glass and polymer manufacturing, building integration), a field of high economic growth and high employment generation. PVIC is intended to solidify leadership in the manufacture of current PV technology and build infrastructure ensuring leadership in the PV technologies of the future.

2 PV IN OHIO

The state of Ohio has a proven history of generating companies capable of producing and selling photovoltaics modules and their components. Building on this track record, PVIC will repeat this success by providing services and expertise to companies possessing PV module fabrication technology promising to lead to a lowering of production costs and to lower cost PV modules than currently available.

Imagining stage products. Ohio's major universities are world leaders in the development and characterization of PV materials and the conceptualization of new PV devices. Over the years, the needed expertise and equipment for these activities have been established through industrial, federal and state competitive funding programs.

Incubating stage products. Concurrently, Ohio universities have developed the infrastructure for incorporating PV materials in thin film form into small-area functional PV devices; hence they possess the resources to take PV from the imaging stage on to next level of development. These resources have been obtained for the "second generation" thin film PV such as amorphous silicon (a-Si), cadmium telluride (CdTe), and copper indium-gallium diselenide (CIGS). These resources can be used to bring the "third generation" technologies from the imagining to the incubating stage; these include inorganic quantum dots, wires, and tubes, organic thin films, and hybrids. PVIC incorporates the infrastructure to overcome barriers to incubation of the third generation technologies that will sustain the PV industry.

Demonstration stage products. Ohio's major universities are world leaders in the fabrication and characterization of small-area PV devices. For the second generation PV devices, the capability is established for scaling up small-area functional PV device processes into large areas. PVIC acts to overcome barriers to demonstration projects by completing the state-of-the-art facility for large-area deposition of PV structures. This facility has the capability of using a variety of different substrates as required by the industry collaborators of PVIC. In addition, PVIC will install large-area characterization tools also required by the industry collaborators. With state-of-the-art deposition and characterization instruments and the expertise to supervise

them, PVIC becomes unique world-wide and is expected to attract new companies to Ohio. Although a majority of the large-area deposition equipment is devoted to second generation thin film modules, some equipment is devoted to third generation demonstration projects, as is all the characterization instrumentation.

(a) **Demonstration of substrates and coatings.** Ohio has long been a leader in the production of glass and polymer articles. Currently, glass, stainless steel, and polymers are used as substrates for the fabrication of PV devices and modules. Companies with plants in Ohio seek to coat glass and polymer substrates with high performance electrically conducting materials, thereby establishing product lines preferred by PV fabricators. These companies possess the necessary resources to fund such a project, but lack the equipment and expertise to fabricate PV devices and modules and thus are not in a position to optimize their coated substrates for the PV market. PVIC provides services and expertise leading to the fabrication of completed PV devices and modules along with comprehensive device characterization. In this way PVIC speeds up the production of coated substrates optimized for the PV fabricators. In summary, PVIC overcomes market entry barriers faced by substrate producers by providing PV semiconductor deposition and completed PV devices and modules for the coated substrate producers.

Ohio has also been a leader in the production of value-added coatings on glass. Currently, coated glass substrates are modified to enhance the performance of PV devices and modules. Optimized coatings for maximum performance are required by all the thin film PV technologies from a-Si to organic semiconductors. Overcoming these problems, by providing these companies with testing facilities ranging from optical properties to durability of the coatings, directly enhances the power output and durability of PV modules.

(b) **Demonstration of modules.** Ohio companies exist with semiconductor deposition technology capable of fabricating photovoltaic modules at a lower cost (for example, using less expensive atmospheric pressure systems), and have sufficient resources to build manufacturing deposition equipment. In order to complete work on preparing the technology for market entry questions of weathering, long term light exposure, laser scribing and field testing need to be addressed so as to provide a convincing set of data to potential customers. PVIC acts to overcome market entry barriers related to PV module durability by building infrastructure providing these services. Weathering, light soak, field testing and other forms of durability testing (as well as scribing technology) are required for all PV technologies.

Market entry. PVIC assists Ohio PV companies to progress rapidly from the demonstrating phase toward the market entry phase -- made possible through world-class expertise and facilities. Problem areas of great interest to PVIC members include bringing to market (i) low-cost high performance PV based on glass substrates for utility scale generation of electricity and (ii) a flexible building-integrated PV roofing product designed for the residential market with higher performance, easier installation, and pleasing aesthetics. Throughout the demonstrating and market entry process, module developers interact closely with system designers, engineers, and installers within PVIC so products brought to market yield lower balance of systems, installation costs, and satisfy customers demands. PVIC staff with broad technical expertise along the full value chain will catalyze this dialog.

Growth and sustainability. PV is a paradigm-changing technology requireing not only overcoming the barriers to the marketplace, but also tracking levels of customer acceptance. With the rapid increase in residential and commercial PV use, consumer acceptance is also likely to change rapidly and this is a key variable in any marketing strategy. Thus, market research will be helpful to identify consumer trends in PV as well as the marketing strategies required by PV value chain companies. The conclusions of this research are expected to impact the nature of the product and may require returning to unanticipated demonstrating to meet the customer's evolving needs. In addition, consumer education is required to counter misconceptions about PV identified in surveys and to emphasize the societal benefits of renewable energy.

3 PVIC ALLIANCE STRUCTURE

PVIC's success in achieving its targeted problem solving assets, intellectual property commercialization and PV production goals will be significantly impacted by organizational and operational structures. The PVIC organizational structure revolves around an alliance between universities, companies and the Battelle Memorial Institute ("PVIC Alliance"). The PVIC Alliance enables a more effective and efficient operational structure crucial to the integration of solar energy research, technology development and intellectual property commercialization. The PVIC Alliance success factors include a synergistic technology pooling from multiple universities for the purpose of licensing "technology packages" of greater value and create new business development growth in photovoltaic materials, components and solar energy products. The PVIC Alliance organizational and operational structures eliminate or minimize many of the biggest challenges companies face in dealing with multiple institutions by including the following elements:

- o A formal alliance agreement establishing protocols for collaborative technology platform research focused on market driven customer needs.
- o Companies negotiate with one entity (PVIC IP management) for multi-institution IP created on laboratory equipment and instrumentation of the PVIC Center.
- o Companies will be able to collaborate with multiple institutions simultaneously through the PVIC Alliance and fund multi-laboratory research

and development projects, knowing that the resulting IP will be pooled into a technology package for a specific commercialization target, and business development objective.
- o Companies will be able to structure projects within PVIC spanning the "solar energy value chain". These projects will involve collaborative projects involving material suppliers, component producers and solar energy device manufacturers with each company having the option to license technology packages for a specific field complementary to their position in the value chain and commensurate with their level of participation in the Center.
- o The executive director of the center will have an Industrial Advisory Board assisting in the development of the R&D agenda, a business plan and a licensing strategy based on pooled intellectual property.

Reporting to the co-directors will be a "IP Strategy and Commercialization Manager" ("IP Manager") responsible for developing an IP strategy aligned with the PVIC Center business plan and provides for an integration and pooling of IP.

An objective of the PVIC Alliance agreement is to have a position below the co-directors to coordinate IP issues, which provides a single person with whom companies interact for IP packages involving multiple institutions.

4 INTERACTIONS WITH PVIC

The primary objective of PVIC is to assist companies in overcoming barriers inhibiting their progress along the Commercialization Framework. Some of the barriers are quite straightforward and only require the collection of a set of data. The data collection may involve an analytical characterization, a material deposition or some other technique within the expertise of PVIC personnel. In other cases the barriers may be more complex and require further analysis. A company may require an involved study in order to better understand the barrier to commercialization inhibiting growth and the introduction of products to the market. In order to meet this objective PVIC can interact with companies in one of three ways:

1) the company can apply to PVIC for services for which IP is not generated and there is not an exchange of IP between the company and PVIC;
2) the company requires further knowledge and the possibility exists for IP generation; or
3) the company is attract to the IP held by PVIC and arranges a license agreement to use the IP in the company's commercialization strategy.

Companies interacting with PVIC would initially do so through the appropriate director (either in Toledo, Dr. Robert Collins, or Columbus, Dr. Robert J. Davis) who would determine whether the company is interested in PVIC services or requires a PVIC research contract or requires a licensing agreement. For PVIC services the company would pay user fees for collection of the information required. For companies requiring more in-depth analysis where IP may be generated on the part of PVIC members a research contract between the company and PVIC would be implemented. In research contracts and licensing agreements issues related to the handling of IP and the scope of the work would be clarified. Conceivably, other possibilities exist for interactions between PVIC and companies. The appropriate co-Director in consultation with the Executive Board of Industrial Advisors will handle each case to ensure the best interests of PVIC members, the company, and the State of Ohio are served.

The PVIC "IP Strategy and Commercialization Manager" ("IP Manager") reports to the co-Directors and is responsible for leading an IP Innovation Council (IPIC) composed of technology transfer/commercialization managers from the PVIC institutions. The IP Manager is responsible for pooling and bundling IP from all institutions for the purpose of accelerating the licensing of "technology systems" that can be commercialized by PVIC member companies, new prospect companies or can be the basis for launching start-up companies. The PVIC IP Innovation Council will meet quarterly to ensure the intellectual management process is focused on creating an intellectual property pipeline aligned with market drivers and customer's needs.

5 COMMERCIALIZATION

The PVIC commercialization strategy supports this mission by focusing on two generations of PV technologies and a portfolio of advanced products that will enable existing companies to grow, and also holds the prospect of creating new start-ups and commercialization alliances that will benefit the job growth targets of PVIC. The foundation of the commercialization strategy is based on a strong proprietary contract research program aligned with market drivers, a PVIC catalyzed "constellation of alliances" throughout the PV product value chain, and the ability of a multi-organizational alliance to create an intellectual property pipeline that meets the needs of industry.

The Center will create and support a portfolio of products as the industrial collaboration projects move closer to commercialization. A nearer-term "portfolio of PV products" includes the following PVIC collaborations and alliances with industry shown in the figure below.

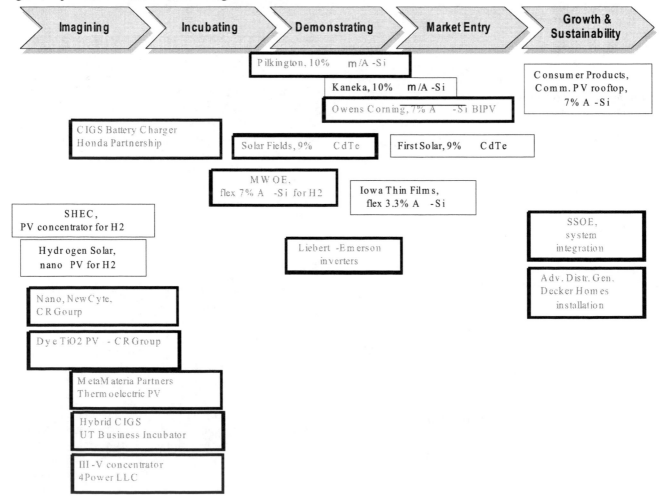

Bulk Heterojunction Organic-Inorganic Photovoltaic Cell Based on Doped Silicon Nanowires

Gary Goncher*, Lori Noice**, and Raj Solanki***

Department of Physics, Portland State University, Post Office Box 751, Portland, Oregon 97207-0751,
*ggoncher@verizon.net, **noice@pdx.edu, ***solanki@pdx.edu

ABSTRACT

Heterojunction photovoltaic devices were fabricated using single crystal silicon nanowires and the organic semiconductor regioregular poly-(3-hexyl thiophene) (RR-P3HT). N-type nanowires were first grown on an n+ silicon substrate by the vapor-liquid-solid (VLS) method. Devices were then fabricated by filling the gap between the nanowires and a transparent PEDOT:PSS-coated ITO electrode with RR-P3HT evaporated from chlorobenzene. Device performance indicates that both silicon and P3HT act as absorbers for photovoltaic response. Initial results show both open circuit voltage and short circuit current are both lower than expected, most likely duedue to the methods of photovoltaic cell construction.

1 INTRODUCTION

Polymeric organic semiconductors have been used to fabricate photovoltaic materials for more than a decade` with steady improvement in both materials and device architectures [1]. Some of the advantages of these materials are ease of fabrication and high optical absorption by thin films. However, charge carrier mobilities are typically orders of magnitude lower than commonly used inorganic semiconductors. In addition, excitons in organic semiconductors have a much high binding energy than crystalline inorganic semiconductors, and thus relatively large electric fields are required for dissociation into charge carriers after photoexcitation. The fields necessary for exciton dissociation (and hence charge carrier formation) exist only at heterointerfaces, and thus organic layers must be kept relatively thin to produce efficient devices.

In order to exploit some of the advantages of organic semiconductors while minimizing the impact of their major disadvantage, poor charge transport, photovoltaic devices have been fabricated using bulk heterojunctions of organic polymers. In this configuration there are interpenetrating networks of materials, minimizing the distance that excitons must travel before dissociating at a heterointerface. A particularly successful version of this type of device uses (6,6)-phenyl C61 butyric acid methyl ester (PCBM) as the donor, resulting in efficiencies exceeding 3% [2]. A natural extension of this idea is the use of inorganic semiconductors as one of the materials. Effective photovoltaic devices have been fabricated using inorganic nanoparticles [3]. The problem that occurs with this type of architecture is that one of the conduction paths requires percolation of charges to reach the contact electrode. In this work we propose the use of doped single crystal nanowires as the inorganic material in an organic-inorganic bulk heterojunction photovoltaic cell. The n-type silicon nanowires are grown on a highly doped n-type silicon substrate, resulting in a continuous conducting pathway for charge carriers in the inorganic phase.

2 NANOWIRE FABRICATION

Silicon nanowires were fabricated via VLS growth mechanism on n-type silicon (111) wafers. The wafers were coated with a thin Au film (~ 3 nm) by sputtering or evaporation, then placed in tube furnace on a Mo block for nanowire growth. Nanowires were grown at 420-450° C using 10% disilane in Ar at a pressure of 2 Torr. Full details of the growth have been previously published [4]. N-type doping of the nanowires was achieved by incorporating 100 ppm phosphine into the gas flow used during growth, producing a doping level of approximately 3×10^{16} [5].

Growth of nanowires via the VLS process occurs at reaction sites comprised of small Au islands. To determine the size and density of Au islands produced on the substrates used in these experiments, an Au film was deposited on a Si wafer, the wafer was heated to the growth temperature used for nanowire growth in the tube furnace, then cooled and examined using atomic force microscopy (AFM). Figure 1 shows an AFM scan taken with a Digital Instruments microscope of an Au film after annealing at 420° C. Au nanoparticle diameters were 46 +/- 17 nm, and particle density on the surface was 3.7×10^{10} particles/cm^2. Silicon nanowire diameters are typically several nm larger than the size of the Au nanoparticles used to nucleate and control the growth in the VLS process [6]. The length of the nanowires used in these experiments was 5-10 µm, produced during a 15 min. growth period in the furnace. The length varies between individual wires, with some dependence on wire diameter.

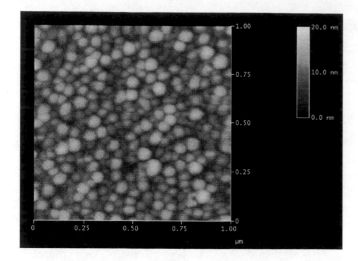

Figure 1. AFM image (tapping mode) of Au nanoparticles on a silicon surface following anneal at 420° C.

The silicon nanowires used in these experiments were single crystalline material with a principle growth axis along the (111) direction, as shown in the HRTEM image in figure 2. The wire tips are typically terminated with Au, and the edges of the wires have an amorphous SiO_2 layer approximately 4 nm thick. To provide better junctions between the n-type nanowires and the p-type organic semiconductor used to create the p-n heterojunction, the oxide was etched in buffered HF immediately prior to placing the nanowire samples in a nitrogen glove box for device fabrication.

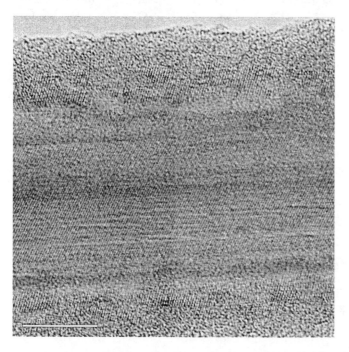

Figure 2. HRTEM image of silicon nanowire showing crystalline core and 3-4 nm amorphous oxide at the edges of the wire.

Quantum effects are important in small silicon nanowires, increasing the bandgap and causing the lowest energy transition to be direct. However, both theory and experiment have shown that these effects are only important for wire diameters below approximately 8 nm [7,8]. Thus the wires used in these experiments, with an average diameter of nearly 50 nm, behave essentially like bulk silicon as far as electronic transitions are concerned. However, the area of the heterojunction is considerably larger than would be the case for planar devices without nanowires. For the wire parameters discussed above, the area increase is approximately 2×10^4 greater than for a planar device. The random orientation of wires has the additional benefit of trapping more light than would a planar or textured surface.

3 ORGANIC-INORGANIC DEVICES

For this work we chose to use regioregular poly(3-hexylthiophene) as the p-type organic semiconductor at the heterointerface due to its high carrier mobility and low bandgap (2.0 eV) compared to other polymeric semiconductors. The interface between silicon and organic semiconductors has been investigated by a number of groups [9-11]. The heterojunction between n-Si and RR-P3HT was found to be a nearly ideal diode with a rectification ratio of greater than 10^3 [10]. Due to the fact that the nanowires used in this study were grown on an opaque n-Si wafer, the contact to the P3HT was made using a transparent ITO conductor on glass. Poly-[3,4-(ethylenedioxy)-thiophene]: poly-(styrene sulfonate) (PEDOT:PSS) has been widely used as a hole transport layer [1], and was used in our devices between the P3HT and ITO electrode.

The structure of the devices fabricated for this study is shown in figure 3. Briefly, nanowires were first grown on an n-type silicon substrate, and transferred to a nitrogen glove box following a brief (10 sec.) HF etch to remove surface oxide. PEDOT:PSS (Bayer Baytron P) was spin coated onto glass with a thin transparent electrode of sputtered ITO and also transferred to the glove box. Several drops of a solution of P3HT in chlorobenzene (5 g/l, deoxygenated by nitrogen bubbling) were applied to the surface of the PEDOT-PSS film, and the n-Si nanowire sample was placed face-down on top of the transparent electrode. The P3HT film was allowed to evaporate slowly over several days, then the sandwich was annealed for 1 hr. at 100° C to allow the P3HT film to form an oriented layer [12]. The device was then sealed with epoxy and removed from the glovebox for testing. Contact to the n side of the cell was made by scratching the back surface of the n-Si wafer to remove oxide and applying a Ga contact; contact to the p side was made by soldering a wire to the ITO using In-alloy solder.

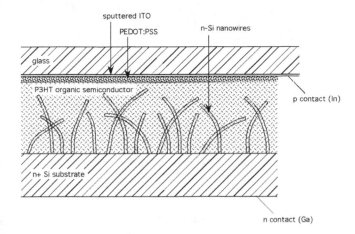

Figure 3. Diagram of the silicon nanowire PV cell structure (not to scale).

Energy levels relative to the vacuum level are shown in figure 4 for the components of the photovoltaic cell (without any adjustments for dipole formation or band bending due to Fermi levels). The HOMO of P3HT is positioned to inject holes into PEDOT-PSS and hence into the ITO electrode, and should accept holes generated by light absorption in the silicon nanowires. The LUMO of P3HT is well above the Fermi level of the n-Si nanowires and electron collection should occur efficiently at the silicon interface. Electrons generated in the nanowires will be collected at the Ga electrode.

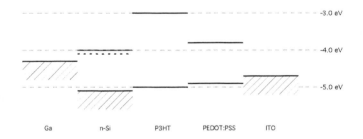

Figure 4. Energy levels of components of the cell.

Diode characteristics were observed for the assembled devices (figure 5). However, a relatively high leakage current was observed in the reverse direction. A likely suspect for this is nanowire penetration of the P3HT and possibly PEDOT:PSS layers during drying of the P3HT layer due to hydrostatic forces. Another possibility for high currents through the device is degradation of the P3HT polymer through exposure to oxygen. Although materials were handled in a nitrogen glove box and encapsulated prior to testing, this failure mechanism is well known for P3HT devices [13]. There may also be carrier recombination at the heterointerface, although P3HT heterojunction devices have been fabricated with very high internal quantum efficiency and low recombination [14].

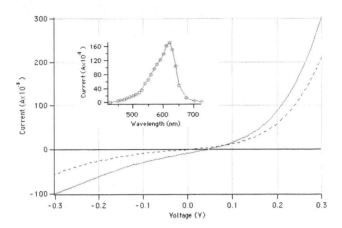

Figure 5. IV curve for silicon nanowire PV cell under 100 mW/cm^2 illumination (solid line) and dark (dashed line). Inset shows photoresponse as a function of wavelength.

The peak in photoresponse of the cell is at 600 nm (figure 5 inset). Since the absorption peak of P3HT is between 500 and 550 nm (2.0 eV bandgap), a substantial fraction of the photovoltaic response must be due to absorption in the silicon nanowires.

Open circuit voltage for the initial set of devices (0.05 V) are lower than expected, due most likely to a low shunt resistance in the cells. High dark currents in the devices indicate that conduction barriers are not present. Oxidation of the Si nanowire surface would limit currents to lower values than those observed.

Low short circuit photocurrents in the devices fabricated ($J_{sc} = 5 \times 10^{-6}$ A/cm^2 at 100 mW/cm^2 illumination) are likely due to the assembly method chosen, which requires a thicker layer of P3HT than would be optimal. Carrier mobilities are rather low in P3HT, and holes generated in the silicon nanowires must be transported across the P3HT film to the ITO electrode. For light absorbed in the P3HT layer, excitons are generated which must diffuse to a heterojunction for dissociation and collection. Hence the layer thickness should be kept very thin (~10 nm) to permit eciton diffusion and carrier collection. We are exploring alternate geometries such as a thin film coating of P3HT over the nanowires and a thicker PEDOT:PSS layer for charge transport to improve device efficiency.

4 CONCLUSIONS

Photovoltaic cells using heterojunctions of n-type silicon nanowires and P3HT organic semiconductor have been assembled and tested. The cells show evidence of carrier generation in both the silicon and polymer sides of the heterojunction. The first cells constructed have low V_{oc} and I_{sc}, but the geometry shows promise for both high light absorption and efficient carrier transport to electrodes. Efforts are underway to modify materials parameters (nanowire doping, polymer thickness, and charge transport layer composition) to improve device characteristics.

REFERENCES

[1] Harald Hoppe and Niyazi Serdar Sariciftci, J. Mater. Res. 19, 1924, 2004.

[2] F. Padinger, R. S. Rittberger, and N. S. Sariciftci, Adv. Funct. Mater. 13, 1, 2003.

[3] Wendy U. Huynh, Janke J. Dittmer, and A. Paul Alivisatos, Science 295, 2425, 2002.

[4] C. A. Decker, R. Solanki, J. Freeouf, J. R. Carruthers, and D. A. Evans, Appl. Phys. Lett. 84, 1389, 2004.

[5] G. Goncher, R. Solanki, J.R. Carruthers, J. Connley, Jr., and Y. Ono, J. Electronic Materials 35, 1509, 2006.

[6] Yi Cui, Lincoln J. Lauhon, Mark S. Gudliksen, Janfang Wang, and Charles M. Lieber, Appl. Phys. Lett., 78, 2214, 2001.

[7] B. Delley and E. F. Steigmeier, Appl. Phys. Lett., 67, 2370, 1995.

[8] D. D. D. Ma, C. S. Lee, F. C. K. Au, S. Y. Tong, and S. T. Lee, Science 299, 1874, 2003.

[9] Evan L. Williams, Ghassan E. Jabbour, Qi Wang, Sean E. Shaheen, David S. Ginley, and Eric A. Schiff, Appl. Phys. Lett. 87, 223504, 2005.

[10] C. H. Chen and I. Shih, J. Mater. Sci.: Mater. Electron. 17, 1047, 2006.

[11] Vignesh Gowrishankar, Shawn R. Scully, Michael D. McGehee, Qi Wang, and Howard Branz, IEEE 4th World Conf. on Photovoltaic Energy Conv., 2006.

[12] Gang Li, Vishral Shrotriya, Yan Yao, and Yang Yang, J. Appl. Phys. 98, 043704, 2005.

[13] G. H. Gelinck et. al., Appl. Phys. Lett. 77, 1487, 2000.

[14] Pavel Schilinsky, Cristoph Waldauf, and Christophe J. Brabec, Appl. Phys. Lett., 81, 3885, 2002.

Cleantech 2007, ISBN 1-4200-6382-0

Molecular Building Blocks for Efficient Solid State Lighting

P.E. Burrows, L.S. Sapochak. A. B. Padmaperuma, H. Qiao and P. Vecchi

Energy Science and Technology Directorate
Pacific Northwest National Laboratory
PO Box 999, Richland, WA 99352, USA, burrows@pnl.gov

ABSTRACT

General illumination consumes 22% of the electricity generated in the U.S. This huge proportion is partly due to the ubiquity of artificial lighting but also the inefficiency of converting electrical energy to light. Incandescent lightbulbs convert a mere 5% of the supplied power into light (most of the rest emerging as heat) whereas the more efficient fluorescent bulbs achieve about 20% efficiency. Improving the efficiency of these light sources is difficult since in all cases the emission of light is essentially a by-product of an energetic excitation process. In contrast, solid state lighting utilizes materials which directly convert electrical energy to light with little production of heat and therefore have the potential for far higher efficiency, with over 70% demonstrated in the infrared. New materials based on direct bandgap semiconductors and organic light emitters may permit this level of efficiency for general lighting. In both cases, however, understanding the nanoscale structure of the material is critical to achieving high efficiency. This is particularly evident in the case of organic molecular compounds, where weak inter-molecular interactions can permit the photophysical properties of a solid to be tuned by changing the chemical structure of the molecular building block.

Keywords: lighting, molecular, organic, solid state

1 INTRODUCTION

Organic light emitting device (OLED) technology has improved in efficiency over the last decade due to the development of new molecular materials.[1] As a result, commercially available products (e.g., small color displays for cellular telephones) based on this technology have become available in the marketplace.[2] A further outcome of this progress is that the efficiency of OLEDs has increased to the point where they are worth considering for solid state lighting applications.[3] OLEDs have unique advantages for lighting applications in that they are large area emitters (as opposed to point sources) which can therefore be used without a diffuser or luminaire, potentially increasing the overall lighting system efficiency.

Realizing this potential has created new challenges to overcome. For example, a white OLED-based lighting solution reaching the efficiencies in excess of 150 lm/W will require devices, which operate at close to 100% internal quantum efficiency (IQE) and close to the

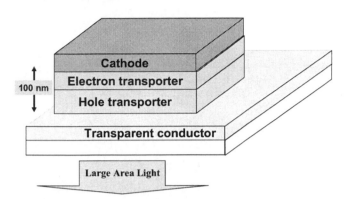

Figure 1: Simplified schematic cross-section of an organic light emitting devices (OLED)

minimum achievable operating voltage. An IQE in excess of 80% has already been demonstrated and various schemes exist[4,5] to improve the optical out coupling at least to 50%. In all cases, however, the most efficient materials on the basis of IQE are guest-host composites consisting of up to 20% of an organometallic phosphor in a charge transporting host material. Electroluminescence (EL) from the dopant results either from energy transfer from the host, direct trapping of charge on the phosphor or a combination of the two processes.[6,7] The best phosphors are based on a chelated heavy metal ion (typically Ir and Pt). Operating voltages of small molecule composites, however, remain high (typically ~ 10 V at the high brightness relevant to lighting) compared to polymer-based OLEDs unless ionic or small molecule dopants are added.[8] Blue electrophosphorescence, which is a necessary component of white light, has been a particular challenge because the triplet excited state of the host material must be higher than that of the dopant in order to prevent quenching of the dopant emission. This requires a host material with even higher triplet energy than the blue phosphor dopant and this is difficult to achieve in a molecule which also forms morphologically stable thin films.

The host material for a blue OLED must have a triplet energy (E_t) > 2.8 eV which requires an extremely short conjugation length. There is a tradeoff , however, between decreasing the extent of the π-aromatic system to increase the singlet and triplet exciton energies and adversely affecting charge transport. Deep blue phosphors have been demonstrated using insulating, wide bandgap host materials based on tetra aryl silanes with charge transport occurring via hopping between adjacent dopant molecules, but at increased operating voltage, and therefore less power

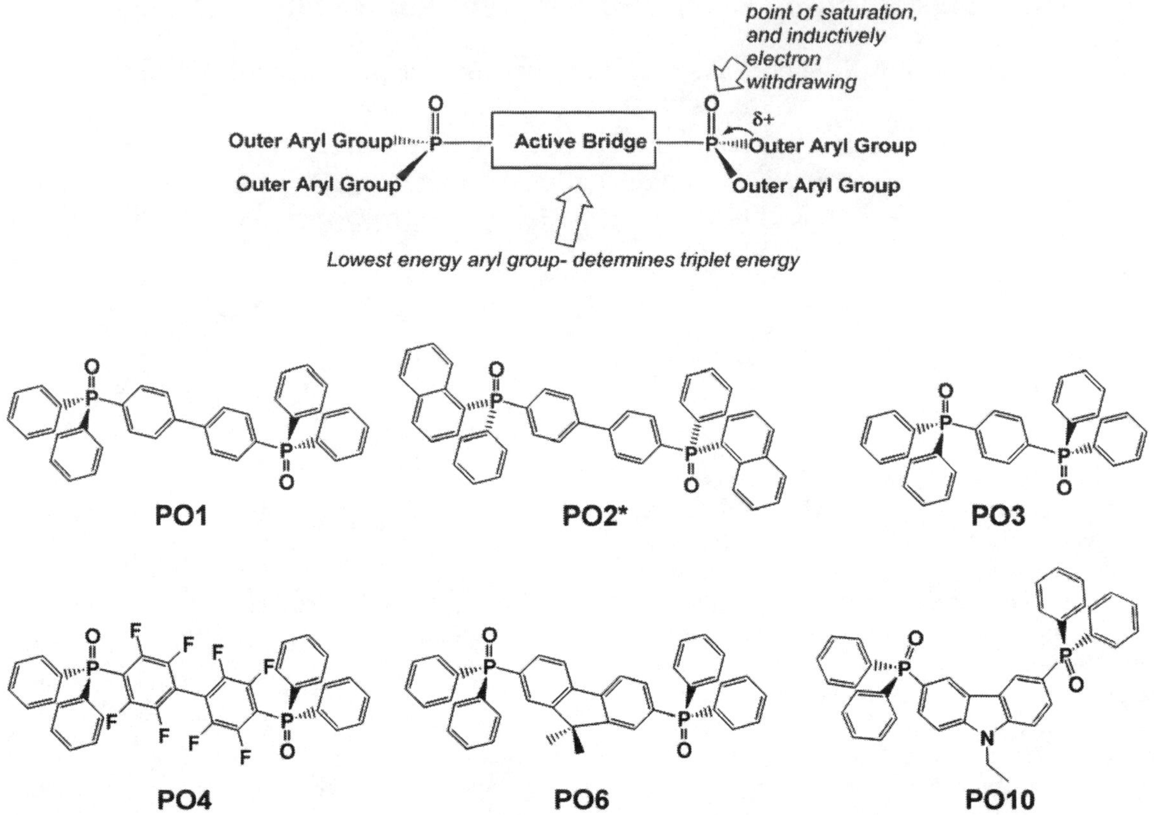

Fig. 2. General design concept for achieving organic charge transporting host materials with high triplet energies and the chemical structures of specific examples are shown. Note that the "active" aryl group in PO2 is the naphthyl ring, rather than the biphenyl bridge. *PO2 is a mixture of optical isomers and no stereochemistry is specified.

efficient devices. In this paper, we present a design strategy for developing organic host materials for blue electrophosphorescence which have the targeted high triplet energies without sacrificing the corresponding charge transport properties required for power efficient OLEDs. This is accomplished by using saturated linkers to connect molecular building blocks with high triplet exciton energy to build larger, tractable molecules in a bottom-up design scheme.

We recently reported that organic phosphine oxides function as wide band gap and charge transporting host materials for the sky blue phosphorescent dopant, iridium(III)bis(4,6-(di-fluorophenyl)-pyridinato-N,C2') picolinate (FIrpic) in OLEDs with peak quantum efficiencies of ~ 8% and low drive voltages.[9,10] Tthe phosphine oxide (P=O) moieties act as a point of saturation (i.e., breaks π-electron conjugation) between the "active" chromophore bridge (biphenyl and 9,9-dimethylfluorene in PO1 and PO6, respectively) and outer phenyl groups, resulting in materials with triplet energies characteristic of the lowest energy aryl group in the molecule (i.e. the active bridge) but with physical properties suitable for device fabrication by thermal sublimation. Here, we describe a general design concept for high triplet energy host materials (Fig. 2) by extending the work to other high triplet energy chromophores including naphthalene, phenyl,

octofluorobiphenyl, and N-ethylcarbazole. Device results using some of these hosts doped with FIrpic are discussed.

2 MOLECULAR BUILDING BLOCKS

The PO compounds were designed to have photophysical properties (e.g. high triplet exciton energy) characteristic of the lowest energy aryl group with the thermal and charge transport properties of a larger molecule. To demonstrate the effectiveness of the design, absorption and emission spectra of PO1, PO2, PO4, and PO10 in CH_2Cl_2 solution are shown in Fig. 3 compared to the absorption and emission properties of the relevant aryl group.

The absorption maximum of PO1 was red shifted from biphenyl by ~ 20 nm (corresponding to an energy shift of 0.4 eV) and was closer to the absorption spectrum of the dibromo derivative (see Fig. 3a). This result is similar to previously reported work using the 9,9-dimethylfluorene bridged PO derivative, PO6. The energy shifts for both PO1 and PO6 are similar to the brominated aryl bridges because both $Ph_2P=O$ and Br substitution impart an inductive electron withdrawing effect along the long axis of the respective bridge. Similar trends were observed for PO4, shown in Fig. 3c. In the case of PO2, which is the only material in this study containing two extended chromophores (the biphenyl bridge and the outer naphthyl

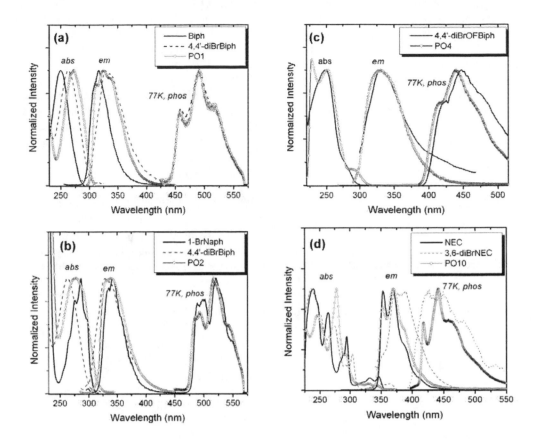

Fig. 3. Absorption and emission spectra (room temperature, CH_2Cl_2) and phosphorescence spectra (77K, CH_2Cl_2) shown for (a) PO1, (b) PO2, (c) PO4 and (d) PO10 compared to their component aryl groups [biphenyl (Biph), 4,4'-dibromobiphenyl (4,4'-diBrBiph), 1-bromonaphthalene (1-BrNaph), 4,4'-dibromooctofluorobiphenyl (4,4'-diBrOFBiph), 3,6-dibromo-N-ethylcarbazole (3,6-dBrNEC) and N-ethylcarbazole (NEC).

groups), the absorption spectrum showed contributions from both aryl groups (see Fig. 3b). The lowest energy absorption band originates from the naphthyl ring, which is also the lowest energy aryl group in the molecule. The absorption spectrum of the carbazole bridged compound, PO10, shown in Fig. 3d, was closer to N-ethylcarbazole (NEC) than the dibrominated derivative (3,6-diBrNEC) because substitution of the NEC chromophore is along the *short* axis of the molecule (as discussed previously by Marsal, et al. for different brominated derivatives of carbazole).[9] Photophysical spectra for PO3 are shown in Fig. 4. This wide band gap phenyl bridged PO compound exhibited an absorption spectrum similar to triphenyl phosphine oxide (TPPO),[10] but with a weak red shifted absorption band likely from the bridging phenyl ring.

3. LIGHT EMITTING DEVICES

To evaluate the usefulness of the PO compounds as electron transporting host materials for blue electrophosphorescence, OLEDs were fabricated using the shortest wavelength commercially available phosphorescent dopant, FIrpic, doped at 5%, 10% and 20% by mass via co-evaporation into a PO (PO1, PO3, PO6 and PO10) host

layer. The device structure was grown on ITO glass (< 15Ω/ □) by thermal evaporation as described previously[11,12] and was composed of, in sequence, 200 Å CuPc / 200 Å α-NPD / 60 Å TCTA / 200 Å 5, 10 or 20% FIrpic in PO / 200 Å PO/ 6 Å LiF / 1000 Å Al. The PO compounds served both as electron transporting host in the emissive layer and as an exciton and hole blocking layer in this device configuration. It is assumed that holes are transported in the doped light emitting layer by hopping conduction between FIrpic molecules. Device results are reported in Table 1 and compared to previous results reported for PO1 and PO6.

All PO materials were effective as hosts for FIrpic giving maximum external quantum and luminance efficiencies of ~ 8% and ~ 21 cd/A, respectively. Furthermore, light emission was observed at very low applied voltage of < 3.9 V at maximum quantum efficiency (at which point the brightness was typically ~ 10 cd/m²) giving luminous power efficiencies ranging from ~ 17 to 25 lm/W. The high brightness results (800 cd/m²) were obtained at 4.8 – 6.3V as compared with > 9V typically reported using carbazole-based host layers.[13] Only FIrpic emission was observed from devices with a 20% FIrpic doping concentration. At 10% doping, a small contribution from TCTA fluorescence is observable. This becomes more significant at the lowest doping concentration of 5%. This is consistent with electron leakage from the light emitting layer into the TCTA and poor hole injection from the TCTA into the light emitting layer at low FIrpic concentrations (because holes are injected directly into FIrpic states due to the low-lying HOMO of the PO molecules), suggesting that the efficiency data could be further improved by incorporation of an electron blocking layer and use of higher work function anode materials or additional hole injection layers.

Table 1. Device properties for OLEDs using PO hosts doped with FIrpic

	Property / Host : % FIrpic	PO1 : 20%	PO3 : 10%	PO6 : 10%	PO10 : 20%
Results at maximum quantum efficiency	$\eta_{ex,max}$(%) [J (mA/cm^2)]	7.8 (0.09)	8.1 (0.23)	8.1 (0.002)	8.3 (0.03)
	$\eta_{c,max}$(cd/A)	20.8	21.5	21.5	22.2
	V @ max QE	3.9	3.6	3.0	3.7
	$\eta_{p,max}$(lm/W)	16.7	18.8	25.1	18.8
Results at "lighting brightness" taken to be 800 cd/m^2	η_{ex}(%)	6.7	7.2	4.4	5.7
	J (mA/cm^2)	4.2	4.1	7.6	5.6
	η_c(cd/A)	18.0	19.3	11.8	15.0
	V_{opt}	5.6	4.8	5.6	6.3
	η_p(lm/W)	10.1	12.7	6.7	7.7

[a]$\eta_{ex,max}$, maximum external quantum efficiency; $\eta_{c,max}$, maximum luminance efficiency; $\eta_{p,max}$, maximum luminance power efficiency; and V_{opt}, operating voltage at a specified current. *Reported at lighting brightness (800 cd/m^2).

4. CONCLUSIONS

The phosphine oxide moiety has been successfully used as a point of saturation in order to build sublimable, electron transporting host materials starting from small, wide bandgap molecular building blocks. The presence of the P=O group is expected to lead to a lowering of the LUMO and HOMO states, which is consistent with the photophysical data presented. This design principle leads to a range of new materials suitable as host materials for blue organic phosphors which generate bright light at a lower voltage than previously published material systems. The low operating voltage is ascribed to a combination of facile electron injection into the PO layer and hole transport by hopping conduction between phosphorescent dopant molecules. All the materials tested give similar external quantum efficiencies although the current density and phosphor loading at peak efficiency and the operating voltage show some variation. These variations do not appear to be correlated with the band gap or frontier orbital energies of the particular host material. We note, however, that the P=O moiety also has a strong order-directing influence on the material and some evidence of molecular-scale aggregation has been previously obtained for PO6. The bridge unit also influences the preferred intermolecular interaction geometry via hydrogen bonding and edge-to-face interactions.[14] These properties are expected to strongly influence electron transport within the layer, which has a concomitant effect on the charge balance in the light emitting layer, and therefore on the device efficiency as a function of injected current. Further study of the nanoscale structure of these materials is therefore necessary to fully understand how to maximize the efficiency and minimize the voltage of devices based on these materials.

[1] See e.g. "The Special Issue on Organic Electronics," *Chem. Mater.* 16(23), 4381-4846 (2004).
[2] W.E. Howard, "Better Displays with Organic Films," in *Scientific American*, Jan. 12, 2004.
[3] Dept. of Energy, Solid State Lighting Website, http://www.netl.doe.gov/ssl
[4] C. Adachi, M. Baldo, M.E. Thompson, and S.R. Forrest, *J. Appl. Phys.* 90(1), 5048-5051 (2001).
[5] B.W. D'Andrade and S.R. Forrest, *Adv. Mater.* 16(18), 1585-1595 (2004).
[6] R.J. Holmes, S.R. Forrest, Y.-J. Tung, C. Kwong, J.J. Brown, J.J.; S. Garon, and M.E. Thompson, *Appl. Phys. Lett.* 82(15), 2422-2424 (2003).
[7] R.J. Holmes, B.W. D'Andrade, S.R. Forrest, X. Ren, J. Li, and M.E. Thompson, *Appl. Phys. Lett.* 83(18), 3818-3820 (2003).
[8] M. Pfeiffer, K. Leo, X. Zhou, J.S. Huang, M. Hofmann, A. Werner, J. Blochwitz-Nimoth, *Org. Elect.* 4, 89-103 (2003) and references therein.
[9] P. Marsal, I. Avilov, D.A. da Silva Filho, J.L. Brédas, and D. Beljonne, *Chem. Phys. Lett.* 392, 521-528 (2004).
[10] P. Changenet, P. Plaza, M.M. Martin, Y.H. Meyer, and W. Rettig, *Chem. Phys.* 221, 311-322 (1997).
[11] P.E. Burrows, A.B. Padmaperuma, L.S. Sapochak, P. Djurovich, and M.E. Thompson, *Appl. Phys. Lett.* 88, 183503-1 – 183503-3 (2006).
[12] A.B. Padmaperuma, L.S. Sapochak, P.E. Burrows, *Chem. Mater.* 18, 2389-2396 (2006).
[13] R.J. Holmes, S.R. Forrest, Y.-J. Tung, C. Kwong, J.J. Brown, J.J.; S. Garon, and M.E. Thompson, *Appl. Phys. Lett.* 82(15), 2422-2424 (2003).
[14] P.A. Vecchi, A.B. Padmaperuma, H. Qiao, L.S. Sapochak, and P.E. Burrows, *Org. Lett.* **8**, 4211 (2006).

Dynamics of Water in Nafion Fuel Cell Membranes: the Effects of Confinement and Structural Changes on the Hydrogen Bonding Network

David E. Moilanen, Ivan R. Piletic and Michael D. Fayer[*]

[*]Stanford University Department of Chemistry
Stanford, CA. 94305, USA, fayer@stanford.edu

ABSTRACT

The complex environment experienced by water molecules in the hydrophilic channels of Nafion fuel cell membranes is studied by ultrafast infrared pump-probe spectroscopy. A wavelength dependent study of the vibrational lifetime of the O-D stretch of dilute HOD in H_2O confined in Nafion membranes provides evidence of two distinct ensembles of water molecules. While only two ensembles are present at each level of membrane hydration studied, the characteristics of the two ensembles change as the water content of the membrane changes. Anisotropy measurements show that the orientational motions of water molecules in Nafion membranes are significantly slower than in bulk water and that lower hydration levels result in slower orientational relaxation.

Keywords: Nafion, water, confinement, dynamics, fuel cells

1 INTRODUCTION

Nafion is the most common membrane separator used in polymer electrolyte membrane fuel cells (PEMFCs) due to its chemical and thermal stability and its high proton conductivity. It is a polymer, consisting of a long chain fluorocarbon backbone with pendant, sulfonic acid terminated, polyether side chains. The extreme difference in the polarity of the fluorocarbon backbone and the sulfonic acid side chains causes segregation of the membrane into hydrophobic and hydrophilic aggregates.[1]

The importance of the hydrophilic regions of Nafion for proton conduction has inspired a great deal of research over the past three decades since its development in the early 1970s. One of the first experiments to look directly at the water inside Nafion membranes was presented by Falk, who used steady state IR spectroscopy of H_2O, D_2O and HOD to study the properties of water in Nafion.[2] IR spectra show that water experiences a range of environments in Nafion and that the hydrogen bonding network in Nafion is weaker than in bulk water.[2, 3] Often, the properties of Nafion are studied at various degrees of hydration from dry to fully hydrated. Scattering experiments show that the hydrophilic domains swell with increased hydration. NMR[4, 5], IR[3], and MD simulations[6, 7] show that the pendant side chains in the hydrophilic domains can rearrange as the hydration increases. Ultimately, the practical question in terms of fuel cell operation is: how does the swelling and rearrangement of the hydrophilic domains affect the mechanism for proton transport through Nafion?

Here, we present an IR pump-probe study of the dynamics of water absorbed in Nafion membranes at four hydration levels. IR spectroscopy of dilute HOD in H_2O or D_2O provides a direct probe of the hydrogen bonding network and dynamics of water. The local environment of an HOD molecule is reported by its linear IR spectrum[8] and its vibrational lifetime. Vibrational relaxation is sensitive to both local fluctuating forces acting upon an excited vibrational mode and the availability of lower frequency accepting modes to dissipate the energy. Changes in the local environment can alter the fluctuating forces and shift the energy levels of the accepting modes, altering the vibrational lifetime. Although it is difficult to pinpoint the cause of a change in the vibrational lifetime, the fact that it is sensitive to local effects allows the identification of multiple ensembles based on differences in the vibrational lifetime at different frequencies.

Global rearrangements of the hydrogen bonding network are reflected in the time dependent orientational anisotropy, which is a measure of the reorientational motions of the water molecules. Reorientation requires the concerted motion of several water molecules and involves the reorganization of the hydrogen bonding network through the breaking and forming of hydrogen bonds. This requirement causes the anisotropy to be sensitive to the characteristics of the hydrogen bonding network as a whole rather than the local environment of an O-D oscillator. The detailed wavelength dependent study of the vibrational lifetime and orientational relaxation presented below provides insights into both the local and long range dynamics of water contained in the hydrophilic domains of Nafion. A detailed description of the experimental setup can be found in the recent work by Moilanen et. al.[9] Samples of Nafion were prepared at several different hydration levels, λ. We define λ as the number of water molecules per sulfonate group.

2 RESULTS AND DISCUSSION

2.1 Vibrational Lifetimes

Recently, the vibrational lifetime of dilute HOD in H_2O absorbed in Nafion was measured for the first time.[3] The measurement of the vibrational lifetime as well as the IR spectrum of water in Nafion provided evidence for multiple ensembles of water molecules in Nafion. While the IR spectra at all hydration levels could be fit using a weighted sum of two fixed spectral components, the vibrational lifetime showed evidence that the characteristics of the two ensembles was changing with hydration. The vibrational lifetime measurements in this earlier study were conducted only at the peak of the absorption spectrum for each sample.

Here we report a complete wavelength dependent study of the vibrational lifetime at each hydration. Figure 1 shows normalized plots of the vibrational lifetime of $\lambda = 3$ Nafion at four frequencies in the 0-1 absorption region.

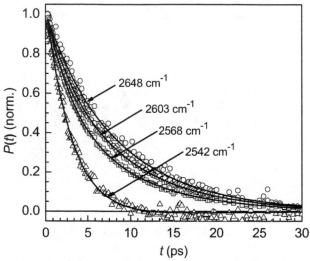

Figure 1: Population dynamics (vibrational lifetimes) at four different frequencies in the 0-1. The solid lines are biexponential fits to the data at each frequency with the time constants fixed at $\tau_1 = 3.2$ ps and $\tau_2 = 8.6$ ps.

It is clear that the vibrational lifetime decay is much faster on the red (low frequency) side of the 0-1 region than it is on the blue (high frequency) side. The vibrational lifetime increases monotonically from the red side to the blue side across the 0-1 transition and is fit well by a biexponential at all wavelengths. Interestingly, the vibrational lifetime can be fit at all wavelengths in the 0-1 (and also in the 1-2) using two fixed time constants. This is strong evidence that water in $\lambda = 3$ Nafion experiences two very different environments. The relative amount of water in each environment can be extracted from the amplitudes of the two exponentials. For normalized population decays we have:

$$P(t) = ae^{-t/\tau_1} + (1-a)e^{-t/\tau_2} \qquad (1)$$

where a is the fraction of water molecules in environment 1 with vibrational lifetime τ_1 and $(1-a)$ is the fraction of water molecules in environment 2 with vibrational lifetime τ_2. In $\lambda = 3$ Nafion, $\tau_1 = 3.2$ ps and $\tau_2 = 8.6$ ps. The solid lines in figure 1 are fits to the data using these two time constants.

Evidence for multiple hydrogen bonding environments has been observed in several other systems including AOT

reverse micelles,[10] and mixtures of water and acetonitrile.[11] Nafion is similar in many ways to AOT reverse micelles. In fact, early models of Nafion invoked a reverse micellar structure for the hydrophilic regions. However, recent experiments have shown that the two environments in AOT reverse micelles, a core type region with bulk water characteristics, and a shell type region are preserved at all reverse micelle sizes studied.[10] In Nafion, both components of the vibrational lifetime change as the hydration level is increased. Table 1 contains the results of global fits to the vibrational lifetime data of the four different hydration levels of Nafion studied. At each hydration level, the relative amount of each component at a given wavelength follows a similar trend to the results for $\lambda = 3$ Nafion shown in figure 1.

	τ_1 (ps)	τ_2 (ps)
$\lambda = 1$	5.1	11
$\lambda = 3$	3.2	8.6
$\lambda = 5$	2.3	6.5
$\lambda = 7.5$	2.0	5.9

Table 1: Population relaxation times for the two components of the vibrational lifetime at different hydrations.

The change in both components of the vibrational lifetime as the hydration level of the membrane increases bears some resemblance to the results for acetonitrile/water mixtures, yet it is clear that the hydrophilic regions of Nafion cannot be viewed as a simple binary solution due to the fluorocarbon backbone of the polymer which restricts the mobility of the sulfonate terminated side chains. While the hydrophilic domains of Nafion may seem similar to AOT reverse micelles in many respects, the change in the vibrational lifetime with changing water content indicates that the hydrophilic domains restructure, in contrast to the interfacial region of AOT. The sulfonate groups in AOT reverse micelles have very little mobility because they are constrained by the close packing of the surfactant molecules. In Nafion, the side chains are clearly not constrained to the extent that they are in AOT and are therefore able to rearrange to some degree. Acetonitrile molecules in binary mixtures of acetonitrile/water have essentially complete mobility to rearrange in order to form the most stable water cluster possible. Clearly, the ability of the sulfonate terminated side chains in Nafion to rearrange is intermediate between the fixed structure of the headgroups in AOT reverse micelles and the free reorganization that is possible in a binary acetonitrile/water solution. The side chains in Nafion have some ability to reorganize to produce more thermodynamically stable water clusters but this restructuring is constrained by the structure of the fluorocarbon backbone and the packing of the polymer's hydrophobic aggregates.

Cleantech 2007, ISBN 1-4200-6382-0

2.2 Orientational Relaxation

Unlike the vibrational lifetime, orientational relaxation, which is measured through the time resolved anisotropy, has a concrete physical connection to the motions of water molecules. The breaking and forming of hydrogen bonds, which is central to the reorientation of a water molecule, is an important step in both translational diffusion and proton transfer. In Nafion, both translational motion and proton transfer are critical for fuel cell performance, and understanding how the orientational motions of water molecules change as the amount of water in the membrane varies is fundamental in modeling these processes.

Anisotropy decays collected at the peak of the absorbance spectrum for each sample are shown in figure 2 along with the anisotropy decay of bulk water for comparison. Clearly, the anisotropy decays become significantly slower as the hydration level of the membrane decreases, and none of the curves (except water) can be fit by a single exponential decay.

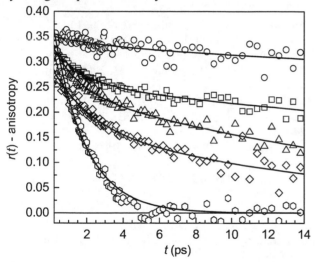

Figure 2: Anisotropy decays at the peak of the absorption spectrum for $\lambda = 1$ (circles), 3 (squares), 5 (triangles), and 7.5 (diamonds). The anisotropy decay of bulk water (hexagons) is shown for comparison.

While several groups have measured the anisotropy decays of bulk water[12-14] and confined water,[10, 15-18] only recently has an MD simulation provided a physically meaningful interpretation of the anisotropy decay for bulk water.[19] Often, the single exponential decay of the experimentally measured anisotropy of bulk water is modeled by the Debye-Stokes-Einstein (DSE) equation. The DSE equation is based on the reorienting molecule taking random, infinitesimal steps. While the data may be fit with this model, infinitesimal diffusion does not take into account an important property of water: its tetrahedral hydrogen bonding network. The directionality of the interactions between water molecules within the tetrahedral framework of the hydrogen bonding network means that

hydrogen bonds must be broken and new bonds formed for complete reorientation to occur.

The biexponential decay of the anisotropy observed for water in AOT has been described using a model based on restricted orientational diffusion or "wobbling in a cone."[10, 17] In the wobbling in a cone model, the motions of the molecules are restricted at short times to lie within a cone of half-angle θ_c. This motion is diffusive in nature, but rather than proceeding to sample the entire sphere, it is restricted by the hard boundary conditions of the cone. After this initial wobbling period, the slow, long time decay of the anisotropy is attributed to full sampling of the rest of orientational space. The final exponential decay constant was interpreted using the DSE equation. However, recent MD simulations have provided evidence for a different interpretation of the long time component of the anisotropy.[19]

The MD simulation of Laage et. al. highlighted the well established tetrahedral nature of the hydrogen bonding network of water.[19] Based on this physical picture for the microscopic structure of water, Laage et. al. found that the orientational motions of water molecules are better described by a jump diffusion model, in which the jumps correspond to the rearrangement of hydrogen bonds among water molecules.[19] From a physical point of view, this description makes sense; the breaking and forming of hydrogen bonds, which changes the orientation of the hydroxyl transition dipole, cannot involve many infinitesimal random steps. Likewise, it is energetically unrealistic to imagine a water molecule breaking a hydrogen bond without being able to immediately form a new hydrogen bond.[20]

In Nafion, the anisotropy decay does not fit to a single exponential. Unlike the vibrational lifetime, biexponential behavior does not imply that the anisotropy is produced by two subensembles.[10] Here the time scales of the two components of the anisotropy decay differ by a large enough amount that it is possible to propose a reasonable physical underpinning for the different mechanisms that give rise to the decays. At short times, the motions of water molecules are restricted by the attractive potential of the hydrogen bond (wobbling in a cone). At later times, large amplitude fluctuations in the orientation occur through a rearrangement of the hydrogen bond network by breaking and forming new hydrogen bonds.

With this physical picture in mind it is possible to combine the wobbling in a cone model applied previously (short time dynamics) with the jump diffusion model (long time dynamics) for water reorientation. A detailed description of this analysis can be found in the recent paper by Moilanen et. al.[9]

There are several trends in the anisotropy parameters reported in Table 2. First, the amplitude of the short time component, A_S, increases with increased hydration leading to larger cone angles for higher hydration levels. This means that the water molecules are able to sample more of

	A_S	τ_w (ps)	θ_c	A_L	τ (ps)
$\lambda = 1$	0.02 ± 0.02	3 ± 3	$20° \pm 4°$	0.33 ± 0.02	∞
$\lambda = 3$	0.08 ± 0.01	1.8 ± 0.2	$31° \pm 2°$	0.25 ± 0.01	63 ± 10
$\lambda = 5$	0.10 ± 0.01	1.3 ± 0.2	$33° \pm 2°$	0.24 ± 0.01	22 ± 2
$\lambda = 7.5$	0.12 ± 0.02	1.9 ± 0.4	$42° \pm 3°$	0.17 ± 0.02	17 ± 3
water				0.34 ± 0.01	2.6 ± 0.1

Table 2: Orientational relaxation parameters: A_S is the short component amplitude, τ_w is the wobbling time constant, θ_c is the cone angle, A_L is the long component amplitude, and τ is the jump diffusion time constant.

their orientational space on a fast time scale. At the lowest hydration level, $\lambda = 1$, virtually no reorientation occurs after the small amplitude short time component. While the time constant for the fast component is similar for all four samples, the long time component, τ, becomes significantly faster for the higher hydration levels. If the reorientational angular displacements that water molecules make during a jump are similar at each of the hydrations, the increase in the decay of the long component indicates that the jump rate is increasing with increased hydration. In the case of $\lambda = 1$, so little water is present in the membrane that it is never possible to find a new hydrogen bond acceptor, so no jumps occur on the time scale of the experiment..

It is clear that within this model, the orientational motions of water molecules are not local phenomena. Complete reorientation requires the breaking and forming of a hydrogen bond, which depends on the concerted motion of several water molecules to lower the energy barrier of the hydrogen bond transition state. Although the local reorientational event requires the participation of water molecules in the first and second solvation shells of the reorienting molecule, the ability of these two molecules to accept a hydrogen bond depends on the dynamics of their own solvation shells. The necessity for cooperative motion means that the anisotropy is sensitive to the hydrogen bond dynamics of the entire cluster of water molecules, not the local environment of a single water molecule. The short time scale wobbling in a cone motion is also determined by the nature of the hydrogen bond network bonded to the wobbling OD hydroxyl group under observation.

3 CONCLUSIONS

Wavelength dependent studies of the vibrational lifetime at four different hydration levels show evidence for two ensembles of water molecules that have characteristics which change with the amount of water in the membrane due to a restructuring of the sulfonate terminated side chains. Unlike the vibrational lifetime, the anisotropy is sensitive to the global structural rearrangements of the changing hydrogen bond network. The orientational motions of water molecules in Nafion membranes are strongly restricted with virtually no reorientation possible at the lowest hydration levels. In fuel cells, the performance of polymer electrolyte membranes like Nafion depends on the dynamical properties of the water inside the membranes. Understanding how the changing environment of the water molecules affects their ability to reorient provides a key tool for unraveling the mechanisms for the transfer and transport of protons in fuel cell membranes.

Acknowledgements: This work was supported by the Department of Energy (DE-FG03-84ER13251). DEM thanks the NDSEG for a Graduate Fellowship.

REFERENCES

[1] Rubatat, L., G. Gebel, and O. Diat. Macromolecules, 2004. **37**(20): p. 7772-7783.

[2] Falk, M.. Canadian J. Chem., 1980. **58**: p. 1495-1501.

[3] Moilanen, D.E., I.R. Piletic, and M.D. Fayer. J. Phys. Chem. A., 2006. **110**(29): p. 9084-9088.

[4] Giotto, M.V., et al.. Macromolecules, 2003. **36**: p. 4397-4403.

[5] Meresi, G., et al.. Polymer, 2001. **42**: p. 6153-6160.

[6] Blake, N.P., et al.. J. Phys. Chem. B., 2005. **109**(51): p. 24244-24253.

[7] Urata, S., et al.. J. Phys. Chem. B., 2005. **109**(9): p. 4269-4278.

[8] Glew, D.N. and N.S. Rath. Can. J. Chem., 1971. **49**(6): p. 837-56.

[9] Moilanen, D.E., I.R. Piletic, and M.D. Fayer. J. Phys. Chem. C., 2007. **Accepted**.

[10] Piletic, I.R., et al.. J. Phys. Chem. A., 2006. **110**(15): p. 4985-4999.

[11] Cringus, D., et al.. J. Phys. Chem. B., 2004. **108**: p. 10376-10387.

[12] Rezus, Y.L.A. and H.J. Bakker. J. Chem. Phys., 2005. **123**(11): p. 114502-1-7.

[13] Rezus, Y.L.A. and H.J. Bakker. Journal of Chemical Physics, 2006. **125**: p. 144512-1-9.

[14] Steinel, T., J.B. Asbury, and M.D. Fayer. J. Phys. Chem. A, 2004. **108**: p. 10957-10964.

[15] Piletic, I.R., et al.. J. Am. Chem. Soc, 2006. **128**(32): p. 10366-10367.

[16] Piletic, I.R., H.-S. Tan, and M.D. Fayer. J. Phys. Chem. B., 2005. **109**(45): p. 21273-21284.

[17] Tan, H.-S., I.R. Piletic, and M.D. Fayer. J. Chem. Phys., 2005. **122**: p. 174501(9).

[18] Tan, H.-S., et al.. Phys. Rev. Lett., 2004. **94**: p. 057405(4).

[19] Laage, D. and J.T. Hynes. Science, 2006. **311**: p. 832-835.

[20] Eaves, J.D., et al.. Proc. Natl. Acad. Sci., 2005. **102**(37): p. 13019-13022.

Determination of Pore Size Distribution of Nafion/Sulfated β-Cyclodextrin Composite Membranes as Studied by ^{1}H Solid-State NMR Cryoporometry

J.–D. Jeon[*], B. H. Lee[**] and S.–Y. Kwak[***]

School of Materials Science and Engineering, Seoul National University, San 56-1,
Sillim-dong, Gwanak-gu, Seoul 151-744, Korea
[*]jdjun74@snu.ac.kr, [**]redboho@kist.re.kr, [***]sykwak@snu.ac.kr

ABSTRACT

Nafion/sb-CD membranes were prepared by mixing 5 wt% Nafion solution with H[+]-form sulfated β-cyclodextrin (sb-CD), and their water uptake, ion exchange capacity (IEC), and ionic cluster size distribution were measured. The water uptake and IEC of the membrane increased with an increase in the sb-CD content. The SAXS experiments confirmed that an increase in the sb-CD content of the membranes shifted the maximum SAXS peaks to lower angles, indicating an increase in the cluster correlation peak. NMR cryoporometry is based on the theory of the melting point depression of a liquid confined within a pore, which is dependent on the pore diameter. The intensity-temperature (*IT*) curves showed that the cluster size distribution gradually became broader with an increase in the sb-CD content due to the increased water content, indicating an increase in the ionic cluster size. This result indicates that the presence of sb-CD results in increases in the cluster size as well as in the water uptake and the IEC.

Keywords: Nafion, sulfated β-cyclodextrin, pore size distribution, cryoporometry, fuel cell

1 INTRODUCTION

Direct methanol fuel cells (DMFCs) are promising candidates for portable power sources and transport applications because they do not require the fuel processing equipment that is essential for polymer electrolyte membrane fuel cells (PEMFCs) [1]. One of the critical problems hindering the commercialization of DMFCs is high methanol permeation rate across proton-exchange membranes [2]. Methanol permeates into the membranes primarily through the ionic clusters, and thus the size distribution of these clusters determines the methanol permeability. Therefore, an improved understanding of the cluster size distribution might help improve the performances of fuel cell membranes.

To our knowledge, there has been no previous direct and precise characterization of the cluster size distribution of Nafion-based membranes using ^{1}H NMR cryoporometry. In this study, we report the ionic cluster size distribution of Nafion-based membranes, which are determined using NMR cryoporometry. In order to reduce the methanol permeability of the Nafion membranes while minimizing the loss of proton conductivity, we add sulfated β-cyclodextrin (sb-CD) into the membranes. Cyclodextrins have a shallow truncated cone shape and a hydrophobic cavity that is apolar relative to the outer surface. Among them, β-CD is the most accessible, lowest-priced and generally the most useful CD, and is used in this study. Sulfated β-CD is very hydrophilic because its external surface has many reactive sites, i.e., sulfonic acid groups. Therefore, the addition of hydrophilic sb-CD into the Nafion membranes assists the transport of protons, since the number of reactive ionic cluster sites is increased. In addition, the presence of sb-CD nanoparticles inside the ionic clusters in the membranes means that the methanol transport pathway is tortuous, resulting in a decrease in the methanol permeability.

The objective of this study is to prepare Nafion/sb-CD composite membranes with various sb-CD contents and to probe the ionic cluster size distributions for the swollen membranes with NMR cryoporometry. The detailed results of this experimental approach are presented and discussed. In the near future, we plan to report the performance, i.e., the proton conductivity and methanol permeability, of these composite membranes in DMFCs.

2 EXPREMENTAL

2.1 Materials

Nafion perfluorinated ion-exchange resin (5 wt% solution in a mixture of lower aliphatic alcohols and water) was purchased from Aldrich Chemicals and used as the membrane material; it has an equivalent weight (EW) of 1100 g for each sulfonic acid group. The Na[+]-form sulfated β-CD (typical substitution: 7 − 11 moles/mol β-CD) was purchased from Aldrich Chemicals. The H[+]-form sulfated β-CD (denoted hereafter as sb-CD) was obtained by recrystallization after adjusting the pH of the Na[+]-form sulfated β-CD solution.

2.2 Fabrication of Composite Membranes

The composite membranes were prepared by the solution casting method. The desired amount of sb-CD was added into 5 wt% Nafion solution, and then stirred at room temperature and degassed by ultrasonication. The sb-CD content of the mixture was 1, 3, or 5 wt% with respect to

Nafion. The resulting mixture was slowly poured into a glass dish in an amount that would produce a formed membrane thickness of approximately 100 μm. The filled glass dish was evaporated at 40 °C for 2 days and then annealed at 120 °C in a convection oven for 2 h. After cooling, the membrane was peeled off the glass dish by the addition of water. The membranes were stored in deionized water so that they were water-saturated. In this study, NC*x* denotes a Nafion/sb-CD composite membrane containing *x* wt% of sb-CD.

2.3 Characterization

The swelling characteristics of the membranes were determined with water uptake measurements. The samples were completely dried under vacuum for 3 days at 30 °C and then weighed. They were then placed in deionized water for a week at 25 °C. Water on the surfaces of the wet samples was removed with filter paper, and then the samples were immediately transferred to a weighing dish and weighed. The water uptake was calculated according to the equation

$$\text{water uptake}\,(\%) = \frac{W_{\text{wet}} - W_{\text{dry}}}{W_{\text{dry}}} \times 100 \qquad (1)$$

where W_{wet} and W_{dry} are the weights of the wet and dried membranes, respectively.

The ion exchange capacity (IEC) indicates the number of milliequivalents of ions in 1 g of the dry polymer. H^+-form samples of similar weight were soaked in 50 mL of 0.1 N NaCl solution for 24 h at 25 °C in order to achieve complete H^+ to Na^+ ion exchange. 10 mL of the released H^+ was then titrated with 0.1 N NaOH solution, in which phenolphthalein was used as an indicator. The IEC was calculated from the titration data with the following equation:

$$\text{IEC}\,(\text{mequiv/g}) = \frac{V_{\text{NaOH}} \times N_{\text{NaOH}} \times 5}{W_{\text{dry}}} \qquad (2)$$

where V_{NaOH} is the amount of NaOH required to neutralize a blank sample, N_{NaOH} is the normality of the NaOH solution, 5 is the ratio of the amount of NaOH required to dissolve the sample to the amount used for titration, and W_{dry} is the weight of the dried sample.

Small-angle X-ray scattering was carried out on a Bruker AXS Nanostar. In the experiments, the X-ray beam was produced with a rotating-anode X-ray generator operated at 40 kV and 35 mA, of which X-ray source was a monochromatized Cu Kα (λ = 1.54 Å) radiation. The samples were immersed in deionized water for a week and then three swollen pieces were placed within a Mylar bag under wet conditions.

In order to carry out the ^{1}H NMR cryoporometry, each sample was immersed in deionized water for a week, and its surface was wiped to remove excess water before the sample was packed into a 10 mm outer diameter NMR tube and sealed. In order to prevent the evaporation and desorption of water from the samples, the spaces above the samples in the NMR tube were filled with hydrophobic perfluorooctane. The NMR cryoporometry measurements were carried out with a Bruker mq20 spectrometer at 0.47 Tesla and a resonance frequency of 19.95 MHz. The spin-echo amplitude from $90°-\tau-180°-\tau$–echo pulse sequence was measured with a pulse separation time, τ, of 10 ms to ensure that the signal was entirely from the liquid present. Since the spin-spin relaxation time of solid ice confined within pores of NC*x* membranes was very short, whereas the corresponding relaxation time of mobile water was long, the mobile water could be detected during 20 ms of total echo time. The signal amplitude was measured as a function of temperature ranging from 210 to 277 K with an interval of 1 K using a Bruker BVT-3000 temperature control unit. All measurements were obtained by increasing the temperature after initially cooling the samples to a low temperature (i.e., 190 K) in order to prevent the complications of supercooling or hysteresis. The warming rate was low enough to achieve equilibrium, which was controlled by waiting above 10 min at each temperature step for NMR signal to stop changing. The signal intensity was corrected for temperature by implementing the Curie law, i.e., the observed signal intensity was multiplied by the factor T/T_{o} (T_{o} = 273 K and T defining actual temperature).

3 RESULTS AND DISCUSSION

3.1 Water Uptake and Ion Exchange Capacity (IEC)

Figure 1 shows the equilibrium percentage sorption of water, obtained by soaking the membranes in water at 25 °C. The water uptake of the membranes was found to increase from 21.4 for NC0 to 24.4% for NC5 with the increase in the sb-CD content. In order to validate whether or not there was any loss of sb-CD during swelling, the weight of the dried membranes before and after water uptake (i.e., swelling) experiments was measured. As a result, there was no significant difference in membrane weight between before and after the experiments, implying that loss of sb-CD did not occur during swelling. This result shows that the water uptake of the membranes can be controlled by varying their sb-CD content.

The ion exchange capacity (IEC) provides an indication of the number of ion-exchangeable groups present in an ion-conducting polymer membrane; these groups are responsible for the conduction of protons and thus the IEC is a reliable measure of the proton conductivity. The IEC of each membrane was determined by using the acid-base titration method. The results are shown in Figure 1 and demonstrate that the IECs of the membranes increase from 0.89 to 0.96 mequiv/g with increase in the water uptake. Thus the introduction of sb-CD with its many reactive sulfonic acid sites into the membranes contributes to an increase in the IEC.

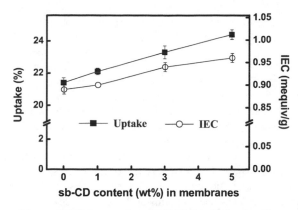

Figure 1: Water uptake and IEC for the composite membranes as functions of sb-CD content.

3.2 Cluster Correlation Peak: SAXS

Under wet conditions, the SAXS results indicate that there are significant changes in the properties of the membranes with variation in the sb-CD content, i.e., with variation in the water content. Figure 2 shows the relative scattering intensity as a function of scattering vector, S, for the hydrated Nafion membrane (NC0) and the hydrated Nafion/sb-CD composite membranes (NC1, NC3, and NC5). The scattering vector, S, is defined by $S = 2\sin\theta/\lambda$. The scattering intensity increased with increase in the water content. This is due to the enhancement of the difference in the electron density by decreasing the electron density of the ionic clusters containing more water content relative to the backbone [3]. It can be also seen that the maximum SAXS peak shifts to lower scattering vector (i.e., scattering angle) as the sb-CD content (i.e., water content) increases. This maximum peak position is inversely proportional to the cluster correlation peak. Therefore, these decreases in scattering angle indicate the expansion of the hydrophilic ionic clusters due to the increased cluster correlation peak. In addition, by revealing these changes in the cluster correlation peak, the SAXS data confirmed the presence of sb-CD particles in the ionic clusters. It has previously been reported that the increase in the cluster correlation peak for Nafion membranes is directly proportional to the volume of absorbed water [4].

3.3 Cluster Size Distribution: NMR Cryoporometry

The theoretical basis for this NMR application is the well-known Gibbs-Thompson equation [5–7], which relates the melting point depression, ΔT_m, of a confined liquid to the pore diameter, D:

$$\Delta T_m(D) = K\frac{1}{D} \tag{3}$$

where K is a constant depending solely on the physical properties of the liquid confined within the porous material. Measurement of the amount of liquid confined within the pores as a function of temperature enables the pore size, D, and its distribution to be estimated when K is known. In this study, water was used as the probe liquid, for which K is approximately 62 K nm. This equation indicates that the difference between the normal and depressed melting temperatures is inversely proportional to a linear dimension of the liquid confined within the pores. The spin-echo signal intensity, V, indicates the amount of liquid water confined within the pores at a particular temperature, T, and thus the volume of the pores with linear dimensions equal to the corresponding the pore diameter, D, can be calculated.

$$\frac{dV}{D} = \frac{K}{D^2}\frac{dV}{dT_m(D)} \tag{4}$$

Therefore, the measurement of $dV/dT_m(D)$ enables the pore size distribution of the samples to be determined, provided K is known.

Figure 3 shows the intensity-temperature (IT) curves of water confined within the ionic cluster pores of the Nafion/sb-CD composite membranes. In the case of the solid phase, the relative signal intensities were denoted as 0, corresponding to totally frozen water. The relative signal intensities from the liquid phase of totally molten water were denotes as 1. These curves show that the signal intensities increase gradually with increasing temperature (below 273 K) due to the gradual melting of the frozen water confined within the ionic cluster pores in the membranes. The abrupt increase in intensity at the melting point of bulk water, 273 K, is due to the melting of the bulk supernatant water.

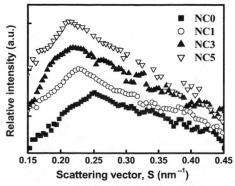

Figure 2: SAXS curves for swollen composite membranes.

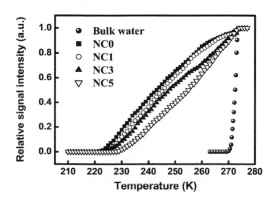

Figure 3: Relative signal intensities versus temperature (*IT* curves) of water confined within the ionic cluster pores of the Nafion/sb-CD composite membranes.

If the amount of liquid confined within the pores of the membranes and K are known, their cluster size distribution can be estimated using equation (4). It has generally been reported that K is in the range $41 - 73$ K nm, when water is used as the probe liquid. We assumed that $K = 62$ K nm to obtain the cluster size distributions. These data were used in equation (4) to obtain the relative pore volume as a function of pore diameter, as shown in Figures 4(a) to (d).

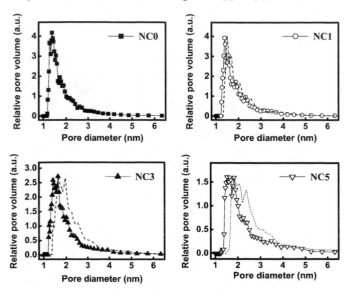

Figure 4: Pore size distribution curves of the Nafion/sb-CD composite membranes. The dashed lines of NC1, NC3, and NC5 are corrected by the melting point depressions, which are caused by colligative effect.

On the other hand, colligative effect will also affect the *IT* curves due to the addition of different sb-CD content. This effect is a phenomenon by which the freezing/melting point of a solution is lowered when more solute is dissolved in the solution. In order to validate the colligative effect of water/sb-CD solutions confined within ionic clusters of composite membranes, the melting point of water/sb-CD solutions was measured by using the NMR spin-echo method. The water/sb-CD solutions confined within ionic clusters of NC1, NC3, and NC5 composite membranes had various sb-CD concentrations of 4, 11, and 17 wt%, respectively. These concentrations were determined by assuming that water (the amount of water measured by the gravimetric method) and sb-CD particles were confined within only ionic clusters of composite membranes. As a result, the melting point depressions of the water/sb-CD solutions with 4, 11, and 17 wt% concentrations compared to pure water were found to be 2, 6, and 8 K, respectively. Therefore, the original *IT* curves of the composite membranes were corrected by adding their melting point

depressions to original temperature. In Figure 4, it is clear that the cluster size distribution of the composite membranes broadens gradually with an increase in their sb-CD content, i.e., there is an increase in the ionic cluster pore size. This is probably due to the expansion of the hydrophilic ionic clusters due to the increased water content, which is caused by the addition of sb-CD with its many sulfonic acid groups into the hydrophilic ionic clusters. This increase in cluster size is in agreement with the trend in the results obtained from the SAXS measurements. Overall, these results show that NMR cryoporometry can be used to determine the ionic cluster size distributions of Nafion-based membranes and that this method provides a means with which to determine pore sizes on the nanometer scale.

4   CONCLUSIONS

Nafion/sb-CD composite membranes were prepared with the solution casting method. The water uptake values were found to increase with an increase in their sb-CD content. The acid-base titration results show that the IEC of the membranes increases from 0.89 to 0.96 mequiv/g with an increase in their sb-CD content, indicating that the introduction of sb-CD with its many reactive sites (sulfonic acid groups) into the membranes results in an increase in IEC. SAXS and ^{1}H solid-state NMR cryoporometry results were obtained to determine the cluster correlation peak and ionic cluster size distributions of the membranes, respectively. The SAXS results show that the cluster correlation peak increases with an increase in the sb-CD content. From the *IT* curves by NMR cryoporometry, it was determined that the cluster sizes of the membranes increase with an increase in their sb-CD content due to their increased water content, i.e., there is an increase in the ionic cluster pore size. Thus the presence of sb-CD with its many sulfonic acid groups in the Nafion membranes leads to increases in their water uptake, IEC, and ionic cluster size. In conclusion, ^{1}H solid-state NMR cryoporometry was found to be a very effective method for characterizing the ionic cluster size distribution of Nafion-based membranes, which is strongly correlated with the performance of membranes for polymer electrolyte fuel cells (PEMFCs).

REFERENCES

[1] T. Yamaguchi, F. Miyata and S.-I. Nakao, Adv. Mater., 15, 1198, 2003.
[2] M. K. Ravikumar and A. K. Shukla, J. Electrochem. Soc. 143, 2601, 1996.
[3] C. L. Marx, D. F. Caufield and S. L. Cooper, Macromolecules, 6, 344, 1973.
[4] M. Fujimura, T. Hashimoto and H. Kawai, Macromolecules, 14, 1309, 1981.
[5] J. W. Gibbs, "The Collected Works of J. Williard Gibbs," Academic Press, New York, 1928.
[6] W. Thompson, Philos. Mag., 42, 448, 1871.
[7] C. L. Jackson and G. B. McKenna, J. Phys. Chem., 93, 9002, 1990.

Hydroxyapatite-imidazole-polymer composite films as a proton conductor under no humidified condition

H. Tsutsumi and Y. Hisha

Department of Applied Chemistry, Faculty of Engineering, Yamaguchi University,
2-16-1, Tokiwadai, Ube 755-8611, Japan, tsutsumi@yamaguchi-u.ac.jp

ABSTRACT

Hydroxyapatite ($Ca_{10}(PO_4)_6(OH)_2$, HAp)-imidazole (Im)-poly(vinylidene fluoride) (PVdF) composite films were prepared by casting the HAp dispersed solution that included Im and PVdF. Their conducting behaviors under no humidified condition and various temperatures were investigated. Two kinds of HAp (a-HAp and n-HAp) with different particle size were used. The average particle size of a-HAp was 1.25 μm in diameter and that of n-HAp was 0.61 μm in diameter. The conductivity for a-HAp-Im-PVdF (1:2:5, by weight) film at 120°C was 8.4 μS cm^{-1} and that for n-HAp-Im-PVdF (1:2:5, by weight) was 298 μS cm^{-1}. The particle size of HAp and the amount of Im and HAp in the films affected their conducting behavior and their mechanical properties.

Keywords: proton conductor, dry condition, polymer electrolyte membrane fuel cell

1   INTRODUCTION

Solid polymer electrolyte type fuel cell (PEFC) is one of the candidates for an energy generator with high performance and low production of carbon dioxide [1]. Higher temperature operation (> 100°C) of PEFC will offer higher energy conversion performance [2]. However, typical solid polymer electrolytes (SPE) for PEFC, such as Nafion™, base on the conduction mechanism through the water molecules in the SPE. To keep the water molecules in the SPE, the operation temperature of the fuel cells with the SPE should be lower than 100°C. SPE for the fuel cells with higher operation temperature (> 100°C) requires other proton conduction mechanism that is not depended on water molecules. Some investigations were performed with various systems, Nafion-imidazole (Im)-phosphoric acid [3], Teflon™-Im-phosphoric acid [3], polymer electrolyte membranes based in the polybenzimidazole (PBI) doped with phosphoric acid [4], and PBI doped with Im or 1-methylimidazole and phosphoric acid [5]. Previously, we reported preparation of SPE films based on hydroxyapatite (HAp), Im, and poly(ethylene oxide) (PEO) and their temperature dependence of conductivity and their proton conduction mechanism without water molecules [6]. However, the HAp-Im-PEO films were not stable over 60°C, because the melting point of PEO is about 60°C. Crosslinking of PEO (c-PEO) with tolylene 2, 4-diisocyanate as a cross-linking reagent was effective to keep the dimension of the film. However, the SPE with c-PEO matrix was not acceptable because of their dimension and mechanical instability over 100°C. In this paper we report that preparation of HAp-Im-poly(vinylidene fluoride) (PVdF) composite films and their conductive performance at the temperature over 100°C. PVdF acts as a binder to improve the thermal and mechanical stability of the composite films.

2   EXPERIMENTAL

2.1   Materials

Poly(vinylidene fluoride) (PVdF), average molecular weight 534,000 (Aldrich) and imidazole (Im) (Wako) were used as received. Other chemicals were also purchased and used as received.

2.2   Preparation of hydroxyapatite

Two types of HAp (a-HAp and n-HAp) were prepared [7, 8]. One type of HAp, a-HAp was prepared according to the reported procedure [7] with some modification. The pH of calcium chloride aqueous solution ($[Ca^{2+}]$ = 0.2 mol dm^{-3}]) was adjusted to 9.01 by addition of Tris-HCl buffer (Solution A). Diammonium hydrogenphosphate aqueous solution ($[PO_4^{-3}]$ = 0.12 mol dm^{-3}) was prepared and the pH of the solution was adjusted to 9.30 by addition of NH$_3$-NH$_4$Cl buffer (Solution B). The solution B (40 ml) was added to 40 ml of the solution A at room temperature. After the addition the mixed solution was stirred for 2 h at room temperature. The white participate was filtered off from the solution and the solid was dried at 80°C for 24 h and then at room temperature for 24 h under dynamic vacuum conditions (*ca.* 0.13 kPa).

Another type of HAp, n-HAp was prepared according to the method in the paper [8] with some modification. Phosphoric acid (2.03 g) was added to 150 ml of water. Dispersion of calcium hydroxide (2.22 g / 150 ml of water) was added into the phosphoric acid aqueous solution at room temperature (addition rate 7.5 ml min^{-1}). After the addition of the Ca(OH)$_2$ dispersion, the reaction mixture was stirred 2h and then the mixture was left at rest for 15 h at room temperature. The participated solid was dried at 80°C for 24 h and then at 80°C for 24 h under dynamic vacuum condition (*ca.* 0.13 kPa). The n-HAp solid was grinded in an agate mortar and the grinded powder was riddled with 200 mesh sieve. Calcinated HAp (c-HAp) was prepared from n-HAp by heating at 800°C, for 2 h in the air.

2.3 Preparation of HAp-Im-PVdF composite

Typical preparation procedure of the composite was as follows. Hydroxyapatite (a-HAp or n-HAp), 0.1 g was dispersed into 3 mL of *N, N*-dimethylformamide (DMF). Imidazole (0.2 g) was added into the HAp dispersed DMF. The HAp dispersed DMF solution containing Im was stirred for 24 h (Dispersed solution C). PVdF (0.5 g) was dissolved in 5 mL of DMF and stirred for 24 h (Solution D). The dispersed solution C was added to the solution D and the HAp dispersed solution was stirred at room temperature for 24 h. The HAp dispersed solution containing PVdF and Im was poured into an Al foil dish and the DMF was removed at 50°C for 24 h under atmosphere pressure and then at 55°C, for 6 h under dynamic vacuum condition (*ca.* 0.13 kPa). The resulted film was presented as HAp-Im-PVdF (1:2:5) composite film. Resulted film was referred as HAp-Im-PVdF (x:y:z) composite film. The x, y, and z in the parenthesis express the weight ratio of each component in the composite film. The HAp particles in the films were dispersed homogeneously under the visual observation.

2.4 Measurements

A composite film for conductance measurement was sandwiched with two stainless steel plates (13 mm in diameter). Conductivity for the composite film was measured with an LCR meter (HIOKI 3531 Z Hi tester, 10-100 mV$_{p-p}$, 1-100 kHz) under various temperature conditions from 50 to 120°C. Infrared spectra of samples were recorded with an FTIR spectrophotometer (IRPrestige-21, Shimadzu). XRD patterns of HAp and the composites films were recorded with an X-ray diffraction meter (XD-D1, Shimadzu, CuKα, λ=0.1542 nm). DSC measurements of the samples were preformed with a differential scanning calorimeter (DSC5100S, Bruker AXS); heating rate was 10 K min^{-1}. The mechanical properties of the composite films under heating condition (100°C) were recorded with a thermal mechanical analysis instrument (EXSTAR6000 TMA/SS, SII). Typical sample size was 10.0 mm × 3.0 mm, 100 μm thickness. All the tests were conducted at a crosshead speed of 50 μm/min and offset load of 50 mN. The percentage strain (λ) was calculated as follows:

$$\lambda = [(L - L_0)/L_0] \times 100 \tag{1}$$

where *L* is the total extension measured from the grip displacement and *L$_0$* is the grip distance. The initial Young's modulus was calculated from the initial slope of the stress–strain curve obtained. SEM observation of HAp particles that were Au sputtered with an Ion Sputter (E101, Hitachi) was performed with an SEM equipment (JSM-T300, JEOL).

3 RESULTS AND DISCUSSION
3.1 Structure of hydroxyapatite

Characterization of a-HAp and n-HAp was performed with XRD, FTIR, and SEM measurements. FTIR spectra of a-HAp and n-HAp showed the peak at 875 cm^{-1} that is attributed to HPO$_4^{2-}$. This suggests that a-HAp and n-HAp are partially calcium deficient hydroxyapatite [9]. Fig. 1 shows the XRD patterns of HAp used in this investigation. The XRD pattern of c-HAp powder indicates that c-HAp is crystalline form. The XRD pattern of a- or n-HAp powder suggests that the structure of a-HAp or n-HAp is mainly hydroxyapatite structure containing partially amorphous phase. The average particle size of a-HAp was 1.25 μm and that of n-HAp was 0.61 μm determined from SEM observations of them.

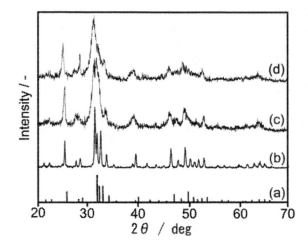

Fig. 1 XRD patterns of (a) JCPSD card (Card No.90432), (b) a-HAp, (c) n-HAp, and (d) c-HAp (Calcinated at 800°C for 2h in the air).

3.2 Temperature dependence of conductivity for HAp-Im-PVdF composite films

PVdF-based composite films were prepared from HAp dispersed DMF solution containing Im and PVdF, and the resulted films with HAp were free-standing, flexible, and opaque. Conductivity for c-HAP-Im-PVdF film was very low, under 10^{-9} S cm^{-1}, which is the limitation on measurement of our equipment. Fig. 2 shows temperature dependence of conductivity for a-HAp-Im-PVdF and n-HAp-Im-PVdF composite films. The conductivity clearly increases with temperature for all the samples because of the higher diffusivity of the Im molecules and the higher flexibility in the chains of the PVdF matrix. Conductivity at 120°C of a-HAp-Im-PVdF (1:2:5) was 8.4 μS cm^{-1} and that of n-HAp-Im-PVdF (1:2:5) was 298 μS cm^{-1}. Conductivity of the SPE film with n-HAp was about 35 times higher than that with a-HAp. At all temperatures, the conductivity for n-

HAp-Im-PVdF films with other composition was also higher than that for a-HAp-Im-PVdF films. This suggests that conductivity for the SPE film depends on the kind of HAp. Optical microscope observation of the films suggested that dispersion of n-HAp particles in the SPE films was not similar to that of a-HAp. The n-HAp particles were well dispersed in the composite film and a-HAp particles were not in the matrix.

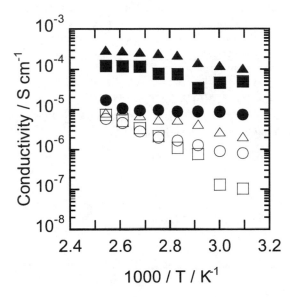

Fig. 2 Temperature dependence of conductivity for HAp-Im-PVdF composite films.

a-HAp-Im-PVdF (x:y:z): △ (1:2:5), ☐ (2:2:5), ○ (5:2:5)

n-HAp-Im-PVdF (x:y:z): ▲ (1:2:5), ■ (2:2:5), ● (5:2:5)

The Arrhenius plots show linear lines irrespective of type of HAp and concentration of Im and PVdF. They can be interpreted by normal Arrhenius relationship, eq 2.

$$\log \sigma = \log \sigma_0 - E_a/RT \qquad (2)$$

where σ_0 is the conductivity at infinite T, R is gas constant, E_a is activation energy, the results are illustrated in Table 1.

Table 1 Parameters of Arrhenius plots calculated from temperature dependence of conductivity for HAp-Im-PVdF composite films

Sample	σ_0 / S cm^{-1}	E_a / kJ mol^{-1}
a-HAp-Im-PVdF(1:2:5)	6.09×10^{-2}	9.15
a-HAp-Im-PVdF(2:2:5)	1.88×10^{-4}	28.0
a-HAp-Im-PVdF(2:5:5)	1.50×10^{-3}	21.3
n-HAp-Im-PVdF(1:2:5)	1.19×10^{-1}	6.96
n-HAp-Im-PVdF(2:2:5)	1.69×10^{-1}	5.81
n-HAp-Im-PVdF(2:5:5)	8.01×10^{-2}	8.25
Im-PVdF(2:5)	1.20×10^{-4}	29.5

The E_a for the composite films, n-HAp-Im-PVdF is 5.81 - 8.25 kJ mol^{-1}. The E_a for hydrated pure Nafion obtained by NMR is 11.0 kJ mol^{-1} [10] and by the macroscopic conductivity measurements is 9.34 kJ mol^{-1} [11]. The E_a for the n-HAp-Im-PVdF is lower than that for Nafion.

The parameter, σ_0 is closely related with the number of charge carriers in the composite film. The σ_0 for the composite film with n-HAp is the one or two orders in magnitude larger than that for the film with a-HAp. Thus, the conductivity for the composite film with n-HAp is higher than that for the film with a-HAp.

3.3 Mechanical property of the composite films

Mechanical property of an SPE film for a PEFC operated under higher temperature (> 100°C) condition is also important for its long-life operation. Fig. 3 describes the stress–strain results at 100°C obtained on the HAp-Im-PVdF films as well as for a pure PVdF film. Specific values of the mechanical properties of HAp-Im-PVdF films are summarized in Table 2. A yielding phenomenon was observed clearly with the PVdF film. The maximum stress appeared at the yield point (36%, 7.1 MPa). The breaking elongation of PVdF film was 102%. Strain-stress curves of a-HAp-Im-PVdF films were difference with that of PVdF. The yield point of a-HAp-Im-PVdF (1:2:5) was 2.8% and 3.9 MPa. The breaking elongation of the film was 18% and the values of other films with a-HAp were 15 and 10%. The n-HAp-Im-PVdF (1:2:5) film showed a yielding phenomenon, however, the maximum stress (6.8 MPa) appeared over the first yield point (21%, 5.7 MPa). The breaking elongation of the film was 99%. Mechanical toughness of the films is reduced by addition of a-HAp particles. Addition of n-HAp particles to the PVdF film almost preserves the mechanical strength of an original PVdF film.

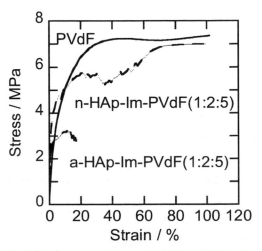

Fig. 3 Strain-stress curves of PVdF and HAp-Im-PVdF composite films at 100°C.

3.4 Effects of dispersion of HAp in PVdF matrix to conductivity and mechanical property

We reported the charge carriers in HAp-Im-PEO or HAp-Im-c-PEO (Crosslinked PEO) are imidazolium ions that are produced by acid-base reaction between imidazole and the phosphate groups on the surface of HAp [6]. We also checked the PVdF-based composite films with FTIR measurements. Infrared spectra of the composite films suggest that the charge carriers in them are imidazolium ions that are produced by acid-base reaction between imidazole and the phosphate groups on the surface of HAp.

Depersiveness of HAp in the PVdF matrix is also concerned with conductivity and mechanical property of HAp-Im-PVdF composite films. Distance (d) between particles (r is radius of the particles) in a matrix is calculated as follows [12]:

$$d = [(4\pi \times \sqrt{2} / 3V)^{1/3} - 2] \times r \qquad (3)$$

where V is volume fraction of the particles in the matrix. When the content of HAp in the a-HAp-Im-PVdF composite film is equal to that in the n-HAp-Im-PVdF film, the estimated distance between an HAp particle and another one in the n-HAp-Im-PVdF composite film is about 2 times shorter than that in the a-HAp-Im-PVdF film. The shorter distance between the HAp particles will provide higher conductivity and enhanced mechanical property for the n-HAp-Im–PVdF composite films. Detail investigation about the effect of dispersibility of HAp on the composite film system is now in progress.

ACKNOWLEGEMENT
This work is partially financial supported by Research for Promoting Technological Seeds (JST).

REFERENCES

[1] K. Sopian, W. R. W. Daud, *Renewable Energy*, **31**, 719 (2006) and references cited therein.

[2] Q. Li, R. He, J. O. Jensen, N. J. Bjerrum, *Chem. Mater.*, **15**, 4896 (2003) and references cited therein.

[3] J. Sun, L. R. Jordan, M. Forsyth, D.R. MacFarlane, *Electrochimica Acta*, **46**, 1703 (2001).

[4] M. Kawahara, J. Morita, M. Rikukawa, K. Sanui, N. Ogata, *Electrochimica Acta*, **45**, 1395 (2000).

[5] A. Schechter, R. F. Savinell, *Solid State Ionics*, **147**, 181 (2002).

[6] H. Tsutsumi, Y. Hisha, Y. Shibasaki, *Electrochemical and Solid-State Letters*, **8**, A237 (2005).

[7] T. Taguchi, A. Kishida, M. Akashi, *Kobunshi Ronbunshu*, **57**, 324 (2000).

[8] R. Kumar, K. H. Prakash, P. Cheang, and K. A. Khor, *Langmuir*, **20**, 5196 (2004).

[9] A. Siddharthan, S. K. Seshadri, T. S. Sampath Kumar, *J. Materials Science: Materials in Medicine*, **15**, 1279 (2004).

[10] G. Ye, N. Janzen, G. R. Goward, *Macromolecules*, **39**, 3283 (2006).

[11] P. Costamagna, C. Yang, A. B. Bocarsly, S. Srinivasan, *Electrochimica Acta*, **47**, 1023 (2002).

[12] K. Chujyo, Y. Harada, *Kolloid Z.*, **201**, 66 (1965).

Table 2 Mechanical property of PVdF and HAp-Im-PVdF composite films at 100°C

Sample [a]	Modulus / MPa	Yield stress / MPa	Yield strain / %	Breaking elongation / %
PVdF	120	7.1	36	102
a-HAp-Im-PVdF (1:2:5)	390	2.8	3	18
a-HAp-Im-PVdF (2:2:5)	440	3.8	15	34
a-HAp-Im-PVdF (5:2:5)	330	3.1	10	16
n-HAp-Im-PVdF (1:2:5)	160	5.7, 6.8	21, 72	99
n-HAp-Im-PVdF (2:2:5)	350	6.0	20	98
n-HAp-Im-PVdF (5:2:5)	590	5.5	9	96

a) Typical sample size is 10.0 mm × 3.0 mm, 100 μm thickness. Cross head rate 50 μm min^{-1}

Block Copolymer Directed Nanoparticle for Fuel Cell Applications

Terry Dermis*, Sundar Mayavan*, Naba K. Dutta*, Namita Roy Choudhury* and Steven Holdcroft**

*Ian Wark Research Institute, ARC Special Research Centre, University of South Australia, Mawson Lakes Blvd., South Australia, Australia
**Department of Chemistry, Simon Fraser University, Canada

ABSTRACT

A simple method for synthesizing nanoparticles (NPs) using a fluorinated block copolymer is proposed. In this method, platinum NPs were introduced, via a reduction method, into a poly(vinylidene difluoride -co- hexafluoropropylene) -b- poly(methyl methacrylate) P((VDF-co-HFP)-b-PMMA) copolymer matrix. NP dispersion was characterized using transmission electron microscopy (TEM) coupled with an energy-dispersive x-ray (EDX) device for spectroscopic analysis, indicating whether NPs were selectively bound to one of the two phase separated copolymer blocks. Ultimately, exploitation of this technology would be a major advantage in the fuel cell market, specifically as the catalytic electrodes of proton exchange membrane fuel cells (PEMFCs).

Keywords: block copolymers, nanoparticles, self-assembly, fuel cell, AFM

1 INTRODUCTION

In recent years, researchers have attempted to develop various electrode structures for commercializing fuel cell systems. The crux of electrode performance would be the best electrocatalyst for a given charge transfer reaction. Platinum (Pt) alloys, and metal oxides have been the most effective electrocatalysts developed so far [1]. However, achieving high catalyst dispersion and subsequent improved access of the fuel towards it (i.e. high particle surface area), is a critical issue. This has led to intense research in developing a catalytic system in which metallic particles are directly deposited on to a block copolymer (BCP) matrix, enabling high dispersion of the catalytic particles whilst maintaining low Pt loadings. Generally, a block copolymer system containing dissimilar blocks A and B, is able to self-assemble into an ordered structure with a specific micro- or nano-domain orientation. The resulting BCP morphology and domain structure is directly related to the chemical nature of the blocks, molecular weight of each polymer block, the type of solvent and substrate used, and annealing times and temperatures [2]. In dilute solutions of selective solvents (i.e. selective to only one block), BCPs can intermolecularly associate to form aggregates. In such block copolymers, phase segregated polymer with formation of meso-domains with size ca. several tens of nanometers is observed. The situation is versatile and complicated since for a block copolymer system a solvent that is good for one block may be neutral, slightly selective, or strongly selective, depending on whether it is good, near Θ, or a non-solvent for the other block/s. The state of a system is principally governed by $\varphi\chi N$, where φ is the polymer volume fraction, χ is the interaction parameter between the individual component, and N is the total degree of polymerization. By controlling appropriately the segment nature and length of each constituent of the block in block copolymers, a wide variety of micro-domain structures of high degree of richness and complexity in solution phase are possible. Solution properties of block copolymers as well as their aggregation properties have long been investigated and have attracted much attention from both theoretical and experimental viewpoints. The knowledge of their solution behavior is of utmost importance for NP synthesis as they tend to act as nanoreactors by stabilizing the NPs within micelles of the diblock [3]. Control over the size and shape of the metal colloids can be achieved by 'growing' the desired colloids within domains of suitable size and shape. Block copolymers [4] and ionomers [5] comprising fluoro/non-fluoro segments have significant potential applications in the field of fuel cells. Recently, Shi and Holdcroft [6] successfully synthesized a series of high molecular weight block copolymers; P(VDF-co-HFP)-b-PMMA and P(VDF-co-HFP)-b-sulphonated PS (polystyrene). The former BCP was used in this study owing to the remarkable characteristics of PMMA and its viability in the proposed application. While a variety of procedures is found in the literature for synthesizing BCP nanocomposites, a convenient way of producing them is through the reduction of metal halides or anionic metal chloride precursors with alcohols, $NaBH_4$, Hydrazine or H_2 in the presence of BCP as a stabilizing agent [7]. Here we report the preparation and properties of these novel block copolymer nanocomposites for fuel cell catalyst applications. Pt based metallic precursors were incorporated into three compositions of block copolymer matrices and then reduced to metallic NPs within the BCP. The effect of copolymer composition and concentration on BCP self-organization, and NP deposition behavior, distribution, and size has been investigated using microscopic and spectroscopic techniques.

2 EXPERIMENTAL
2.1 Materials & Preparation

Poly(vinylidene difluoride-*co*-hexafluoropropylene)-*b*-poly(methyl methacrylate) (PVDF-*co*-HFP)-*b*-PMMA), synthesized at Simon Fraser University, Burnaby, Canada [6], was used as the model diblock copolymer without further ionization. Three samples of the copolymer were used possessing differing MMA molecular weights which contributed to dissimilar overall compositions for each of the samples (Table 1). Tetrahydrofuran (THF, reagent grade), was used as the selective solvent, where 0.1 wt%, 0.01 wt% and 0.005 wt% solutions were made from each of the samples. Pre-cleaned silicon wafers were used as substrates for applying thin films of the 0.01 and 0.005 wt% BCP/THF solutions, using a spin coater set at 4000 rpm.. These were then dried with high purity N_2 gas before being characterized using microscopic techniques.

The procedure for preparation of the BCP - stabilized colloid was as follows: $PtCl_4$ (Sigma Aldrich) (0.015 wt %) and P(VDF-*co*-HFP)-*b*-PMMA (0.1 wt %), were dissolved in THF and stirred overnight to form a pale yellow homogeneous solution. Sodium borohydride ($NaBH_4$) (Sigma Aldrich, ~1.5x the amount of $PtCl_4$ to assure complete reduction) was then added quickly with vigorous stirring. The color of the solution changed quickly from pale yellow to dark brown in a few seconds indicating that Pt(IV) ions were reduced to Pt^0 in aqueous solution. This dark brown solution was stirred overnight to give the BCP-stabilized metallic colloids. Each procedure was conducted under a N_2 atmosphere, where exposure to an open air atmosphere was kept to a minimum.

Table 1 : Molecular weight/compositions of each block in the three copolymer samples

Composition (%)	P(VDF-co-HFP) Mn^2 (g/mol)/PDI*	Block Mn^2 (g/mol)/PDI	PMMA Mn^3 (g/mol)
15/85	22k/1.50	147k/1.40	125k
65/35	22k/1.50	34k/1.41	12k
88/12	22k/1.50	25k/1.83	3k

*PDI = Polydispersivity Index

2.2 Characterization

BCP phase separation and morphology was analyzed by AFM using a NanoScope III Multimode SPM with controller from Digital Instruments, Veeco Metrology Group, Santa Barba, CA. All measurements were performed in tapping mode under ambient air using single-crystal silicon cantilevers. Data evaluation was performed with the NanoScope software version 5.30 (Digital Instruments, Veeco), and all images were flattened to the first order.

UV–Visible absorption spectra were recorded using a Varian Cary 1E UV/Vis spectrophotometer with a 10-mm quartz cell. The scanning range was from 200 to 800 nm. Baseline correction was made with respect to the THF solvent.

Transmission electron microscopy (TEM) was used to obtain micrographs of the Pt particles distributed among the block copolymer matrix, and possibly showing the resulting phase separated structure of each composition. A Philips CM200 operated at an accelerating voltage of 200kV and microscope fitted with an EDAX DX-4 energy-dispersive X-ray (EDX) system was used. A drop of dilute block copolymer/Pt^0 aqueous solution was dispensed onto carbon films supported by 200 mesh copper grids, where the solvent was allowed to evaporate before analysis took place.

3 RESULTS AND DISCUSSIONS

AFM analysis was conducted on the substrate cast diblock copolymer thin films. This technique was employed to examine confinement phenomena, such as the nanoscale patterns, formed due to the dissimilarity of the two block segments. Fluorinated and non-fluorinated polymer segments were expected to form highly ordered structures, dictated by the copolymer structure and molecular weights of each phase, substrate, film thickness, and selective solvent. Since the Si wafer is hydrophilic, it was expected that the PMMA segment would have a higher affinity for this surface in which it would reside to, whereas P(VDF-*co*-HFP), having a lower surface energy, would have a higher affinity for the free surface [8].

Figure 1 (a) shows definitive spherical type structures of one phase existing among the matrix of the other phase. The spheres are assigned to the P(VDF-*co*-HFP) phase, since the block ratio for this is of the lower order and, as in thin film geometries, the interfacial interactions impose restrictions on the local A-B segmental concentration profiles, hence, the lower surface energy of the fluorinated phase has caused these spheres to reside at the free surface. Annealing is expected to result in integration of like-segments, reducing the number of unfavorable A-B enthalpic contacts, in an attempt to decrease the overall free energy and the A-rich and B-rich domains may grow, in order to increase the surface to volume ratio. In an effort to maintain constant segmental density throughout this process, the copolymer chains must stretch beyond their equilibrium conformations. Also, due to the preferential interaction between the PMMA component and the Si substrate a higher degree of phase segregation of PMMA from the P(VDF-*co*-HFP) phase is probable, than if they were in the bulk.

The 65/35% composition, illustrated in Figure 1 (b), shows a unique phase separated structure of the two segments. Again, asymmetric equilibrium features evolve and due to molecular weight/compositional differences in the blocks, the system minimizes its free energy by creating a discontinuous gyroid type structure. Here the P(VDF-*co*-HFP) phase is seen to construct this gyroidal network in a 'threaded-bead necklace' (TBN) type organization, where these bead-type domains seem to run along striations bringing about the complex gyroidal symmetry. This phase image is representative of a very dynamic system of which a level of ordering will result with time or under the influence of temperature.

Figure 1 (c) represents the 88/12% block copolymer composition. Here, the molecular weight of the P(VDF-*co*-HFP) is much higher than that of PMMA, resulting in the

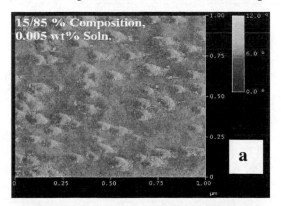

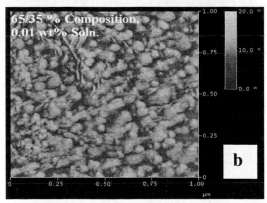

Figure 1 : (a), (b), and (c) represent 1 x 1 micron scanned AFM phase images of the various BCP systems cast on Si wafer substrates. Spherical domain diameters for each composition are: ~60-70 nm, ~40-60 nm, and ~25 – 45 nm (PMMA) and ~140 – 180 nm (P(VDF-co-HFP)) for the above images (a), (b), and (c), respectively.

PMMA block forming small spherical domains aligned in a cross-hatch type ordered morphology. It is noticed that the smaller PMMA domains, where present, segregate the larger P(VDF-*co*-HFP) domains due to their affinity for the Si substrate. It is also observed that there is attachment of both P(VDF-*co*-HFP) and PMMA segments suggesting continuity, which is expected since corresponding blocks will segregate in an attempt to lower the overall free energy of the system.

All three images reveal that the copolymer exhibits a hierarchy of patterns, dependent on the BCP composition, and bulk solution concentration. The origin of pattern formation is thought to be associated with short-range and long-range van der Waals intermolecular forces [8].

The formation of P((VDF-*co*-HFP)-*b*-PMMA/Pt colloids via reduction of $PtCl_4$ with $NaBH_4$ was studied by UV-Vis spectroscopy. Once $NaBH_4$ was added to the mixture a new

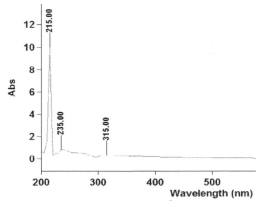

Figure 2 UV/Vis spectrum of Pt^0/15/85% BCP solution after 24 hrs mixing under N_2 atmosphere.

absorption band at 215 nm appeared and increased in intensity with time, which is characteristic of the colloidal Pt [9]. The resultant solution was mixed for 24 hrs in which UV/Vis results confirmed full reduction of $PtCl_4$ to Pt^0 after 24 hours (Figure 2).

Figure 3 (a) and (b) show representative TEM images of the distribution of NPs within BCP nano-domains. If observed closely a contrast is seen along the contours where the Pt NPs are residing. This contrast mechanism in all TEM images possibly results from thickness and density fluctuations between the two phases. Low structural detail and poor contrast of the BCPs is accounted for due to the inability to stain either blocks with RuO_4 [10]. It is, however, observed that the metal particles are quite uniform and well distributed. This relates well with the assumption that the Pt colloids primarily reside to the hydrophilic PMMA phase and stabilization and association of Pt nanoparticles can be controlled by using this unique fluoro non-fluoro block copolymer. The average size of the Pt particles in all images (88/12 % composition not shown here) are about 2 – 3.5 nm in diameter which is much smaller than that of the commercial catalyst (dia ≈ 3.8 nm) [11]. We also conclude that the NPs are stable amongst the BCP matrix since all TEM analysis was conducted 24 hours after full reduction had been confirmed by UV/Vis spectrophotometry measurements. It can be established from the weight percentages of the elements (shown in Table 2), detected by EDX analysis, that the Pt particles are residing to the PMMA phase since insignificant amounts of fluorine have been detected. Furthermore, silica was detected in all samples, and this can be accounted for from inevitable cross-contamination of all apparatus used. Hence, a well dispersed organization among the NPs, stabilized by the copolymer matrix, due to phase separation of the dissimilar blocks, can be seen.

Table 2 : Weight percent distribution of elements found from TEM-EDX analysis for each composition. (Analysis performed on each TEM image shown in Figure 3).

Comp. (%)	C	O	F	Na	Si	Cl	Pt
15/85	40.5	5.8	0	3.2	3.4	0.7	46.4
65/35	17.3	0.9	0.3	0.2	1.9	0.2	79.2
88/12	32	2.2	0	0.2	1.8	0.4	63.5

4 CONCLUSION

AFM images indicated highly ordered BCP morphology with distinct patterns dependent on block compositions and copolymer concentration. TEM results demonstrate that stable, well dispersed nanocomposites can be achieved through selective incorporation of the NPs onto specific nano-domains of the phase-separated BCP. This method allows preparation of simple or complex highly ordered structured thin films with deposited NP arrays for use as electrocatalyst layers within a PEMFC membrane electrode assembly.

5 ACKNOWLEDGEMENTS

The authors wish to acknowledge the Australian Research Council for funding of this work through Discovery Grant. T. Dermis also gratefully acknowledges the funding provided by the Ian Wark Research Institute through an Honours scholarship.

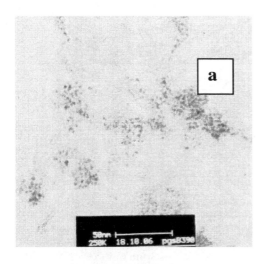

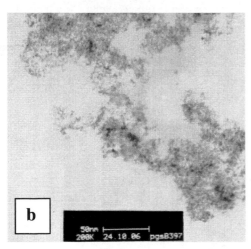

Figure 3: Representative TEM images of (a) 15/85% and (b) 65/35% compositions with incorporated Pt nanoparticles. (Scale bar = 50 nm).

6 REFERENCES

[1] A. H. Brian C. H. Steele *Nature* **2001**, *414*, 345.
[2] T. P. Lodge, *Macromolecular Chemistry and Physics* **2003**, *204*, 265-273.
[3] L. Zhang, H. Niu, Y. Chen, H. Liu and M. Gao, *Journal of Colloid and Interface Science* **2006**, *298*, 177-182.
[4] L. Rubatat, Z. Shi, O. Diat, S. Holdcroft and B. J. Frisken, *Macromolecules* **2006**, *39*, 720-730.
[5] R. A. Weiss, A. Sen, C. L. Willis and L. A. Pottick, *Polymer* **1991**, *32*, 1867-1874.
[6] Z. Shi and S. Holdcroft, *Macromolecules* **2004**, *37*, 2084-2089.
[7] R. M. R. H. Bönnemann, *Eur. J. Inorg. Chem* 2455-2480.
[8] P. F. Green and R. Limary, *Advances in Colloid and Interface Science* **2001**, *94*, 53-81.
[9] Z. Zhou, S. Wang, W. Zhou, L. Jiang, G. Wang, G. Sun, B. Zhou and Q. Xin, *Physical Chemistry Chemical Physics* **2003**, *5*, 5485-5488.
[10] J. S. Trent, J. I. Scheinbeim and P. R. Couchman, *Macromolecules* **1983**, *16*, 589-598.
[11] S. W. Zhenhua Zhou, Weijiang Zhou, Luhua Jiang, Guoxiong Wang, Gongquan Sun, Bing Zhou and Qin Xin, *Phys. Chem. Chem. Phys.*, **2003**, *5*, 5485 - 5488.

Biofilm Based Microbial Fuel Cell

M. Willert-Porada*, K. Lorenz, R. Freitag**, V. Jerome**, and S. Peiffer***

*Chair of Materials Processing, University of Bayreuth, monika-willert-porada@uni-bayreuth.de,
**Chair of Bioprocess Engineering, ruth.freitag@uni-bayreuth.de,
*** Chair of Hydrology, University of Bayreuth, D-95440 Bayreuth, Germany

ABSTRACT

Biofilms of natural anaerobic microbial consortia externally grown on Gas Diffusion Electrodes (GDL) in a bioreactor, were investigated within a Membrane-Electrode-Assembly with respect to lifetime and electricity production upon variation of biodegradable materials with an oxygen demand of 1000mg/l as simulated waste water. A remarkable differentiation and plasticity of the films is observed, to resist the toxicity of 0.5 mg/cm² Pt used on the inner side of the anode for smooth hydrogen combustion and to recover bioactivity upon alteration of the biodegradable material. Such robust biofilms are intended to provide a new technology for water purification and electricity production from industrial and community wastewater.

Keywords: microbial fuel cells, water purification, electricity

1 INTRODUCTION

Opposite to Polymer Electrolyte Membrane fuel cells, PEM-FC, Microbial fuel cells, MFC`s are not yet commercialized, although such fuel cells could generate electricity and pure water using ubiquitous bacteria and biodegradable waste, instead of expensive catalysts and hydrogen or synthetic fuel generated by electrolysis or petrochemical processes [1].

Future utilisation of bacteria for energy production is not limited to MFC`s; devices like e.g., a bioelectrochemically assisted microreactor, BEAMR, or Bacterial Electrolysis BE, offer further possibilities for Hydrogen or electricity production utilising the electron transfer processes in bacteria and organic materials as "fuel" [2].

The working principle of a MFC is to offer the bacteria externally a better acceptor for the electrons transferred during metabolic processes as compared to the internal electron acceptor. In such case, the electrons can be used for electricity generation. The device to be build must ensure a close contact of the enzymatic active centre of the bacteria with some means for electron transfer and an electro acceptor in close contact. As shown in Fig. 1, such an environment is provided by a Membrane Electrode Assembly used in PEM-FC, because the electrode materials are made of carbon, with a high degree of biocompatibility. The biofilm is at the anode, air oxygen is the electron acceptor, water is the reaction product at the cathode, CO_2 at the anode.

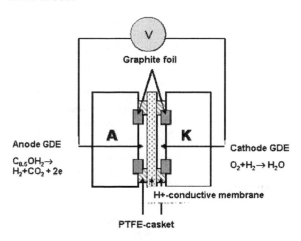

Figure 1: MFC-set up

The difficulty of MFC`s, BEMR or BE technology as compared to direct methane production from biomass [3], is based on the lack of deep knowledge about the physiology of bacterial consortia directly interfaced with an artificial, technical system. The variety of communication among such consortia is remarkable: some evidence is given for "wireless" and "wired" communication as well as for processes called "quorum sensing", with partial destruction of the biofilm under stress [1,4].

Therefore, the study presented in this paper is focussed on the investigation of the electricity production as an indicator of plasticity and differentiation within a biofilm-based natural bacterial consortia upon fuel cell operation with varying biodegradable materials as fuel.

2 EXPERIMENTAL

2.1 Biofilm growth kinetics

The biofilms were grown in a media consisting of 5 g yeast extract, 10 g Pepton from Casein, 5 g NaCl per 1 l de-ionized water. All parts of the reactor were steril (20 min at 121 °C in an autoclave). Sterile filters (0,2 m pore size, Sartorius) were used in all silicone tubing. The apparatus is shown in Fig.1. Two different substrate materials based on carbon, 2x2 cm² or 0,5x2 cm² in size were used: Gas Diffusion Electrodes, GDE commonly applied for Polymer

Electrolyte fuel cells (ETEK, o,5 mg Pt/cm² , type A-6-ELAT-SS) and Gas Diffusion Layer, GDL (GDL 10 BB, SGL, Germany). The growth of bacteria in the media was followed by pH-measurements, Optical Density (OD) measurements, (1 OD = $8*10^8$ cells/ml), and by optical microscopy after staining (BacLight Bacterial Viability Kit, Molecular Probe). The biofilms were characterized by fluorescence confocal laser microscopy, living cells were stained with CFDA (5-Carboxyfluorescein-diacetate), defect cells by Propidium iodide (PI). In addition, cryo-SEM was used to see the roughness of the biofilm on the electrodes.

2.2 Membrane Electrode Assembly Measurements

Voltage-current measurements with the MEA shown in Fig. 1 were performed under Nitrogen atmosphere with IM6e, Zahner, under constant feed flow. As biodegradable substance acetate, glucose and starch solution were employed, with 1000mg/l CSB (chemical oxygen demand). Gas analysis of the liquid feed at the anode GDL was performed by GC (Agilent 6890).

3 RESULTS

The biofilm growth kinetic is characterized by thickness measurements of the film using fluorescence laser scanning microscopy. As shown in Fig 2, different bacteria are present, mainly belonging to Clostridia. The red stain indicates living bacteria.

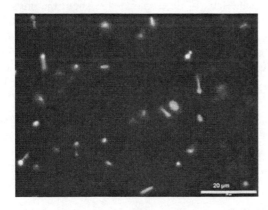

Figure 2. Laser fluorescence image of a typical biofilm: red stain indicates living bacteria

The thickness measurement are related to the electrode material used and to the duration of biofilm growth, as shown in Fig. 3. and Fig.4.

It is clearly seen, that even a very small amount of Pt inhibits significantly the biofilm growth. The biofilm bacteria grow predominantly at the opposite side on the carbon fibre composite, in large colony, as also seen from the cryo-SEM image in Figure 4.

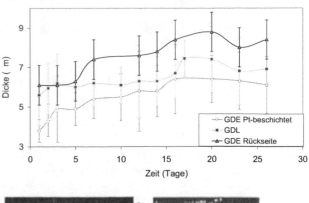

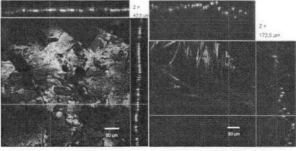

Figure 3: Average biofilm thickness as function of growth time and carbon substrate material (top); florescence image of morphology and density of living cells (red stained)

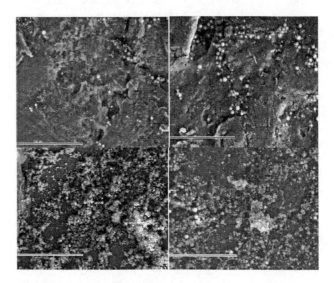

Figure 4: Cryo-SEM image of the Pt-side of a biofilm-GDE-anode after 2 days (right) and 1 week (top) as compared to a GDL as substrate: the amount of bacteria colonies is significantly depleted at the Pt-side of the GDE and increased at GDL as substrate, 100 m scale bar

Roughness measurements of the substrate indicate, that the depletion at the Pt-containing side of the anode GDE could also be caused by the smoother surface and reduced number of large pores at this side of the electrode. Biofilm

However, the bioactivity of such Pt-containing GDE-anode in the MEA-measurement is still significantly higher as compared to a biofilm grown on a Pt-free GDL, as shown in Figure 5.

One can assume, that the contact between the bacterial enzyme system and the electrode is comparably good in both cases, but in the presence of Pt the hydrogen produced by the bacteria undergoes faster "cold combustion" as compared to the Pt-free system, therefore the overall reaction yield is higher.

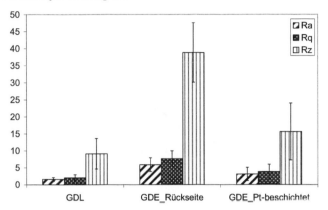

Figure 5: Roughness of the substrates for biofilm growth

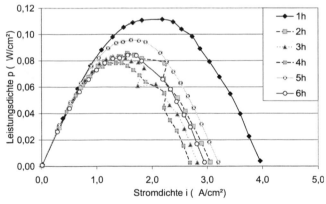

Figure 6: Power production of a biofilm as function of time elapsed from media change, feed contains 1000 mg/l acetate

The MEA measurement reveal, that it needs several hours for the biofilm-bacteria to accommodate their bioactivity to a changing biodegradable organic substrate. The biofilm is grown on the anode in a different media as compared to the one used in the MEA-measurement. At the beginning of the measurement, some residue of this media is still present in the porous substrate material, therefore bioactivity is high, as reflected by the voltage-current curve shown in Figure 6. As soon as the bacteria "recognize" the change in media, their activity drops significantly. After a certain period of time, it "recovers", therefore we assume, that some reorganization of the microbial colonies has taken place.

This behavior is independent of the feed used, however the time and extend of recovery differs, as showm in Figure 6 for acetate and figure 7 for glucose and starch. Because the concentration of the biodegradable material was kept constant in each case with respect to the CSB, a direct comparison is justified.

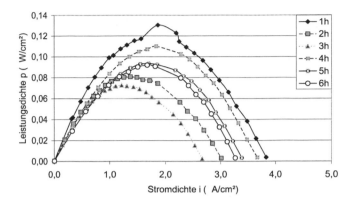

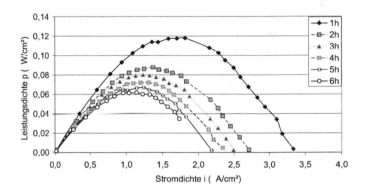

Figure 7: Power production of a biofilm as function of time elapsed from media change, feed contains 1000 mg/l glucose (top) or starch (bottom)

The results of gas analysis measurements in the anode-liquid confirm bacterial metabolism as the source of electricity production, as shown in Table 1.

	CO_2 der Gasphase (ppm)		H_2 der Gasphase (ppm)		CH_4 der Gasphase (ppm)	
Glukose	672,67	x	x	x	38,43	x
Glukose in MBZ	731,80	739,47	x	1,93	40,41	39,56
Glukose GDL MBZ	716,10	666,56	x	1,71	39,95	39,29
Stärke	601,45	x	0,75	x	41,91	x
Stärke MBZ	740,81	906,49	x	2,63	43,24	40,57
Stärke MBZ 4 W	771,99	934,30	x	1,26	40,34	39,22
Acetat	1723,25	x	x	x	38,19	x
Acetat MBZ	862,34	785,86	x	2,18	39,29	38,52
	1h	4h	1h	4h	1h	4h

x = nicht gemessen

Table 1: GC-gas analysis results at the anode liquid.

More quantitative measurements as well as a full analysis of errors upon MEA-measurements have been reported in [6]. In future, membrane materials will be introduced, which have an increased biocompatibility as compared to the ones used in the present study [7,8], in order to increase the bioactivity of the biofilm at the inner side of the anode and study the maximum possible energy output but releasing the limitation of biodegradable material to 1000mg/l, which was intentionally applied in the work reported here.

4 REFERENCES

[1] B.E. Logan, J.M. Regan, Electricity-producing bacterial communities in microbial fuel cells. Trends Microbiol. 14 (2006) 512-518.

[2] B.E. Logan, S. Grot, A bioelectrochemically assisted microbial reactor that generates hydrogen gas, Patent application 60/588,022(2005).

[3] A. Weiss, V. Jerome, R. Freitag (2006) Molecular characterisation of an established biogas plant – isolation of genomic DNA suitable for compilation of a metagenome database. J. Chromatogr. A, submitted

[4] M.J. Kirisits, M.R. Parsek, Does Pseudomonas aeruginose use intercellular signaling to build biofilm communities? Cell. Microbiol. 8 (2006) 1841-1849.

[4] C.A. Fux, J.W. Costerton, P.S. Steward, P. Stoddly, Survival strategies of infectious biofilms. Trends Microbiol. 13 (2005) 34-40.

[5] W. Teughels, N. van Assche, I. Sliepen, M. Quirnen. Effect of materials characteristics and/or surface topography on biofilm development. Clin. Oral Imp. Res. 17 (2006) 68-81.

[6] K. Lorenz, Biofilm basierte Katalysatoren fuer PEM-BZ, Diplomarbeit Univ. Bayreuth, July 2006

[7] F. Bauer, M. Willert-Porada, Comparison Between Nafion and Nafion Zirconium Phosphate Nano-composite in Fuel Cell Applications, Fuel Cells 3-4, 2006a, p. 261-269

[8] A. Wojtkowiak, Biomimetische Konzepte zur Herstellung von Ionomermembranen für PEM-BZ, Diplomarbeit Univ. Bayreuth, January 2006.

Flame spray synthesis of visible light active nanocrystalline bismuth oxide based photocatalysts

Kranthi Kumar Akurati,* Andri Vital,* Felix Reifler,** Axel Ritter,** Thomas Graule*

* EMPA, Materials Science and Technology, Laboratory for High Performance Ceramics, Ueberlandstrasse 129, CH-8600 Duebendorf, Switzerland. E-mail: kranthi.akurati@empa.ch, andri.vital@empa.ch, thomas.graule@empa.ch

** EMPA, Materials Science and Technology, Laboratory for Functional Fibers and Textiles, Lerchenfeldstrasse 5, CH 9014, St. Gallen, Switzerland. E-mail: felix.reifler@empa.ch, axel.rittler@empa.ch

ABSTRACT

$BaBiO_3$ nanoparticles have been synthesized by dissolving Ba and Bi precursors in a suitable solvent and spraying into the high temperature acetylene flame using an atomizing gas. Resulting powders were characterized by nitrogen physisorption (measuring specific surface area), x-ray diffraction (phase composition), transmission electron microscopy (size, shape and morphology of the particles), whilst UV-vis diffuse reflectance spectroscopy analyzed with the Kubelka-Munk function has been used to study the visible light absorption of the photocatalyst and the optical band gaps. Specific surface area of the nanoparticles has been varied by changing the flow rate of the of the precursor solution that has significant influence on the combustion enthalpy density (CED) of the flame. Rate of degradation of formaldehyde under visible light illumination (>400 nm) has been used as the measure of the photocatalytic activity (PCA) of the particles whose specific surface area ranges from 5 to 50 m^2/g. Clear dependence of the specific surface area and crystallinity of the particles on the PCA has been observed which signifies the advantages of nanoparticles.

Keywords: nanoparticles, flame spray, photocatalysis, visible light, Barium oxide, bismuth oxide

INTRODUCTION

Photocatalysis phenomenon has attracted considerable interest in recent years because of its usage in water purification, environmental cleaning, solar energy conversion and generation of alternative energy resources. Development of a practical photocatalytic system focuses on the cost effectiveness of the process. Usage of the expensive solar concentrators and artificial UV irradiation for photocatalytic reactions has negative influences on the cost effectiveness. Instead, more practical and inexpensive step is to employ renewable solar energy as the illumination source to activate the photocatalyst for photocatalytic decomposition reactions. To date, TiO_2 has undoubtedly proven to be the most effective photocatalyst. However, owing to the large band gap of TiO_2, it is not suitable for the usage as visible light active photocatalyst. Narrow band gap (<2.6 eV) bismuth oxide based materials, MBi_2O_4 (M = Ba, Sr, Ca), has been reported [1] as an efficient visible light active photocatalysts and are synthesized by the conventional solid state reaction methods that leads to particles having very low specific surface area (<1 m^2/g). Since photocatalysis is a surface phenomenon, less surface area of the large particles hinders the number of active sites where the photocatalytic reaction can take place. Hence there is a need to synthesize the visible light absorbing photocatalysts with high specific surface area. In the present study, flame spray synthesis has been employed to synthesize $BaBiO_3$ nanoparticles. High temperatures prevailing in the flame result in the nanoparticles with high degree of crystallinity. Reduction of the particle size is also associated with the increased absorption of the visible light irradiation. Moreover, the ability of the synthesis of Ba-Bi-O nanoparticles by flame spray synthesis at high production rates has been shown.

EXPERIMENTAL PROCEDURE

Fig 1 shows the schematic of the experimental set-up of the flame spray synthesis. It consists of a syringe pump to feed the precursor mixture, an external mixing gas-assisted nozzle and the powder collection unit. Liquid precursor is stored in a separate vessel and the flow rate is adjusted with the syringe pump. The nozzle consists of a central opening (2.8 mm) incorporated in a capillary tube (1.05 and 1.59 mm internal and external diameter, respectively) through which the precursor and fuel mixture is fed to the flame. The spacing between the capillary tube and central opening is used to feed the atomizing gas (oxygen) that forms fine droplets of the liquid precursor mixture. The liquid is ignited by six supporting premixed flamelets produced by C_2H_2 (13 l/min) and O_2 (17 l/min). Openings (1.3 mm) contributing to the supporting flames are located at 3.25 mm from the centre of the nozzle and all gas

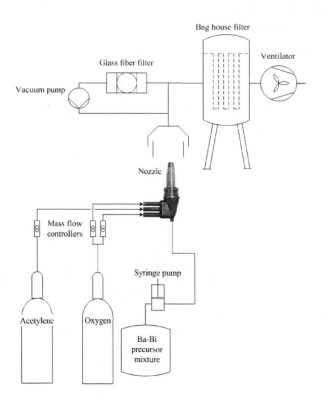

Fig. 1: Experimental set-up for the synthesis of BaBiO₃ nanoparticles

flow rates are controlled by the mass flow controllers (Bronkhorst HI-TEC, Netherlands). Due to the high exit velocities of the process gases, particles are collected in a bag house filter (Friedli, Switzerland) and the representative samples of about 1 g are collected on a glass fiber filters (Type GF50, Schleicher and Schuell, Germany), via a by-pass, using a vacuum pump.

As a precursor source of Bi, bismuth nitrate pentahydrate (Bi(NO₃)₃. 5H₂O, purity>98.5%, Sigma Aldrich, Switzerland) was used by dissolved in water with 15 vol% HNO₃. Barium acetate [Ba(CH₃COO)₂] was used as a precursor source for Ba and was dissolved in distilled water. The CED in this study is defined as the ratio of the total liquid precursor mixture plus acetylene-oxygen combustion enthalpy to the total gas flow in the system. Changing the combustion enthalpy density is associated with the variation of oxygen concentration (Lambda, λ) in the process. Lambda is defined as the ratio of the actual fuel-to-oxygen ratio of the reactants to the stoichiometric fuel-to-oxygen ratio.

CHARACTERIZATION

The specific surface area (SSA) of the product powder was determined from a five-point N₂ adsorption isotherm obtained from BET (Brunauer–Emmett–

Teller) measurements using a Beckman-Coulter SA3100. Prior to BET analysis, the powder samples were degassed at 200°C for 180 min under flowing N₂ atmosphere to remove adsorbed H₂O from the surface. Assuming monodisperse, spherical primary particles, the BET- equivalent primary particle diameter d_{BET} is calculated by

$$d_{BET} = \frac{6}{(\rho_B) * SSA}$$

Where ρ_B– density of BaBiO₃ (7.88 g/cm³). The primary particle size, shape and morphology of the particles were investigated by transmission electron microscopy (TEM). Powder samples were dispersed in isopropanol (purity > 99.5%, Fluka, Switzerland) and a few drops of the dispersion were allowed to dry on carbon-coated copper grids (Plano GmbH, Germany). The TEM analysis was performed on a Philips CM30 electron microscope operating at 300 kV.

X-ray diffraction (XRD) was used for identification of the crystal phases and determination of the average crystallite size. Diffraction measurements were performed with a PANalytical PW 3040/60 X'Pert PRO instrument using Ni-filtered Cu-Kα radiation of wavelength 1.5418 Å. A 2θ scan range from 5 to 80°, a scanning step size of 0.01° and a scintillation counter detector was used. Curve fitting and integration was carried out using proprietary software from Philips X'Pert high score plus.

The PCA of the as-synthesized powders was evaluated by the degradation of the formaldehyde using TL 20W visible light lamps (Philips, λ_max – 450 nm, 400-500 nm range).

Results and discussion

Characterization:

BET

Fig 2 shows the specific surface area (SSA) of the WO₃/TiO₂ nanocomposites synthesized as a function of flow rate of the precursor solution. From the figure it is apparent that the SSA of the BaBiO₃ particles decreased with increasing flow rate of the precursor. The temperature of the flame increases with the increasing the flow rate of precursor that enhances the combustion enthalpy density (CED) [2]. An increase in the flame temperature and combustion enthalpy density enhances the sintering rate of the particles facilitating them to grow to large sizes and concomitantly decrease the SSA. In addition, decrease of precursor concentration in the flame reduces the initial particle number concentration, resulting in lower coagulation rates and smaller primary particles. This is similar to the conven-

Cleantech 2007, ISBN 1-4200-6382-0

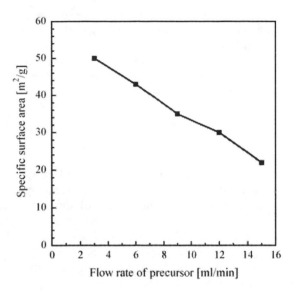

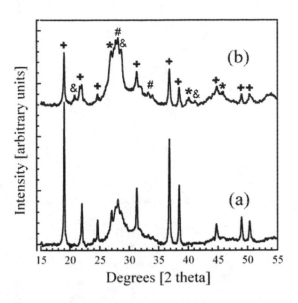

Fig. 2: Specific surface area (SSA) of the BaBiO₃ nanoparticles synthesized with various flow rates of precursor solution.

Fig. 3: XRD pattern of the particles synthesized with 6 ml/min precursor flow rate, (a) using water as the solvent to dissolve the Ba and Bi precursors and (b) water mixed with organic solvent to dissolve the precursors. Symbols +, *, # and & corresponds to $Ba(NO_3)_2$, $BaCO_3$, Bi_2O_3 and $BaBiO_3$ phases, respectively.

tional flame aerosol process where the surface area of the particles decreases with increasing temperature as reported by several authors [3, 4]. It is also in agreement with other FSP studies reported earlier. [5, 6]

XRD

Fig. 3a and 3b shows the XRD pattern of the particles synthesized using water and the mixture of water plus organic solvent, respectively as the solvent for the Ba and Bi precursors. The melting point of bismuth nitrate pentahydrate and barium acetate is 30 and 460°C, respectively [7]. XRD pattern of the particles synthesized using only water as the solvent shows several undecomposed products of barium, i.e. $Ba(NO_3)_2$ and $BaCO_3$. In addition weak reflections corresponding to BaBiO₃ is also seen suggesting that the Ba and Bi precursors did not decompose homogeneously. Water cools the high temperature flame drastically and sufficient heat is not available to completely decompose the high melting point barium acetate precursor. Presence of the individual peaks of Bi_2O_3 suggests that the entire bismuth nitrate has been decomposed. Hence it can be said that, only the decomposed Ba-acetate precursor contributed to the formation of BaBiO₃ phase.

Fig. 3b shows the XRD pattern of the particles synthesized by using an organic solvent in addition to the water used for dissolving the Ba and Bi precursors. Reflections corresponding to the $Ba(NO_3)_2$ phase has

been significantly reduced and intensity of the reflections corresponding to BaBiO₃ phase increased. Usage of more amount of heat generating organic solvent assists in the complete decomposition of barium acetate precursor and resulting in the formation of the stoichiometric BaBiO₃ phase. Also the SSA of the stoichiometric BaBiO₃ particles is significantly higher than the particles synthesized by the partial decomposition of the precursor mixture.

TEM

Fig. 4 shows the TEM image of the BaBiO₃ particles synthesized with 6 ml/min flow rate of the precursor mixture. Particles have aggregated morphology with the sinter necks in between.

PHOTOCATALYTIC ACTIVITY

Fig. 5 shows the degradation behavior of the formaldehyde using visible light irradiation. No detectable degradation of formaldehyde occurs without BaBiO₃ or with visible light irradiation alone. The degradation behavior is studied by gas chromatography. Figure 5 shows the photocatalytic activity of the BaBiO₃ particles synthesized at various flow rates of the precursor mixture. The degradation is compared with the commercial Degussa P25-TiO₂ which shows

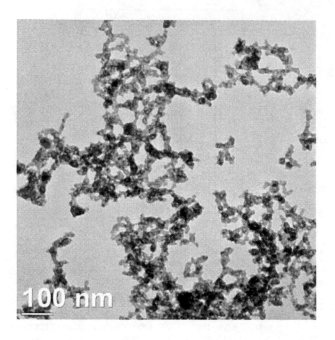

Fig. 4: TEM image of the BaBiO₃ particles

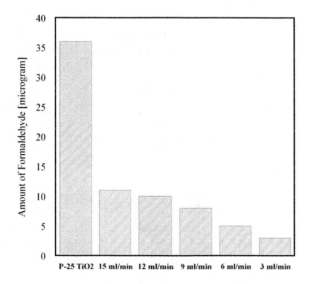

Fig. 5: Degradation profile of the formaldehyde after the visible light irradiation for 3 hours.

excellent activity towards the degradation of formaldehyde under UV irradiation. However, under the visible light irradiation P25 TiO₂ shows poor activity due to its inability to absorb visible light by virtue of its wide band gap [8]. PCA of the BaBiO₃ particles increased with the decrease of the precursor flow rate. As shown in the previous sections, synthesis of the particles with decreasing the precursor mixture flow rate accompa-

nied with the increase of the SSA. Clear dependence of the specific surface area of the particles on the PCA has been observed which signifies the advantages of nanoparticles. Photocatalysis being a surface phenomenon, improvement in the PCA is attributed to the increased number of active sites where the photocatalytic reaction can take place.

CONCLUSIONS

Crystalline nanoparticles of BaBiO₃ have been successfully synthesized by flame spray synthesis. Influence of the melting points of the individual solid precursors to produce the mixed oxides is shown. As-synthesized BaBiO₃ nanoparticles shows activity towards the degradation of the formaldehyde under the visible light irradiation and the activity increases with the decrease of the particle size.

ACKNOWLEDGEMENTS

The authors would like to acknowledge the EC and the Swiss BBW for their support of the FP5-Project Photocoat (EU contract No G5RD-CT-2002-00861; BBW project No 01.0571-1).

REFERENCES

[1] Tang, J., Zou, Z. and Ye, J. Angew. Chem. Int. Ed. 43, 4463, 2004.

[2] Mueller R., Maedler L. and Pratsinis S. E., *Chem. Eng. Sci.* 58, 1969, 2003.

[3] Wegner K. and Pratsinis S. E., *AIChE J.* 49, 1667, 2003.

[4] Mueller R., Kammler H. K., Pratsinis S. E., Vital A. Beaucage G. and Burtscher P., *Powder Technol.*, 140, 40, 2004.

[5] Mueller R., Jossen R., Kammler H. K. and Pratsinis S. E., *AIChE J.* 50, 3085, 2004.

[6] Maedler L. and Pratsinis S. E., *J. Am. Ceram. Soc.*, 85, 1713, 2002.

[7] www.sigmaaldrich.com

[8] N. Daude, C. Gont, C. Jonamin, Phys. Rev. B 15, 3229, 1977.

Nanoparticles Formed by Complexation of Poly-γ-glutamic Acid and Lead Ions

M. Bodnar[*], A.-L. Kjøniksen[**], J. F. Hartmann[***], L. Daroczi[****], B. Nyström[**] and J. Borbely[*#]

[*]Department of Colloid and Environmental Chemistry, University of Debrecen
H-4010 Debrecen, Hungary, jborbely@delfin.unideb.hu
[**] Department of Chemistry, University of Oslo, Blindern, N-0315 Oslo, Norway
[***] ElizaNor Polymer LLC, Princeton Junction, New Jersey 08550, USA
[****] Department of Solid State Physics, University of Debrecen, H-4010 Debrecen, Hungary
[#]BBS Nanotechnology Ltd., H-4225 Debrecen 16. P.O.Box 12.

ABSTRACT

The present investigation describes the preparation and characterization of novel biodegradable nanoparticles based on complexation of poly-gamma-glutamic acid (γ-PGA) with bivalent lead ions. The prepared nanosystems were stable in aqueous media at low pH, neutral and mild alkaline conditions. The solubility and size of these nanoparticles in the dried and swollen states will be described and discussed. The correlation of size of particles, pH of the solutions, concentration and the ratio of compound polyelectrolytes have been studied.

It was found, that the size of individual particles was in the range of 40-100 nm measured by TEM. The low and high pH values in mixtures with high concentrations of γ-PGA and Pb^{2+} ions favored the growth of large complexes. The γ-PGA nanoparticles, which are from a biodegradable biomaterial with high flocculating and heavy metal binding activity, may be useful for various water treatment applications in aqueous media.

Keywords: poly-γ-glutamic acid, lead binding, complexation, nanoparticles

1 INTRODUCTION

Recently, reduction of water consumption by industry has been an important challenge. Enhanced ultrafiltration with polymers is a feasible method to remove metal ions from diluted wastewater streams [1,2]. Biomacromolecules, biodegradable polymers as biomaterials have an important role in a wide range of industrial fields such as water treatment [3, 4]. For separation of toxic heavy metal ions, including lead ions, several natural polymers have been investigated [5, 6]. The most valuable properties of these biopolymers are their biocompatibility, biodegradability, and flocculating activity for metal ions [7, 8].

Flocculation of polyelectrolytes in the presence of bivalent ions is an important process, and is widely used in water treatment technologies. For separation of toxic heavy metal ions, natural poly-γ-glutamic acid (γ-PGA) and other natural polymers have been investigated [3, 5]. The aggregate size distribution in flocculants was studied and it

was found that it mainly depended on the pH and concentration of the electrolytes [9].

The present paper reports the formation of complexes of poly-gamma-glutamic acid (γ-PGA) with bivalent lead ions. The solubility and size of these nanoparticles in the dried and swelled states were investigated. In aqueous solution the average size of the particles varies strongly depending on pH and the concentrations of γ-PGA and Pb^{2+}. The γ-PGA – lead complexes may form separated spherical colloid particles or aggregates in aqueous media depending on the pH, and the concentrations of γ-PGA and lead ions in the mixture.

In our research work stable colloid particulate systems were performed based on complexation of γ-PGA with lead ions. Solubility of the systems has been surveyed by turbidity measurements. The sizes of swelled complexes in aqueous solutions have been determined by means of dynamic light scattering (DLS). TEM micrographs made the visual observation of the dried nanoparticles possible. It was studied the correlation of size of particles, pH of the solutions, concentration and the ratio of compounds.

2 EXPERIMENTAL SECTION

2.1 Materials

Poly-γ-glutamic acid ($M_W = 1.2 \times 10^6$) was prepared in our laboratory by using the biosynthetic methods described earlier [10, 11]. Briefly, poly-γ-glutamic acid was produced from *Bacillus licheniformis*, strain ATCC 9945a, which was maintained on 1.5% (w/v) Bouillon-agar slants to produce appropriate cultivation conditions. γ-PGA was precipitated by addition of acetone and filtered. The γ-PGA was re-dissolved in water, dialyzed against distilled water and freeze-dried.

Lead(II)-nitrate was purchased from Sigma-Aldrich Co., Hungary, and was used as received without further purification.

2.2 Characterization

Turbidimetry. The transmittances of γPGA-Pb^{2+} mixtures of different composition and pH were measured with a temperature controlled Helios Gamma (Thermo Spectronic, Cambridge, UK) spectrophotometer at a wavelength of 500 nm. The apparatus is equipped with a temperature unit (Peltier plate) that gives a good temperature control (25 ± 0.05 °C) over an extended time. The turbidities τ of the samples can be determined from the following relationship: $\tau = (-1/L)\ln(I_t/I_0)$ where L is the light path length in the cell (1 cm), I_t is the transmitted light intensity, and I_0 is the incident light intensity.

Transmission Electron Microscopy (TEM). A JEOL2000 FX-II transmission electron microscope was used to characterize the size and morphology of the dried γ-PGA nanoparticles. For TEM observation, the nanoparticles were prepared from the reaction mixture, the pH of the solution was 3.0. The sample for TEM analysis was obtained by placing a drop of the colloid dispersion containing the γ-PGA nanoparticles onto a carbon coated copper grid. Mean diameters and the size distribution of diameters were obtained from measured particles visualized by TEM images and then analyzed by using SPSS 11.0 program file.

Dynamic Light Scattering (DLS). The beam from an argon ion laser (Lexel laser, model 95), operating at 514.5 nm with vertically polarized light, was focused onto the sample cell through a temperature-controlled chamber filled with refractive-index-matching silicone oil. The sample solutions were filtered through 5 μm filters (Millipore) directly into precleaned 10 mm NMR tubes (Wilmad Glass Company) of highest quality.

2.3 Formation

Synthesis of γ-PGA nanoparticles with lead ions. γ-PGA solution (c = 6 mmol, pH = 2.6) and $Pb(NO_3)_2$ solution (c = 3.125 mmol, pH = 2.3) were produced and used for preparation of γ-PGA nanoparticles by lead ion complexation. $Pb(NO_3)_2$ solution was added to the γ-PGA solution dropwise. The mixture was diluted to 50 ml and the pH was adjusted to the desired pH value with 0.1 M sodium hydroxide solution. The reaction mixture was stirred at room temperature. Formation of γ-PGA particles with bivalent lead ions at diverse stoichiometric ratios and concentrations were made.

Sample	γ-PGA solution	$Pb(NO_3)_2$ solution
γPGA-Pb 1	25 ml	16 ml
γPGA-Pb 2/1	12.5 ml	16 ml
γPGA-Pb 2/2	12.5 ml	8 ml
γPGA-Pb 2/3	12.5 ml	4 ml
γPGA-Pb 3	6.25 ml	4 ml

Table 1. Reaction conditions for the formation of γ-PGA nanoparticles by lead ion complexation

3 RESULTS AND DISCUSSION

3.1 Formation of nanoparticles

The carboxylic groups of linear γ-PGA chains were cross-linked with lead ions, and formed stable nanoparticles. Complexation was observed and separated spherical particles or aggregates were obtained depending on pH and the γPGA-Pb^{2+} composition. The size and the stability of the particles depended on the pH and the concentrations of γ-PGA and Pb^{2+} ions. In neutral and alkali media, the carboxylic groups are deprotonated, and the repulsive electrostatic forces between the negatively charged parts affect the physicochemical properties of these particles. In acidic media, γ-PGA is an uncharged macromolecule, and this may lead to smaller sizes of the γPGA-Pb^{2+} particles unless the concentrations of the components in the mixture are sufficiently high to form intermolecular complexes.

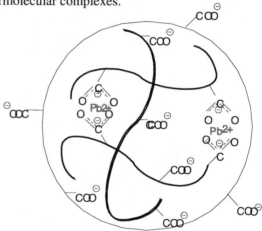

Figure 1. Schematic illustration of γ-PGA-nanoparticles formed through the interaction with lead ions

3.2 Turbidimetry

The pH dependence of the turbidity at a fixed γ-PGA/Pb^{2+} ratio (3.0) and for mixtures with a constant γ-PGA concentration were investigated.

The general trend that appears in Figure 2 is that the value of the turbidity rose as the pH decreased, and this effect became stronger when the concentrations of the components increased. This is compatible with the idea that interpolymer associations were formed under acid conditions, and this behavior should have be strengthened as the concentrations of γ-PGA and Pb^{2+} increased. At a constant γ-PGA concentration, a higher level of Pb^{2+} addition leaded to a stronger upturn of the turbidity at low pH values, which suggested that higher Pb^{2+} concentrations promoted interchain associations.

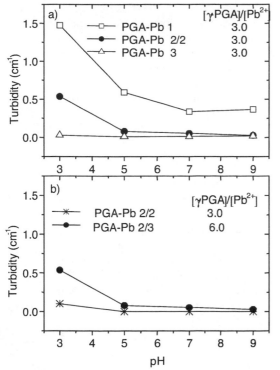

Figure 2. Effects of pH on the turbidity at the mixture conditions indicated.

3.3 Particle size by TEM

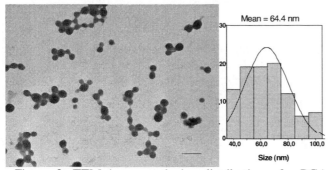

Figure 3. TEM image and size distribution of γ-PGA nanoparticles formed in the sample γPGA-Pb 2/2 by bivalent lead ions. The bar in the Figure is 200 nm.

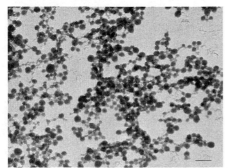

Figure 4. TEM image of complexes of γ-PGA nanoparticles formed in the sample γPGA-Pb 1 by divalent lead ions. The bar in the Figure is 200 nm.

TEM micrographs provide visual evidence of the morphology and the size as well as the size distribution of the dried nanosystems. Figure 3 and 4 represent the nanoparticles formed by complexation of γ-PGA with lead ions at pH 3.

The elemental analysis can exhibit the heavy metal content of nanoparticles. Figure 5 shows the high lead content of dried sample, focusing on the nanoparticles. The spectrum confirms that the complexation of the polymer with lead ions was successfully obtained. The copper content of the sample ensues from the copper grid, as the holder of TEM samples.

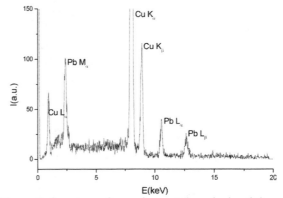

Figure 5. Spectrum from elemental analysis of the sample γPGA-Pb 1.

3.4 Particle size by DLS

Figure 6 shows the pH dependences of the hydrodynamic radii, derived from the fast ($R_{h,f}$) and the slow ($R_{h,s}$) relaxation mode, for γPGA-Pb^{2+} mixtures with a fixed value of the ratio and at different values. At a constant ratio, the value of $R_{h,f}$ increased as the concentration of the components rose, and the upturn of $R_{h,f}$ at low pH was most pronounced at the highest concentration of the components. This finding stressed again that low pH and high levels of γ-PGA and Pb^{2+} favored the formation of intermolecular associations. At high pH, the growth of $R_{h,f}$ was stronger at the highest concentration of the components, which probably reflected the formation of intermolecular γPGA-Pb^{2+} complexes. The parameter $R_{h,f}$ monitored some average size of single particles and small particle clusters. This means that the growth of $R_{h,f}$ was an indicator of that the number of small aggregates in the system increased. The values of $R_{h,s}$ were significantly higher than the corresponding values of $R_{h,f}$. A prominent feature was the strong upturn of $R_{h,s}$ at low values of pH when the concentrations of γ-PGA and Pb^{2+} were high, which again suggested that acid conditions promote the growth of large clusters.

The results reveal that $R_{h,f}$ was almost independent of pH and the effect of moderate Pb^{2+} additions was modest on the value of $R_{h,f}$ (the average value of $R_{h,f}$ is approximately 40 nm). The behavior of $R_{h,s}$, which characterizes the size of the large species, was quite different especially at the low lead ion concentration. If the proportion of lead ions is very small, the lead ions cannot react with the carboxylic groups optimally at higher values of pH. This is expected to lead to a situation where an increase in pH will give rise to the formation of large-scale complexes that have a broad distribution of sizes. At higher levels of lead ions, intermolecular complexes between γ-PGA and Pb^{2+} can be formed at higher pH, and this was signalized by the enhanced values of $R_{h,s}$.

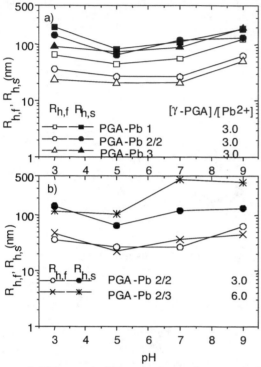

Figure 6. Effects of pH on the hydrodynamic radii from the fast (Rh,f) and the slow (Rh,s) relaxation modes at the compositions of the mixture indicated.

4 CONCLUSION

In this work, the complexation between γ-PGA and Pb^{2+} ions at different concentrations of the components and at various pH values has been studied with the aid of several experimental methods. Depending on the concentrations of the polymer and the lead ions, flocculation of large complexes may occur. High concentration of the components and low pH value promoted the formation of huge aggregates. The turbidity experiments have supported the hypothesis that high concentrations of the components and low pH favored the growth of clusters.

The TEM measurements on the γPGA-Pb^{2+} mixtures in the dry state have shown that individual nanoparticles were formed together with some clusters of particles.

The DLS experiments also have shown the effects of pH and composition on the size and size distribution of the clusters. At high concentration of the components in the mixture, low pH values produced association complexes, whereas at high pH, where the polymer is charged,

intermolecular γPGA-Pb^{2+} complexes (with a broad distribution of sizes) were formed. An interesting feature was found when there is a deficiency of Pb^{2+} ions in the sample. In this case large clusters with a wide size distribution were formed at high pH values.

Acknowledgement. This work was supported by RET (Grant of the Regional University Knowledge Center) contract number (RET-06/432/2004) and by ElizaNor Polymer LLC, USA.

REFERENCES

[1] R. Molinari, S. Gallo, P. Argurio, Water Res., 38, 593, 2004.
[2] J. Llorens, M. Pujola, J. Sabate, J. Membrane Sci., 239, 173, 2004.
[3] M. Taniguchi, K. Kato, A. Shimauchi, X. Ping, H. Nakayama, K. I. Fujita, T. Tanaka, Y. Tarui, E. Hirasawa, J. Biosci. Bioeng., 99, 245, 2005.
[4] U. S. Ramelow, C. N. Guidry, S. D. Fisk, J. Hazard. Mater., 46, 37, 1996.
[5] D. Solpan, M. Torun, Colloid Surface A, 268, 12, 2005.
[6] L. Qi, Z. Xu, Colloid Surface A, 251, 183, 2004.
[7] M. Taniguchi, K. Kato, A. Shimauchi, X. Ping, K. I. Fujita, T. Tanaka, Y. Tarui, E. Hirasawa, J. Biosci. Bioeng., 99, 130, 2005.
[8] H. Yokoi, T. Arima, J. Hirose, S. Hayashi, Y. Takasaki, J Ferm Bioeng., 83, 84, 1996.
[9] C. Rattanakawin, R. Hogg, Colloid Surface A, 177, 87, 2001.
[10] A. Krecz, I. Pocsi, J. Borbely, Folia Microbiol., 46, 183, 2001.
[11] M. Borbely, Y. Nagasaki, J. Borbely, K. Fan, A. Bhogle, M. Sevoian, Polym. Bull., 32, 127, 1994.

Cleantech 2007, ISBN 1-4200-6382-0

Nano hetero nuclear fuel structure

L. Popa-Simil, I.L. Popa-Simil

LAVM LLC, Los Alamos, NM 87544

ABSTRACT

The direct energy conversion of nuclear energy requires nano-structured nuclear fuel The principle of the radiation energy direct conversion is the usage of the knock-on electrons produced by the moving particle interaction with the lattice. These electrons transfer their energy to other electrons creating a shower of low energy and high current. A structure created of independent nanowires of two different materials insulated in a thin dielectric coating may use as electron showers energy harvesting device. The dimensions of the conductors are in the range of few tens of nm while the insulators are of several nm, offering a breakdown voltage of few milivolts. When the mowing particles interacts with them it generates consistent e-showers when riches the high electron density conductor which are passing through insulators and stops in the low electron density conductor polarizing it negatively. These two types of conductors are connected to an external load transferring the accumulated charge. To obtain a higher voltage, in a uniform radiation field bimetal nanobeads insulated together and self-organizing have to be created around the radiation source. The advantages consist in transforming the actual nuclear reactors into high power solid-state batteries with no heat exchangers and turbines.

Keywords: nano structure, self-organized, nuclear fuel,

1 INTRODUCTION

The direct energy conversion attempts started by 1940th [1, 2] when nuclear power civil applications started. They were followed by the fast development of the semiconductor devices and the research oriented to establish their behavior in various radiation fields. The 1950th up to 1970th period is dominated by the attempt of using various semiconductor structures to harvest electric power from the radiation field energy [3-8], marking the beginning of the beta-voltaic as a technical domain. The development of electronic devices, and ion beam applications inspired the devices developed afte the capacitor model [9, 10]. Other researches were looking in the potential applications of the MEMS devices as instruments to harvest nuclear power [11, 12], or in various effects possible to occur in special semiconductors and junctions [13, 14], which seems promising for the nuclear power conversion. Most of these previous attempts of nuclear energy harvesting were exhibiting low overall energy efficiency or incompatibilities with the nuclear structures.

2 THE HETERO-NANO STRUCTURE CONCEPT

The concept development starts in early 1980th when during ion beam based experiments various manifestations of direct energy conversion occurred as annoying and in few cases jeopardizing the experiments [15]. Later during the recoil implantation and IBA experiments [16] these phenomenon have been understood and special care have been taken to prevent its unwanted manifestations.

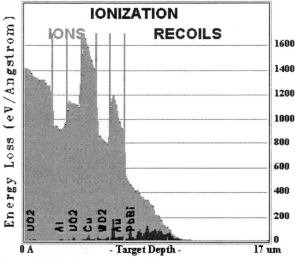

Fig. 1 – SRIM 2006 simulations for a multiplayer target

The final developments are explained by using the Monte Carlo simulation [17] for a multi-layered material

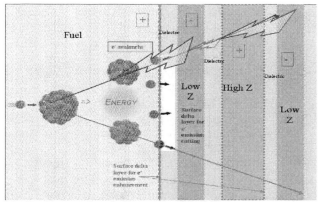

Fig.2 – Energy harvesting schematic diagram

shown in Fig. 1. The sandwich material structure is made of various layers with different electron densities. The nuclear particles stopping in these strata are mainly ionizing the material along its path. The ionization energy transferred from the nuclear particle (photon, X or gamma, electron, ion, or neutral nucleus) is further used to create knock-on electrons and resonant X lines.

Fig. 2 shows a potential energy harvesting structure looking like Young's [10] and Ritter's [9] patents but applied from the fission products generated in the near by layer. The fission product carries 80% of the energy as kinetic energy shared between the two fission products. According to the simulation in Fig. 1 the range of the fission products in near by matter is less than 20 microns while the stopping time is less than 100 ps. During this time the knock-on electrons generating by the Couloumbian collisions are generating showers, which are interacting further with the atoms in lattice, until all the energy is degraded into heat. During the electron showers generation

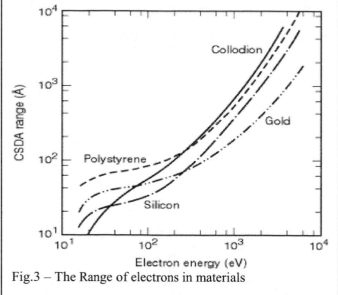

Fig.3 – The Range of electrons in materials

and displacement the polarization of the material occurs and the electrons are coming back after a path similar to the Brownian displacement further exciting the phonon modes. To prevent the material heating, there is possible to alternate the layers and offer electrons conductive paths. As shown in Fig. 3 the free electron path in materials [18] is about 1 to 10 nm, comparable with the equivalent Debye length for the free electrons in conduction bands. It shows that for a insulator material thickness the electron may have a ballistic behavior for a thickness up to 10 nm which may offer a breakdown voltage of few mV. If we connect in parallel the nano-layers made from various materials as in Fig. 2 is possible to obtained a capacitor like structure loaded by the nuclear particles induced current and discharged electrically. The concept represents an enhancement from the Young's patents [10, 19], being applied to all the kind of nuclear particles from photons to fission products. In divergence to the previous art, the low "Z" concept is good for educational purposes only, in reality what matters is the real electron density, and the Fermi level concept is not directly applicable to the

structure because the high heterogeneity in the energy discharge, all the process taking in average 20 ps. As a synthesis, for each fission act delivering about 25 pJ, about 2.5nA at 10mV electric energy is obtained.

3   THE NANO-STRUCTURE

The previously presented intuitive structures looking similar to super-mirrors or super-capacitors do not withstand the time and radiation fields, layers diffusing into each-other. The new development is improving this feature by using a nano-grain structure or nano-wires connected by polarity and immersed into insulator, creating a structure

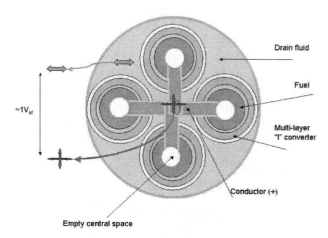

Fig. 4 – Spherical nano-hetero structure

similar to the MOS-FET, but generically called "Cici". The "C" conductor is a high yield generator of electron shower that passes ballistic through the "I" insulator and stops into the "c" conductor, which becomes negatively polarized. The "c" conductor is a low-yield electron shower generator, deflected back by the "i" insulator. Quantum interactions as those predicted by Klimov [14] may be stimulated in order to obtain the carrier multiplication.

Fig. 4 shows the structure in a energy harvesting voxel having a spherical symmetry. As shown in Fig. 1 towards the end of the range a lot of energy is given to recoil, inflicting high damage to the lattice. To avoid the radiation damage and fission products accumulation a micro-nano-hetero-structure may be designed. A liquid metal, with or without harvesting properties, surrounds the onion foil harvesting structure. Finally this structure generates about 50 microwatt per micro-bead.

3.1   Power extraction system

To carry out this power a local DC/AC converter is required. Fig. 5 shows the principal diagram of such a converter. The charge accumulated in the "Cici" structure is transferred to a buffer capacitor set that has its contacts switched by a MEMS device on an inductive load, representing the primary of a voltage multiplier transformer. The MEMS switch is controlled from a central unit, and is also providing status data, used for the reactor diagnosis and control.

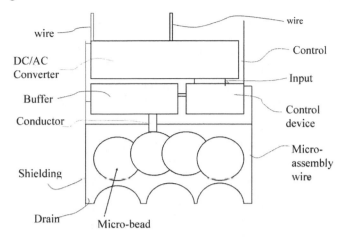

Fig 5 – DC/AC MEMS converter

The micro-assembly device spans over several cubic millimeters generating several watts alternate current.

3.2   The serial nano-hetero structure

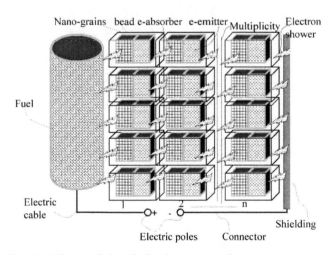

Fig. 6 – Bi-material, polarized nano-powder system

The system shown in Fig. 5 is starting the conversion from few mV that requires a smart MEMS switch to produce the voltage multiplication by D class wave simulation, or high efficiency autotransformer. Another option is presented in Fig. 6 where the voltage multiplication is achieved by using a new type of bi-material nano-grains, powder like, immersed into an insulator drainage liquid. This polarized powder acts like a serial connection, of several tens to hundreds of capacitors, allowing higher output voltages.

This multi-layer powder may be contained in screening, metallic coatings, making intermediary capacitor systems in order to reduce the probability of breakdown or short-circuit which will produce the fuel heating. Several micro-

beads may be connected in series or parallel in order to achieve the best power extraction combination.

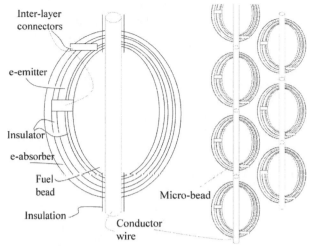

Fig. 7 Multi-bead connection system.

Fig. 7 shows an example of high voltage serial beads, placed in a variable geometry structure that allows higher burnup factors. The nano-structure extraction efficiency may be maximized, such as the residual heating energy to be at the limits of acccptablc opcrating temperatures. This parameter dictates on the power density this structure may carry. Has to be understood that only the energy which was nor extracted as electricity is producing heat. The theoretical limits of a near 100% efficiency "Cici" mm^3 structure made by 10% carrier conductor and 90% capacitive structure is about 5 kw/mm^3, being several thousands higher than the actual power densities smaller than 1Kw/cm^3 based on thermal floe power extraction.

3.3   Electric system structure

To carry this power outside the reactor and to deliver into grid a pyramidal structure of impedance adapters and voltage multiplier system have to be build as in Fig.8.

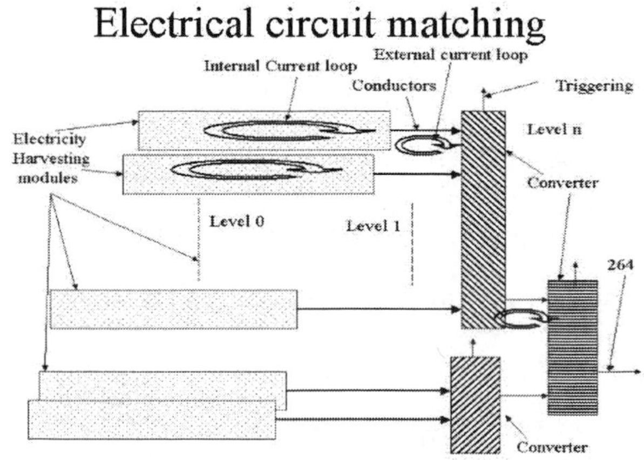

Electrical circuit matching

Fig. 8 – Pyramidal electric power extraction system

This system has to mitigate between the direct conversion efficiency and the electricity conversion efficiencies. This new nuclear fuel represents a great challenge for the actual nuclear reactors concepts, representing a drastic volume reduction, finally reflected in costs.

4 THE HARVESTING STRUCTURE

The nano-micro-hetero structure may be build using actinides in its structure, and becoming an evaluated nuclear fuel, but may be also build without fissile materials in its structure. In this case, it may be used to harvest the energy of external radiations like that coming from radioactive materials, ion beams, fusion reactors.

Element of Actinic panel

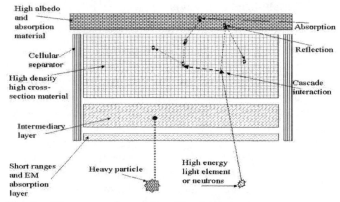

Fig. 9 – The energy harvesting tile structure

Fig. 9 shows the block diagram of a multi-radiation energy-harvesting device designed for fusion and space application. The first layer is very thin, being mainly an active coating collecting heavy nuclei and X ray, followed by a medium density layer, for medium energy radiation and a n energy harvesting and amplification layer, by using actinide based materials. A cryogenic version of this device is possible opening the way for high efficiency, high power density device.

5 CONCLUSIONS

The direct conversion nuclear fuels have the capability to reduce the size of the nuclear reactor down to a trailer, by cutting all the heat exchangers, turbine and associated equipments.

The nano-hetero-structures open the way for high power density, high conversion efficiency and portable nuclear power applications.

REFERENCES

1. Harmut Israel Kallmann, E.K., *Device for measuring the intensity of a radiation of slow neutrons by means of ionization chamber*. US 2288718, 1940.
2. Ernest G. Linder, *Method and means for collecting electric energy of the nuclear reactions*. US2517120, 1946.
3. Volney C Willson, *Generator of Power*. US2728867, 1955.
4. Eric Gustav Karl Schwarz, *Secondary emission type of nuclear battery*. US2858459, 1958.
5. Schuyler M. Christian, *Radiation responsive voltage sources*. US2847585, 1958.
6. WILLIAM E. PARKINS, *NUCLEAR BATTERY INCLUDING PHOTOCELL MEANS*. US3483040, 1969.
7. OLSEN K., *Semiconductor nuclear battery*. U.S. Pat. No. 3,706,893, 1972.
8. Adler Karl, D.G., *Miniaturized nuclear battery*. US3934162, 1976.
9. Ritter James C., *Radioisotope photoelectric generator*. US4178524, 1979.
10. YOUNG R.D., H.J.P., LIGHT G.M., SEALE S.W. Jr.,, *Charged-Particle Powered Battery*. WO9748105A1, 1997.
11. Amil Lal, L.L., Hang Guo,, *Tiny Atomic Battery Could Run For Decades Unattended*. TECH SPACE, 2002.
12. Anghaie S., S.B., Knight T.,, *Direct Energy Conversion Fission Reactor Gaseous Core Reactor with Magnetohydrodynamic (MHD) Generator*. DE-FG03-99SF21894, 2002. **Final Report**.
13. Nyevorov V., *The direct conversion of nuclear power into electric current for various applications*. Basic & Applied Research Bureau, 2006. **Proposals**(1): p. 5.
14. Klimov V.I., *Mechanisms for photogeneration and recombination of multiexcitons in semiconductor nanocrystals*. J. Phys. Chem., 2006. **B**(110): p. 16827.
15. POPA-SIMIL L., e.a., *Multi-thermocouple structure for measuring the target's temperature field during beam power deposition*. Internal Report, 1986. **NIPNE-HH, Accelerator division**(Special projects): p. 2.
16. POPA-SIMIL L., e.a., *Machining regime influence on rougness and carbon concentration in surfaces*. Vacuum, 1994. **99**(1).
17. Zigler James, *PARTICLE INTERACTIONS WITH MATTER, SRIM - The Stopping and Range of Ions in Matter*. IBA, 2006. **ibm**(web).
18. NIST, *Electrons interactions with matter*. web, 2005. **www.nist.gov**.
19. YOUNG Robert D., H.J.P., LIGHT Glenn M., SEALE Stephen W. Jr.,, *CHARGED-PARTICLE POWERED BATTERY*. US5861701, 1999.

Development of a novel material for hydrogen storage

E. Titus*, Gil Cabral, J.C. Madaleno, M.C. Coelho, T. Shokuhfar and J.Gracio

Department of Mechanical Engineering, University of Aveiro, 3810-193, Portugal

ABSTRACT

Nanocomposites of polymer and carbon nanotubes (CNTs) are fascinating and progressing area of hydrogen storage research. The dispersion of the CNTs in polymer matrices is an important issue while making the nanocomposites. In-situ polymerization is a better approach for synthesizing homogeneous polymer CNT composites. However, the dispersion of CNT in the monomer solution is still problematic. In this paper, we report a novel approach for dispersing multiwalled (MW) CNTs directly into polyaniline (PANI)/ ethyl alcohol solution and preparation of uniform composite of PANI/MWCNT with the aid of nickel catalyst. The possibility of the preparation of uniform PANI/CNT suspension retaining the structure of both PANI and CNT is successfully demonstrated for the first time.

Key words: Carbon nanotubes, Polyaniline, Hydrogen storage, Composites

INTRODUCTION

A novel hydrogen storage medium is highly desirable for the next generation clean fuel. Metal hydrides are already showing good promise for future hydrogen storage. However, their relatively low gravimetry capacity is an obstacle for maximum hydrogen storage. Conducting polymers are safe, highly stable, light weight and economic materials. PANI is a unique type of conducting polymer in which the charge delocalization can, in principle, offer multiple active sites on the polymer backbone for the adsorption and desorption of hydrogen, involving weakening of the H-H bond followed by adsorption of this hydrogen onto the adjacent nano-fibrous network. However, the major disadvantage of polyaniline is its complicated processing due to its degradation below the melting point. In addition, major difficulties have been encountered in attempts to dissolve the material. The controversy on the solubility of polyaniline is dated back to 1910. Willstatter and Dorogi reported that an oligomeric (eight-monomer chain compound) aniline was largely insoluble [1]. Green et al repeated their experiments and claimed solubility of this non-polymeric material in 80% acetic acid, 60% formic acid, pyridine and concentrated sulfuric acid [2]. Later, Angelopoulos and co-workers and Wang et al. reported only partial solubility of polyaniline, in its emeraldine base form, in N-methylpyrrolidone (NMP), dimethylformamide (DMF), tetrahydrofuran (THF), benzene and chloroform [3]. The strategies to render the conducting polymer solution processible, were progressed significantly in the following years also. A significant improvement in this field occurred when camphorsulfonic acid (CSA) [4], dodecylbenzensulfonic acid [5] or alkylene phosphates [6] have been proposed as dopants. Composites of conjugated polymers are fascinating because of their potential for enhanced properties that are difficult to attain separately with individual components. It is reported that PANI can store as much as 6 –8 wt% of hydrogen [7], however, the reproducibility is problematic due to the complicated processing [8]. In recent years many research work have been reported on PANI composites with CNTs [9]. The report demonstrates the compatibility between PANI and CNT. However, the exact nature of the composite still remains unclear and further study is required.

We report the synthesis of new PANI/CNT composite by the solubilization process using nickel catalyst. We recommend this material for hydrogen storage due to its combined propeties. Fourier Transform Infrared (FTIR) analysis was done to study the nature of the material individually and in composite.

MATERIALS AND METHODS

The CNTs were purchased from Nanocyl, Belgium, and the purity of the sample is 95%. Polyaniline in emeraldine base form was purchased from sigma Aldrich. MWCNTs were functionalised using HNO_3/H_2O_2 medium and the PANI/MWCNT uniform solution (PANI/MWCNT/nickel composite) was prepared with the aid of nickel. The PANI/MWCNT/nickel composite was later filtered and dried at 60^0C. Nickel used was in the form of powder. FTIR analysis was carried out using Nicolet NEXUS bench machine with 128 scans and resolution of 1 cm^{-1}. The IR samples were prepared by mixing the samples with KBr powder.

RESULTS AND DISCUSSION

FTIR spectra (400-4000 cm⁻¹) for CNT (a), PANI (b) PANI/CNT (c) and PANI/CNT doped with nickel (d) are shown in Figure 1.

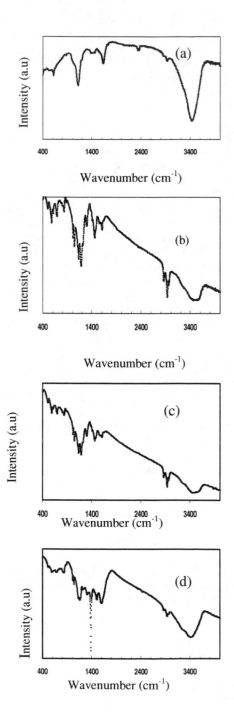

Figure 1

The characteristic vibrational modes of MWCNT, C=C (~1650 cm⁻¹) and O-H (~3400 cm⁻¹) are apparent in spectrum (a). The C=C vibrations occur due to the internal defects and the O-H vibration is associated with the oxidation of the sample during sample purification. The high purity of CNTs is also important in CNT-composite preparation. The spectrum for the PANI (b) consists of major peaks centered at around 3500, 1579, 1473 and 1309 cm⁻¹ which are typical of PANI. The spectrum shows quinoid band at 1579 cm⁻¹ and benzene ring band at 1473 cm⁻¹. The spectrum also displays a small band at 1309 cm⁻¹ which is associated imine group. The band of deprotonated PANI is visible at 1189 cm⁻¹ and out of plane bending of benzene ring appears at 835 cm⁻¹. The N-H stretching band at 3500 and the C-H stretch band at 2854, 2929, 2960 cm⁻¹ respectively are correctly observed in the spectrum. The spectrum of CNT after dispersion in PANI (figure 1c) with out nickel did not exhibit any major change from PANI spectrum. This shows no interaction between CNT and PANI. However, PANI/CNT doped with nickel shows few changes in the spectrum (figure 1d). The benzene ring band was found to be shifted to 1509 cm⁻¹. This is expected due to the interaction PANI with CNT. The major shift is observed for NH peak of PANI (3500 cm⁻¹), which is shifted to lower wavenumber (3426 cm⁻¹). The most interesting feature is the appearance of an additional band at 1400 cm⁻¹ which is attributed to the nickel related peak. We confirmed the position of this peak by doping of CNT alone with nickel (figure 2).

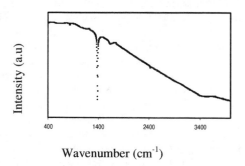

Wavenumber (cm⁻¹)

Figure 2

The presence of nickel in PANI/MWCNT/Ni composite is further confirmed by EDS analysis (figure 3). The schematic of the possible interaction is shown in figure 4. The SEM image of the PANI/MWCNT/Ni composite is shown in figure 5.

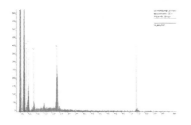

Figure 3

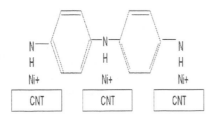

Figure 4

Figure 5

CONCLUSION

We assume nickel plays as an intermediate role in forming the composite. The interaction between PANI and CNT after nickel doping is confirmed by the shift in IR signal of NH vibrations. However, the exact role of nickel in forming uniform suspension of PANI/MWCNT has to be studied further. We propose this material for hydrogen storage due to the successful incorporation of nickel and CNT onto polymer matrix.

References:

1. Willstatter, R., et al., "Uber Anilinschwarz. II.", Berichte der Deutschen Chemischen Gesellschaft, 42, 2147, 1909
2. Green, A. G., "Aniline-Black and Allied Compounds. Part I.", J. Chem. Soc. 97, 2388, 1910
3. Angelopoulus, M., et al., "Polyaniline Processability from Aqueous Solutions and Effect of Water Vapor on Conductivity", Synthetic Metals, 21, 21, 1987
4. Y. Cao, P. Smith and A.I. Heeger Synth Met, 48, 91, 1992
5. Y. Cao, G.M. Treacy, P. Smith and A.J. Heeger Appl Phys Lett 60, 2711, 1992
6. I. Kulszewicz-Bajer, J. Sobczak, M. Hasik and J. Pretula, Polymer, 37, 25, 1996
7. Cho SJ, Song KS, Kim JW, Kim TH, Choo K. Fuel Chem Div Prepr 47, 790, 2002
8. Panella B, Kossykh L, Dettlaff-Weglikowska U, Hirscher M, Zerbi G, Roth S. Volumetric measurement of hydrogen storage in HCl-treated polyaniline and polypyrrole. Synth Met 151, 208, 2005
9. Cochet M, Maser WK, Benito AM, Callejas MA. Chem Commun 40, 1450, 2001

*Corresponding author: email: elby@mec.ua.pt

Active Coatings Technologies for Customized Military Coating Systems

J.L. Zunino III

U.S. Army Corrosion Office, U.S. Army RDE Command,
Bldg 60 Picatinny, NJ 07806-5000 jzunino@pica.army.mil

ABSTRACT

The main objective of the U.S. Army's Active Coatings Technologies Program (ACT) is to develop technologies that can be used in combination to tailor coatings for utilization on Army Materiel. ACT is divided into several thrusts, including the Smart Coatings™ Materiel Program, Novel Technology Development, as well as other advanced technologies areas. The goal of the ACT Program is to conduct research leading to the development of multiple coatings systems for use on various military platforms, incorporating unique properties such as self-repair, selective removal, corrosion resistance, sensing, ability to modify coatings' physical properties, colorizing, and alerting logistics staff when tanks or weaponry require more extensive repair. A partnership between the U.S. Army Corrosion Office at Picatinny Arsenal, NJ along with researchers at several universities are developing active coatings systems via novel technologies such as nanotechnology, Micro-electromechanical Systems (MEMS), meta-materials, flexible electronics, electrochromics, electroluminescence, etc.

Keywords: Active Coatings Technologies, Smart Coatings, Army, nanotechnology, sensors, MEMS

INTRODUCTION & PROGRAM DRIVERS

The Army is transforming into a lighter yet more lethal "objective force," all while fighting a war in Afghanistan and Iraq. The Army's Future Combat System is the heart of the Objective Force. Its new platforms must be deployable, 70% lighter and 50% smaller than current armored combat systems, while maintaining equivalent lethality and survivability[1]. To meet the lighter yet more lethal requirements of the Future Combat Systems, our scientists and engineers need to capitalize on new technologies and breakthroughs in the scientific arena.

The coatings we apply to our tanks, helicopters, and other weapon systems need to better protect their structures and crew since design margins are significantly tighter resulting in much less room for error for these lighter vehicles.

The U.S. General Accounting Office (GAO) estimates that the total cost for DOD corrosion related problems alone is $20 billion per year, $4 billion of which is related to painting and de-painting operations[2]. The coatings applied to weapon systems today lack the ability to self-correct when environmental conditions and circumstances change, nor do they have the ability to tell the user of potential anomalies such as corrosion, damage or adhesion problems.

A crucial impact of corrosion is related to safety. The U.S. Army had forty-six mishaps, thirteen serious injuries, and nine fatalities directly related to corrosion between 1989-2000. This does not include all the indirect impacts or figures beyond 2000. Now that we are at war, there is a far greater impact of corrosion on safety[3].

The U.S. Army Corrosion Office at Picatinny Arsenal, NJ is addressing issues of military coatings systems by developing coatings capable of collecting, analyzing, managing and adapting to data from the environment in real-time. If an anomaly such as a scratch or degradation from corrosion is detected within the coating, embedded sensors will analyze the data and initiate a response. The response may result in the coating self-healing if a crack exists or the coating's color patterns may change via electroluminescence and/or electrochromics to visually display corroded areas on the tank, if desired.

There have been major advances in the Active Coatings Technologies program. These advances include researching into MEMS and Nano devices in order to create coating systems that will self correct and show where areas of weakness are. These devices will be crucial in this coating process; helping the Army move into the lighter realm. Self correcting coatings will significantly cut costly repairs. The research is ongoing and will continue looking at smaller electronics that will better help the Warfighter.

Current and future efforts under this program include the integration and powering of these sensing packages into a multi-layered Smart Coatings™ system capable of "thinking and reacting" autonomously. The resulting Smart Coating™ materiel will ultimately aid the Army/ Future Combat System and DOD by; 1) developing a novel multi-functional coating system for enhanced protection of tactical equipment; 2) decreasing life-cycle costs while increasing readiness i.e. reducing equipment down time; 3) reducing the logistics burden of the Soldier; 4) controlling materiel signature & footprint via masking techniques; and 5) reducing the potential hazards associated with current painting/de-painting operations.

DISCUSSION

A vast array of novel technologies has been introduced over the past few years. Advancements in nano-technologies, MEMS, polymers, composites, flexible electronics, and numerous other areas allow the

Department of Defense to improve and create faster, lighter, and more lethal systems.

The U.S. Army is attempting to take these technologies and implement them into an active coatings system though the Active Coatings Technologies Program, thus creating numerous technologies that can be combined to develop customizable coatings solutions to meet military user requirements. These technologies give one the ability to work at the molecular level, atom by atom, to create large structures with fundamentally new molecular organization and yield advanced materials that will allow for longer service life and lower failure rates.

Nanostructured materials yield extraordinary differences in rates and control of chemical reactions, electrical conductivity, magnetic properties, thermal conductivity, strength, and fire safety. The small size allows for numerous systems and functions to be incorporated together and embedded into materials such as metals, polymers, paints/films, composites, etc.

The Army Corrosion Office has assembled a team including university support from the New Jersey Institute of Technology (NJIT), Clemson University, University of New Hampshire, Pennsylvania State University, University of Massachusetts, South Carolina Research Authority, as well as other military and industry representatives. This team is developing multilayer, modular active coatings with numerous functionalities such as self-repair, visual display, artificial intelligence, self-management, sensing package, and corrosion inhibitors, that can be customized as needed. An illustration of the Smart Coatings™ System is depicted in Figure 1.

To date, several working prototype modules have been developed under this program. Some of the key areas of research within the modules include color modifying coatings, flexible electronics, wireless sensor packages, nanotube development, intelligent nano-clays, alternative fuel/power sources, de-painting/self-repair, material modification, and other military capabilities.

Color modification methods include using electrochromics, electroluminescence, intelligent clays (i-clays), single and multi-wall carbon nanotubes, and chemical additives to control and adapt color change capabilities on demand.

Self-Repair
Visual Display
Novel Materials
AI Network
Sensing Package
Power
Corrosion Inhibitor

Figure 1: Smart Coatings™ System
(PATENT App: 11/307611
02-14-06 U.S. Army & NJIT)

The program is also developing flexible electronic capabilities for sensing, communication, data collection/storage, and power alternatives. Flexible electronics have been developed at NJIT that demonstrate the capabilities of several different types of the flexible sensors, some of which can detect strain in the material, scratches on the surface, corrosion, pressure, flow, and temperature change.

Numerous tasks are currently working with nanotubes including their functionalization, development, and production. Single-walled carbon nanotubes (SWCNT) are being implemented into coatings and inks to initiate self-healing, active switching, sensing, color modification, and other functionalities. Work is also being performed to develop cost effect methods of development and scale-up of advanced Nanomaterials and their production. Nanotubes are also being utilized for power/fuel cells development and electroluminescence. For example, photovoltaic power sources are currently being developed. One method includes the use of p-n junction SWCNT coatings as photovoltaic modules with the bottom layer functioning as a proton exchange membrane (PEM) fuel cell. This cell will provide power for active coatings capabilities while electroluminescence will serve as a modifiable display. Also, photovoltaic coatings have been developed that demonstrate solar cells functionalities as well as active display capabilities.

Solubility and polymer wrapping of SWCNTs have provided us the ability to functionalize these tubes. Beyond that, chemistries are being developed that enable the production of single-walled nanotubes with precise but tunable dimensions (properties).

The creation and development of intelligent nanoclays (i-clays) to detect corrosion & chem/bio via color changes or luminescent properties within the Smart Coatings™ System is also underway. These i-clays can be incorporated into inks, paints, composites, etc. to add functionalities to current and future coating systems.

Self-repair and de-painting research is on going using micro-encapsulation techniques, nano particles, micro-etching, and MEMS. These technologies are being developed into a micro distribution systems mimicking the body's vascular and healing capabilities.

These technologies, tasks and modules are designed to be integrated into customizable coating systems tailored with desired functionalities to meet user requirements. The ability to have an active, adaptive coatings system that acts more like a living entity than a typical paint job allows the coating to be utilized on both current and future weapon systems, military and civilian applications.

Major advances in the ACT program thus far include the development of military grade active sensing packages to detect damage (corrosion, substrate integrity, etc) and environmental conditions (i.e. radiation, chemicals, temperature, gases, strain, etc).

The sensors and technologies developed under the program can be used with Army Materiel. A potential application is corrosion of the main support members of the Chinook Helicopter (Figure 2). The Chinook Helicopter has structural support members beneath floorboards that are susceptible to corrosion. Typically, bi-weekly inspections include 3 man days to remove floor boards, 2 man days of inspection, and 3 man days to replace floorboards. Utilizing the flexible sensor array and ribbon cables that are accessible in the cockpit or can transmit data wirelessly, technicians can utilize the sensors to determine the structural integrity of the support members without bi-weekly visible inspections. By enabling visible inspections to be conducted remotely, time and cost savings can be realized.

Another problem these technologies can help resolve exists on Heavy Expanded Tactical Truck (HEMTT). The fuel tanks on these trucks are only half filled since the structural integrity and safe load capacity of a given truck is not known due to material degradation. This means that twice the number of "runs' occur to deliver the required amount of fuel; costing valuable resources and time. Using the sensors contained in active coatings the safe load capacity can be known instantly. With both the Chinook and HEMTT, the data can be transmitted remotely so the number of visual inspections can be greatly reduced and conditioned based maintenance can be performed, saving time and money.

Fuel Version - Only 1/2 filled due to unknown load limit caused by material degradation

Figure 3. Fuel version of HEMTT with corroded supports

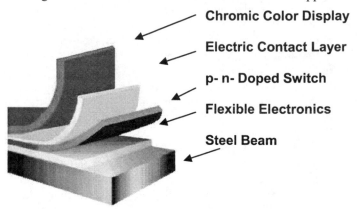

Chromic Color Display

Electric Contact Layer

p- n- Doped Switch

Flexible Electronics

Steel Beam

Figure 5: Schematic of Smart Coatings™ Prototype

The Army Corrosion Office, along with the New Jersey Institute of Technology, has developed these prototype systems modeling the supports in the HEMTT, Chinook and Blackhawk Helicopters (Figure 4). The embedded electronics can potentially sense changes in the beam in real-time and alert the user of anomalies such as damage, corrosion, cracks, etc. The data can be collected and saved or used as an early warning system giving real-time status of the supports. This can be done through a visual display on the beam or the information can be transmitted to the cockpit and/or control center wirelessly.

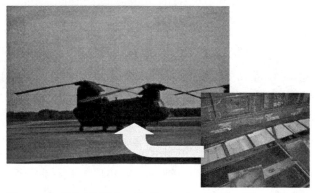

Figure 2: CH-47 Chinook Corrosion Under Floor Boards

Several of the technologies developed thus far have been incorporated into "proof-of-concept" prototypes. These prototypes were successfully demonstrated at the U.S. Army Corrosion Summit in 2005 & 2006. One prototype (Figure 4) demonstrates the ability to sense a change in the environment, analyze the change, and alert the user of the anomaly through color changes on the substrate. Figure 5 illustrates a schematic of that prototype system.

Another prototype (Figure 6) demonstrates a wireless version of the system in which the data is collected, wirelessly transmitted, and analyzed. The information can illustrate where and how structural changes, damage, material degradation, and material loss affect the integrity of beams, supports, and other structures (Figure 7).

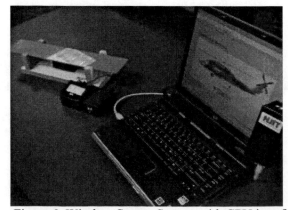

Figure 6: Wireless Sensor System with CPU interface

Figure 7: Display Illustration of Damage on CH-47D

Current and future efforts under this program include more advanced sensor packages collecting real-time data of operational environments and conditions, as well as the integration and powering of these sensing packages into a multi-layered Smart Coatings™ system capable of "thinking and reacting" autonomously. These advancements will allow the actions to be repeatable so that the coating will resemble a living entity that has the capacity for self-sustainment.

The overall goal is to develop systems to be utilized on current military systems and to transition technologies to the Future Combat System (FCS).

CONCLUSIONS

The need to protect our current and future military assets is obvious. It is in the Department of Defense's best interest to use the latest technologies to advance the protection of these assets. The current and future advances made in nanotechnology and MEMS are leading to the development of novel materials and systems that ultimately will allow the military to advance into the twenty-first century and beyond.

Corrosion, material degradation, and coating failures are a serious cost driver for our military. Current coatings on military systems are not capable of self-sustainment, or alerting the user of potential anomalies that can cost the DOD billions of dollars per year as well as the loss of equipment and lives.

The Active Coatings Technologies Program, including the Smart Coatings™ Materiel Program, is helping to address this issue by integrating state–of-the-art technology into and on military systems. Through its research and development, the program will 1) develop a novel multi-functional coating system for enhanced protection of tactical equipment; 2) decrease life-cycle costs while increasing readiness i.e. reducing equipment down time; 3) reduce the logistics burden of the Soldier; 4) control materiel signature & footprint via masking techniques; 5) reduce the potential hazards associated with current painting/de-painting operations.; and
 6) correspond with Army's Transformation Strategy to help safeguard our national and international interests.

The goal of this program is to develop those technologies needed to create active coating systems to aid in Army Transformation. This transformation will result in new and modernized weapons systems fielded globally that are capable of meeting current and potential challenges.

Active Coatings Systems will increase survivability & readiness while decreasing life-cycle costs, maintenance and potential hazards. The resulting systems will ultimately aid the Army/ Future Combat System and DOD in 1) developing a novel multi-functional coating system for enhanced protection of tactical equipment; 2) decreasing life-cycle costs while increasing readiness i.e. reducing equipment down time; 3) reducing the logistics burden of the Soldier; 4) color modification techniques; 5) reducing the potential hazards associated with current painting/de-painting operations.

REFERENCES

[1] *Future Combat System (FCS)*: Article. www.globalsecurity.org.

[2] *Cost of Corrosion Report to Congress Jan 02:* National Association of Corrosion Engineers (NACE) 2002.

[3] Research performed on the *Technology Demonstration for the Prevention of Material Degradation Program,* U.S. Army Corrosion Office, U.S. Army ARDEC-RDECOM, Picatinny Arsenal, NJ.

[4] Research performed on the *Smart Coatings™ Materiel Program,* U.S. Army Corrosion Office, U.S. Army ARDEC-RDECOM, Picatinny, NJ.

[5] J. Zunino III, et al. *U.S. Army Development of Active Smart Coatings™ System for Military Vehicles,* NSTI, Nanotech 2005.

[6] J. Zunino III, et al. *Development of Active Sensor Capabilities for the U.S. Army's Active Smart Coatings™ System,* 2005 Tri-Service Corrosion Conference. Department of Defense & NACE.

Design and Manufacturing Concepts of Nanoparticle-reinforced Aerospace Materials

M. Kireitseu[*], G. Tomlinson[*], and V. Basenuk[**]

[*]University of Sheffield, Dept. of Mechanical Engineering, RR-UTC
Mappin Street, Sheffield S1 3JD, the United Kingdom, indmash@yandex.ru
[**]Institute of mechanics and machine reliability NAS of Belarus, Dynamics laboratory

ABSTRACT

Nanoparticle-based vibration damping shows the effect that molecule-level mechanism can have on the damping and that nanoparticles/fibres/tubes-reinforced composite materials can provide enhanced strength and vibration damping properties over the broader operational conditions. It is particularly worth noting that carbon nanotubes can act as a simple nanoscale spring. The mechanisms involved in such materials need to be understood and the relevance to damping identified. The focus in this paper is directed toward the development of the next generation of vibration damping systems, providing a road map to manufacturing technology and design solutions. The research work concentrates on an investigation related to nanoparticle-reinforced materials extensive dynamic characterization and modelling of their fundamental phenomena that control relationships between design and damping properties across the length scales.

Keywords: nanoparticle, damping, fan blade, design

1 INTRODUCTION

Vibrations and noise exist in almost every aspect of our life and are usually undesirable in engineering structures [1]. Vibrations are of concern in large structures such as civil (airbus A 380) or Rolls-Royce powered military aircrafts [2, 3], as well as small structures such as electronics [4]. Manufacturers have stated that for the next generation of transportation we need light-weight, cost-effective and reliable vibration-absorbing materials [5, 6].

Recent breakthroughs in automotive, maritime and aircraft design have resulted in the successful use of synthetic parts and novel designs in applications previously considered too demanding for non-traditional materials, comprising 20-25% of a typical car body weight [7]. Composites are also, besides aluminium, the most important materials for aerospace applications. Due to the opportunities they present for weight saving, their share has reached more than 15 % of the structural weight of civil aircraft, and more than 50% of the structural weight of helicopters and military aircrafts [8]. In addition to their high mass specific stiffness and strength, the high potential of composites for additional functionality is another reason for their success. Defined anisotropic behaviour, the possibility to integrate sensors or actuators, high structural damping, and superior fatigue performance are typical advantages. Motivation to the research work could be outlined as follows:

1. Nanoparticle/tube/fibre-reinforced composite material is a relatively new vibration damping technology entailing placement of numerous nanoparticles inside vibrating material structure that has wide applications in areas of transportation (aerospace, auto, rail, maritime) and electronics. Carbon nanotubes are particularly cost-decreasing material for potential large scale industrial applications.

2. Manufacturers such as Rolls-Royce, Boeing, Airbus require in a long term 10 time stronger and 2 time lighter components and materials as well as their 3 to 5 year goal is 30-40% enhancement of these operational parameters that can be achieved by advanced design and manufacturing based on about 50% nanotechnology content.

3. The CNT-reinforced material damping technology and related energy dissipation phenomenon is complex because of the variety of mechanisms involved and there is the gap between engineering applications/structures and nanotechnology leading to the next generation engineering.

It is now accepted that nanotechnology may considerably enhance strength/damping behaviour and reduce noise of engineering structures through the utilisation of nanomaterials that dissipate a substantial fraction of the vibration energy that they receive [7]. Carbon nanotubes are particularly promising cost-decreasing reinforcement material [8, 13]. Boron nitride or silicon carbide nanotubes are another possible candidate for aerospace material reinforcement [2, 9]. The benefits that may be achieved by both carbon nanotubes in lower temperature damping applications and boron-nitride/silicon carbide nanotubes in high temperature aerospace applications may be very significant. The present paper will outline some preliminary efforts made by the institutions in this direction under EU programmes. The key goal of this communication is to provide a route map in the future research and outline a project concentrated on the next generation industry-oriented nanotechnology-based solutions for enhanced vibration damping/dynamics performance.

2 DESIGN CONCEPTS

2.1 Damping nanomaterials

Selection of engineering material for the future nanoparticle-reinforced composite material is available from three types of solids that are metals, polymers and some ceramics. Selected matrix will mostly determine mechanism of

Cleantech 2007, ISBN 1-4200-6382-0

energy dissipation [10, 11]; however, small CNT volume (1-5%) may greatly affect material's behaviour due to extreme nanotube properties over traditional materials [5-12].

Nanotube-metal matrix composite materials are still rarely studied [6, 13]. Most used metallic alloys are hard or soft materials such as titanium/nickel and aluminium/bronze respectively. While titanium is stiff, light-weight and good for high temperature applications such as turbine fan blades, car panels are made of light-weight and cheaper aluminium and its alloys. Produced metal foams have excellent energy absorbing/damping properties over bulk material [14] and most likely are used for the next generation engineering. The CNT-reinforced metal materials are generally prepared by standard powder metallurgy melting [16] or ultrasonic liquid infiltration method [15], but good mixing and dispersion of the nanotubes should be achieved. In this respect it would be worth producing CNT-reinforced metallic foam composites and investigate their damping/dynamics performance.

Carbon nanotube-reinforced ceramic-matrix composite materials are a bit more frequently studied [17]; most successful efforts were made to obtain tougher ceramics (SiC, Al_2O_3, etc.) [37]. The composites can be processed by 1) mechanically mixed nanotubes with the matrix and then sintering [13, 37], melting [20] or spraying [18] of the particle mixture; 2) CVD deposited CNT-based thin films on SiC substrate (up to 50 μm) [19]; and 3) electrochemical processing such as micro-arc oxidizing of metal substrate in an liquid electrolyte with added nanoparticles [21]. Some ceramic coatings are successfully used to enhance damping of titanium fan blades [18] and therefore, would be recommended as a candidate matrix material, but dispersion and orientation of nanotubes in the matrix is yet out of some control. Bulk properties other than mechanical are also worth being investigated.

Nanotube-polymer composites are now intensively studied [22-29], notably epoxy- and polymethylmethacrylate (PMMA)-matrix composites [25]; however, their damping behaviour is rather contradictory result than plausible information (see table 1). The ability of the polymer molecular chains to form large-diameter helices around individual nanotubes favours the formation of a strong bond with the matrix [36]. Selection of related manufacturing technologies is available from some well-known in the aircraft industry that are 1) manually or ultrasonically melt mixing and extrusion of nanotubes and polymer-layered silicate [12, 22], 2) CNT-reinforced resin by using so-called calandering technique [23, 24], 3) polymerization of carbon fibre by interfacial polymerization [26, 27].

2.2 Design of nanoscale-reinforced fan blade

Advanced damping material and fan blade design concepts could be introduced as follows:

1) Particle micro-balloons are currently being used for enhanced damping performance of fan blade over bulk material core or rigid stiffeners in a hollow blade [18].

Nanoparticle-reinforced micro-balloons may be added separately or used to create syntactic foam (a two-part epoxy adhesive filled with balloons) such that the density of the foam is about 1000 kg/m^3 (Fig. 1).

Motivation: Volume and weight of filler material should be minimized. In large civil engines, the blades are hollow and usually have stiff rib-like metallic structures in order to increase the rigidity and maintain cross-sectional profile of the blade. The filled fan concept is to replace this metal structure with CNT-reinforced foam simultaneously acting as a strengthener and a damping element.

2) CNT-reinforced damping coatings: both single layer and multi-layered sandwich structure.

Motivation: Coating has considerable adhesion and adds significant damping to titanium fan blades. Ceramic coating is desirable in high temperature applications, but its damping level is lower than that of polymeric ones. Another problem is fracture and fatigue of hard coatings on dynamic blade.

Syntactic foam fillers are frequently used to stiffen hollow fan on large civil aircraft engines. These typically comprise a two-part epoxy compound filled with glass micro-balloons such that the density of the foam is well below 1000 kg/m^3 and the concept of the cavity-filled fan blade has been introduced [8-10]. Hollow fan blades currently enclose a strong, stiff metallic structure to maintain the cross-sectional profile of the blade when subjected to the large static and dynamic forces experienced in normal operation. In the cavity-fill fan concept, the polymeric core replaces this metal structure simultaneously acting as both strengtheners and damping elements – an example is shown in Figure 1. It would be expected that incorporating CNT may affect not only the damping performance, but also the integrity under static, impact and fatigue loads. While hard balloons are desirable for maximising stiffness and damping, CNT-reinforced polymeric ones may be better for the damping and integrity.

Further improvements can be made to both the materials and to the blade design. In its simplest form, the CNT reinforcement concept is simply a fan blade coated or filled with a suitably selected damping material. Hollow fan blade may be filled with CNT-reinforced material and coated with CNT-reinforced layer on the top [12]. The blades are composed of a titanium sheet and composite internal / external manifolds with nanoparticle reinforcement.

Due to the complex whole structure and advanced design, these blades should have low natural frequencies and bending-torsion coupled mode shapes that could potentially lead to aeroelastic instabilities. Increasing the damping levels in these blades will improve the fatigue life and reduce aeroelastic instability concerns. The vibratory modes of interest include the first and second bending modes as well as first torsion mode. Due to the geometric constraints of the blade shape and large internal manifolds very little room is available for damping treatment placement; however, it is expected that large volume-to-surface ratio of carbon nanotubes may overcome the problems. Some possible outcomes would be also extending the temperature range over which some polymeric damping materials is present and finding ways to increase the modulus without sacrificing density and integrity. It is anticipated that significant weight, thickness and manufacturing cost reductions could be achieved in this way.

3 EXPERIMENTS

SWNT-reinforced polymeric matrix composites are prepared using an extrusion procedure. The polymer is made of polymethylmethacrylate polymer samples. SWNT particles produced by Dynamic Enterprises, UK (2 nm in diameter and 80-90 vol% sample purity) are added in concentrations of 10% by weight (including impurities) along with a surfactant (polyoxyethylene) to aid in dispersion. The mixture is put into a mold and subjected to a vacuum for 25 min and curing. Scanning electron microscope (SEM) is used to evaluate the nanotube dispersion and orientation in the polymeric matrix. CNT orientation was controlled by pressure rates at extrusion.

Hollow titanium alloy (6%W, 4%V) samples were then sandwiched with the composite material and glued by an adhesive epoxy resin. Resonance frequencies, the mode shape and damping at each mode were determined by laser vibrometry at standard vibration shaker tests. Heating of clamped sample was provided by two 1000W electrical "Philips" bulbs. The clamping block is fixed so that friction losses and extraneous damping is minimized. The data acquisition and control of the electro-dynamic system is based on Computer Measurement System. The deformation rates were 200 rad/s for shaker. The vibrometry procedure yields not only the resonance frequency, but also the mode shape.

Material	Young modulus at 25°C, GPa	Loss factor at 25°C
Epoxy resin Polyurethane at 850Hz Epoxy adhesive filler	3.1-3.4 [12, 19] 0.3 [19] 3 and 0.00275 at 850Hz [19, 34]	0.01 [12]* 0.58 [19]** 0.4 [19]
1-5 wt% CNT-matrix 50-60 wt% CNT-matrix	3.2-3.6 [23, 24], 7.1-7.5 [35]	0.08 [12] (* - 8 times increase)
MWNT-reinforced thin SiC ceramic film (no matrix) at 850Hz	284 [19] Bending stiffness + 30% [19, 34]	0.3 [19] (** - 2 times decrease) +200% [34]

Table 1: Damping properties of CNT-reinforced polymeric materials.

In this paper, the concepts of CNT-reinforced titanium alloy sandwich have been introduced and CNT-reinforced composites were applied so as to investigate their applicability to aerospace components (fan blade). Modal strain energy numerical methods [18] for predicting damping was validated experimentally under non-rotating conditions.

Loss factor exceeding $\eta=1$ have been demonstrated on fan blades under non-rotating conditions showing the design potential of the concept. CNT-reinforced composites affect not only the damping performance/strength, but also the integrity under static and cyclic loading. While CNT-reinforced ceramic material may be desirable for stiffening, polymeric ones were better for damping and integrity. The results (fig. 1) clearly show that the analytical methods used are reliable and that significant levels of damping can be achieved in fan blades using the cavity fill concept.

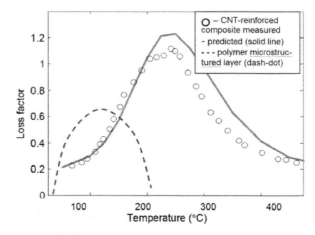

Figure 1: Comparison between measured and predicted modal properties under non-rotating conditions, showing a very close correlation.

Size of reinforcement nanoparticles may vary from 1 to 100 nm in diameter and length. The number of CNT walls and their size affect on stress concentration in the composite [29] and thus short and even round particles are the strongest ones (diamonds etc.), but longer fibres are flexible and may be worth for damping while CNT may particularly act as a simple nanoscale spring and a crack trapping nanomaterial blocking the holes in the composites [30, 31]. Such a damping phenomenon could be multiplied by a factor of billions when CNTs are dispersed in a material.

Orientation and geometry (waviness) of CNT particles may affect mechanisms of energy dissipation/fracture mechanics and maximum stiffness is achievable at 90^0 longitudinal CNT orientations [25, 29]. Notably those open-end CNTs do not collapse/failure/buckle due to higher stress concentrations while many authors [12, 22-31] have used closed-end CNT-reinforced composites. Thus isolated single-walled open-end nanotubes (SWNT) may be desirable for the future damping applications due to significant load-bearing ability in the case of CNT-matrix interactions. Defects of carbon particles limit the performance [23-27].

CNT dispersion should be optimized for damping at 1-5% volume content because of carbon fibre conglomeration at higher volumes [32, 35], but 60 wt% CNT concentration in polymer was modelled [35]. The main disadvantage of CNT dispersion is that it involves a large uncertainty to desired damping effect due to nanoparticles.

Cleantech 2007, ISBN 1-4200-6382-0

4 CONCLUSION

At some preliminary point it was validated that a concept of using CNT as vibration damping oscillators where CNT acts like a nano-shock-absorber and loss factor exceeding $\eta=1$ have been demonstrated on clamped specimens simulating fan blades under non-rotating conditions showing the design potential of the concept; however, a proof of manufacturing concept is required.

Future work should be concentrated on improvements to both the filler materials and the blade design. Efforts should be focused on ecology-oriented manufacturing & design, assessment of life cycle, durability, utilization and repair possibilities, comprising several steps for tailoring the nanoparticle-matrix interface, dispersion and orientation for specific application. It is anticipated that significant weight, thickness and manufacturing cost reductions could be achieved in this way. Selection of nanoparticles just lightly depends on its price, but application outcomes.

ACKNOWLEDGMENTS

The support of research work by the Royal Society in the U.K. and Marie-Curie Fellowship Ref. # 021298-Multiscale Damping 2006-2008 at the Rolls-Royce Centre in Damping, the University of Sheffield in the United Kingdom is gratefully acknowledged. It should be noted however that the views expressed in this paper are those of the authors and not necessarily those of any institutions.

REFERENCES

[1] T.R. Chung, Journal of Materials Science 36, 5733 – 5737 (2001).

[2] C.E. Harris, M.J. Shuart, and R.Hugh, "Survey of Emerging Materials for Revolutionary Aerospace Vehicle Structures & Propulsion Systems" (NASA Langley Res. Center, Hampton, USA, 2002), p.175.

[3] R.Chandra. Journal of Sound and Vibration 262, 475–496 (2003).

[4] J.J. Hollkamp and R.W. Gordon, Smart Mater. & Struct. 5(5), 715-223 (1996).

[5] P.R. Westmoreland (ed.), "Applications of Molecular and Materials Modelling" (NSTI, USA, 2002), p.180.

[6] R.W. Siegel, E. Hu, M.C. Roco (eds.), "Nanostructure Science and Technology: A Worldwide Study" (IWGN, NSTC, USA, 1999), p. 250

[7] C.Q. Ru, "Encyclopedia of nanoscience and nanotechnology" (Amer. Scientif. Publish., USA, 2003), p. 520.

[8] P.F. Harris, "Carbon Nanotubes and Related Structures" (Cambridge Univ. Press, Cambridge, 1999), p.540.

[9] T. Talay, "Systems Analysis of Nanotube Technology" (NASA, Washington, D.C., 2000), p. 240.

[10] R.DeBatist, J. de Physique (Paris), Colloque C9, 44(12), 39-45 (1983).

[11] A. Kelly, Proceedings of the UK Royal Society A 344, 287–302 (1970).

[12] X. Zhou, E. Shin, K.W. Wang, C.E. Bakis, Compos. Sci. Technol. 71, 1825–1831 (2004).

[13] B. Bhushan (ed.), "Handbook of Nanotechnology" (Springer-Verlag, New York, USA, 2004), p. 1220.

[14] S.A.Nayfeh, M.J.Verdirame, K. Varanasi, Journal of Sound and Vibration 214, 320-325 (2001).

[15] Y.Deming, Y. Xinfang, P. Jin, Journal of Material Science Letters 12, 252–263 (1993).

[16] Y.K. Favstov, L. Zhuravel, L.P.Kochetkova, Metal Sci. and Heat Treatment 45(11–12), 16–18 (2003).

[17] D.M.Wilson, J.A. DiCarlo, H.M.Yun, ASM Engineered Materials Handbook, Vol.1 (2001), p. 340.

[18] S. Patsias, C. Saxton, M. Shipton, Materials Science and Engineering A 370, 412-416 (2004).

[19] N. Koratkar, B. Wei, P.M. Ajayan, Compos. Sci. Technol. 63, 1525–1531 (2003).

[20] G.R. Tomlinson, D. Pritchard and R. Wareing, Proceedings of I. Mech. Eng. Part C 215, 253-257 (2001).

[21] M. Kireitseu, J. of Particulate Sci. and Technol. 20(3), 20-33 (2003).

[22] S. Peeterbroeck et al., Compos. Sci. Technol. 68, 1627–1631 (2004).

[23] F.H. Gojny et al., Chemical Physics Letters 370, 820–824 (2003).

[24] F.H. Gojny et al., Compos. Sci. and Technol. 64, 2363–2371 (2004).

[25] C.A. Coopera ct al., Compos. Sci. and Technol. 62, 1105–1112 (2002).

[26] E.C. Botelhoa et al., Comp. Sci. and Technol. 63, 1843–1855 (2003).

[27] S.B. Sinnott, J. Nanosci. Nanotechnol. 2(2), 113–123 (2002).

[28] V.T. Bechel, R.Y. Kim, Compos. Sci. and Technol. 64, 1773–1784 (2004).

[29] J. Sandler, M.S.P. Shaffer, A.H. Windle, et al., Phys. Rev. B 61, 301-305 (2002).

[30] J.L. Rivera, C. McCabe, and P.T. Cummings, Nanoletters 3(8), 1001-1005 (2003).

[31] C. Li and T.W. Chou, Phys. Review B 68, 403-405 (2003).

. Qizn, E. Dickey, R. Andrews, T. Rzntell, Appl. Phys. Letters 76(20), 2868-2870 (2000).

[32] Y.N. Hu, C.Y.Wang, Key Eng. Materials 259-260, 141-145 (2004).

[33] N. Koratkar, B. Wei, P.M. Ajayan, Adv. Mater. 14(13-14), 997–1000 (2002).

[34] G.M. Odegard et al., Compos. Sci. and Technol. 64, 1011–1020 (2004).

[35] A.H. Barber et al., Compos. Sci. and Technol. 62, 856–862 (2004).

[36] G.D. Zhan et al., Appl. Phys. Letters 83, 1228-1231 (2003).

Titanium Dioxide Nanostructured Coatings: Application in Photocatalysis and Sensors

JA Byrne, JWJ Hamilton, TA McMurray, PSM Dunlop, V Jackson A. Donaldson, J Rankin, G Dale, D Al Rousan

Nanotechnology and Advanced Materials Research Institute,
University of Ulster at Jordanstown, Northern Ireland, United Kingdom, BT37 0QB
Tel. +44 28 90 36 89 41   Email: j.byrne@ulster.ac.uk

Abstract

This paper presents some of the research work taking place at the University of Ulster investigating preparation, characterisation and application of nanostructred TiO$_2$. Four exemplars are used to demonstrate the potential applications of these materials i.e. photocatalytic disinfection of water containing chlorine resistant microorganisms, photocatalytic 'self-cleaning' of surfaces contaminated with protein, transducers for electrochemical biosensors and finally new opportunities presented by electrochemical growth of TiO$_2$ aligned nanotubes.

1.0 Introduction

Nanostructured titanium dioxide has been widely researched for application towards the photocatalytic treatment of purification of water and air, "self-cleaning" and superhydrophilic coatings for surfaces, and dye-sensitised voltaic cells. Nanoparticle TiO$_2$ films present a large surface area to geometric area ratio, which is useful in water and air purification and dye sensitised cells. In addition, these films also give high surface area desirable for electrochemical sensor applications. Titanium dioxide is found in three crystal forms, brookite, rutile and anatase, the latter of which is the most suitable for photocatalytic applications. Anatase TiO$_2$ is a wide band gap semiconductor (3.2 eV) and absorbs photons with $\lambda < 387$ nm. Band gap excitation produces electron hole pairs, which can take part in electrochemical reactions at the interface and result in the production of radical species. This photocatalytic action has been reported to degrade organic pollutants (in water and air) to CO$_2$ and H$_2$O, and kill a wide range of microoganisms. Furthermore, photocatalytic films can degrade protein material (including temperature stable proteins) adhered to their surface, and could find application in, for example, the sterilisation of surgical devices.

Nanostructured TiO$_2$ thick films present a high surface area, which is desirable for electrochemical sensor and biosensor applications. TiO$_2$ is biocompatible, non-toxic, and chemical stable under conditions found within the body, making it a suitable material for implantable biosensors. Furthermore, TiO$_2$ can be used for the electrochemical detection of hydrogen peroxide in the presence of oxygen. This makes nano-structured TiO$_2$ electrodes suitable for non-mediated biosensors utilising oxidase enzymes e.g. glucose oxidase.

Self-assembled titanium oxide nanotube arrays with maximum packing density can be formed by the anodic oxidation of titanium metal [1]. Such materials may prove to have enhanced properties for photocatalytic, sensor and other applications.

2.0 Photocatalytic disinfection of water

Cryptosporidium poses significant problems to the drinking water industry. It is ubiquitous in surface water, difficult to remove by conventional drinking water treatment processes and if ingested can cause serious illness [2]. The cost-effective removal of *Cryptosporidium* from drinking water sources remains one of the industries greatest challenges [3]. There have been many outbreaks of cryptosporidiosis all over the world associated with consumption of contaminated drinking water [4]. The largest outbreak occurred in Milwaukee, USA in 1993 with the death of 104 AIDS patients and an estimated 403,000 people becoming ill [5].

The bactericidal effect of TiO$_2$ photocatalysis has been widely reported [6,7] however a limited number of studies have reported the effectiveness of photocatalysis against chlorine resistant organisms. In this paper we present the photocatalytic inactivation of *Cryptosporidium parvum* oocysts

TiO₂ powder (Degussa P25) was electrophoretically immobilised onto Ti alloy substrates [8] and placed in a quartz water-jacketed reactor with the illumination source focused on the coated area (125W HPR lamp (Philips), mainline emission 365nm). An oocyst suspension was prepared by diluting fresh oocysts (Moredun Scientific) in saline solution to achieve a working concentration of 2×10^4 oocysts per cm^3. The reactor was thermostatically controlled at $20 \pm 2°C$ and agitation of the 10 cm^3 of oocyst suspension was provided by a small magnetic stirrer. Air sparging was achieved using a small aquarium pump, flow rate of 900 cm^3 min^{-1}. The reactor was allowed to reach equilibrium under dark conditions for 15 min prior to irradiation. A 100 μL sample was removed and the electrode illuminated. Samples were removed every 60 minutes thereafter for a period of 240 minutes. Analysis for oocyst damage was preformed using the vital dye exclusion protocol developed by Robertson et al [9].

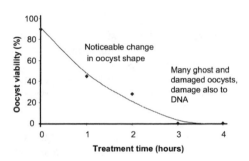

Figure 1. Photocatalytic inactivation of Cryptosporidium parvum oocysts.

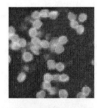

Before treatment:
Time = 0 mins
Vital Dye Exclusion
Assay indicating 90% oocyst viability

After photocatalysis:
Time = 240 min
Vital Dye Exclusion
Assay shows extensive damage to oocysts

During the first hour observable changes in oocyst shape were observed associated with a decrease in viability (see figure 1). Following three hours treatment viability decreased to zero and fragmentation of the oocyst bodies was evident along with the presence of many ghost oocysts (oocysts missing their DNA).

Further studies are being carried out confirm the loss of infectivity via in-vitro infectivity and an examination to elucidate the mechanism of disinfection will be undertaken.

3.0 Self-cleaning coatings for surface decontamination

Decontamination of surfaces is an area of current interest. Health care acquired infections (HAIs) cost the Health care sector billions of pounds every year and cause patient discomfort, prolonged hospital stays, and even death. Conventional approaches to the decontamination and sterilisation of re-usable surgical devices may not be wholly effective. While photocatalytic coatings have been reported to be 'self-cleaning' and even commercialised for this purpose e.g. Pilkington Activ self cleaning glass [10], there are few published reports dealing with photocatalytic decontamination of protein from surfaces.

In this work TiO₂ thin films were prepared using a sol gel route. Titanium IV butoxide was hydrolysed under controlled conditions with acetic acid as a catalyst. The resulting sol gel was spin coated onto glass slides and then annealed. Raman spectroscopy analysis and glancing angle XRD confirmed the presence of anatase crystal phase. Samples were contaminated with fluorescein-isothiocyanate labelled bovine serum albumin (FITC-BSA) as a model contaminant. A series of 1 μL drops of different concentrations of FITC-BSA were casted onto slides and allowed to dry. The glass slide was two-thirds coated with TiO₂, using the sol gel route, so that control spots (no TiO₂) were on the same slide. Slides were irradiated for 0 h, 6 h, 12 h and 24 h with a UVA source. Humidity was not controlled in these experiments. The fluorescence intensity measured (Fluoromax – P) and the relative protein concentration was calculated from a calibration response. Figure 2 shows the decrease in fluorescence intensity with increasing UV exposure time. These results demonstrate the photocatalytic degradation of FITC-BSA. Experiments are ongoing investigating the degradation of other protein species which are resistant against physico-chemical or biochemical treatment.

4.0 TiO₂ electrodes as transducers for biosensors utilising oxidase enzymes

Commercially available glucose biosensors utilise glucose oxidase (GOD) as the biorecognition component which is wired to the transducer

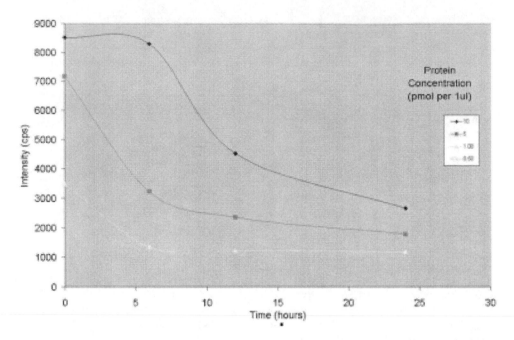

Figure 2. Photocatalytic degradation of fluorescein-isothiocyanate labelled bovine serum albumin on TiO_2 coated glass

using a mediator. However, mediated glucose biosensing may not be suitable for in-vivo sensing as mediators may be toxic or simply lost into the blood. Non-mediated glucose biosensing is possible using materials which can either be directly wired to the enzyme or which can selectively detect the H_2O_2 product in the presence of oxygen. There is an opportunity to produce a wide range of biosensors, utilising oxidase enzymes.

The use of mesoporous titanium dioxide electrodes has been reported previously for the amperometric detection of glucose via electro-reduction of released hydrogen peroxide. [11]

Electrophoretic coating may be used to produce porous nanocrystalline TiO_2 electrodes [8,12]. These electrodes have been tested for the electrochemical reduction of H_2O_2 in the presence of O_2 and the response is independent of O_2 at potentials more positive than –0.4 V vs the saturated calomel electrode (SCE). Titanium foil samples were coated with nanoparticle TiO_2 by the electrophoretic method. Electrical contact was made to an area of the Ti foil not coated with TiO_2 using copper wire and conducting epoxy. The contact and any remaining uncoated foil area were insulated using a negative photoresist.

All electrochemical analyses were carried out using a three-electrode electrochemical cell, with pH 6 phosphate buffer as the supporting electrolyte. The reference electrode was a

saturated calomel (SCE), and the counter electrode was a platinum disc. Amperometric detection was carried out using a BAS LC-4C amperometric detector connected to a Lloyd instruments PL3 x-y plotter. All potentials are reported against SCE. Glucose oxidase was added to the buffer (air sparged) in free suspension, and the electrode response was measured at –0.4 V. Standard additions from a stock solution of glucose were made to the cell and the steady state current measured. The steady state current as a function of glucose concentration is shown in figure 3. The GOD oxidises the additions of glucose to produce H_2O_2, which is then reduced at the TiO_2 electrode. The electrode response was directly proportional to the glucose concentration over the 2.0 to 20.0 mM glucose.

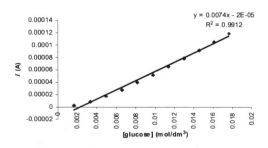

Figure 3. Electrode response to standard additions of glucose. 10 mg GOD was present as free enzyme in 30 cm³ pH 6 phosphate buffer. The TiO_2 electrode was held at a fixed potential of -0.4 V.

5.0 Electrochemical growth of TiO₂ nanotubes on titanium metal foil

In this work the effect of HF concentration and anodisation potential were investigated. Anodisation was carried out in a one compartment cell with a titanium foil anode platinum foil cathode. Constant potential conditions were employed. The total cell volume was 100 cm³ and electrode separation was 20 mm. The salient parameters investigated were HF concentrations 0.005%-0.5% w/v and cell potentials in the range 5.0 to 30 V. Above concentrations of 0.15% HF the Ti foil dissolved rapidly. The optimum HF concentration was found to be 0.05 % w/v.

In this system, only cell potentials above 15 V produced the nano-structuring effect and potentials greater than 30 V destroyed the nanotube formation. Sample morphology was examined using an FEI quanta SEM at an accelerating voltage of 30 kV and a beam current of 47 pA. The mean tube diameter was ca. 80 nm for the samples prepared at a cell potential of 25 V in 0.005%w/v HF (see figure 4)

Work is ongoing investigating the potential uses of TiO₂ nanotubes for photocatalysis, biosensing and other applications

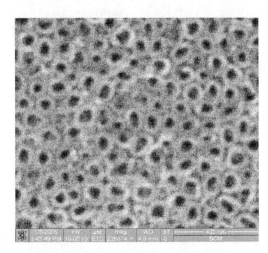

Figure 4. SEM of TiO₂ nantubes grown electrochemically on Ti foil.

7.0 Acknowledgements

We would like to thank the following; DEL NI for funding Rankin, Dale, Al Rousan, and Donaldson. J Dooley and C Lowrey, UU, for analysis of Cryptosporidium. University of Edinburgh for FITC-BSA and analysis. Dept of Health UK, R&D Office HPSS NI, Invest Northern Ireland and European Commission for funding. Degussa for samples of P25. Philips lighting, Netherlands, for light sources.

7.0 References

1. D Gong, G.A. Grimes, O.K. Varghese, W. Hu, R.S. Singh, A. Chen, E.C. Dickey, "Titanium oxide nanotube arrays prepared by anodic oxidation," *J.Mater.Res.*, 2001, **16**, 3331-3334.
2. Boucher, "Cryptosporidium in Water Supplies. Third Report of the Group of Experts" (Department of Health/Department of the Environment, Published by HMSO, 1998).
3. Badenoch, "*Cryptosporidium* in Water Supplies. First Report of the Group of Experts" (Department of Health/Department of the Environment. Published by TSO Ltd, 1990)
4. H. V. Smith, J. B. Rose, "Waterborne cryptosporidiosis: Current status", *Parasitology Today*, 1998, **14**, 14-22
5. W. R. Mackenzie, N. J. Hoxie, M. E. Proctor, M. S. Gradus, K. A. Blair, D. E. Peterson, J. J. Kazmierczak, D. G. Addiss, K. R. Fox, J. B. Rose, J. P. Davis, "A massive outbreak in Milwaukee of *Cryptosporidium* infection transmitted through the public water supply", *New England Journal of Medicine*, 1994, **331**, 161-167
6. H. Zheng, P. C. Maness, D. M. Blake, E. J. Wolfrum, S. L. Smolinski, W. A. Jacoby, "Bactericidal mode of titanium dioxide photocatalysis," *Journal of Photochemistry and Photobiology A-Chemistry*, 2000, **130**, 163-170.
7. D. M. Blake, P. C. Maness, Z. Huang, E. J. Wolfrum, J. Huang, W. A. Jacoby, "Application of the photocatalytic chemistry of titanium dioxide to disinfection and the killing of cancer cells," *Separation and Purification Methods*, 1999, **28**, 1-50
8. J. A. Byrne, B. R. Eggins, N. M. D. Brown, B. McKinney, M. Rouse, "Immobilisation of TiO₂ powder for the treatment of polluted water," *Applied Catalysis B-Environmental*, 1998, 17, 25-36
9. L.J. Robertson, A.T. Campbell, H.V. Smith, "Viability of Cryptosporidium parvum oocysts: Assessment by the dye permeability assay," *Applied and Environmental Microbiology* 1998, **64** (9) 3544
10. Ashton V., "Windows on the future." *Chemistry in Britain*, 2002, **38**(6): 26-28.
11. S Cosnier, C Gondran, A Senillou, M Gratzel, N Vlachopoulos, "Mesoporous TiO₂ Films: New Catalytic Electrode Materials for Fabricating Amperometric Biosensors Based on Oxidases", *Electroanalysis*, 1997, **9(18)**, 1387-1392.
12. J A Byrne, B R Eggins, S Linquette-Mailley and P S M Dunlop, "The effect of hole acceptors on the photocurrent response of particulate titanium dioxide", *Analyst*, 1998, **123**, 2007-2012.

CNT Composites for Aerospace Applications

S Bellucci[1], C Balasubramanian[1,2], P Borin[1,3], F Micciulla[1,3], G Rinaldi[4]

[1] INFN-Laboratori Nazionali di Frascati, Via E. Fermi 40, 00044 Frascati, Italy
[2] Department of Environmental, Occupational and Social Medicine, University of Rome Tor
  Vergata, Via Montpellier 1, I-00133 Rome, Italy
[3] University of Rome "La Sapienza", Department of Aerospace and Astronautics Engineering Via
  Eudossiana 18, 00184 Roma, Italy
[4] University of Rome "La Sapienza", Department of Chemical Engineering and Materials, Via
  Eudossiana 18, 00184 Roma, Italy
bellucci@lnf.infn.it, bala@lnf.infn.it, borin@pd.astro.it, starfederico@inwind.it,
gilberto.rinaldi@ingchim.ing.uniroma1.it

ABSTRACT

Carbon nanotubes were synthesized by thermal arc plasma process after optimization of the synthesis parameters. These samples were then analysed by Scanning and Transmission electron microscopes (SEM and TEM), in order to establish the morphology of the nanostructures. Atomic Force Microscopy (AFM) and Electron diffraction studies were also carried out before using the sample for the composite material preparation. Composites of epoxy resin with curing agent as well as a mixture of graphite and carbon nanotubes were prepared with varying proportions of the mixture. The electrical resistivity of the material was studied under varying pressure and voltage conditions. Preliminary results of these studies present interesting features which are reported here.

1. INTRODUCTION

The study of nanotubes has advanced tremendously in a relatively short span since its first discovery in 1991 by Iijima [1]. The properties of these nanostructures are so unique and enhanced that it is finding applications in various spheres of life – right from bio-medical to optical and to space applications [2-12].

Essentially two families of carbon nanotubes exist: SWNT or (single wall nanotubes), that are constituted by only one rectilinear tubular unity and the other MWNT (multi wall nanotubes}, that are constituted by a series of coaxial SWNT. Though generally both the types have high aspect ratio, high tensile strength, low mass density, etc. the actual values could vary depending on whether it is SWNT or MWNT. Of the two types SWNT is better suited for mechanical applications.

Owing to their exceptional morphological characteristics, electric, thermal and mechanical, carbon nanotubes yield a material particularly promising as reinforcement in the composite materials with metallic matrixes, ceramics and polymers.

The key factor in preparing a good composite rests on good dispersion of the nanotubes, the control of the bonding between nanotubes and matrix, the density of the composite material [2]. Besides, the type of nanotubes (SWNT, MWNT) the synthesis mode (arc discharge, laser, CVD) etc are important variables since they determine the perfection of the structure and the reactivity of the surface.

2. EXPERIMENT

Carbon nanotubes were synthesized in a DC arc plasma system in helium atmosphere at a pressure of 600 torr. Arc was struck between two electrodes consisting of a high purity graphite rod and a block of graphite. The discharge is typically carried out at a voltage of 20V and a current in the range of 80 – 100 A. Some amount of the evaporated carbon condenses on the tip of the cathode, forming a slag-like hard deposit. The deposit essentially in the cathode consists of bundles of carbon nanotubes mixed with small quantity of amorphous carbon.

The as-synthesised samples were characterized by means of SEM, TEM and AFM. Figures 1 and 2 show SEM images.

Cleantech 2007, ISBN 1-4200-6382-0

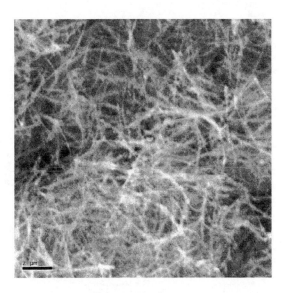

Fig.1 : SEM images of CNT with arc discharge.

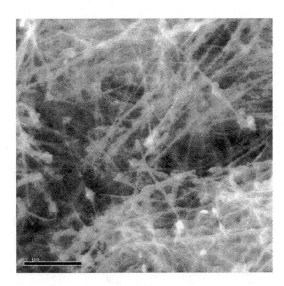

Fig. 2: SEM image of CNT which shows straight and long nanotubes

2.1 Nanotubes Composites

Due to the unique properties of carbon nanotubes they are being widely studied as a constituent of composite material. CNT based composite materials are increasingly being considered for mechanical, electrical and space applications. Even studies on biosensor composites based on functionalized nanotubes and nanoparticles are reported [9-12]. They are also being studied for the suitability and applications in aerospace and aeronautical fields. A prospective application in aerospace that we are studying is the improvement of electrical properties of composites made from carbon nanotubes and epoxy resin [13-15]. To start with it was decided to mix the epoxy resin with graphite. The purpose was to make a light, thin and mechanically strong composite material to cover electric circuits against external electromagnetic interference. This is very important for air and space crafts.

The epoxy resin that was used is a commercial Shell product Epon 828. Two types of curing agent were used along with the resin; mainly A1 curing agent and PAP8 agent. Also some of the resin+curing agent samples were mixed with 20 wt% of graphite and these were used for the analysis of the electrical resistivity studies. We stress that the first curing agent possesses polar groups in its chemical composition, whereas the second agent contains benzene groups. As a consequence, the mechanical properties of composites where the PAP8 agent has been used turn out to be improved [16]. However, the stability of the mechanical properties, under varying pressure conditions, as well as the corresponding resistivity behavior, has not been investigated yet. In the present work, we fill up this gap, in the part concerning the electrical transport properties.

The composite was made by manually mixing the micron sized (particle size ~ 20 microns) graphite powder in the resin+curing agent. Care was taken to avoid air bubbles in the mixture. The experiments were performed in two stages : Initially two types of resin with curing agents were used to find the one most suitable for the earlier defined applications. In the second stage this resin was mixed along with the CNT to study the change/ enhancement of the electrical property.

In order to comply with the standard specification of the U.S. military authorities, we tested the electrical properties of the composite materials, making use of "Y" shaped electrical circuits having two parallel lines as the tail of the "Y" with 1 mm gap between them and a length of about 2.5 cms. The circuits were made on a PC base with silver print and the two arms of the "Y" were connected to the picoammeter and the high voltage supply. The composite mixtures were spread, like thin films, on the circuit and electrical resistance tests were carried out using Keithley 6485 Picoammeter with short circuit protection.

The current through the sample was recorded for three different applied DC voltages, namely – 200, 500 and 1000 V. The resistance and the resistivity were then calculated. The experiment was repeated under three different pressures – atmospheric, 10^{-2} and 10^{-6} mbar. The low pressure measurements gave indirectly the effect of moisture on the resistivity values of the samples. The plots in figures 3, 4, 5 and 6 show the resistivity *vs* applied voltages for various samples under varying voltage and pressure conditions.

3. RESULTS AND DISCUSSION

A. Studies with resin and graphite

Analysing the data it is observed that the resistivity of samples with curing agent A1 is found to be a few times lower than the samples with curing agent PAP8. It is important to note that the absolute change in resistivity is less over a wide voltage range of 200 to 1000 volts for the sample with A1 curing agent (as seen from figures 3 & 4), whereas for the sample with PAP8 curing

agent the resistivity changes marginally more with increasing voltage.

Notice that the resistivity data were collected with the same samples at two different times of the year (i.e. July 2005 and September 2005), in order to have a rough estimate of the influence of climatic and environmental conditions on their performance. It appears, from a preliminary analysis of our data, that the stability of composites employing PAP8 agent is jeopardized by the addition of graphite. In the case of composites with A1 curing agent the behavior is quite opposite, i.e. the stability of the material increases as graphitic additions are included. This seems to favor the use of A1 curing agent from the point of view of the optimization of the aerospace applications sought for.

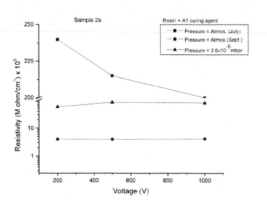

Fig. 3: Plot of resistivity vs. voltage for the sample Resin+A1 with no graphite added

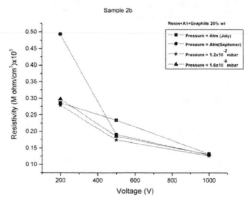

Fig. 4: Plot of resistivity vs. voltage for the sample Resin + A1+graphite added.

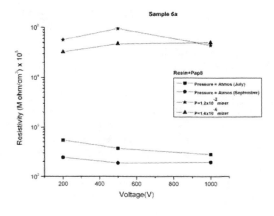

Fig. 5: Plot of resistivity vs. voltage for the sample Resin + PAP8

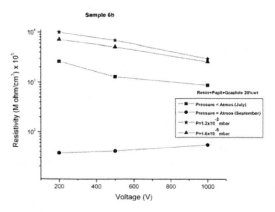

Fig. 6: Plot of resistivity vs. voltage for the sample Resin + PAP8 + graphite added

3.1 Variation of resistivity with pressure/humidity

It is expected that when the ambient pressure is decreased while doing the resistivity measurements the humidity also gets decreased resulting in higher resistivity values. The resisitivity values for all the samples show some variation when done in atmosphere as compared to when done in low pressure. However, this variation gets reduced when graphite is added to the resin.

From the plots above we observe that, for the first sample (i.e. sample with curing agent A1 – figures 3 and 4), when we work in different pressure conditions, the resistance changes very little Instead in the second sample, the resistance undergoes remarkable variations when we work in different pressure and humidity conditions, as seen in (figures 5 and 6). This feature might constitute a drawback for the use of the corresponding curing agent PAP8 for composite devices working under standard aerospace conditions, where the values of the pressure can undergo substantial variations.

3.2 Variation of resistivity with graphite addition

It is observed that the resistivity change is very large – near about 3 orders of magnitude when 20% graphite is added to the resin + A1 curing agent, whereas for the PAP8 curing agent the increase in resistivity due to addition of graphite is comparatively only marginal, of about 3 to 5 times.

These above results when considered in totality gives a broad spectrum wherein we find that the resin + A1 curing agent + graphite seems to be an ideal candidate for applications in various pressure ranges as well as voltage ranges. The Resin + A1 + graphite has the lowest changes in the resistivity values for voltages from 200 to 1000 V and also for a pressure difference of atmospheric to 10^{-6} mbar

B. Studies of Resin with CNTs

Resistivity measurements were performed for composites with A1 resin in combination with carbon nanotubes (shown in Fig. 1). Composites were made replacing graphite with CNTs. The quantity of CNTs added was 0.5 wt% of the resin mixture. Figure 7 shows the plot of resistivity vs. voltage for this sample. As can be observed the resistivity values changes drastically with the addition of a small quantity of CNTs. The Resin A1 with no graphite or CNT has a resistivity in the range of few tens of M ohms ($\times 10^5$) /cm^3 whereas when 0.5 wt% of CNT is added the resistivity reduces by a factor of 10^3 to values ranging from 0.01 to 0.04 Mohms ($\times 10^5$)/cm^3. Also when these values are compared with the composite of resin A1 with graphite (refer Fig. 4), we observe that the resistivity 20 wt% of graphite is ten times higher than the addition of a small fraction of CNT.

Fig. 7 : Plot of resistivity vs voltage for composites of Resin A1 with CNT

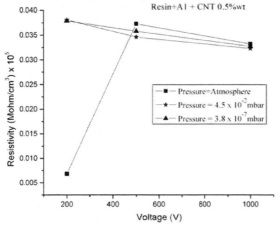

4. CONCLUSIONS

Resistivity studies were performed on the composite material made from a resin with different curing agents (namely A1 and PAP8). Comparison was also made of composite of graphite against composites made from carbon nanotubes. The results of these studies present interesting features which are useful in choosing the ideal composition and ratio of the composite material for use in shielding of electrical circuits of space vehicles from radiations of the outer space.

We can deduce that the PAP8 curing agent is very sensitive to the humidity variations over a long time period and cannot optimize the performance of the circuits. Instead the A1 curing agent in combination with 20 wt% graphite microparticles is not very sensitive to these variations and makes the behaviour of an electrical circuit more stable.

The same resing A1 when combined with carbon nanotubes, in place of graphite powder, yielded resistivity values which were orders of magnitude better than either with plain resin or with large quantity of graphite powder.

REFERENCES

[1]   Iijima S., *Nature* (London) **354**, 56 (1991).

[2]   Ahn C H et al. 2004 Springer Handbook of Nanotechnology, Bharat Bhushan Editor.

[3]   S. Bellucci, *Phys. Stat. Sol. (c)* **2**, 34 (2005).

[4]   S. Bellucci, *Nucl. Instr. Meth. B* **234**, 57 (2005).

[5]   S. Bellucci, *Atti XVII Congresso AIV*, Ed. Compositori (2005), p. 61, ISBN 88-7794-495-1.

[6]   S. Bellucci, C. Balasubramanian, F. Mancia, M. Marchetti, M. Regi, F. Tombolini, *3rd International Conference on Experimental Mechanics*, Proc. of SPIE Vol. 5852 (2005), p. 121, ISSN 0277-786X, ISBN 0-8194-5852-X.

[7]   S. Bellucci, *CANEUS 2004-Conference on Micro-Nano-Technologies*, Nov. 2004, Monterey, CA, USA, AIAA paper 2004-6752.

[8]   S. Bellucci, C. Balasubramanian, G. Giannini, F. Mancia, M. Marchetti, M. Regi, F. Tombolini, *CANEUS 2004-Conference on Micro-Nano-Technologies*, Nov. 2004, Monterey, CA, USA, AIAA paper 2004-6709.

[9]   Bottini M., Bruckner S., Nika K., Bottini N., Bellucci S., Magrini A., Bergamaschi A., Mustelin T., *Toxicology Letters*, **160**, 121 (2006).

[10]   M. Bottini, L. Tautz, H. Huynh, E. Monosov, N. Bottini, M. I. Dawson, S. Bellucci, T. Mustelin, *Chem. Commun.* **6**, 758 (2005).

[11]   M. Bottini, C. Balasubramanian, M. I. Dawson, A. Bergamaschi, S. Bellucci, T. Mustelin, *J. Phys. Chem. B* (in press).

[12]   M. Bottini, A. Magrini, A. Di Venere, S. Bellucci, M. I. Dawson, N. Rosato, A. Bergamaschi, T. Mustelin, *Journal of Nanoscience and Nanotechnology* (in press).

[13]   Lau K-T and Hui D 2002 "*The revolutionary creation of new advanced materials - carbon nanotube composites*", Composites: Part B 263-277.

[14]   M.J. Biercuk et al., *Appl. Phys. Lett.*, **80**, 2767 (2002).

[15]   S. Narasimhadevara et al., "Processing of carbon nanotube epoxy composites", Masters Thesis, University of Cincinnati.

[16]   G. Rinaldi, *"Materiali per l'ingegneria"*, Ed. Siderea, Roma.

Advancements in the Supercritical Water Hydrothermal Synthesis (scWHS) of Metal Oxide Nanoparticles

Edward Lester[*a], Paul Blood[a,b], Jun Li[a,b] and Martyn Poliakoff[b]

[a]School of Chemical Environmental and Mining Engineering (SChEME)

[b]Clean Technology Research Group, School of Chemistry

University of Nottingham, University Park, Nottingham, NG7 2RD, UK

Edward.Lester@nottingham.ac.uk

ABSTRACT

Supercritical Water Hydrothermal Synthesis (scWHS) is a relatively simple and environmentally friendly process for the production of potentially valuable metal oxide nanoparticles. Previous problems with blockages forming in the original T piece reactor were overcome by redesigning the reactor using image analysis and computational fluid dynamics. An optimised reactor, termed the *Nozzle Reactor*, has been developed which can be run continuously and is able to produce a range of different metal particles including titania, ceria, zirconia, copper oxide, YAG, hematite, magnetite and silver. The reactor also shows a dramatic improvement in process reproducibility ($\pm$ 5m^2/g for BET surface area) and in control of particle size. Preliminary evidence suggests that the reactor could eventually lead to the ability to good control of particle properties, such as size, composition and shape, through the manipulation of process variables.

Keywords: supercritical water, metal oxides, image analysis, reactor design.

1 INTRODUCTION

Nanoparticulate metal oxides and metals are finding increasing applications in areas as diverse as sun blocks, electro-conductive printing inks, electronic displays, pigments and catalysts. Many of the routes to such particles involve relatively noxious chemicals, are not easily scalable, have a complex and time-consuming sequence of stages, or may require expensive precursors. These methods include sol-gel[1] (aerogels and xerogels), metal-atom aggregation in cryogenic inert gas matrices[2], thermal or ultrasonic decomposition of metal carbonyls, reduction of metal ions[3] semiconductor particles, zeolites and inverse micelles[4]. By contrast, Supercritical Water Hydrothermal Synthesis (scWHS) offers a relatively simple route which is inherently scalable and chemically much more benign. Mixing supercritical water with a metal salt causes precipitation of a metal oxide after a hydrolysis and then a dehydration stage. The difficulty with continuous operation of the scWHS process is in the maintaining of complete control over mixing, in order to prevent blockages, and over product quality, in terms of particle size and size range. Experiments to visualise fluid mixing were undertaken in order to understand this mixing process more fully, and also to explain why the T-Piece reactor originally used at Nottingham was prone to blockages[5].

2 MODEL MIXING

The two fluids were carefully selected to model the properties of the two streams in the continuous scH$_2$O reactor. Methanol (Pseudo-scH$_2$O) was chosen to represent scH$_2$O whereas the aqueous metal salt stream was modelled by a 40 % w/w aqueous sucrose solution (Pseudo-metal salt). The density and viscosity ratios between these two fluids are similar to those between scH$_2$O and the pressurized aqueous metal solution stream in the actual supercritical system. Methylene Blue dye was used at a near-saturation concentration of 12 ppm to colour the sucrose solution. All modelling experiments were initiated by feeding the blue sucrose solution into a flowing MeOH environment thus recreating the scenario in the supercritical reactor (the metal salt is introduced into a flowing scH$_2$O environment).

The results illustrate the path taken by the pseudo-metal salt and highlight any zones of strong mixing/turbulance within the supercritical water reactor. The reactor model of the T-piece reactor was constructed out of polished transparent acrylic resin, with a constant internal tube diameter of 3.8 cm. Corresponding flow rates in the pseudo-rector were calculated so that the inlet Reynolds numbers of the Methanol and Sucrose streams were close to those exhibited in the actual scH$_2$O system.

Figure 1 shows how the system was arranged and Figure 2 shows the mixing in a T piece that approximated to the actual supercritical water reactor.

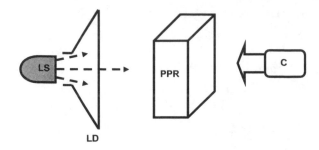

Figure 1: The modeling setup used to video fluid mixing in the pseudo reactor: Key: LS – light source (500W), LD – light diffuser, PPR – Perspex pseudo reactor, C – camera.

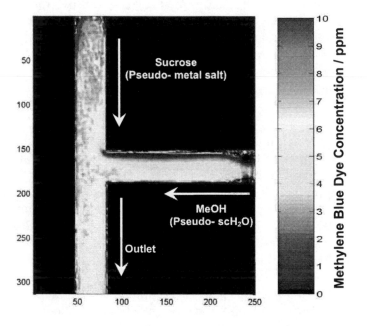

Figure 2: The mixing regime inside the T piece reactor

This experiment demonstrated clearly why blockages were occurring readily in the T piece system. Note fluid partitioning in the side arm where the supercritical water is introduced and the premixing in the metal salt inlet.

Many other experiments were carried out on different geometries and different flow rates including T piece, Y piece, tangential side swirl and countercurrent tangential swirl reactors. Each reactor had numerous orientations and flow arrangements. Each experiment highlighted the difficulties of mixing two fluids with significant differences in density and therefore buoyancy. The optimal mixing geometry would clearly need

- *Instantaneous strong and uniform mixing of two reactant streams – to give steady state*

operation and to aid in the instantaneous formation of many small metal oxide nuclei which is desirable for small particle formation

- *Very short average residence time combined with a narrow residence time distribution* - to minimise the subsequent particle growth and narrow particle size distribution

- *Minimal heating of aqueous metal salt stream prior to the reactor, followed by immediate and rapid heating within the reactor* – to prevent precipitation/deposition of metal salts in the pipes prior to the reactor

- *Strong net downstream flow/eddies for the rapid transport of product particles out of the reactor* – to prevent particle accumulation within the reactor and to minimise subsequent particle growth

3 THE NOZZLE REACTOR

Figure 3 illustrates the nozzle reactor [6,7]; with its pipe-in-pipe concentric arrangement in which the internal pipe has an open-ended nozzle with a cone attached. The supercritical water is fed downwards through the internal pipe and out the end of the 'Nozzle'; the aqueous metal salt steam is fed counter-currently upwards through the outer pipe. The reactor outlet is situated upwards through the outer pipe.

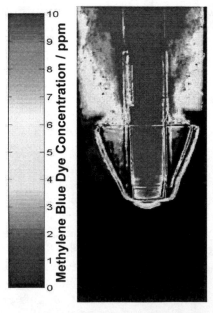

Figure 3: The nozzle reactor mixing supercritical water (down the innerpipe) with metal salt (up the outer pipe)

As the two reactant fluids are introduced, the mixing is instantaneous and strong. The resultant turbulent macro-mixing eddies are streamlined downstream to the outlet of the reactor. No contamination, and therefore reaction, occurs within the reactor inlets, nor are there any areas of poor net flow ('stagnant zones'). The strong downstream macro-eddies in the Nozzle Reactor are advantageous for two reasons: i) they result in uniform and high net flow through the reactor to its outlet and, therefore, a relatively short residence time; ii) the strong downstream eddies also aid in transporting of the particles out of the reactor. Both of these advantages prevent accumulation in the reactor and minimise particle growth.

The metal stream can also be kept below 40°C up until it is contacted with the supercritical water stream within the reactor; hereby preventing precipitation of the metal salt prior to the reactor, which is a common problem during scWHS. In the Nozzle Reactor, there is no upstream mixing inside either inlet since the density difference between the two reactants can only induce eddies streamed one-way, towards the reactor outlet. Therefore, the metal salt stream will remain 'cold' until it is mixed with the supercritical water within the reactor; the resultant mixing is almost instantaneous, resulting in rapid heating of the metal salt. This fast heating/fast mixing scenario, may have potential in many other supercritical water processes. The efficiency of this heat transfer profile can be increased with the addition of extra cooling and heating. A cooling jacket can be attached to the metal salt inlet to prevent the conduction of heat down the pipes from the reactor and a band heater to the area surrounding the nozzle to maintain and control the reaction temperature.

4 NANOPARTICLES FROM THE NOZZLE REACTOR

4.1 Iron Oxide
Hematite (Fe_2O_3) nanoparticles can exhibit superparamagnetic behaviour and improved coercivity. Their potential applications include M.R.I. contrast agents, recording media, catalysts, pigments, targeted drug delivery and sensors. With iron nitrate as the metal salt precursor at a concentration of 0.05M, the size of product can be controlled between 5nm to 50nm by adjusting the water feed temperature. Figure 4 shows the relationship between particle size and operating temperature.

4.2 Copper Oxide
Copper oxide (mixed CuO and CuO_2) particles can be used as an antimicrobial agent, in wood preservation, conductive inkjet printing and in pigments. By using copper formate as the metal salt precursor at a concentration of 0.01M, nanoparticles around 50nm can be produced continuously (see Figure 5).

4.3 Cobalt Oxide
Co_3O_4 has potential application in biomedicine, catalysis and recording media[8]. Operating at 415°C using cobalt acetate tetrahydrate as the organic metal salt precursor. Figure 6 shows a TEM image of the particles around 20-30nm.

4.4 Cadmium Sulfide
CdS has potential applications in photochemical catalysis, non-linear optics, infrared detectors, and photosensitizing sensors. Using cadmium nitrate and thiocarbohydrate as the two precursors and operating at 200°C, 10nm particles can be produced (Figure 7)

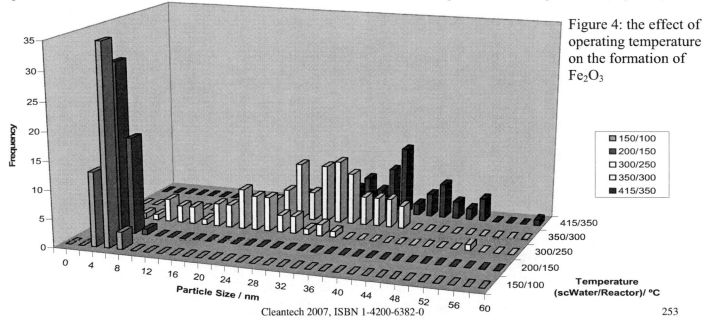

Figure 4: the effect of operating temperature on the formation of Fe_2O_3

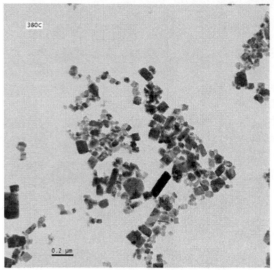

Figure 5: copper oxide nanoparticles. Note lower concentrations can be used to reduce particle size.

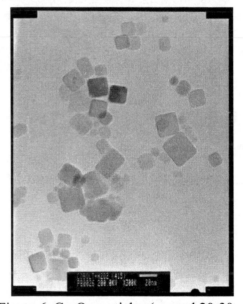

Figure 6: Co$_3$O$_4$ particles (around 20-30nm)

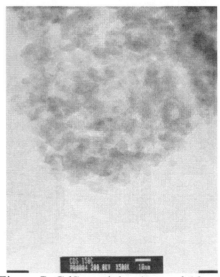

Figure 7: CdS particles (around 10nm)

5 CONCLUSIONS

The Nozzle Reactor is an optimised design specifically developed for the supercritical water hydrothermal synthesis of nanoparticles. The nozzle reactor takes advantage of differential buoyancy forces to produce an ideal steady-state mixing profile.

The application of the nozzle design on the actual scWHS rig has resulted in a significant improvement in both product reproducibility and process reliability. We are in the position where we can investigate the process chemistry and the influence of process variables effectively. It is capable of producing highly useful and valuable nanoparticles such as silver, copper oxide, cobalt oxide, YAG, CdS and iron oxide in a clean, flexible and relatively cheap process.

The large selection of particles produced since the implementation of the nozzle design is perhaps a hint of this process's commercial potential. Further work is required into the real application, treatment and capping of these particles.

6 ACKNOWLEDGEMENTS

Particular acknowledgement is made to the Engineering and Physical Sciences Research Council and to the Strategic Technology Group at ICI, Wilton.

7 REFERENCES

[1] I. Kartini, P. Meredith, X. S. Zhao, J. C. D. da Costa, G. Q. Lu, *Journal of Nanoscience and Naotechnology* 2004, *4*, 270

[2] K. J. Klabunde, C. Mohs, *Chemistry of Advanced Materials*, Wiley, New York, 1998.

[3] P. Mulvaney, H. A., *J. Phys. Chem.* 1990, *94*, 4183.

[4] Y. Wang, N. Herron, *J. Phys. Chem.* 1988, *92*, 4988.

[5] P. J. Blood, J. P. Denyer, B. J. Azzopardi, M. Poliakoff, E. Lester, *Chem. Eng. Sci.* 2004, *59*, 2853.

[6] E. Lester, P. Blood, J. Denyer, D. Giddings, B. Azzopardi, M. Poliakoff accepted for publication in J. Supercritical Fluids doi:10.1016/j.supflu.2005.08.011

[7] E. Lester, B. Azzopardi, Countercurrent mixing device, World Patent WO2005077505 - 2005-08-25

[8] L. Cote, A. S. Teja, A. P. Wilkinson, Z. Zhang, *J. Mater. Res.* 2002, *17*, 2410

Monometallic nano-catalysts for the reduction of perchlorate in water

D. M. Wang, H. Y. Lin and C. P. Huang[*]

Department of Civil & Environmental Engineering
University of Delaware, Newark, DE, 19716
[*]Corresponding author: huang@ce.udel.edu

ABSTRACT

Perchlorate removal in a clean, cost-effective and publicly acceptable approach is one of the important issues in current drinking water treatment practice. Catalytic membrane (CM) was prepared by coating monometallic catalysts of the nano-size onto supports by chemical or electrochemical method. The support materials were stainless steel and graphite. Nano-catalysts were made of transitional metals from the first, the second and the third row of the periodic table. The CM was characterized by surface analysis techniques including SEM, XPS and BET. The CM was used as cathode where the reduction of perchlorate occurred through hydrogen atoms that were generated on the CM surface. All experiments were performed at ambient conditions. It was found that for the first time perchlorate could be reduced readily by hydrogen atoms in the presence of catalyst. At a maximum perchlorate concentration of 100 mg/L, it is possible to achieve a > 90% removal in 8 h using nano-catalysts such as Sn, Ti, and Co. The rate constants were between 5.1 and 9.6 μM-L^{-1}-hr^{-1} among the 18 different monometallic nano-catalysts tested. Chloride was the major end product, whereas a small quantity of chlorite was observed in the presence of Co catalyst. Membrane coated with metallic nano-catalysts at different mass showed different reduction rate, e.g., the optimum surface coverage for Sn was 0.6 to 0.7 mg per gram stainless steel membrane.

Keywords: monometallic catalyst, nano-size, perchlorate reduction, electrochemical, hydrogen atoms

1 INTRODUCTION

Perchlorate salts, such as ammonium, potassium, and sodium, are important national strategic chemicals [1]. Perchlorate in the environment is mostly contributed from production of explosives, pyrotechnics and blasting formulations [2, 3] in addition to naturally occurring to a limited extent [4].

Perchlorate ion persists in the environment for many years [5, 6]. It is now known that perchlorate is present in groundwater and surface water of more than 20 US states. The detection of perchlorate at elevated concentrations in drinking water supplies in Nevada, Utah, and California in the late 1990s has prompted interest and concern of this new chemical species in the environment.

Perchlorate is a stable ion that may affect humans and other animals through multiple pathways of exposures. The toxicological mechanism through which perchlorate exerts its effects have been reviewed in some detail [7, 8, 9]. The U.S. EPA named perchlorate a contaminant of concern after studies linked the chemical to thyroid disorders and other potential health problems.

Two major technologies were currently suggested for the treatment of large-volume water that contains perchlorate: ion exchange and biological processes or their combinations. Other emerging or studied technologies such as membrane separation, electrodialysis, adsorption and chemical process are not being applied at field treatment facilities.

None of these treatment technologies is feasible for real-world applications due to process cost, operation difficulties, relatively low removal efficiency, difficult in maintenance, low process flexibility, low public acceptance and subsequently low political buyout. More researches are needed for a cost-effective, high efficient and public-acceptable removal process.

2 EXPERIMENTAL

2.1 Materials

Perchlorate solution was synthesized by dissolving proper perchlorate salt into distilled water. Distilled water was prepared in the laboratory by a water-purification system (Mega-Pure System, Model MP-290). Sodium perchlorate (> 95%), ammonium perchlorate (> 98%), potassium chlorate (> 98%), sodium chlorite (> 80%) and sodium chloride (>99.5%) were purchased from Sigma-Aldrich (Allentown, PA). All chemicals used for the preparation of catalytic membranes, were supplied by different companies, namely, cadmium (II) perchlorate hydrate (> 99.9%), chromium (III) perchlorate hexahydrate, cobalt (II) nitrate (> 98%), cupric sulfate (> 99%), lead chloride (> 98%), manganese chloride (> 97%), molybdenum (II) acetate (> 98%), palladium chloride (>98%), platinum chloride (> 98%), rhodium (III) chloride hydrate (> 99%), ruthenium chloride hydrate (> 98%), scandium perchlorate (40%), stannous (II) chloride(> 99%), titanium(III) oxide(> 99.9%), vanadium (II) chloride (95%), zirconium(IV) chloride (>99.5%) from Sigma-Aldrich Company (Allentown, PA); Nickel (II) perchlorate (> 99%) from Johnson Matthey Company, (Wayne, PA) and zinc chloride (> 99%) from the Fisher Scientific International Inc.. All chemicals were used as received.

Stainless mesh (openings 0.1 mm, thickness 0.25mm), was purchased from InterNet (Anoka, MN) and TWP Inc. (Berkeley, CA). Ion-exchange membrane (AMX) was purchased from Tokuyama Soda Inc. (Burlingame, CA).

2.2 Preparation of Catalytic Membrane

The raw membrane was first cut into small pieces (40 mm x 40 mm) upon receive in the laboratory. The membranes were then hanging one by one on a mixing motor rotating at a low speed of < 100 rpm while being immersed in the sulfuric solution of 0.01M for cleaning. After cleaning, the membranes were rinsed with DIW water. Each membrane was dried in a dryer for 1 hour and weighted before coating of catalyst. The metal ion solutions were prepared by dissolving the appropriate chemicals in DIW water at a concentration range of 0.01 mM to 0.1 M depending on the solubility. The washed membrane was hung again on the same motor, and immersed in the metal ion solution completely. An electroplating system was set up with the membrane as cathode and graphite as anode. When the motor began to rotate at a speed of 20 rpm, the potential was applied on the system with a constant current of 20 to 100 mA. After electroplating, the catalytic membrane was washed and dried before weighing. The catalyst density was determined from the amount of metal deposited on the membrane and the weight of the tare membrane. The images of the catalytic membrane was observed using SEM (JOEL 7400F) as to reveal the morphology and the degree of surface coverage of the metallic catalysts on surface of the inorganic membrane.

2.3 Reduction of perchlorate on membrane system

The main reduction chamber of the system was a glass connector (ACE Glass Company, Wilmington, DE). The catalytic membrane was attached at the opening bottom of the glass connector. The catalytic membrane behaved as a cathode, which was prepared according to procedures described above. An anode (e.g., iron rod) was placed in the glass connector as a counter electrode. The electrochemical reaction producing hydrogen atoms was driven by a power supply (Model FB1000, Fisher Scientific International Inc., USA). A second power supply (Model WP705B, Vector-VID) was used to facilitate the transport of perchlorate ion from the bulk to the catalytic membrane. The first power supply supplied the constant current for the generation of hydrogen atoms at the cathodic catalytic membrane and the second power supply provided constant potential for perchlorate ion transport.

Reduction of perchlorate. The reaction system described above was used with the addition of an ion-exchange membrane outside the catalytic membrane. The purpose of the ion-exchange membrane adjacent to the catalytic membrane at the opening bottom of the glass connector was to increase the concentration of perchlorate at the vicinity of the catalytic membrane. The volume between these two membranes was estimated to be one mL. The initial concentration of perchlorate was 10 ppm, and the experiment was performed at ambient conditions.

Effect of catalyst mass weight. Effect of surface coverage of catalysts was performed by investigating the difference of perchlorate removal for different mass weight of catalyst coated on unit weight of support. As described in section 2.2, catalytic membranes with different weight of catalysts per unit weight of support were prepared by applying different electroplating time (0.5 to 10 minutes) at the same constant current. Surface of the membranes was pictured by SEM instrument (JOEL 7400F).

Effect of catalyst type. In order to identify the effective catalysts, different catalytic membranes were prepared using metals from the first, the second, and the third row of the periodic table. A total of 18 monometallic catalysts of the transition metal in the periodic table were coated on the stainless steel mesh support. Fifty liters of perchlorate solution (10 mg/L) was used as the initial solution after purging with hydrogen gas for 30 minutes. The volume of the solution for each experiment was 1.2 liter during the screening of catalysts. The changes in perchlorate concentrations in addition to reaction intermediates, namely chlorate, chlorite and especially chloride were monitored and pH values recorded.

2.4 Analytical methods

Perchlorate, chlorate, chlorite, and chloride were analyzed using a Dionex ion chromatograph system with a GP50 pump, conductance detector and an EG 40 effluent generator. The separation part was a 4-mm Dionex AS-16 anion-exchange analytical column and guard column. The flow rate of sodium hydroxide effluent was 1.20 mL-min^{-1} with a concentration gradient to assure satisfying separation and detection limit. For all experiments with an initial perchlorate concentration of larger than 0.001 mM, a 25-μL-injection loop was used for perchlorate analysis. For concentration of less than 0.001 mM, the injection loop was changed to a 1000- μL.

3 RESULTS AND DISCUSSION

3.1 Surface analysis of the membrane

Figure 1 shows clearly that the surface of morphology of catalysts is different. For catalysts such as Sn and Mn, the surface area is supposed to be small due to its layered-structure on the membrane. For catalysts such as Sc, Co, Cu and Cr, the spongy structure will yield a large surface that provides active sites for the reaction. In the mean time, catalysts such as Pd and Mo may not have enough active sites due to the lack of mass on the membrane.

Figure 1 SEM images of different catalytic membranes.

3.2 Reduction of perchlorate

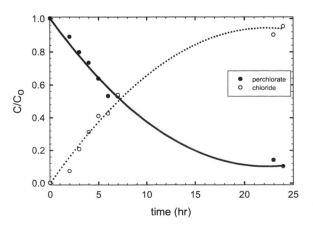

Figure 2 Reduction of perchlorate by catalytic membrane.

The reduction of perchlorate was obvious as shown in Figure 2. This is the first time that perchlorate can be reduced by hydrogen at reasonable rates. The decrease of the concentration of perchlorate was sharply at the first several hours indicated the reduction rate was rapid. It can be found it was a zero order reduction at high concentration of perchlorate. With the concentration became low, the rate turned out to be a first order.

From the analysis of the chemicals in the solution, chloride was the main end product of the reduction. However, both chlorate and chlorite were the intermediate since their peak appeared in the ion chromatography during the analysis. Their concentrations were so low that cannot be shown in Figure 2.

3.3 Effect of catalyst mass

It is speculated that the amount of catalyst on the stainless steel mesh support can also affect the rate of perchlorate reduction. As shown in Figure 3, the surface coverage of catalyst on the membranes was totally different at different amount of surface catalyst, e.g., Sn.

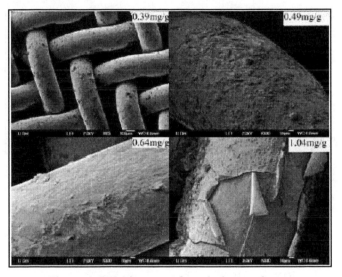

Figure 3 SEM images of catalytic membranes coated with different mass of Sn.

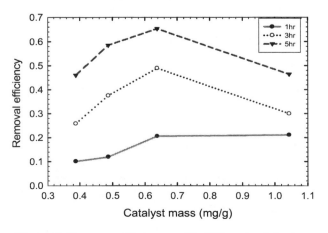

Figure 4 Remove efficiency with different catalyst mass.

Results shown in Figure 4 clearly indicate that there was an optimal catalyst density range at 0.6 to 0.7 mg/g as exemplified by Sn. The perchlorate removal efficiency increased with the increase in the catalysts density till 0.65 mg/g, but decreased abruptly as the catalyst density became greater than 0.7 mg/g. Further study is necessary to gain insight into the mechanisms of catalyst density on the perchlorate removal efficiency.

3.4 Effect of catalyst type

The abundance of electrons in the d orbital renders the transition metals good catalytic properties. It is well known that the bonding strength is closely related to the number of d electrons. The bonding strength plays an important role in the catalytic activities of a catalyst. In general, the catalytic activity of these metals can be correlated with the electronic configuration of the d-orbital as "percentage d characteristic" of the metallic bond based on Pauling's valence bond theory or with the strength of the metal adsorbate bond.

Element		Z	d-electron	$k^{\#}$ $\mu M\text{-}L^{-1}\text{-}hr^{-1}$	k'^{*} $mM\text{-}L^{-1}\text{-}hr^{-1}\text{-}mg^{-1}$
IIIA	Sc	21	1	5.9	11.8
IVA	Ti	22	2	11.0	2.5
	Zr	40	2	5.5	1.5
VA	V	23	3	7.4	2.6
VIA	Cr	24	5	9.9	16.5
	Mo	42	5	9.3	13.5
VIIA	Mn	25	5	7.0	2.2
VIIIA	Co	27	7	10.4	1.4
	Ni	28	8	6.8	1.8
	Ru	44	7	9.2	3.4
	Rh	45	8	7.3	4.6
	Pd	46	10	7.6	2.9
	Pt	78	10	7.6	18.2
IB	Cu	29	10	8.6	1.0
IIB	Zn	30	10	6.6	2.3
	Cd	48	10	8.9	22.2
IVB	Sn	50	10	9.8	3.4
	Pb	82	9	5.1	0.7

Table 1 List of Electron Characteristics and Rate constants.

Table 1 lists the atomic number (Z), d-electron, rate constants and specific rate constants. From Table 1, it is seen that among the 18 metallic catalysts tested, Sc, Ti, Cr, Mo, Pt and Cd were the most promising.

4 SUMMARY

On-site hydrogen atoms produced electrochemically by using membrane coated with monometallic catalysts can readily reduce the perchlorate to chloride. The reduction rate was different for different catalyst and/or mass weight of the catalysts.

5 ACKNOWLEDGEMENT

This research was supported by the U.S. Department of Defense through the Strategic Environmental Research and Development Program (SERDP), #ER-1430.

REFERENCES

[1] Urbansky, E. T. (1998). "Perchlorate chemistry: Implications for analysis and remediation." Bioremediation Journal 2(2): 81-95.

[2] Herman, D. C., and Frankenberger, W. T., (1998). "Microbial-mediated reduction of perchlorate in groundwater." Journal of Environmental Quality 27(4): 750-754.

[3] Greer Monte A; Goodman Gay; Pleus Richard C; Greer Susan E. (2002), Health effects assessment for environmental perchlorate contamination: The dose response for inhibition of thyroidal radioiodine uptake in humans. " Environmental health perspectives 110(9), 927-37.

[4] Susarla, S., Collette, T. W., Garrison, A. W., Wolfe, N. L., and McCutcheon, S. C., (1999). "Perchlorate identification in fertilizers." Environmental Science and Technology 33(19): 3469-3472.

[5] Fisher J; Todd P; Mattie D; Godfrey D; Narayanan L; Yu K., (2000),"Preliminary development of a physiological model for perchlorate in the adult male rat: a framework for further studies. " Drug and chemical toxicology 23(1), 243-58.

[6] Urbansky, E. T., (2002), "Perchlorate as an environmental contaminant. " Environmental Science and Pollution Research International 9(3), 187-192.

[7] Bradford, C., Rinchard J., Carr, J., and Theodorakis, C., (2005). "Perchlorate affects thyroid function in eastern Mosquitofish (Gambusia holbrooki) at environmentally relevant concentrations." Environmental Science and Technology 39: 5190-5195.

[8] Landrum, M., Canas, J. E., Coimbatore, G., Cobb, G. P., Jackson, W. A., Zhang, Baohong, and Anderson, T. A., (2005). "Effects of perchlorate on earthworm (Eisenia fetida) survival and reproductive success." Science of the Total Environment: in press.

[9] Park, J.W., Rinchard, J., Liu, F., Anderson, T. A., Kendall, R. J., and Theodorakis, C. W., (2005). "The thyroid endocrine disruptor perchlorate affects reproduction, growth, and survival of Mosquito.sh." Ecotoxicology and Environmental Safety: in press.

Functionalized Nanoporous Ceramic Sorbents for Removal of Mercury And Other Contaminants

S. V. Mattigod, G. E. Fryxell, R. Skaggs, K. E. Parker

Pacific Northwest National Laboratory
POB 999, Battelle Boulevard, Richland, Washington, USA, shas.mattigod@pnl.gov

ABSTRACT

A new class of high-performance nanoporous sorbents has been developed for heavy metal removal that overcomes the deficiencies of existing technologies. These novel materials are created from a combination of synthetic nanoporous ceramic substrates that have specifically tailored pore sizes (2 to 10 nm) and very high surface areas (~1000 m^2/g) with self-assembled monolayers of well-ordered functional groups that have high affinity and specificity for specific types of free or complex cations or anions. These sorbents known as SAMMS™ (Self-Assembled Monolayers on Mesoporous Silica) are hybrids of two frontiers in materials science: molecular self-assembly techniques and nanoporous materials. One form of SAMMS™ containing monolayers of mercaptopropyl-trismethoxy silane has shown exceptional sorptive properties for mercury and other soft cations such as silver, cadmium, and lead. Another form of SAMMS™ with a functional monolayer consisting of ethylenediamine-Cu(II) complex (Cu-EDA) specifically adsorbs tetrahedral oxyanions such as arsenate, selenate, molybdate, chromate and pertechnetate even in the presence of high concentrations of sulfate. Separation of actinides can be addressed by SAMMS™ material synthesized with a set of monolayer functionalities consisting of hydroxypyridinones, acetamide and propinamide phosphonates. These nanoporous sorbents offer a better choice for efficient and cost-effective removal contaminants from diverse waste streams.

Keywords: nanoporous sorbent, self-assembled monolayers, heavy metals, oxyanions, radionuclides

1 INTRODUCTION

Silica based nanoporous materials using liquid crystal templating was achieved about a decade ago [1, 2]. This has led to a number of applications of these nanoporous materials in diverse fields such as catalysis, sensor technology, and sorbents. Due to their relatively high surface areas (~500 – 1000 m^2/g), nanoporous materials as sorbents offer a significant advantage over conventional sorbents, and to be effective in this role, it is necessary to functionalize the pore surfaces of these materials..

We have developed a method to activate the pore surfaces of the silica based nanoporous materials so that these materials could be used as effective sorbents. This process consists of synthesizing within pores, self-assembled monolayers of adsorptive functional groups that are selected to specifically adsorb specific groups of contaminants. Molecular self-assembly is a unique phenomenon in which functional molecules aggregate on an active surface, resulting in an organized assembly having both order and orientation. In this approach, bifunctional molecules containing a hydrophilic head group and a hydrophobic tail group adsorb onto a substrate or an interface as closely packed monolayers. The tail group and the head group can be chemically modified to contain certain functional groups to promote covalent bonding between the functional organic molecules and the substrate on one end, and the molecular bonding between the organic molecules and the metals on the other. For instance, populating the head group with alkylthiols (which are well-known to have a high affinity for various soft heavy metals, including mercury) results in a functional monolayer that specifically adsorbs heavy metals such as, Hg, Ag, Cd, Cu, and Pb). If the head group consists of Cu-ethylenediamine complex (Cu-EDA) the monolayer will with high specificity adsorb oxyanions (As, Cr, Se, Mo). Additional monolayers with head groups we have designed include acetamide (APH) and propinamide (PPH) phosphonates for binding actinides (Am, Pu, U, Th), The functionalized monolayer and substrate composite was designated as SAMMS™ (Self-Assembled Monolayers on Mesoporous Silica). Detailed description of the functionalization techniques have been published previously [3-5]. Various self-assembled monolayer functionalities and the contaminants they were designed to target are shown in Table 1.

SAMMS™ Type	Contaminant
thiol	Ag, Cu, Cd, Hg, Pb
Cu-EDA	As, Cr, Se, Mo, ^{99}Tc
APH, PPH	^{241}Am, ^{237}Np, ^{239}Pu, ^{238}U

Table 1. SAMMS™ Technology developed for Contaminant Removal

2 ADSORPTION DATA

2.1 Thiol-SAMMS™

Heavy metal adsorptive properties of thiol-SAMMS was tested by contacting known quantities of sorbent with a fixed volume of 0.1M $NaNO_3$ solution containing of the metal of interest (Ag, Cd, Cu, Hg Pb). The initial concentrations of these metals ranged from 0.05 to 12.5 meq/L and the solution to sorbent ratio in these experiments ranged from ~200 – 5000 ml/g. The suspensions were continually shaken and allowed to react under ambient temperature conditions (~ 25 °C) for approximately 8 hours. Next, the sorbent and the contact solutions were separated by filtration and the residual metal concentrations in aliquots were measured by using inductively-coupled plasma mass spectrometry (ICP-MS).

The data from the adsorption experiments indicated that thiol-SAMMS adsorbed the heavy metals with significant affinity (Table 2). The predicted adsorption maxima were 0.56, 0.72, 1.27, 4.11 and 6.37 meq/g for Cu, Pb, Cd, Ag, and Hg respectively. The calculated distribution coefficients were $4.6x10^1 – 1.8x10^5$, $2.2x10^2 – 8.6x10^3$, $2.2x10^2 – 1.9x10^4$, $1.2x10^3 – 8.7x10^5$, and $1x10^3 – 3.5x10^8$ ml/g for Cu, Pb, Cd, Ag, and Hg respectively. Such selectivity and affinity in binding these heavy metals by thiol-SAMMS can be explained on the basis of the hard and soft acid base principle (HSAB) [6-8] which predicts that the degree of cation softness directly correlates with the observed strength of interaction with soft base functionalities such as thiols (-SH groups). According to the HSAB principle, soft cations and anions possess relatively large ionic size, low electronegativity, and high polarizability (highly deformable bonding electron orbitals) therefore, mutually form strong covalent bonds. The order of adsorption maxima observed in this experiment appears to reflect the order of softness calculated by Misono et al [9] for these heavy metals.

Heavy Metal	Adsorption Max (meq/g)	K_d (ml/g)
Ag	4.11	$1.2x10^3 – 8.7x10^5$
Cd	1.27	$2.2x10^2 – 1.9x10^4$
Cu	0.56	$4.6x10^1 – 1.8x10^5$
Hg	6.37	$1.0x10^3 – 3.5x10^8$
Pb	0.72	$2.2x10^2 – 8.6x10^3$

Table 2. Heavy Metal adsorption characteristics of thiol-SAMMS

2.2 Cu-EDA SAMMS™

Oxyanion adsorption characteristics of Cu-EDA SAMMS™ was evaluated by contacting the sorbent with 3 meq/L of Na_2SO_4 solution containing either chromate (~0.02 – 18 meq/l) or arsenate (~0.01 – 29 meq/l) ions. Solution to sorbent ratio in these experiments ranged from 100 – 500 ml/g. After 12 hours of contact, filtered aliquots of solution were analyzed by inductively-coupled plasma atomic emission spectrometry (ICP-AES).

Results of the arsenate and chromate adsorption tests indicated that Cu-EDA SAMMS very effectively adsorbed both these oxyanions from Na_2SO_4 solution. The predicted adsorption maxima from the Langmuirian fit to the data were 2.13 and 2.08 meq/g for arsenate ($HAsO_4$) and chromate (CrO_4) respectively. The calculated distribution coefficients were $2.4x10^2 – 1.0x10^4$ and $1.7x10^2 – 1.0x10^6$ ml/g for arsenate and chromate respectively. The bonding mechanism of these oxyanions was studied by Kelly et al [10,11] using X-ray Adsorption Fine Spectroscopy (XAFS) which indicated that bonding was monodentate in nature. The copper ion was bonded directly to the oxyanion by a shared oxygen. The adsorption process changed the coordination of copper from octahedreal to trigonal bipyramidal geometry. The bonding did not alter the tetrahedral symmetry of $HAsO_4$ ion whereas, the symmetry of the CrO_4 ion was distorted with two short and two long Cr-O bond distances.

2.3 Acetamide- Propinamide Phosphonate SAMMS

The actinide-specific APH- and PPH-SAMMS™ were tested by contacting 100 mg quantities of each these sorbents with 10 mL portions of solutions containing 2 x 10^6 counts per minute (CPM)/l of Pu(IV) in a matrix of acidified (pH=1) 1M $NaNO_3$ with separately spiked (0.01M) Pu-complexing ligands such as, phosphate, sulfate, ethyelenediaminetetraacetate (EDTA), and citrate. After 1-4 hours of reaction, the solution was separated by filtration, mixed with Ultima Gold™ scintillation cocktail and the residual alpha activity of Pu(IV) was measured by using a liquid scintillation counter (2550 TR/AB Packard Instruments, Meriden, Conn).

Plutonium adsorption data (Table 3) indicated that both APH and PPH SAMMS™ adsorbed this actinide with high specificity. However, on average, APH SAMMS performed slightly better in adsorbing Pu(IV) from solution. In this experiment, the presence of complexants with differing chelating strengths did not significantly affect Pu(IV) adsorption by these nanoporous sorbents. Considering that the complexation constants of Pu(IV) with these ligands vary in the order EDTA>Citrate>phosphate >>sulfate>nitrate[12] very high adsorption affinity shown by these sorbents indicate that the CMPO ligand (APH and PPH) functionality based adsorption substrates are capable of chelating Pu(IV) much more strongly than these ligands. Additionally, the APH SAMMS has been shown to adsorb Pu(IV) very rapidly with bulk of the sorbate removed from solution under one minute [13].

Ligand	APH SAMMS	PPH SAMMS
	---------K_d (ml/g)---------	
Nitrate	2.12×10^4	1.58×10^4
Nitrate+Phosphate	2.04×10^4	1.75×10^4
Nitrate+Sulfate	1.98×10^4	1.82×10^4
Nitrate+EDTA	2.05×10^4	1.56×10^4
Nitrate+Citrate	2.31×10^4	1.87×10^4

Table 3. Adsorption of Pu(IV) by APH- and PPH-SAMMS

3. SUMMARY AND CONCLUSIONS

Tests conducted on a new class of functionalized nanoporous materials showed that these are very effective in removing heavy metals, oxyanion and radionuclide contaminants from aqueous waste streams. The thiol-functionalized SAMMS^{TM} designed for heavy metal removal adsorbed 0.56 to 6.36 meq/g of Ag, Cd, Cu, Hg and Pb with very high selectivity (K_d 4.6×10^6 to 3.5×10^8 ml/g).

Sorption tests conducted on Cu-EDA SAMMS^{TM} showed that this material very effectively adsorbed arsenate and chromate with loading approaching ~2.1 meq/g with distribution coefficient values ranging from 1.7×10^2 to 1.0×10^6 ml/g.

Other forms of SAMMS^{TM} (APH and PPH) developed for specifically adsorbing radionuclides indicated that they were quite effective in removing Pu(IV) from solutions containing various complexing ligands such as, EDTA, citrate, phosphate, sulfate, and nitrate. Distribution coefficients as high as 2×10^4 ml/g indicated that these nanoporous substrates are very effective scavengers for actinides such as Pu.

Self-assembled monolayers of selected functionalities on nanoporous silica substrates can achieve very high contaminant loadings with relatively high specificities. These novel class of sorbent materials show great promise for effective removal of wide variety of targeted contaminants from aqueous waste streams.

4 ACKNOWLEDGEMENTS

This work was supported by the Office of Science Office of Biological and Environmental Research of the U. S. Department of Energy and the IR&D funds from Battelle. Pacific Northwest National Laboratory is operated for the U. S. Department of Energy by Battelle under contract DE-AC06-76 RLO 1830

REFERENCES

[1] Beck, J. S., J. C. Vartuli, W. J. Roth, M. E. Leonowicz, C. T. Kresge, K. D. Schmitt, C. T-W Chu, D. H. Olson, E. W. Sheppard, S. b. McCullen, J. B. Higgins, and J. L. Schlenker. *J. Am. Chem Soc.*114, 10834-10843, 1992.

[2] Kresge, C. T., M. E. Leonowicz, W. J. Roth, J. C. Vartuli, J. S. Beck. *Nature.* 359, 710-712, 1992.

[3] Feng, X., G. E. Fryxell, L. Q. Wang, A. Y. Kim, J. Liu, and K. M. Kemner. *Science.* 276, 865, 1997.

[4] Fryxell, G. E., J. Liu, A. A. Hauser, Z. Nie, K.F. Ferris, S.V. Mattigod, M. Gong, and R.T. Hallen. *Chem. Mat.* 11, 2148-2154, 1999.

[5] Birnbaum, J. C., B. Busche, Y. Lin, W. J. Shaw, and G. E. Fryxell. *Chem. Comm.* 1374-1375, 2002.

[6] Pearson, R. G. *J. Chem. Educ.* 45, 581-587, 1968.

[7] Pearson, R. G. *J. Chem. Educ.* 45, 643-648, 1968.

[8] Hancock R. D. and A. E. Martell. *J. Chem. Educ.* 74 644. 1996.

[9] Misono, M., E. Ochiai, Y. Saito, and Y. Yoneda. *Jour. Inorg. Nucl. Chem.* 29, 2685-2691, 1967.

[10] Kelly, S., K. Kemner, G. S. Fryxell, J. Liu, S. V. Mattigod, and K. F. Ferris. *Jour. Synchrotron Rad.* 8, 922-924, 2001.

[11] Kelly, S., K. Kemner, G. E. Fryxell, J. Liu, S. V. Mattigod, and K. F. Ferris. *Jour of Phys Chem.* 105, 6337-6346, 2001.

[12] Cleveland, J. M. *The Chemistry of Plutonium.* Am. Nucl. Soc. LaGrange Park, Illinois. 1979.

[13] Fryxell, G. E., Y. Lin, H. Wu. *Studies in Surf. Sci. Cat.* 141, 583-589, 2002.

Thermostability of TiO$_2$ nanoparticle and its Photocatalytic Reactivity at Different Anatase/Rutile Ratio

Yao-Hsuan Tseng[*], Hong-Ying Lin[**], Chien-Sheng Kuo[***], Yuan-Yao Li[***], and Chin-Pao Huang[**]

[*]Energy and Environment Laboratories, Industrial Technology Research Institute, Taiwan R.O.C., yaohsuanTseng@itri.org.tw
[**]Department of Civil and Environmental Engineering, University of Delaware, USA
[***]Department of Chemical Engineering, National Chung Cheng University, Taiwan , R.O.C.

ABSTRACT

Thermostability study of commercial P25 TiO$_2$ nanoparticles was carried out by ascending annealing temperature from 400 to 1100 °C. The thermostability of TiO$_2$ structure was measured by X-ray diffraction (XRD). Anatase-Rutile phase transition occurred only when temperature exceeds 600°C. Rutile weight fraction increased from 25 to 100 % between 400 to 840 °C. Phase transition activation energy was calculated by using Arrenhius plot to be 27 kJ/mol. Transmission electron microscope (TEM), Brunauer-Emmett-Teller (BET), and dynamic light scattering (DLS) techniques were applied to determine the size of grown particles. Results of particle size analysis using TEM imaging method were the same as those using BET instruments up to 840 °C. BET measurements tend to overestimate the particle size at temperature greater than 840 °C. In contrast, DLS overestimate the size of TiO$_2$ particles due to agglomeration in solution. Mean TiO$_2$ particle sizes grew from 25 to 450 nm when temperature increased from 400 to 1100 °C. Furthermore, the photocatalytic reactivity in the degradation of dye decreased with the increase of particle size and rutile fraction.

Keywords: photocatalyst, thermostability, methylene blue

INTRODUCTION

Since the discovery of the photoelectrochemical splitting of water on titanium dioxide electrodes [1], the semiconductor-based photocatalysis has been extensively studied. Titanium dioxide has become one of the most popular photocatalysts due to its non-toxicity, low cost, strong redox ability and wide applications, such as air purification, water treatment, deodorization, self-cleaning, and antibacterial coating [2]. The TiO$_2$ absorbs photons, and evolves active oxygen species, such as OH radical and O$_2^-$ ion, by reaction with H$_2$O and O$_2$ that are adsorbed on the TiO2 surface. The high oxidation potential of active oxygen species is responsible for the decomposition of many pollutants [1,2].

Much has been reported on the photocatalytic reactions of TiO$_2$. Majority of the research on the photocatalysis of TiO$_2$ focus on the bandgap modifications and effect of physical properties such as particle size and crystal structures on photocatalysis [2]. No information is available on the thermostability and its effect on the photocatalysis. In this present study, we investigated the thermostability of P25 nanoparticle and its photocatalytic activity. The degree of phase transformation was observed using XRD and the activated energy of Anatase-Rutile phase transition was determined by the Arrhenius equation accordingly. The particle size of TiO$_2$ photocatalyst was characterized by TEM, specific surface area by the BET, and DLS techniques. The anatase-rutile fraction ratio was determined as a function of temperature.

EXPERIMENTAL

Preparation and characterization of TiO$_2$

Degussa P25 TiO$_2$ was heat-treated in a conventional muffle furnace at 773, 873, 933, 1053, 1113, 1243, and 1373 K for four hours. The crystal phase and surface area were measured with a powder X-ray diffraction (Rigaku D max B) with a CuK radiation source and a N$_2$-gas adsorption analyzer (NOVA 2000), respectively. The particle size was determined using the TEM (Jeol 2000) and DLS (Malvern 3000HS) in distilled water medium. This is necessary as to minimize the degree of aggregation.

Photocatalytic decoloration of methylene blue

A volume of 100 mL of the aqueous solution of methylene blue (Co = 6.25 nM) was introduced into a batch reactor with 10 mg of TiO2, which was pre-dispersed with a 100-W ultrasonic tip for 2 minutes. The decoloration reaction was carried out at 288 K under UVA irradiation provided by two 20-W black lamps. The distance between the solution and the lamp was 10 cm. After centrifugation at 10000 rpm for 15 minutes, the absorbance of the methylene blue solution at 660 nm was measured using a UV/VIS spectrometer (HP 8452A) to determine the residual concentration.

Cleantech 2007, ISBN 1-4200-6382-0

RESULTS AND DISCUSSION

Characterization of TiO$_2$

In this work, the anatase-rutile phase composition was determined by XRD. Based on the XRD data it is possible to calculate the distribution of anatase and rutile using the following empirical equation [3]:

$$W_R = I_R/I_0 = I_R/(0.88I_A + I_R) \qquad (1)$$

where W_R is the rutile weight fraction in percent, I_A and I_R are integrated diffraction peak intensity from anatase (101) and rutile (110), respectively. I_o is the total integrated (101) and (110) peak intensity. The variation of anatase-rutile weight fraction (in percent) is shown in the Table 1. The obvious phase transition only occurred about 873 K. The anatase phase decreased with increase of temperature, and disappeared at temperature above 1113 K. Assuming that the phase transformation is a first-order reaction [4], the activation energy can be calculated by the Arrhenius equation as shown in Figure 2. The activation energy was 27.0 kJ/mol with a correlation coefficient value of 0.968, which indicated that mass transfer was the limiting step for the phase transition reaction.

Table 1. The effect of annealing temperature on the anatase-rutile weight fraction, specific surface area, and apparent reaction rate constant

Temp. (K)	W_R (%)	d_{DLS}/d_{TEM}	Surface area (m^2/g)	k_{app} (10^{3}min^{-1})
P25*	27	7.26	53.1	-
773	28	7.30	52.2	53.0
873	47	6.26	44.5	45.2
933	51	6.84	32.9	34.4
1053	89	4.63	15.0	21.8
1113	100	4.33	9.8	16.9
1243	100	2.96	3.4	11.8
1373	100	2.29	2.4	10.9

k_{app}: methyl blue degradation rate constant
*: original TiO$_2$ (P25)

Figures 3 and 4 show the particle size distribution of the TiO$_2$ particles determined by TEM, DLS, and BET surface area techniques. Rapid grain growth and formation of rutile phase appeared at temperature above 1053 K. The particle size can be calculated from the specific surface area by using the following equation with the assumption of spherical and nonporous particles [5]:

$$d_{BET} = 6000/\rho A \qquad (2)$$

where d_{BET} is the calculated particle size (nm), ρ is the density of TiO$_2$ (g/cm^3), and A is the specific surface area (m^2/g). Results from TEM and BET measurements were consistent with each other in the temperature range of ambient to 1113 K. The plausible reason is that there was significant aggregation of TiO$_2$ at high surface energy at elevated temperature [6]. As a result, the specific surface area was underestimated at the temperature above 1053 K. In contrast, DLS overestimates the size of TiO$_2$ particles due to agglomeration in the solution. The degree of aggregation (d_{DLS}/d_{TEM}) increased with decrease in particle size due to the high surface energy of small particles.

Photocatalytic activity

Figure 5 shows the conversion of methylene blue in the presence of various TiO$_2$ photocatalysts. The decoloration reaction did not proceed in the absence of both TiO$_2$ and UV light. The decoloration reaction followed a pseudo-first order rate law, which was consistent with previous literatures [7].

Figure 4 also shows the effect of annealing temperature on the apparent decoloration rate. The reaction rate decreased with increase in the annealing temperature due to grain growth of TiO$_2$. Several reports have indicated that anatase phase has greater photocatalytic activity than the rutile phase due to the difference in band-gap energy ($E_{g,A} = 3.2 > E_{g,R} = 3.0$) and electron-hole recombination rate ($r_{rec,R} > r_{rec,A}$) [8]. Results of this present work show that the surface area and the aggregation effect also play important roles on the photocatalytic characteristics. The samples prepared at 873 K and 933 K had similar anatase fraction, but the reaction rate at 873 K was larger than at 933 K because of the difference in specific surface area. Moreover, both the anatase fraction and the surface area of TiO$_2$ at 773 K are much larger than those at 1373 K, but the reaction rate of TiO$_2$ at 773 K was only five times that of TiO$_2$ prepared at 1373 K. It may be in part attributed to the fact the secondary particle of TiO$_2$ at 1373 K was only four times that at 773 K.

Maira and coworkers also reported that the photoactivity of TiO$_2$ in term of trichloroethylene degradation was strongly dependent on the degree of aggregation of TiO$_2$ [9]. Therefore, the anatase fraction is not the most important factor affecting the photocatalytic decoloration of methylene blue. The decoloration reaction is only related to decomposing the chromophores group of the dye; rutile phase alone is sufficient to decolorize the dye through its oxidation activity. The large surface area (small primary particle size) and small secondary particle size are more important than the anatase fraction of TiO$_2$ in determining the decoloration rate.

ACKOWLEDGMENT

One of us, Yao-Hsuan Tseng, wishes to thank Dr. S.W. Yue, Director, Center of Environmental Safety and Health Technology at the Industrial Technology Research Institute,

Taiwan, for his support and encouragement and the award of a Personnel Training grant.

REFERENCES

1. A. Fujishima, K. Honda, *Nature*, 238, 37 (1972).
2. A. Fujishima, T. N. Rao, and D. A. Tryk, *J. Photoch. Photobio. C.*, 1, 1 (2000).
3. J. F. Porter, Y. G. Li, and C. K. Chan, *J. Mater. Sci.*, 34, 1523 (1999).
4. R. D. Shannon and J. A. Pask, *J. Am. Ceram. Soc.* 48, 391 (1965).
5. T. Ohno, M. Akiyoshi, T. Umebayashi, K. Asai, T. Mitsui, and M. Matsumura, *Appl. Catal. A.*, 265, 115 (2004).
6. Y. Sun, A. Li, M. Qi, L. Zhang, and X. Yao, *Mater. Sci. & Eng.*, 86, 185 (2001).
7. A. Houas, H. Lachheb, M. Ksibi, E. Elaloui, C. Guillard, and J. Herrmann, *Appl. Catal. B*, 31, 145 (2001).
8. B. Kraeutler and A. J. Bard, *J. Amer. Chem. Soc.*, 100, 5985 (1978).
9. A. J. Maira, K. L. Yeung, C. Y. Lee, P. L. Yue, and C. K. Chan, *J. Catal.*, 192, 185 (2000)

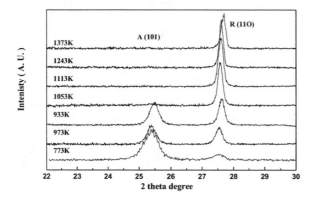

Figure 1 The XRD patterns of phase transition of TiO_2 particle

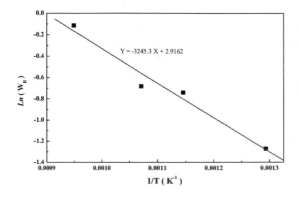

Figure 2 The Arrhenius plot of phase transition of TiO_2 particle

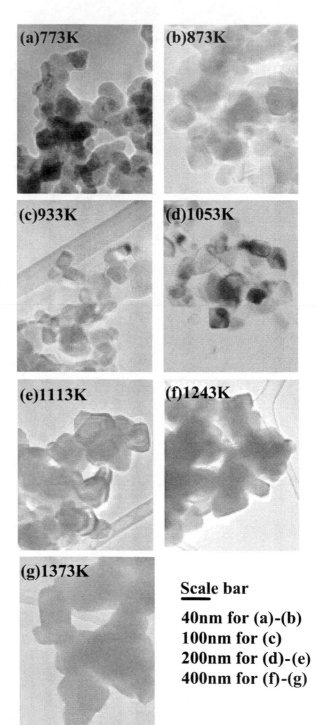

Figure 3 TEM images of variant TiO_2 particles

Scale bar

40nm for (a)-(b)
100nm for (c)
200nm for (d)-(e)
400nm for (f)-(g)

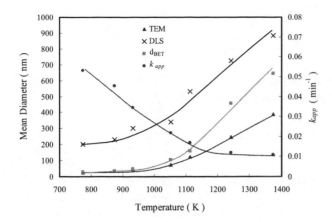

Figure 4 The effect of annealing temperature on the and
particle size apparent rate

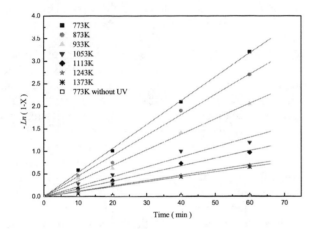

Figure 5 Conversions of methylene blue vs. time with
variant TiO$_2$ catalysts; 100 mL of 6.25 nM
methylene blue, 10 mg of TiO$_2$, 15°C, 2 sets of
20 W black lamp

Destruction of Organophosphate Agents by Recyclable Catalytic Magnetic Nanoparticles

L. Bromberg and T.A. Hatton*

Department of Chemical Engineering, Massachusetts Institute of Technology,
Cambridge, MA 02139, USA, tahatton@mit.edu

ABSTRACT

Organophosphorus (OP) pesticides and warfare agents are catalytically hydrolyzed in aqueous media by suspensions of magnetite (Fe_3O_4) nanoparticles modified with poly(1-vinylimidazole-*co*-acrolein oxime-*co*-acrylic acid). The oxime- and imidazole-modified magnetite particle serves as a nano-sized particulate carrier with nucleophilic groups immobilized on its surface. The oxime-modified magnetite nanoparticles are colloidally stable within a wide pH range and are readily recovered for reuse from the aqueous milieu by high-gradient magnetic separation methods with no loss of catalytic activity.

Keywords: oxime, imidazole, catalysis, organophosphate agent, poly(1-vinylimidazole-co-acrolein oxime-co-acrylic acid), high gradient magnetic separation.

1   INTRODUCTION

Organophosphates (OPs) are the most widely applied insecticides today. They are used in agriculture, the home, gardens, and veterinary practice. Whilst widely used in agriculture, they are also utilized against household and catering establishment pests, as well as against head lice in humans and a number of ectoparasites in domestic animals. The aerial application of OPs is used to control cereal and vegetable pests. Approximately 40,000 metric tons of various OPs are applied annually to agricultural crops in the U.S. Hence, development of decomposition agents for acutely toxic OP pests as well as warfare agents is of significant utility in both civilian and military fields. Notably, hydrolysis of the P-O bond in the OPs by nucleophilic agents yielding phosphoric acids as the main OP metabolite allows for drastic reduction of the OPs toxicity, because the phosphoric acids are not known to be efficient inhibitors of the acetylcholine esterase. We have recently discovered a group of catalytic decomposition agents, friendly to the environment and inexpensive, which could be readily dispersed in an aqueous medium, be it an aqueous reservoir or bodily fluid, maintain their ability to decompose OP at neutral pH, and be capable of being efficiently removed from the medium by a high-gradient magnetic separation (HG-MS)[1]. Such agents comprised magnetite (Fe_3O_4) nanoparticles modified by a powerful α-nucleophile, oxime group, as in a common antidote, 2-pralidoxime (PAM), or its polymeric analog, poly(4-vinylpyridine-N-phenacyloxime-co-acrylic acid). In the present work, we adopted analogous approach toward the OP decomposition agents, but designed a novel copolymer acting as both an efficient nucleophilic agent on the magnetite surface as well as a stabilizing agent enabling an aqueous stability of magnetite suspensions in a broad range of pH. The presence of the oxime group in the novel copolymer was achieved by the use of acrolein that is readily converted to the acrolein oxime by oximation with hydroxylamine. However, the use of polyacrolein, per se a water- and common solvents-insoluble polymer, was impossible for our intended application. Therefore, we utilized a seeded copolymerization of acrolein with an amphiphilic copolymer of 1-vinylimidazole (VIm) and acrylic acid (AA), to arrive at an acrolein copolymer that was subsequently oximated. An analogous free-radical polymerizations of acrolein in basic aqueous conditions initiated by persulfates and using polypyrrole as a seed agent has been reported previously [2]. Application of VIm has been rationalized in terms of imidazole's well-known nucleophilicity that could be weaker than the oxime's, but still sufficient to catalyze ester hydrolysis in the enzyme-like fashion [3]. In organophosphate-hydrolytic enzymes, histidine imidazole serves as a base, deprotonating a water molecule and generating the attacking hydroxide ion that produces the hydrolysis [4]. Furthermore, vinylimidazole is hydrophilic even at pH above its pK_a and poly(1-vinylimidazole) has been shown to maintain colloidal stability of magnetic iron oxide particle up to pH 10 [5]. Ion-exchange equilibria and complexation between the positively charged imidazole-modified particles and organophosphoric acids resulting from the OP hydrolysis can contribute to the acid metabolite recovery along with the particles, using HGMS. Finally, AA was chosen as a minor component of the copolymer to further stabilize magnetite particle at neutral pH. The resulting copolymer was capable of stabilizing magnetite nanoparticles, which catalyzed hydrolysis of a model OP nerve agent, diisopropyl fluorophosphate (DFP), and could be recovered from aqueous media by the HGMS, as described below.

2   EXPERIMENTAL

2.1   *Polymer Synthesis*: Copolymer of N-vinylimidazole (VIm), acrolein (Ac), and acrylic acid (AA) [p(VIm-AcOx-AA)] was synthesized by seeded, free-radical copolymerization using potassium persulfate as an initiator, followed by oximation. A three-necked flask containing a deoxygenated solution of 0.1 mol (9.9 mL) of VIm, 0.025 mol (1.82 mL) of AA, and 15 mg of $K_2S_2O_8$ in deionized water (50 mL) was deaerated by nitrogen purge and kept at 70 °C for another 0.5 h. Then 0.1 mol (5.9 mL) of Ac were added to the reaction mixture via a syringe, followed by addition of 10 mg/mL aqueous $K_2S_2O_8$ solution (10 mL) and stirring. The reaction mixture was kept under nitrogen blanket at 70°C overnight. The resulting black, viscous copolymer [p(VIm-Ac-AA)] solution was repeatedly washed by acetone and freeze-dried. The aldehyde groups of the acrolein in the copolymer were converted to acrolein oxime by reacting the p(VIm-Ac-AA) copolymer (15 g) with 200 mL of a freshly prepared anhydrous methanol solution containing 0.15 mol (10.4 g) of hydroxylamine hydrochloride and 0.15 mol (6 g) of sodium hydroxide. The methanolic solution of the polymer and hydroxylamine was kept under reflux while stirring at 70°C for 2 days and then methanol was evaporated and the resulting solids were repeatedly washed with excess water and acetone on a paper filter. The resulting polymer was exhaustively dialyzed against 10 mM aqueous H_2SO_4 solution using a Spectra/Por® membrane (MWCO, 3.5 kDa). The purified p(VIm-AcOx-AA) samples were lyophilized and stored dry at −20 °C until further use. $(C_{37}H_{54}N_{12}O_6)_x$, found (calc): C 58.14 (58.25); H 7.61 (7.13); N 22.39 (22.03). ^{1}H NMR (400 MHz, CD_3OD): d 1.6 (m, 2 H, CH_2- in the main chain), 2.2, 3.5 (m, 1 H, CH- in the main chain), 7.25 (m, 2H, imidazole), 7.36 (s, 1H, oxime), 7.7 (m, 1H, imidazole). FTIR (KBr): 3370 (bonded N-H, OH stretch), 2918, 2850 (N-H...N= stretch), 1695 (C=O), 1670 (aldoxime C=N), 1650 (aldoxime C=N), 1560, 1435,1371, 1266, 1094, 938, 732 cm^{-1}. Weight-average MW by GPC in N,N-dimethylacetamide was 128.5 kDa, polydispersity index 1.82. The structure of the p(VIm-AcOx-AA) copolymer is depicted in Scheme 1.

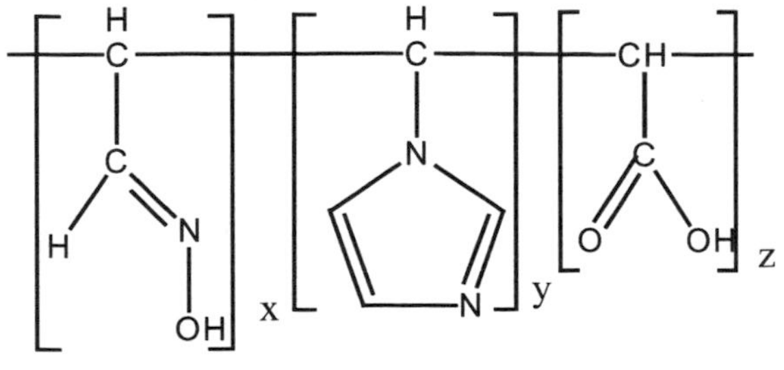

$$x : y : z = 4 : 4 : 1$$

Scheme 1. p(VIm-AcOx-AA) copolymer.

2.2. *Particle Synthesis*: Magnetic nanoparticles were produced by chemical coprecipitation of iron(II) and iron (III) chlorides. Namely, 1.88 g (7.0 mmol) of $FeCl_3.6H_2O$ and 0.69 g (3.5 mmol) of $FeCl_2.4H_2O$ were added to 40 mL of deionized water and the solution was deaerated by nitrogen purge in a stirred 250-mL three-necked flask and

temperature of the flask contents was brought to 80 °C. An aqueous/methanol (1:1 v/v) solution of the p(VIm-AcOx-AA) copolymer (6.25 wt%, 40 mL total, pH adjusted to 6) was added to the flask and the resulting mixture was equilibrated at 80 °C while stirring under nitrogen purge. Then the nitrogen purge was ceased and the contents of the flask were at once added to 80 mL of a 28% ammonium hydroxide and the mixture was vigorously stirred for 10 min. The resulting precipitate possessed strong magnetic properties and was thus separated from the liquid by decantation using a magnetic separator. The precipitate was dried and resuspended in deionized water [1]. The suspension was dialyzed against excess deionized water (membrane MW cut-off, 3.5 kDa) and lyophilized. The composition of the oxime-containing particles were assessed by elemental analysis as follows. $[(C_{37}H_{54}N_{12}O_{14}(Fe_3O_4)_2]_x$, found (calc): C 35.75 (36.25); H 4.09 (4.44); Fe 27.12 (27.33); N 13.74 (13.71), oxime group content 3.2 mmol/g.

2.2 Kinetic measurements

Kinetics of the DFP decomposition were measured at 25 °C with an Orion 96-09 combination fluoride electrode (Thermo Electron Corp.) and a Model 45 Dual Display Multimeter (Fluke Corp.) connected to a PC with FlukeView Forms software for data processing. The electrode was immersed in a stirred 9-mL aqueous sample and the electrode potential-time output was recorded continuously. No significant changes in pH, set initially at 7.0, were observed in any of the runs. The electrode was calibrated in independent series of experiments using aqueous solutions of sodium fluoride.

2.3 Magnetic Separation and Reuse

High-gradient magnetic separation (HGMS) experiments were performed using a cylindrical plastic column with an internal diameter of 7 mm and a length of 20 cm packed with 3.6 g of type 430 fine-grade stainless steel wool (40-66 mm diameter) placed inside a quadrupole magnet system comprising four nickel-plated Neodymium Iron Boron 40 MGOe permanent magnets sized 18x1.8x1.8 cm each (Dura Magnetics, Inc.). The flux density generated inside the packed column was ca. 0.73 Tesla.

Magnetic washing of the particles was performed as described previously [1]. The resulting suspension was subjected to the kinetic experiment using electrode detection of the fluoride ions generated by the DFL decomposition. The process of particle recovery and reuse was repeated in two sequential cycles.

3 RESULTS AND DISCUSSION

Effect of pH on electrokinetic mobility and particle size of the magnetite modified with the p(VIm-AcOx-AA) copolymer is shown in Fig.1. As is seen, the particles remained stable, with the number-average hydrodynamic diameter varying in the range 125-150 nm. No particle precipitation and sedimentation was observed up to pH 11.5. The ζ-potential was significant and positive at pH<8, reflecting the positive charge on imidazole moieties, and turned negative above pH 8, when the contribution from the negatively charged carboxyls of AA turned entire particle negative as the imidazole moieties were no longer charged at these pH. The pK_a of the poly(1-vinylimidazole) is 4.9 [6], and the pI of the VIm and AA copolymers (VIm: AA mol ratio 4:1) has been reported at 6.5 [7]. Attachment of the poly(1-vinylimidazole) to maghemite particles resulted in the particles with pI 7.4 [5], reflecting a shift to higher pH required for the ionization of the immobilized imidazole. Even larger shift to pH ≥ 8 is observed herein.

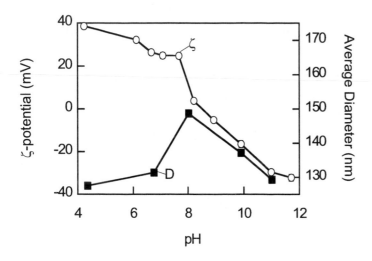

Figure 1: Effect of pH on ζ-potential and hydrodynamic diameter of magnetite particles modified with p(VIm-AcOx-AA). T=25°C; 10 mM Tris buffer with pH adjusted by 1 M NaOH or HCl.

Hydrolysis of DFP to produce the fluoride ion was monitored by the ion-selective fluoride electrode. Typical electrode response curves are shown in Fig.2.

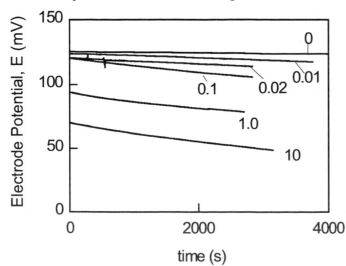

Figure 2: Potential-time response of fluoride-selective electrode to 0.6 mM DFP concentration in the presence of varying concentrations of magnetite particles modified with p(VIm-AcOx-AA). T=25°C; 10 mM Tris, pH 7.0. Numbers stand for particle concentrations in mg/mL. Addition of DFP to a suspension of oxime-containing particles, resulted in the rapid appearance and accumulation of the fluoride ions, as is seen from the response of the ion-selective electrode. The electrode potential was converted to the time-dependent fluoride concentration (C_t) readings, using electrode calibration curves in sodium fluoride solutions [1]. The initial slope of the C_t vs t kinetic curves corresponds to the initial rate of the DFP hydrolysis (v_o). The observed rate constant for the DFP hydrolysis (k_{obs}) was obtained from the experimental data: $-\ln(1 - C_t/[DFP]_o) = k_{obs}t$, where $[DFP]_o$ is the initial concentration of the substrate. Fig.3 shows typical kinetic curves in terms of the above equation. It further illustrates the particle recovery results.

Recovery of the nanoparticles from the aqueous solutions was achieved in a series of magnetic filtration experiments in each of which, following the successful catalytic hydrolysis of DFP by the oxime-coated nanoparticles, the suspension of particles was passed through the HGMS filter placed inside the magnet device. The magnetic particles were trapped in the filter and subsequently recovered by removing the steel wool-packed column from the magnetic environment and passing fresh water through the filter; these particles were then reused in the hydrolysis of a fresh batch of DFP solution (Fig.3). This cycle of the DFP hydrolysis and particle filtration and collection was repeated twice. As is seen in Fig.4, essentially complete recovery and reuse of the particles was possible. The k_{obs} constants for DFP hydrolysis were determined in three cycles to be unchanged at $(2.6 \pm 0.07) \times 10^{-5}$.

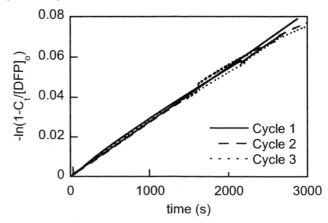

Figure 3: Catalytic stability of magnetic nanoparticles modified by p(VIm-AcOx-AA) indicated by essentially unchanged kinetic profiles observed for DFP hydrolysis

Cleantech 2007, ISBN 1-4200-6382-0

when the particles are recovered, washed, and reused. T=25°C, pH 7.0, 10 mM Tris; initial DFP concentration in all cycles, 0.6 mM. Cycle 1 denoted the first use of the particles, Cycle 2 the first magnetic recovery and resuspension of the particles at 1 mg/mL, and Cycle 3 the second magnetic recovery and resuspension of the particles at 1 mg/mL. The straight line illustrates the average slope of the kinetic curves used to calculate the k_{obs}.

The observed rate of DFP hydrolysis in 10 mM Tris buffer at pH 7.0 in the presence of particles was markedly higher than the k_{obs} of the spontaneous hydrolysis, even at very small particles concentrations (Fig.4). The apparent second-order hydrolysis rate constant, $k'' = v_o /[Catalyst]_o[Substrate]_o$ with p(VIm-AcOx-AA)-modified magnetite was 2-5 times higher than with the previously reported [1] magnetite modified with poly(4-vinylpyridine-N-phenacyloxime-co-acrylic acid), at similar initial oxime and substrate concentrations.

reduce the concentration of the phosphoric acids (the OP metabolites), from water.

REFERENCES

[1] L. Bromberg and T.A. Hatton, Ind.Eng.Chem.Res., 44, 7991-7998, 2005.

[2] T. Basinska, D. Kowalczyk, B. Miksa, and S. Slomkowski, Polym. Adv. Technol., 6, 526-533,1995.

[3] T. Kunitake, F. Shimada, and C. Aso, J. Am. Chem. Soc., 91, 2716-2723, 1969.

[4] M. Harel, A. Aharoni, L. Gaidukov et al., Nat. Struct. Mol. Biol. 11, 412-419, 2004.

[5] M. Takafuji, S. Ide, H. Ihara, and Z. Xu, Chem. Mater., 16, 1977-1983, 2004.

[6] T. Roques-Carmes, F. Membrey, A. Deratani, M.R. Böhmer and A. Foissy, J. Colloid Interface Sci., 256, 273–283, 2002.

[7] K. Ogawa, A. Nakayama, and E. Kokufuta, Langmuir, 19, 3178-3184, 2003.

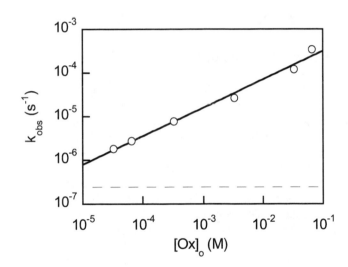

Figure 4: Dependence of the observed DFP hydrolysis kinetics constants (k_{obs}) on the effective initial concentration of the oxime groups ($[Ox]_o$) in suspensions of magnetite modified with p(VIm-AcOx-AA). T=25°C; 10 mM Tris, pH 7.0, $[DFP]_o$=0.6 mM. Broken line shows k_{obs} for the spontaneous hydrolysis of DFP in the buffer solution alone.

4 CONCLUSIONS

Novel, recyclable oxime- and imidazole-modified magnetite nanoparticles have been reported. The particles of hydrodynamic diameter of 125-150 nm are stable in aqueous suspensions in a wide range of pH, catalyze the hydrolysis of the OPs and can be recovered by high gradient magnetic separation and reused without a significant loss of hydrolytic activity. Studies are underway to clarify whether the recovery of the particles can also

Enhancement of Oxygen Transfer in Fermentation by Use of Functionalized Magnetic Nanoparticles

B. Olle[*], L. Bromberg, T.A. Hatton, D.I.C. Wang

Department of Chemical Engineering, Massachusetts Institute of Technology

77 Massachusetts Avenue, Cambridge, MA 02139

[*]bernat@mit.edu

ABSTRACT

Enhancement of oxygen mass transfer has been observed in the presence of colloidal dispersions of magnetite (Fe_3O_4) nanoparticles coated with oleic acid and a polymerizable surfactant. These fluids improve gas-liquid oxygen mass transfer up to 6 fold in an agitated, sparged, cell-free reactor and show remarkable stability in high-ionic strength media over a wide pH range. A combination of experiments using physical and chemical methods has been used to show that the mass transfer coefficient k_L and the volumetric mass transfer coefficient k_La are enhanced in the presence of nanoparticles. An increase of 40% in the oxygen uptake rate has been achieved in *Escherichia Coli* fermentation at a 5.5-L scale by using 0.6% w/w particles dispersed in fermentation media. The enhancement in mass transfer is directly translatable into increased fermenter productivity.

Keywords: magnetic nanoparticles, nanofluids, oxygen mass transfer enhancement, fermentation, mass transfer coefficient

1. INTRODUCTION

Maintaining an adequate oxygen supply to aerobic cell cultures has been a long-standing problem in fermentation technology. This problem is particularly amplified in high cell density cultures and in large scale operations, in which insufficient oxygen transfer rates limit cell growth and ultimately process productivity.

Our approach to reducing the oxygen transport limitation consists of adding functionalized magnetic nanoparticles to the fermentation medium; these materials consist of particles that have a magnetic core and two coatings. The magnetic core facilitates recovery of the fluid after the fermentation by passing it through a magnetic field; the first coating, made of oleic acid, confers high-oxygen storing capacity and the outer layer coating, made of surfactant (Hitenol-BC Polyoxyethylene alkylphenyl ether, Montello, Inc., Tulsa, OK), confers colloidal stability in water to the particle. Using magnetic nanoparticles presents several advantages compared to previous approaches [1, 2], including large interfacial areas and the possibility to readily recover the particles by High-Gradient Magnetic Separation [3, 4]. In addition, the nanoparticles are non-volatile and are synthesized with benign, low-cost chemicals, which makes their use attractive from environmental and economic standpoints [5].

2. EXPERIMENTAL METHODS

2.1 Nanoparticle Synthesis and Purification

A mixture of 94 g of ferric chloride hexahydrate (97% $FeCl_3 \cdot 6H_2O$) and 34.4 g ferrous chloride tetrahydrate (99% $FeCl_2 \cdot 4H_2O$) in 100 g of water was stirred at 80 °C under nitrogen sparging for 30 min in a round bottom flask. Subsequently, 100 g of potassium oleate (40 wt% paste in water, $CH_3(CH_2)_7CH=CH(CH_2)_7COOK$) was added, and the mixture was stirred for an additional 30 min. A 100 mL of an aqueous solution containing 28% ammonium hydroxide (NH_4OH) was added to the mixture, after which the solution immediately turned black because of the formation of magnetite. The reaction continued at 80 °C under stirring and nitrogen sparging for 30 min, after which it is assumed that oleic acid had completely coated the magnetite aggregates. Following the coating of the magnetite aggregates, 100 g of Hitenol–BC and 5 g of ammonium persulfate (>98% $(NH_4)_2S_2O_8$) were added to the reaction mixture. The reaction continued at 80 °C, under nitrogen sparging and vigorous stirring for 30 min. The solution was then cooled to room temperature and remained in the oven at 80 °C overnight, after which most of the residual ammonium hydroxide evaporated. The dispersion was dialyzed against distilled water (14,000 kDa MWCO dialysis membrane) in a 20 L container under mild stirring for 2 days to remove unreacted potassium oleate, Hitenol–BC, ammonium persulfate, and other salts and metal ions. Finally, the dialyzed solution was kept in the oven overnight at 80 °C, after which its solids contents were measured. The final solid contents were typically between 15 to 25% in weight. This synthetic procedure yielded magnetic nanoparticles with an average number diameter between 20 and 25 nanometers (nm).

2.2 Nanoparticle Characterization

Nanoparticles were characterized by Dynamic Light Scattering (DLS) using a Brookhaven BI-200SM instrument at a measurement angle of 90°. The particle sizes cited here correspond to number averages. All measurements were recorded in quadruplicate and reported as average values. DLS measurements showed that particles have a hydrodynamic diameter of 20 nm over the pH range relevant to fermentation (around 7). Zeta potential measurements were recorded using a Brookhaven ZetaPals Zeta Potential Analyzer. Measurements were recorded in quadruplicate and reported as average values. Particles showed a strongly negative zeta potential of approximately -30 mV around pH 7.

3. RESULTS

3.1 Mass Transfer in Cell-Free Systems

3.1.1 Physical Method: Mass Transfer in a Stirred Beaker System

Mass transfer was characterized by a physical method in an agitated beaker. Experiments were conducted in a system that had a fixed and known gas-liquid contact area. The results thus obtained yielded information on the value of the mass transfer coefficient, k_L. This experimental setup was used to avoid problems associated with interfacial area and gas holdup. The experimental system was a cylindrical, 10.3 cm diameter beaker, containing 500 mL. The liquid was agitated by a 4.5 cm diameter, 4-bladed, axial-flow impeller (pitched-blade with each blade measuring 3 cm wide and 1.5 cm long). Dissolved oxygen was measured using a dissolved oxygen sensor (YSI 5010). The temperature was regulated at 37±0.5 °C with a water bath. Oxygen response curves were obtained by first sparging nitrogen until the dissolved oxygen concentration fell to zero and then monitoring the increase in dissolved oxygen concentration due to exposure of the liquid free surface to the room air. To ensure a constant gas-liquid interfacial area no air sparging was used during the second step.

Oxygen mass transfer into the aqueous liquid phase is enhanced in the presence of nanoparticles, as shown in Figure 1. The time required to reach saturation is reduced by approximately 25% in the presence of a nanoparticle mass fraction of $\phi = 0.005$ (0.5% w/w). Further reductions are attained at larger particle

holdups, but the effect is less pronounced above $\phi = 0.01$. Values of the mass transfer coefficient k_L can be extracted from the data in Figure 1 by performing a mass balance on oxygen in the liquid phase

$$\frac{dC_{O_2,\text{bulk}}}{dt} = k_L a\left(C_{O_2}^* - C_{O_2,\text{bulk}}\right) \qquad (1)$$

where a is the specific interfacial area, and $C_{O2,\text{bulk}}$ and C_{O2}^* are the liquid phase concentrations of oxygen in the well-mixed bulk and at saturation respectively.

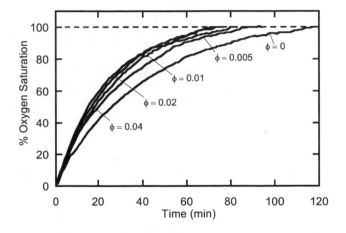

Figure 1: Response curves of dissolved oxygen in a stirred beaker at increasing nanoparticle holdup

The absolute enhancement in k_L, defined as

$$E = k_{L,\text{nanoparticles}} / k_{L,\text{control}} \qquad (2)$$

is presented in Figure 2 as a function of the nanoparticle holdup at different agitation rates. Enhancement increases rapidly at low particle holdups and slowly at larger holdups (above $\phi = 0.01$ approximately). It can also be observed that enhancement is greater at a lower agitation rate.

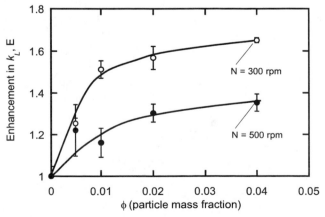

Figure 2: Enhancement in k_L as a function of nanoparticle holdup in at different agitation rates

3.1.2 Chemical Method: Sodium Sulfite Oxidation

A laboratory scale aerated and agitated fermentor was used for the determination of volumetric mass transfer coefficients (k_La) of oxygen. The sodium sulfite oxidation method was used. In the presence of a Cu^{2+} or Co^{2+} catalyst, sodium sulfite is oxidized according to the following reaction

$$Na_2SO_3 + \frac{1}{2}O_2 \xrightarrow{Cu^{2+} \text{or } Co^{2+}} Na_2SO_4 \tag{3}$$

The reaction rate can be adjusted by changing the catalyst concentration so that oxygen transport from the gas to the liquid becomes the limiting step. The oxygen uptake rate (OUR) is calculated by measuring the effluent gas composition with a mass spectrometer and performing a mass balance on oxygen of the gas phase through the reactor as follows

$$OUR = \frac{F_{N_2 \text{ in}}\left[\left(\dfrac{C_{O_2}}{C_{N_2}}\right)_{\text{in}} - \left(\dfrac{C_{O_2}}{C_{N_2}}\right)_{\text{out}}\right]}{V} \tag{4}$$

where OUR is the oxygen uptake rate, $F_{N2,in}$ is the flowrate of nitrogen entering the reactor, C_{O2} and C_{N2} are the concentrations of oxygen and nitrogen entering or exiting the reactor, and V is the working volume. The volumetric mass transfer coefficient can then be determined as

$$k_La = \frac{OUR}{C_{O_2}^* - C_{O_2,\text{bulk}}} \tag{5}$$

where C^*_{O2} is the average liquid phase saturation concentration in equilibrium with the inlet and outlet gas. Experiments were performed in a 5.5 L (working volume) stirred tank reactor (Biolafitte fermentor system, model BL 20.2), with an agitator with 2 Rushton 4-bladed turbine impellers. A 0.67 M sodium sulfite solution was loaded into the reactor, and then a $1 \cdot 10^{-3}$ M solution of copper sulfate catalyst was added. The pH was initially adjusted to 8.0 with sulfuric acid. Temperature for the experimental runs was maintained at 37±0.5 °C. Results in Figures 3a and 3b show that a 4 fold enhancement of k_La can be obtained at a ϕ = 0.0025. At higher particle concentrations k_La is still further enhanced, up to 6 fold, but diminishing returns are obtained as the particle fraction approaches ϕ = 0.01

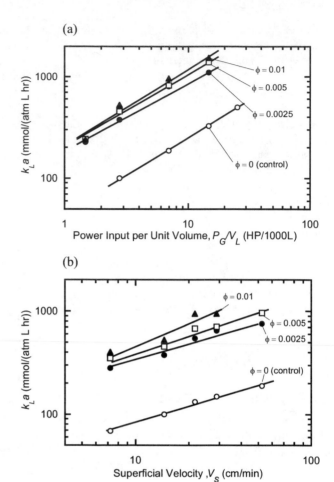

Figure 3: (a) k_La as a function of power input per unit volume at several nanoparticle mass fractions and at a superficial gas velocity of Vs= 14.5 cm/min (b) k_La as a function of superficial velocity at several nanoparticle mass fractions and at a power input per unit volume P_G/V_L of 2.8 HP/1000 L.

3.2 Mass Transfer in Fermentation

Fermentations at a 5.5 L scale in the presence of nanoparticles were conducted to characterize the enhancement of oxygen mass transfer in biological media and to quantify the enhancement of cell growth. The results were compared to a control run conducted in the absence of nanoparticles. The experiments were conducted without controlling dissolved oxygen levels. In this fashion, dissolved oxygen levels fell down to approximately zero (at around 6 hours for the control run) as cells proliferated and thereafter the cultures grew under oxygen mass transfer limitations. During the oxygen limited growth phase it was apparent that oxygen uptake rates were higher in the presence of nanoparticles, as shown in Figure 4. An increase of 40% in the oxygen uptake rate was achieved in *Escherichia Coli* fermentation by using 0.6% w/w particles dispersed in fermentation media. Higher cell growth rates were maintained in the presence of nanoparticles when the cultures grew under an oxygen transfer–limited regime, *e.g.,* after 6 hours in the control run, or after 8 hours for $\phi = 0.0057$. This was supported by dry cell weight profiles, as shown in Figure 5. It can be observed in both Figures 4 and 5 that a nanoparticle concentration of $\phi = 0.02$ does not yield oxygen uptake rates or cell concentrations significantly higher than $\phi = 0.0057$.

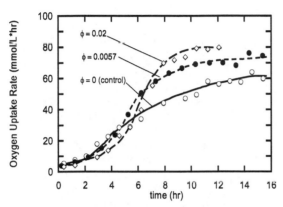

Figure 4: Oxygen uptake rate profiles during fed–batch fermentations of *Escherichia Coli* conducted at mass fractions of nanoparticles of $\phi = 0$, $\phi = 0.0057$, and $\phi = 0.02$.

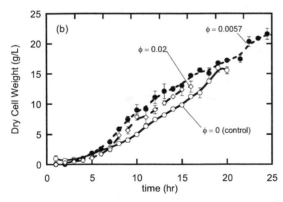

Figure 5: Dry cell weight during fed-batch fermentations of *Escherichia Coli* conducted at mass fractions $\phi = 0$, $\phi = 0.0057$, and $\phi = 0.02$ of oleic acid–coated nanoparticles

CONCLUSIONS

Colloidal dispersions of magnetite (Fe_3O_4) nanoparticles coated with oleic acid and a polymerizable surfactant enhance gas-liquid oxygen mass transfer up to 6-fold in a cell-free system. The dispersions show remarkable stability in high-ionic

strength solutions and over a wide range of pH and can therefore be used in fermentation media. An increase of 40% in the oxygen uptake rate has been achieved in *Escherichia Coli* fermentation by using 0.6% w/w of particles. The enhancement in mass transfer is directly translatable into increased cell concentrations.

REFERENCES

1. Junker, B., T. Hatton, and D. Wang (1990). "Oxygen-transfer enhancement in aqueous perfluorocarbon fermentation systems.1. Experimental-observations." Biotechnol Bioeng **35**(6): 578-585.
2. McMillan, J.D. and D.I.C. Wang (1987). "Enhanced oxygen-transfer using oil-in-water dispersions." Ann NY Acad Sci **506**: 569-582.
3. Ditsch, A., S. Lindenmann, P.E. Laibinis, D.I.C. Wang, and T.A. Hatton (2005). "High-gradient magnetic separation of magnetic nanoclusters." Ind Eng Chem Res **44**(17): 6824-6836.
4. Moeser, G.D., K.A. Roach, W.H. Green, T.A. Hatton, and P.E. Laibinis (2004). "High-gradient magnetic separation of coated magnetic nanoparticles." AIChE J **50**(11): 2835-2848.
5. Bromberg, L., J. Yin, D. Wang, and T.A. Hatton, Olle B., *Bioprocesses enhanced by fluoropolymer-coated magnetic nanoparticles and methods related thereto.* 2004: US Patent Application.

Nanostructured perovskite-based oxidation catalysts for improved environmental emission control

Purnesh Seegopaul, Mahbod Bassir, Houshang Alamdari, André Van Neste
Nanox Inc.
4975 rue Rideau, Suite 100, Québec, (Québec) G2E 5H5 Canada
Phone : 418-692-1131 Fax : 418-692-1165 Website : http://www.nanoxnps.com

ABSTRACT

Global concerns over environmental pollution have resulted in increasingly stringent regulations to control the levels of critical air pollutants, such as, carbon monoxide (CO), nitrogen oxide species (NOx), volatile organic compounds (VOC) and particulate matter (PM). These pollutants are removed by heterogeneous catalysis and the platinum group metals (PGM) remain the catalysts of choice but this situation is now complicated by the requirement for higher performance at lower costs while the PGM are experiencing escalating prices. A solution to this problem is the use of nanostructured perovskite-based Nanoxite™ catalysts engineered with unique structural features and high surface areas that enable higher catalytic efficiency at lower temperatures without sacrificing durability performance. In fact, Nanoxite is a "catalytic washcoat" product in that it simultaneously functions as the emission control catalyst while providing the bulk of the washcoat. As a result, both the PGM level and the amount of conventional washcoat materials are simultaneously reduced. Each powder particle possesses a hierarchical structure where larger micron sized particles hold the < 40 nanometer size perovskite grains. This desired arrangement facilitates easy powder handling and eliminates reactivity typically associated with discrete Nanograin materials. These perovskite-based catalyst formulations are applicable to both diesel engine and stationary emission control with respect to CO / VOC oxidation and the management of NOx and PM.

Keywords: nanostructured, perovskite, Nanoxite, diesel, emission control, CO/VOC oxidation.

1 INTRODUCTION

Significant pollution reduction has been achieved over the last few decades as a result of the implementation of catalyst technology. The Manufacturers of Emission Controls Association (1) reported that automotive catalyst technology has cut pollution by more than 1.5 billion tons in the USA since the mid-1970s. While PGM continue to be the standard catalysts in emission control, both environmental and economic conditions have spurred the search for alternative materials. Over the last few years, Pt pricing jumped from US\$14/g to US\$33/g. Another complicating factor is the steadily tightening of environmental regulations with respect to critical air contaminants suite of CO, NOx, VOC and PM. These tougher standards require more efficient catalysts and lead to higher levels of PGM, with the resulting cost increases. As a result, there is a deep interest to both lower the level of PGM usage and implement alternative non-PGM or significantly reduced PGM catalyst formulations.

Several different approaches are being evaluated and these include the use of active nanomaterials, computer modeling to allow strategic placement of PGM particles for reduced usage rate, platinum (Pt) - palladium (Pd) combination, Pd-loaded perovskites, and improved precious metal thrifting. Recently, researchers at Carnegie Mellon University (2) have quantified the possible impact of nanotechnology on the reduction of PGM usage to meet current and future standards. They estimated that the use of nanotechnology could reduce the annual PGM demand by 139 tons in 2030 while still meeting the environmental standards.

An attractive alternative is based on simple perovskite type material formulations. It is well established that perovskites with the general formula, $ABO_{3\pm\delta}$, exhibit catalytic activity with respect to oxidation reactions, with the performance linked to the nature and valence states of the A and B ions. However, despite many years of research, application of perovskite-based catalysts has been limited as a result of both non-competitive performance from un-optimized material structures and high levels of sulfur in the fuel streams.

With respect to material optimization, Nanox has made major improvements with its Nanograin Nanoxite™ oxidation catalyst formulations that exhibit superior performance over existing commercial technologies at significantly reduced PGM levels. Using its robust, patented Activated Reactive Synthesis (ARS) Technology, Nanox has established the capability to engineer these perovskite-based catalysts with features that enable higher conversion efficiencies of critical air pollutants at lower temperature.

Diesel fuel regulations are changing with the mandated decrease in the sulfur levels. In 2006, the amount of sulfur in diesel for on-road applications will drop to 15 ppm in the USA. Similar restrictions are being placed elsewhere in the world, with off-road reduction to follow. This is a dramatic change from the earlier high levels and facilitates the use of base metal type chemistries.

This paper will review some of the advances made by Nanox for the use of perovskite-based catalysts for environmental emission control. Production technology, powder properties, wash coat capability and performance data will be highlighted to support the applicability of these lower cost catalysts, high performance products as diesel oxidation catalysts (DOC), and oxidation of CO and VOC. The work reported here covers a specific formulation based on lanthanum and cobalt, with doping of iron (Fe) and rare earths in the classical ABO_3 structure.

2 PRODUCTION TECHNOLOGY

Numerous methods have been presented to produce perovskite materials over the years and these techniques include precipitation, sol-gel, thermal processing, and decomposition processes (3). However, not all these techniques are capable of generating catalysts with the required structural features for high catalytic activity. The catalytic activity of perovskites is related to several parameters that include ionic conductivity, oxygen mobility in the lattice, reducibility, and oxygen sorption (4). In order to achieve a formulation that satisfy these conditions, Nanox has pioneered the ARS approach based on a thermo-mechanical technique that combines synthesis and activation to capture the critical features for catalysis.

ARS is a simple production process with strategic advantages that are summarized in Table 1 below. The process steps simply involve conventional raw materials mixing, synthesis, activation, drying and particle separation. Scalability and manufacturability are easily attainable with this technology.

MANUFACTURING	FEATURES
Capability	Varying grain size, activity
Flexibility	Composition control
Manufacturability	Process Control, automation
Safety	Aqueous, Large particles

Table 1: Activated Reactive Synthesis Technology Features

The engineered hierarchical structural arrangement allows free flowing, dust free operation while the handling of discrete nanograins is eliminated. The larger micron-sized particles hold the perovskite nanograins, as evident in Figure 1.

This structure imparts stability and easy handling while being compatible with wash coating technology in place today. This compatibility eliminates the need for any change in wash coating production processes. The sub-particle arrangement facilitates easy mixing with other wash coat additives typically used in these formulations. Shelf life data confirms that there is no activity degradation over time with standard powder storage protocols.

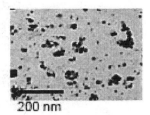

Figure 1: Hierarchical Particle & Grain Arrangement

Table 2 shows some typical properties of the basic lanthanum based formulation. Typically, the grain size exceeds 50 nm while the surface area is above 50 m^2g^{-1}.

PROPERTIES	TYPICAL DATA
Chemical Composition	Lanthanum-based oxides
Structure	Crystalline
Appearance	Black Powder
Bulk Density	0.8 g cm^{-3}
Primary Grain Size	< 50 nanometres
Surface Area (BET)	> 50 m^2 g^{-1}
Mesh Size	- 200

Table 2: Typical Properties of a LaCo-based Formulation

Another advantage of the production technology is linked to the capability of producing stoichiometric materials that are free of residual raw materials and impurities that degrade activity. Figure 2 below shows a typical X-ray diffraction (XRD) profile of a lanthanum cobalt based perovskite.

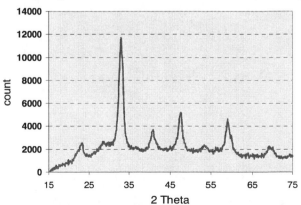

Figure 2: XRD Profile of the LaCo-based Perovskite

Perovskite formulations made with ARS have been compared with similar perovskite formulations produced by other traditional techniques. Temperature programmed desorption studies confirm a clear differentiation of the Nanoxite catalyst especially in terms of oxygen desorption and as a result, these unique features of the Nanoxite facilitate the superior catalytic performance. as shown in the comparative data summarized in Figure 3. The data is not surprising as the Nanoxite powders exhibit better oxygen species distribution and vacancies levels that influence diffusion with the resulting higher activity.

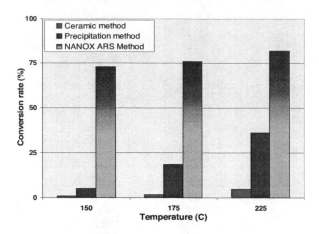

Figure 3: Catalytic Activity of Production Methods

3 WASHCOAT PREPARATION

Coating on substrates is easily completed on cordierite and metallic monoliths with aqueous 50% solids slurry, typically comprised of a mixture of 75% Nanoxite-EC1, 23% high specific surface stabilized ceria-zirconia and 2% of alumina. This solid phase is milled with water at pH 5.5 to 5.8, with loading is at 2.0 g/in^3 with at least 1.5 g/in^3 achievable in a single coating pass. Figure 4 shows a micrograph of the perovskite formulation coated onto a cordierite core (400 cpsi).

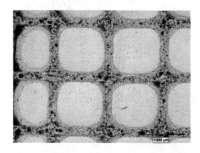

Figure 4: Cordierite Monolith with Perovskite Washcoat

4 PERFORMANCE AND APPLICATIONS

Nanoxite evaluation protocol involves the use of a gas feed stream typically composed of the following: 2000 ppm CO; 200 ppm propylene; 10% H_2O; 20% oxygen, varying levels of SO_2, and inert gas. Detection is completed by FT-IR system directly connected to the quartz micro-reactors at a GHSV of 30,000 h^{-1}. The catalysts are first "degreened" at 400 °C in the test gas stream prior to measurement of catalytic activity.

Figure 5 shows the conversion efficiencies of the Nanoxite coated onto cordierite substrate (400 cpsi) at a coating thickness of 2 g per cubic inch (ci), without any PGM. The data also reveal that low temperature conversion of both CO and propylene are possible in the absence of the PGM.

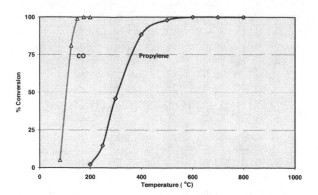

Figure 5: CO & Propylene Conversion

In durability studies, these Nanoxite catalysts maintain the high activity levels after aging for 200 hours at 450 °C in the test gas stream. The performance summarized in Figure 6 shows a stable performance profile, with the actual activity measured at 150 °C over the aging duration. The performance supports the use of these improved formulations as diesel oxidation catalysts and CO/VOC conversion.

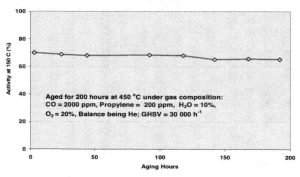

Figure 6: Hydrothermal stability of Nanoxite catalyst

Some applications may require the presence of a limited amount of Pt and/or Pd. These perovskite-based materials are quite compatible with the PGM that are easily added to the washcoat slurry to be incorporated into the perovskite-PGM type composite.

One of the challenges faced by base metal type chemistries is the presence of sulfur in the gas stream. Base metal type catalysts, as with PGM, are poisoned by sulfur dioxide. The current Nanoxite formulation has shown resistance to levels of < 15 ppm SO_2 which can translate to around 600 ppm S in the diesel fuel, assuming a 40:1 air:

feed ratio. We have also shown that addition of low levels of Pt enhances the stability with respect to sulfur poisoning.

The perovskite-based catalyst has a bi-functional role in that while it provides the catalytic conversion power, it also occupies the bulk of the washcoat layer, hence the term "catalytic washcoat". As such, the activity is proportional to the thickness of the perovskite catalytic washcoat layer. This allows the "tuning" of activity profile through thickness control and the incorporation of smaller amounts of PGM, if needed.

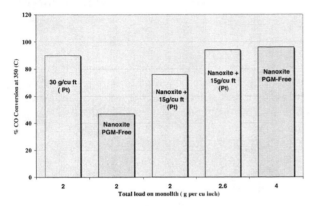

Figure 7: Effect of Nanoxite catalyst loading

Figure 7 shows the effects of this washcoat thickness variation and also the compatibility of perovskite-PGM combination with respect to the conversion of carbon monoxide. These results were obtained a much higher space velocity of 120,000 h^{-1}, and with a washcoat composition of 75% Nanoxite with 25% alumina, further strengthening the value proposition of these lower cost catalysts. The data shows that a high Nanoxite coating thickness of 4 g/ci yielded a conversion efficiency higher than that obtained for a Pt catalyst loading of 30 g per cubic foot (cf). In addition, Nanoxite loading at 2.8 g/ci with Pt at 15 g/cf also showed better performance than the high Pt level of 30 g/cf. Significant cost reduction is expected through the decrease in the Pt usage.

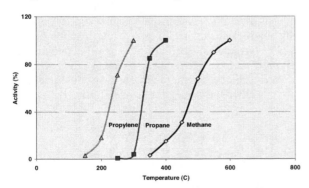

Figure 8: Catalytic Conversion of Hydrocarbons

Perovskite-based catalysts have also shown good conversion with respect to saturated hydrocarbons, such as, methane and propane. Figure 8 shows the profile of both saturated and unsaturated hydrocarbons, with the conversion efficiency higher for propylene as compared to methane. It is to be noted that this competitive conversion levels have been achieved in the absence of PGM.

5 CONCLUSIONS

The evaluation data presented in this paper provides clear and compelling incentives for the introduction of these perovskite-based catalysts in environmental emission control. Conversion of CO and VOC species are easily attained in the absence of the platinum group metals. In particular, the Nanoxite catalysts exhibited performance suitable for use as diesel oxidation catalysts (DOC) and CO/VOC oxidation.

The catalysts are manufactured from standard, conventional raw materials via robust, capable manufacturing process. The particles are strategically engineered with a hierarchical structure that confers stability while being compatible with existing industrial wash coating practices and demonstrating higher catalytic activity.

Perovskite-based catalysts are not new. These materials have been researched for decades but failed to make any significant penetration into the emission control catalyst markets. The major reasons for this situation can be simply linked to a few factors, namely, the un-optimized structures, hydrothermal stability and high levels of sulfur in the gas streams. Nanox has now changed this status with the development of its Nanoxite line of nanocrystalline perovskite-based catalyst formulations.

With the tightening of environmental emission regulation, escalating Pt pricing and demand for higher performance at lower costs, the timing for the broad use of these lower cost catalysts is now. Continuing developments are in progress to expand the applications of these materials to improved NOx management and PM reduction.

REFERENCES

[1] Manufacturers of Emission Controls Association, "Clean Air Facts", www.meca.org/galleries/default-file/advancedfact/pdf. Washington.

[2] S. M. Lloyd, L. B. Lave, and H. S. Matthews, Environ. Sci. Technol. 39, 1384, 2005.

[3] L. G. Tejuca and J. L. G. Fierro, Eds. "Properties and applications of perovskite-type oxides", Marcel-Dekker, 1–25, 1993.

[4] I. Rossetti and L. Forni, Applied Catalysis B: Environmental, 33, 345, 2001.

The authors acknowledge the contribution of Mr. Eric Deschenes, Mr. David Provencher, Mr. Philip Robert and Mr. Yves Rouleau.

Nano-Engineering Of Magnetic Particles For Biocatalysis And Bioseparation

C.H. Yu[a], C.C.H. Lo[b], C.M.Y. Yeung[a], K. Tam[c] and S.C. Tsang[a*]

[a]Surface and Catalysis Research Centre, School of Chemistry, University of Reading, Whiteknights, Reading, RG6 6AD, UK
[b]Center for Nondestructive Evaluation and Ames Laboratory, Iowa State University, Ames, Iowa 50011,USA
[c]AstraZeneca, Mereside, Alderley Park, Macclesfield, Cheshire SK10 4TG, UK

*Corresponding authors, Professor S.C.E Tsang Tel: +44-118-378-6346; fax: +44-118-378-6591 *E-mail address*: s.c.e.tsang@reading.ac.uk

ABSTRACT

Magnetic nanoparticles encapsulated in a thin coating as magnetic separable *nano-vehicle* for chemical species is a hot but challenging area. The facilitated separation of a small magnetic body carrying biologically active species is of a tremendous interest however; the stability of the magnetic body remains a key issue. We report new syntheses of silica encapsulated magnetic nanosize particles as magnetic separable carriers in large quantities based on simple synthetic techniques. The major advantage of using nano-size magnetic particles as carriers is that they display an excellent mass transfer coefficient (high surface area to volume ratio) comparable to soluble species but can still be easily separated from liquid using magnetic interaction with an external applied inhomogeneous magnetic field (i.e. 50MGOe). It is shown that the external coating surfaces can isolate and protect the magnetic core from destructive reactions with the environment where a wide range of conditions for fine chemical catalysis can be made possible. The functionalized surfaces could also offer anchoring sites for the immobilization of active chemical species of interests (enzymes, DNA oligos and antibodies). Most of these applications require nanoparticles covered with appropriate surface chemical functionalities where a strong magnetic core is essential for the separation of each particles from solution.

Keywords: magnetic nanoparticles, biocatalysis, bioseparation, silica encapsulated iron oxide

1 INTRODUCTION

The synthesis of magnetic nanoparticles has been intensively investigated not only for their electrical, optical and magnetic properties but also for their other technological applications including magnetic assisted bio-separation and bio-catalysis. A number of magnetic nanoparticles have recently been applied in biotechnology areas [1]. In biomedicine, magnetic nanoparticles can be used as labeling or imaging reagents when tagged with biological entities. These approaches are very attractive to industry as the magnetically tagged bio-molecules can be recycled easily using magnetic means from solution (minimize the waste production through regeneration [2].)

In general, the most common synthetic methods for the synthesis of magnetic materials with nanometer-scale dimensions can be classified into three categories. They are physical vapour deposition, mechanical attrition and chemical routes from solution. In both the vapour phase and solution routes, the particles are assembled from individual atoms to form nanoparticles. Alternatively, mechanical attrition involves the fracturing of larger coarse-grained materials to form nanostructures. The chemical routes from solution are widely used for the fabrication of nanoparticles. They often provide the best methods for production of nanoparticles due to their enhanced homogeneity from the molecular level design of the materials and, in many cases, cost effective in bulk quantity production. The solution chemical routes provide mild reaction conditions and require less expensive equipments. They also allow control of particle size and size distribution, morphology, and agglomerate size through the individual manipulation of the parameters that determine nucleation, growth, and coalescence. Surface modification of the particles during synthesis or post-synthesis is easily accomplished, and provides additional functionality to the nanoparticle.

Here, we report a single-step solution based chemical synthesis of silica encapsulated iron oxide nanoparticle using microemulsion technique, which can be adopted in bio-catalysis and bio-separation areas [3]. The silica coated iron oxide nanoparticle is also shown to cover with surface hydroxyl groups which facilitate a strong attachment of protein such as Bovine Serum Albumin (BSA) onto its surface.

2 EXPERIMENTAL

2.1 Synthesis

Chemicals: Iron (II) chloride tetrahydrate ($FeCl_2$ $4H_2O$), iron (III) chloride hexahydrate ($FeCl_3$ $6H_2O$), ammonium hydroxide (35 wt%), cetyltrimethylammonium bromide (CTAB) , dried toluene, and tetraethyl orthosilicate (TEOS). All aqueous solutions were prepared with deoxygenated water. Potassium dihydrogen orthophosphate (99%, Aldrich). Sodium chloride (99%) was received from Fisher. Bovine serum albumin (fraction V, 99%), deionized water was used for all experiments.

Synthesis of Silica Coated Fe₃O₄ Particles in Microemulsion: 7.3 g CTAB was added into 180 mL dried toluene in a 250 mL round bottom flask with stirring for 4 h. In general, the higher Wo value used, the larger of the nanoparticles size we observed. In this case, the mole ratio of water to surfactant (Wo =[water]/[CTAB]) of 20 was fixed. Thus, Fe (II)/Fe (III) salt were dissolved in 7.2 mL deionized water by adding the Fe (II)/Fe (III) solution dropwise into the round bottom flask constantly flushing with nitrogen for 2h. After further four hours, ammonium hydroxide was added dropwise to the same flask with a continuous nitrogen purge. The color changed from light yellow to dark brown without precipitation. 6.951 g TEOS was directly added into the microemulsion and the mixture was allowed to age for 5 days to encourage hydrolysis and condensation of the silica precursor. After aging, the resulting nanoparticles were collected as a precipitate when ethanol was added into the solution. The precipitate was then washed with excess ethanol and re-dispersed in toluene. This step was repeated for five times in order to remove the surfactant from the precipitate (until the FTIR showed no trace of surfactant). At last, the precipitate was washed with acetone and left in air for drying at room temperature overnight. The formation of iron oxide nanoparticle from the chemical reaction of ammonium hydroxide with Fe(II)/Fe(III) species is shown in the equation 1, below and Scheme 1 graphically summaries the preparative procedure.

$$FeCl_2 + 2FeCl_3 + 8NH_3 \cdot H_2O \rightleftharpoons Fe_3O_4 + 8NH_4Cl + 4H_2O$$

Equation 1: Iron oxide nanoparticle prepared from ammonia precipitation of Fe(II) and Fe(III) according to chemical reaction

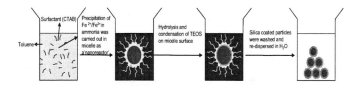

Scheme 1 Synthetic procedure for preparing silica encapsulated iron oxide nanoparticles

Directly immobilization of BSA onto silica@iron oxide
5 mL 315µM of BSA solution was added into 63.6 mg of silica@iron oxide in a 10mL vial and the mixture was agitated at room temperature overnight. The solid immobilized BSA was separated from solution by applying an external magnetic field (50MGOe).

2.2 Characterization

Figure 1 shows a typical X-Ray Diffraction (XRD) pattern of iron oxide in silica. There are three strong peaks

identified from this spectrum corresponding to the lattice spacing of 2.951, 2.509, and 1.475. These match with either the structures of Fe_3O_4 or γ-Fe_2O_3 as compared with an XRD database. A very broad diffraction peak from 9-23° (2θ) corresponding to the poorly crystalline silica is also observed. The average particle size can be derived using Deby-Scherrer equation from the full width at half maximum (FWHM) of the strongest peak. It is noted that the patterns match well with both crystalline magnetite (Fe_3O_4) or maghemite (γ-Fe_2O_3) phases. However, assignment to one of this phase or to the mixture of them based entirely on XRD proofs very difficult since their closely related structures (with almost identical patterns) and nano-metric regime (peak broadening)

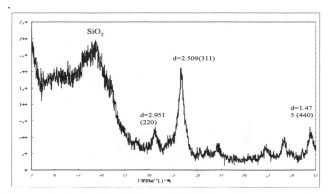

Figure 1: Normalized X-ray diffraction spectrum patterns of silica encapsulated iron oxide in 6.46 nm.

Figure 2 shows a typical high resolution transmission electron microscopic (HRTEM) image of the silica encapsulated iron oxide nanoparticle. It is noted that the image indeed reveals the highly crystalline structure of the iron oxide core (with a lattice spacing of 2.5± 0.1 Å) in an amorphous coating .

Figure 2: HRTEM image of the highly crystalline structure magnetite/ maghemite in silica.

Figure 3a and 3b show general and enlarged vibration saturation magnetization (VSM) responses of the silica encapsulated iron oxide nanoparticle powder upon application of various external magnetic fluxes. The measurements were collected at room temperature. It is evident that the powdered material shows no magnetic hysteresis with both the magnetization and demagnetization curves passing through the origin, which clearly indicates the superparamagnetic nature of the material. This means that the magnetic material can only be aligned under an applied magnetic field but will not retain any residual magnetism upon removal of the field [3]. Thus, this technique appears to be able to prepare magnetically nanoparticles for magnetic separation purposes.

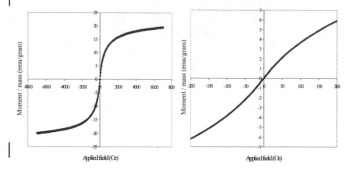

Figure 3: General and enlarged VSM plots of silica encapsulated iron oxide

Figure 4 displays the elemental mapping of the isolated particles of the silica encapsulated iron oxide. After taking the correction of the response factor for each element into account, the atomic ratios of the particle are found to be Fe : O : Si =19.93 : 71.96 : 8.11 with a standard deviation of ± 0.2 %. (excluding carbon analysis because the use of carbon filmed holder which also affects the oxygen analysis).

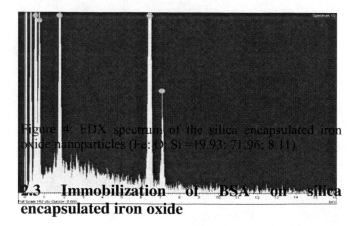

Figure 4: EDX spectrum of the silica encapsulated iron oxide nanoparticles (Fe : O : Si =19.93: 71.96: 8.11)

2.3 Immobilization of BSA on silica encapsulated iron oxide

The ability of the silica encapsulated iron oxide for binding protein is clearly illustrated in the experiment of BSA attachment to the particle. Figure 5 shows the FTIR spectral characteristics of silica encapsulated iron oxide with and without the BSA attachment. Both spectra show an absorption peak near 960 cm^{-1} suggesting the presence of hydroxyl groups. It is noted that the fresh silica encapsulated iron oxide sample appears to contain more hydroxyl groups from its spectrum. It is evident that the appearance of new absorption peaks at 1648 and 1540 cm^{-1}

which are the characteristic features of the BSA, is seen from the sample upon the BSA immobilization. A similar FTIR spectrum was obtained in a previous work who used naked iron oxide for the BSA immobilization [5,6]. It is particularly noted that the high peak near 1648 cm^{-1} corresponds to the Amide I of the BSA [7]. This peak has been regarded as a good indicator whether the immobilized BSA still remains functional as compared to the native free form. If the immobilized BSA is somehow denatured, its secondary structure will be altered leading to a significant shift in this Amide I region [8]. Therefore, the FTIR analysis clearly suggests that the BSA is successfully immobilized onto the silica encapsulated nanoparticle probably via the hydroxyl linkage and the attached protein likely remains functional (no peak shift).

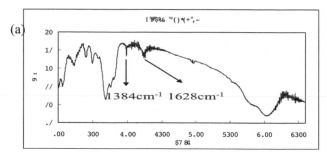

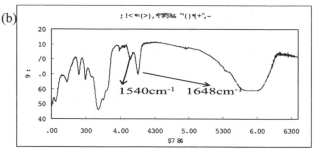

Figure 5 FT-IR of (a) silica encapsulated iron oxide and (b) BSA-bound silica encapsulated iron oxide

3 CONCLUSION

To summarize, we describe a one-step chemical preparative procedure for the synthesis of silica encapsulated iron oxide nanoparticles in solution using microemulsion method. Characterization suggests the material matches with the magnetite or maghemite phases in silica and the sample shows superparamagnetic properties. The material is also shown to cover with hydroxyl groups which assist for the immobilization of BSA thereupon. It is known that BSA is a protein of 66,000 Dalton molecular weight and with approximately 4 × 14 nm diameter. Each albumin molecule has at least 6 distinct binding sites for drugs and endogenous compounds. As a result, this new magnetic immobilized protein may find applications in magnetic drug delivery/administration. Nevertheless, the important point is that the reported technique may be employed for the tagging of a wide variety of biological entities for biomedical or biochemical applications useful for magnetic switch, recording storage and magnetic bio-separations applications.

Cleantech 2007, ISBN 1-4200-6382-0

REFERENCES

[1] F. C. Meldrum, B. R. Heywood, S. Mann, Science, 257, 522, 1992.

[2] S. C. Tsang, V. Caps, I. Paraskevas, D. Chadwick, D. Thompsett, Angewandte Chemie-International Edition, 43, 5645, 2004.

[3] H. H. Yang, S. Q. Zhang, X. L. Chen, Z. X. Zhuang, J. G. Xu, X. R. Wang, Analytical Chemistry, 76, 1316, 2004

[4] X. Gao, C. H. Yu, K. Y. Tam, S. C. Tsang, Journal of Pharmaceutical and Biomedical Analysis, 38, 197, 2005.

[5] Q. Zhang, H. F. Zou, H. L. Wang, J. Y. Ni, Journal of Chromatography A, 866, 173, 2000.

[6] J. Y. Yoon, J. H. Lee, J. H. Kim, W. S. Kim, Colloids and Surfaces B-Biointerfaces, 10, 365, 1998.

[7] T. W. Xu; R. Q. Fu. Chemical Engineering Science, 59, 4569, 2004.

[8] G. Reiter; N. Hassler; V. Weber; D. Falkenhagen; U. P. Fringeli. Biochimica Et Biophysica Acta-Proteins and Proteomics, 1699, 253, 2004.

Gas Sensor Arrays by Supersonic Cluster Beam Deposition

E. Barborini[*], M. Leccardi[*], G. Bertolini[*], O. Rorato[*], M. Franchi[*], D. Bandiera[*], M. Gatelli[*], K. Wegner[*]
A. Raso[**], A. Garibbo[**], C.Ducati[***], P. Piseri[****], P. Milani[****]

[*]Tethis s.r.l., Piazzetta Bossi 4, I-20121 Milan, Italy, emanuele.barborini@tethis-lab.com
[**]Selex-comms, via Negrone 1/A, 16153 I-Genova, Italy
[***]Materials Science & Metallurgy, University of Cambridge, Pembroke Str., Cambridge CB2 3QZ, UK
[****]CIMAINA and Dipartimento di Fisica, Università di Milano, Via Celoria 16, I-20133 Milan, Italy

ABSTRACT

Supersonic cluster beam deposition was employed to produce nanostructured thin films of transition metal oxides to be used as gas sensors. Due to high nanoparticle beam collimation, patterned deposition was easily obtained by using hard masks, achieving sub-micrometric resolution. To exploit hard mask patterning micro-machined substrates having an array structure were developed in order to deposit materials with different properties on each single element of the array. Results on the deposition of nanostructured TiO_2, WO_3, SnO_2 and on the detection of volatile organic compounds (VOC) and gases related to environmental pollution (such as NO_x) are reported.

Keywords: gas sensors, thin film, nanostructure, aerosol deposition, clusters

1 INTRODUCTION

Films of metal oxides kept at high temperature change their electrical conductivity in presence of reactive gases [1]. This process is generally reversible and the original conductivity is restored after desorption of the reactive species. In fact, at a certain temperature the chemical composition of the metal oxide surface reaches an equilibrium state with the chemical composition of the atmosphere to which it is exposed. Modifications in chemical composition of the atmosphere change the equilibrium conditions and the surface chemistry, as a consequence. The electrical properties of the film, such as conductivity, follow these changes. Gas sensors based on these phenomena have been developed, reaching market requests in 1968. The most important materials for gas sensing applications are simple metal oxides, such as SnO_2, TiO_2, WO_3, In_2O_3, ZnO.

Recently, the use of nanostructured thin films as sensing materials attracted interest due to the possibility to enormously increase the specific surface area of the films, that is the fundamental parameter for most applications relying on gas-solid interactions. Deposition of nanoparticles on sensor substrates by wet chemistry methods such as screen printing or doctor blading typically results in rather thick films and is prone to crack development upon evaporation of the solvent. These drawbacks can be circumvented by direct gas phase nanoparticle synthesis and deposition. Among various deposition techniques, supersonic cluster beam deposition (SCBD) turned out to be very promising as nanoparticles of controlled and selected size are directly deposited on substrates without the need of additional treatment and coating steps [2, 3]. Another advantage is the possibility to obtain porous nanostructured films on every kind of micro-machined platform at room temperature and in ultra-clean conditions. A broad variety of substrates is available, including suspended silicon membranes. Soft landing and limited diffusion are characteristics of the deposition process, causing the film to grow accordingly to a highly porous structure at the nanoscale. By using hard masks with micrometric resolution, patterned films can be obtained [4].

Here, we report on the gas sensing properties of nanostructured thin films of transition metal oxides made with the SCBD technique.

2 EXPERIMENTAL

Supersonic cluster beam deposition was employed to produce nanostructured thin films of TiO_2, WO_3, SnO_2. The cluster beam was generated by a pulsed microplasma cluster source (PMCS) [2, 5]. The operation principle of the PMCS is based on the ablation of a metallic rod by a plasma jet (He or Ar), ignited by a pulsed electric discharge in high vacuum conditions. After ablation, metallic atoms thermalize into inert gas and condense to form clusters that are entrained by the gas flux towards the PMCS exit nozzle. The nanoparticles are extracted from the PMCS by supersonic expansion and deposited on a substrate intersecting the beam [2, 3]. Following nozzle expansion, the cluster beam is directed through a set of aerodynamic lenses [6] in order to achieve beam stability, high collimation, high in-axis intensity, and high deposition rates.

The growth of nanostructured films takes place at room temperature on substrates exposed to the cluster beam. Due to the high collimation of the beam, patterned depositions were easily obtained by hard mask method. Figure 1 schematically shows the deposition technique. Soft landing and limited diffusion cause the film to grow to a highly

Cleantech 2007, ISBN 1-4200-6382-0

porous structure. Exposition to air causes the oxidation of the films. The proper oxide stoichiometry is reached during post-deposition high temperature annealing in air. Besides stoichiometry adjustment, annealing is needed to fix the nanostructure of the sensing materials in order to avoid any further modification during sensor operation.

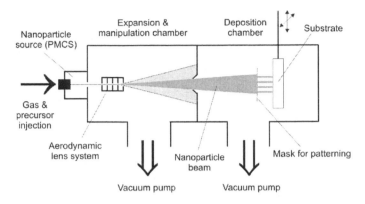

Figure 1: Schematic of the supersonic cluster beam deposition (SCBD) unit with nanoparticle source, particle size selection / beam formation zone and mask for patterned deposition.

3   RESULTS AND DISCUSSION

Figure 2 shows the nanostructure of an as-deposited tungsten oxide film in transmission electron microscopy (TEM).

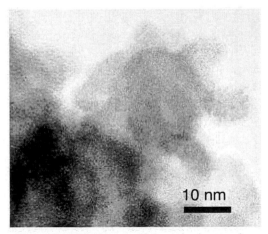

Figure 2: TEM image of deposited amorphous tungsten oxide nanoparticles of 5 to 10 nm diameter.

The film has an open and porous structure at the nanoscale attributed to particle impact with low kinetic energy. Due to their small size and low mass, nanoparticles acquire very little kinetic energy, even when accelerated to supersonic velocity. No lattice fringes are discernible in the as-deposited nanoparticles shown in Figure 2, indicating an

amorphous state. By annealing of the film, the amorphous grains rearrange into a polycrystalline nanostructure.

The gas sensing properties of the films were evaluated with respect to various gaseous species interesting for environmental monitoring, such as CO, NO_x and SO_2, as well as volatile organic compounds (VOC), such as ethanol. Measurements were carried out in a temperature-controlled test cell under well defined atmosphere. The composition of the gaseous atmosphere could be controlled by precision mass-flow meters in a 5-line mixing system. Thereby, individual gaseous compounds could be added at trace impurity level to an inert carrier gas or oxygen/nitrogen mixtures. An electrometer was used to measure the current across the films at a fixed voltage of 5 V during the test sequence, consisting in the injections of suitable amounts of gas into test cell. The injection protocol and the signal acquisition were automatically processed by a PC. Figure 3 shows the response of nanostructured WO_3 film to NO_2 at concentrations less than 10 ppm. The temperature of the sensor was kept constant at 200 °C.

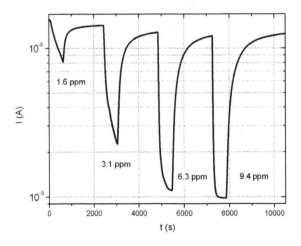

Figure 3: Response of a sensor with nanostructured WO_3 layer to different concentrations of NO_2.

The four peaks of current decreasing correspond to the injection in the test cell of 1.6, 3.1, 6.3, and 9.4 ppm of NO_2. Time sequence was 10' on-time, 30' off-time.

By exploiting hard mask patterning, we developed micro-machined substrates having a 4x4 array structure in order to deposit different oxides on each single element of the array. A thin film heater and a Pt thin wire thermometer were integrated on the back and on the front side of the substrate to control the operation temperature of the sensor. Batch deposition of micro-machined platforms was carried out in order to obtain multi-element sensors to interface neural network analysis algorithms.

4 CONCLUSIONS

Porous nanostructured metal oxide films for gas sensing applications can be prepared by supersonic cluster beam deposition and post-deposition thermal treatment. Using hard mask patterning, it is possible to simultaneaously prepare arrays with individual sensing elements. Alternatively, each array element can be coated with a different oxide resulting in gas sensors with different performance. The use of such a combinatorial approach can be of great help for the understanding of the mechanisms underlying gas selectivity and for the efficient and inexpensive realization of microsensor arrays, e.g. for environmental monitoring.

REFERENCES

[1] T. Seiyama, A. Kato, K. Fujiishi, and N. Nagatani, Anal. Chem, 34, 1502, 1962.

[2] E. Barborini, I.N. Kholmanov, P. Piseri, C. Ducati, C.E. Bottani and P. Milani, Appl. Phys. Lett. 81, 3052, 2002.

[3] I.N. Kholmanov, E. Barborini, S. Vinati, P. Piseri, A. Podesta', C. Ducati, C. Lenardi, and P. Milani, Nanotechnology 14, 1168, 2003.

[4] E. Barborini, P. Piseri, A. Podestà, P. Milani, Appl. Phys. Lett. 77, 1059, 2000.

[5] E. Barborini, P. Piseri and P. Milani, J. Physics D: Applied Physics 32 , L105, 1999.

[6] P. Liu, P.J. Ziemann, D.B. Kittelson and P.H. McMurray, Aerosol Sci. Technol. 22, 293, 1995.

Cleantech 2007, ISBN 1-4200-6382-0

Preparation and Characterization of Various Nanofluids

W.C. Williams, I.C. Bang, E. Forrest, L. W. Hu[*], J. Buongiorno

Massachusetts Institute of Technology
Department of Nuclear Science and Engineering
77 Massachusetts Avenue, Cambridge, MA 02139
[*]lwhu@mit.edu

ABSTRACT

As part of an effort to evaluate water-based nanofluids for nuclear applications, preparation and characterization has been performed for nanofluids being considered for MIT's nanofluid heat transfer experiments. Three methods of generating these nanofluids are available: creating them from chemical precipitation, purchasing the nanoparticles in powder form and mixing them with the base fluid, and direct purchase of prepared nanofluids. Characterization of nanofluids includes colloidal stability, size distribution, concentration, and elemental composition. Quality control of the nanofluids to be used for heat transfer testing is crucial; an exact knowledge of the fluid constituents is a key to uncovering mechanisms responsible for heat transport enhancement.

Keywords: nanofluids, preparation, characterization

1 INTRODUCTION

Nanofluid is the common name of any sol colloid involving nanoscale (less than ~100nm) sized particles dispersed within a base fluid. It has been shown previously that the dispersion of nano-particulate metallic oxides into water can increase thermal conductivity up to 30-40% over that of the base fluid and anomalously more than the mere weighed average of the colloid [1]. An increase in thermal conductivity, and therefore heat transfer coefficients, without a noticeable increase in friction pressure losses would be a significant method for improving the heat transfer performance of any thermal system. The aim of nanofluid research at MIT is to develop the facilities and methodologies for the creation and characterization of water-based nanofluids and to investigate them as convective heat transfer media in single and two-phase flow.

2 PREPARATION OR PURCHASE OF NANOFLUIDS

Three main types of nanofluids are studied here: those made from purchased powders mixed with water, those made from chemical precipitation in the liquid, and those which are purchased. It is very important to understand fully the constituents of any nanofluid under investigation in order to draw significant conclusions as to the heat transfer phenomenon.

2.1 Preparation

Our first attempt to make nanofluids was with nanoparticle powders available from Sigma-Aldrich mixed with a base fluid. In order to properly disperse a powder into a liquid system, the particles must be wetted by the medium, large aggregates must be broken down, particles must be homogenized in the medium, and finally reagglomeration must be prevented. Since the purity of the nanofluid is important, attempts were made to mix the particles directly with water with no additives. The metallic oxide nanopowders are not chemically reactive in atmosphere; however they do tend to form loose micro-sized agglomerates. The most effective method of breaking and evenly dispersing the powder in a fluid is through application of ultrasonic vibration (high speed stirring also works well). Using this methodology the nanofluids were created using the two oxide nanopowders (ZrO_2 and Al_2O_3) and ultrasonic vibration was applied for 12+ hours. The resulting nanofluids initially looked promising, but were not stable with time. Although some particles remained dispersed, the majority formed larger agglomerates and settled out of the liquid.

Further investigation in the literature found that the stability of nanofluids has multiple variables. As stated previously the small particle size gives the potential for the particle to escape settling due to gravity. The Brownian motion of the fluid keeps the particles aloft. However this small particle size dramatically increases the surface to volume ratio of the system. Full stability is only acquired through the separation of the two phases and the minimization of the surface area. The fully suspended system is then either unstable or metastable at best.

An energy barrier must be created between the particles in order to prevent them from passing from the unstable to stable energy states. This is known as the metastable state or "colloidally stable" [2]. The energy barrier must be strong enough to repel the majority of particles impinging on one another due to Brownian motion. The total interaction between particles is thus governed by a balance between van der Waals attraction and short range, electrostatic, and steric repulsions.

The repulsion barrier is typically created by surface charges on the particles. With like charges on the particle surfaces, the particles will repel one another thus preventing agglomeration. Surface charge is typically created by addition of acid or base to the fluid in order to create surface ionization and thus surface charges (electrostatic), or addition of like charged chemical surfactants which attach to the particle surfaces (steric). Both methods have been pursued and have proven effective in this study. Addition of HCl acid in low concentrations to the previously mentioned Al_2O_3 and ZrO_2 nanofluids created by ultrasonication created much more stable nanofluids. Various nanofluids were successfully made using oxide particles dispersed in ethylene glycol, ethanol, methanol, and isopropanol without addition of acids or bases. However the settling did slowly occur over longer periods of time.

Ultrasonic dispersion with surfactants was also somewhat successful. Attempts were made to disperse alumina nanoparticles in ethanol, methanol, isopropanol, and FC-72. Results good dispersion of the particles in ethanol, methanol, isopropanol, but stability was not long lasting. However the FC-72 was found to not wet the particles and therefore no dispersion could be achieved. After further investigation it was found from other works [3] that van der Waals attraction is highly dependent on the dielectric constant of the base fluid. Therefore the overall stability of the colloid or nanofluid depends greatly on the dielectric constant as shown in Figure 1. Dielectric constants are given in Table 1.

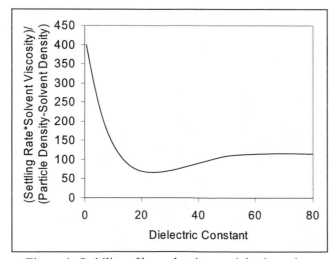

Figure 1: Stability of bare alumina particles in various fluids [3]

The effect of the dielectric constant on van der Waals forces and hence stability was seen first hand in the dispersion of alumina particles. Water, ethylene glycol, methanol, ethanol, and isopropanol all made colloids which were not very stable. If settling is normalized as in Figure1, then ethanol nanofluid was the most stable and water

nanofluid the least stable. Finally FC-72 would not form a colloidal state at all; the low dielectric constant even prevented wetting.

Water	80
Ethylene Glycol	37
Methanol	32.7
Ethanol	24.5
Isopropanol	17.9
FC-72	1.7

Table 1: Dielectric constants of various fluids

The other method for creating nanofluids tried in this study is through chemical precipitation with the addition of surfactants (steric). Making of a Fe_3O_4 nanofluid is described as follows: To begin, excess oxygen is removed from deionized water by bubbling through nitrogen. Then, a 1:2 molar ratio of $FeCl_2 \cdot 4H_2O$ and $FeCl_3 \cdot 6H_2O$ are added to the water. The mixture is stirred and heated to 80°C and becomes yellowish in color and the polymer/surfactant is added to the mixture until it is foaming.

In order to precipitate the particles a solution of Ammonium Hydroxide ~28% is added and the solution immediately turns black with the iron oxide nanoparticles. This solution is allowed to stir and fully react at 80°C for 30 minutes. The resulting nanoparticles are cleaned with acetone and separated from the liquid by using an electromagnet (particles are magenetite). Centrifuging is also possible. The cleaning process is repeated with deionized water, acetone, and the magnet several more times. The final particles are then heated to boil off the excess acetone and can then be dispersed in water. The final dispersion has been found to be stable indefinitely.

Attempts were made to disperse the separated iron oxide particles in to various other fluids. However the results were poor. The surfactant can have an adverse effect depending on the dielectric constant as well. It can almost invert the curve as shown in Figure 1, thus making very high and very low dielectric constant fluids more stable and moderate values less stable [3].

2.2 Purchase

Several companies were approached for samples in this project: Nyacol, Sigma-Aldrich, Nanophase Technologies, Applied Nanoworks and Meliorum Technologies to name a few. These companies produce various nanofluid products with varying quality. The major candidates for this project which are purchased are water-based with zirconia and alumina particles, however some other fluids have also been investigated, like silica, gold, and platinum in water. Gold in ethylene glycol was also available for purchase. Many powders are also available which come with surfactant coatings which allow for easy dispersion.

3 CHARACTERIZATION

It is important to be able to fully characterize the nanofluids under inspection for heat transfer enhancement. The first steps are to quantify the composition, size and loading of the nanoparticles and search for impurities in the nanofluids. Tools utilized to characterize and qualify nanofluids for this study include neutron activation analysis (NAA), inductively-coupled plasma spectroscopy (ICP), transmission electron microscopy (TEM) imaging, thermogravimetric analysis (TGA) and dynamic light scattering (DLS).

3.1 Transmission Electron Microscopy (TEM)

TEM is the primary technique to verify single particle dimensions and to identify agglomerations of particles. The electron beam can be used to see features on the nanometer level. A major drawback to the use of TEM is that samples must be dried out of solution in order to be attached to the carbon matrix. Therefore the particles are not exactly in nanofluid state and agglomeration might occur during drying. However TEM can be used in connection with dynamic light scattering to acquire exact sizing in nanofluid form. An example of one of the nanofluids (Sigma-Aldrich Al_2O_3 in water) can be seen in Figure 2. It is not known by this image alone as to the time of agglomeration, before or after drying of the base fluid. The mean particle size appears to be in agreement with the specifications of Sigma-Aldrich.

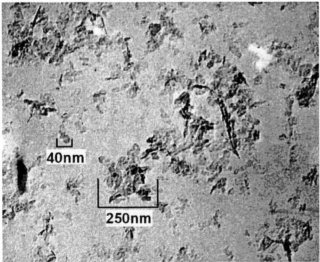

Figure 2: TEM image of Sigma-Aldrich Al_2O_3 nanofluid

3.2 Dynamic Light Scattering (DLS)

DLS can also be used to size particles in a dilute solution. When light from a laser is passed through a colloid the light will scatter uniformly from particles which are less than ~250nm. The time measurement of this scattered light intensity can be used to determine the size of the particles in the colloid. The time variation of the intensity can be correlated to the diffusion of the particles and, through the Stokes-Einstein equation and the fluid viscosity, the hydrodynamic radius of the particles which are moving.

However one of the main drawbacks of DLS is the measurement of very polydisperse systems. The intensity of the light scattered is proportional to the diameter of the particle. Therefore the sizes measured are correct, but the distribution of the particles in these size bins needs to be weighted according to their diameters. The scattering should follow Rayleigh's law for scattering and thus intensity is proportional to the diameter to the sixth order. If the DLS results, from the same Sigma-Aldrich Al_2O_3 nanofluid as viewed with the TEM, are corrected in this way then the majority of the particles are in the 40nm range with a few larger agglomerations around 200-300nm, which reinforces previous findings and the technical specifications of the fluid. The pH of the Sigma-Aldrich is around 4 in order to maintain the particles in dispersion. If the pH is allowed to rise to 7, the agglomerations were seen to increase significantly. This is expected from DLVO theory. Therefore care must be taken to maintain the nanofluid pH stable during dilution for the DLS measurement.

3.3 Neutron Activation Analysis (NAA)

Particle concentration and contamination in the nanofluid solution is another problem of characterization. MIT's research reactor was used to perform neutron activation analysis for the determination of nanoparticle and other impurity concentrations in the nanofluids. NAA is a method that uses the gamma decay emissions of the samples after having undergone neutron irradiation. Irradiation times depend on the materials under investigation and the half life of the gamma decay modes. Around sixty elements detectible using NAA.

Two zirconia nanofluid samples were irradiated, because of the long zirconium half life the samples were irradiated for 6 hours. Both samples were supplied as 10 percent zirconia by weight in water from the vendors. Samples will be designated as A and B. The results found that both nanofluids contained zirconium, sample A has 0.1027 gm/gm or 10.27 wt% zirconia and sample B has 0.0274 gm/gm or 2.74 wt% zirconia. It appears that sample B is dramatically different than the specifications. It was found that there was more zinc than zirconium in sample B. The zinc was found at 0.026 wt% in sample B, which is significantly contaminated. Sample A only has 6.5e-5 gm/gm which is insignificant. It is believed that the sample B specifications are either unreliable or that the batch is overly contaminated. Trace amounts (ppm levels) of other elements (K, Cl, Na) were found which are assumed as additives for pH control.

Some alumina nanofluids were also considered. Nyacol at 20 wt% (NY), Meliorum (ML) and Sigma-Aldrich (SA2) at 10 wt% were the three measured fluids. For the alumina both short and long irradiations were required. The long (6 hours) was for the majority of elements and the short (5 mins) was for the aluminum due to the short half life. The results of the long showed that the samples all contained nearly insignificant levels (ppm) of Na, K, and a few other common elements like Zn.

The results of the shorts gave the loading of the aluminum and thus alumina for the nanofluids. It was found that the NY had 19.37 wt%, specs gave 20wt%. The ML had 9.74 and the SA2 had 8.68 wt% for 10wt% in the specs. These seem to be in good agreement with the specifications of the vendors. The discrepancy of the SA2 is assumed to be due to settling of the nanofluid, because it was an older sample. It is concluded that NAA can be used for quality control of nanofluids. The major limiting factors of NAA is the time and cost of the procedure. It also involves handling of radioactive materials.

3.4 Inductively-Coupled Plasma Spectroscopy (ICP)

A more cost and time effective approach was also used for determining the nanoparticle loading and composition, i.e., the ICP spectrometer. Tests could be done using the ICP to determine concentration and contamination characterization of nanofluids much more effectively. ICP spectrometry uses the light spectrum released from injecting materials into an extremely high temperature plasma. Roughly sixty elements are detectible by ICP. Two samples were tested using the ICP: Al_2O_3 (Sigma-Aldrich) and Fe_3O_4 in water. One key to the analysis is that the nanofluid samples need to be diluted to stay below the concentration of calibration standards. Therefore the Al_2O_3 which was initially at 10wt% was diluted to 0.05wt% (200:1) and the Fe_3O_4 which was initially at 3.5wt% was diluted to 0.035wt% (100:1), both using deionized water.

It was found that the Al_2O_3 was at a concentration of 0.047wt% which is within ~5% of the specified value. Sodium and potassium were also found, which is consistent with the NAA results. These are used to stabilize the nanofluid using the pH method. The Fe_3O_4 was found to have a concentration of 0.034wt% which is within ~2% of the specified value. The sodium of the stabilizing polymer could also be seen by the ICP. These results are determined to be valid. The advantage of ICP is its speed. It was also found to be more accurate than the NAA with respect to the alumina sample. However, the nanoparticles may not be effectively introduced in the nebulizer if large agglomerations occur. Therefore care must be taken to prevent agglomeration when diluting pH stabilized nanofluids.

3.5 Thermogravimetric Analysis (TGA)

Thermogravimetric analysis of nanofluids provides another means of analyzing different components of a nanofluid by weight. This may be desirable when one needs to know an accurate weight percentage of nanoparticles in solvent, or when the amount of polymer (added to improve the nanoparticle stability) in a nanofluid must be known. Thermogravimetric analysis works by heating a small sample on an extremely sensitive balance in a high temperature furnace. The weight vs time curve combined with knowledge of the boiling point of the species in the samples provide the sample composition. The specific equipment used was the Perkin Elmer TGA7. Samples, typically less than 40mg, are placed on a platinum balance and can be heated to over 1000 degrees Celsius.

Results showed for the Fe_3O_4 nanofluid that the surfactant polymer did not completely "cook off" until fairly high temperatures, around 700 or 800 degrees Celsius. It was found the polymer makes up about ¼ of the weight of coated Fe_3O_4 nanoparticles which were found at 3.5wt%. The weight loading of a nanofluid can easily be determined in this fashion. A sample of the 20wt% Nyacol Al_2O_3 nanofluid was also tested. The results showed that the particles consisted of 21.8wt%, which is slightly higher than the specified value. It is believed that this extra weight could possibly be due to the left over hydroxide groups on the surface of the particles. A possible remedy for this would be to push the temperature up to the highest point and allow it to remain there for a longer period of time.

4 CONCLUSIONS

A description of the methodology of preparation and characterization of nanofluids being done at MIT has been given. Insights were found as to the important mechanisms involved in nanofluid creation and stability. A description of the techniques to quantify and characterize nanofluids was also given. These results and methodologies are being incorporated into MIT's project on the determination of nanofluid heat transfer phenomenon.

REFERENCES

[1] S.U.S. Choi, et al. "Anomalous thermal conductivity enhancement in nanotube suspensions," *Applied Physics Letters*, 79(14) pp. 2252-2254. (2001).

[2] D. H. Everett, *Basic Principles of Colloid Science*, Chapter 2, The Royal Society of Chemistry, (1988).

[3] S. Krishnakumar and P. Somasundaran, "ESR investigations on the stabilization of alumina dispersions by Aerosol-OT in different solvents," *Colloids and Surfaces*, A 117 pp. 37-44. (1996).

High performance iPP based nanocomposites for food packaging application

M. Avella, M.E. Errico, G. Gentile

Istituto di Chimica e Tecnologia dei Polimeri, ICTP-CNR
Via Campi Flegrei, 34 Pozzuoli (NA), Italy, mave@ictp.cnr.it

ABSTRACT

High performance iPP based nanocomposites filled with innovative calcium carbonate nanoparticles ($CaCO_3$) were prepared and structure-properties relationships investigated. In particular nanoparticles characterized by high specific surface area (>200 m^2/g) and elongated shape were tested as reinforcement nanophase. In order to promote polymer/nanofillers interactions, $CaCO_3$ were coated with two different surface modifiers, polypropylene-maleic anhydride graft copolymer (iPP-g-MA) or fatty acids (FA). Morphological analysis permitted to assess that the presence of iPP-g-MA promotes a stronger adhesion between polymer/$CaCO_3$ with respect to that achieved by using FA as surface modifier. Mechanical analysis evidenced that Young's modulus increases as a function of nanoparticles content and coating agent nature. Finally, it was observed that the $CaCO_3$ nanoparticles presence drastically reduces the iPP permeability to both oxygen and carbon dioxide.

Keywords: nanocomposites, polypropylene, barrier properties, interfacial adhesion, mechanical properties.

1   INTRODUCTION

Isotactic polypropylene (iPP) is one of the most widely used plastic materials in the food packaging field [1]. Unfortunately polypropylene shows high gas permeability, which results in a poor protection of the packaged food thus limiting its usage [2]. One of the most used methods to improve polypropylene drawbacks is to add a second component such as a polymer in blend or in multilayer, micrometric fillers etc [3, 4]. Nowadays, nanocomposites based on polypropylene matrix constitute a major challenge for industry since they represent the route to substantially improve iPP mechanical and physical properties [5-7]. The enhanced properties are presumably due to the synergistic effects of nanoscale structure and interaction of fillers with polymers.

In this research, high performance iPP based nanocomposites filled with low-cost innovative calcium carbonate nanoparticles ($CaCO_3$) were prepared and structure-properties relationships investigated. Peculiarities of the innovative tested nanoparticles are the very high specific surface area (>200 m^2/g) and the elongated shape. This latter aims to merge the advantages of the well-assessed $CaCO_3$ know-how [8-11] with the properties enhancement attainable, for instance, by addition of clay platelets. In fact, in this way it should be possible to simulate the nanoclay behavior, that reduces gas permeability of polymers according to a tortuous path model, in which the platelets obstruct the passage of gases and other permeants through the polymeric matrix. Moreover, in order to promote polymer/nanofillers interfacial adhesion, $CaCO_3$ coated with polypropylene-maleic anhydride graft copolymer (iPP-g-MA) or fatty acids (FA) as surface modifiers were tested.

Structure-properties relationships were studied performing morphological, mechanical analysis and barrier tests.

2   EXPERIMENTAL SESSION

2.1   Materials

Isotactic polypropylene (iPP), Moplen X 30 S (Mn = 4.69 ×10^4 g/mol, Mw = 3.5 ×10^5 g/mol and Mz = 2.06 ×10^6 g/mol) was kindly supplied by Basell Polyolefins (Ferrara-Italy).

Elongated $CaCO_3$ nanoparticles (250 nm in length and 50 nm in thickness) coated with two different coupling agents, polypropylene-maleic anhydride graft copolymer (iPP-g-MA) and fatty acids (FA) were kindly supplied by Solvay Advanced Functional Minerals (Giraud-France). Nanoparticles are coded as C-PPMA and C-FA respectively

2.2   Methods

iPP/$CaCO_3$ nanocomposites were prepared by mixing the components in a Brabender-like apparatus (Rheocord EC of HAAKE Inc., New Jersey, USA) at 200°C and 32 rpm for 10 min. The mixing ratios of iPP/CaCO3 (wt/wt) were: 100/0, 99/1, 97/3.

Plain iPP and iPP/$CaCO_3$ nanocomposites were successively compression-molded in a heated press at 200°C for 2 min without any applied pressure. After this period, a pressure of 100 bar was applied for 3 min, then the press platelets were rapidly cooled to room temperature by cold water. Finally, the pressure was released and the mold removed from the plates

Morphological analysis was performed by using a scanning electron microscope (SEM), Cambridge Stereoscan microscope model 440, on cryogenically fractured surfaces. Before the observation, samples were covered with a gold layer.

Tensile tests were performed on dumb-bell specimens (4 mm wide and 15 mm long) by using an Instron machine (model 5564) at room temperature and a cross-head speed of 10 mm/min. Young Modulus (E) was calculated from these curves in accordance to the ASTM D256 standard (average 10 samples tested).

Permeability tests were performed in a gas-membrane-gas instrument based on measurement of the downstream pressure increase at a constant upstream side-driving pressure. The apparatus and experimental procedures were similar to those reported elsewhere [12-13]. In each experiment, sufficient time was allowed to ensure attainment of steady-state permeation. The measurements were carried out at a pressure of 1 atm and at a temperature of 30°C.

The permeability was computed from the slope of the linear, steady-state part of the curve representing the permeated gas volume as a function of time. The gas diffusivity was calculated from the 'time lag' determined from the intercept of the steady-state permeability curve on the abscissa.

3   RESULTS AND DISCUSSION

iPP based nanocomposites were prepared by melt mixing. Elongated CaCO$_3$ nanoparticles coated with iPP-g-MA or FA as surface modifier were tested as nanophase. Particular interest was focused on the influence of surface modifier nature on the polymer/nanoparticles interaction and consequently on the final material properties.

3.1   Morphological Analysis

In order to evaluate nanofiller dispersion into polymeric matrix and the interfacial adhesion between the two components, morphological analysis was performed on fractured surface of nanocomposites. As an example, SEM micrographs of nanocomposites containing 3% (wt/wt) of C-FA and C-PPMA are shown in figs. 1 and 2, respectively. As it is possible to observe, both C-FA and C-PPMA appear homogeneously dispersed into iPP allowing to affirm that organic surface modifiers prevent the nanoparticles agglomeration. Nevertheless, the strength of the polymer/nanoparticles interfacial adhesion depends upon the surface modifier nature. In fact in the case of C-FA nanoparticles, SEM micrograph reveals few areas in which debonding phenomena occurred after that a mechanical stress was applied, fig. 1. On the contrary, C-PPMA nanoparticles are completely welded to the iPP phase. This result underlines a stronger adhesion between polymer/C-PPMA with respect to that obtained by using C-FA nanoparticles. It can be hypothesized that the presence iPP-g-MA modifier agent is responsible for strong physical interactions between iPP-g-MA molecules and polymeric matrix entanglements. Interactions via entanglements can be only promoted if the molar mass of the surface modifier is quite high. It is known that the alkyl chains of fatty acids

are too short to interact with matrix chains via the above described mechanism. Nevertheless, the presence of fatty acids reduces the polarity and the absorption surface energy of nanoparticles preventing their agglomeration.

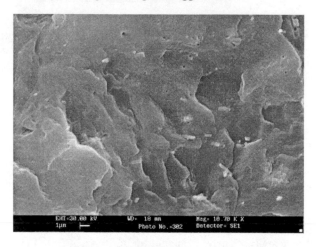

Figure 1: SEM micrograph of iPP/ C-FA 3% (wt/wt) fractured surface.

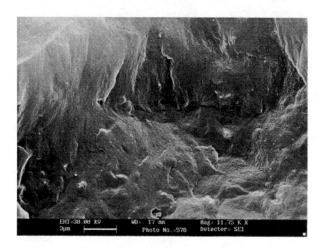

Figure 2: SEM micrograph of iPP/ C-PPMA 3% (wt/wt) fractured surface.

3.2   Mechanical properties

Results of tensile tests performed on iPP based nanocomposites are resumed in fig. 3. As shown, the presence of nanoparticles increases Young's modulus and the extent of this improvement is strictly correlated to the surface modifier nature. In fact, CaCO$_3$ coated with iPP-g-MA gave rise to a more pronounced modulus increase up to 30% with respect to neat iPP value and almost 20% higher than those obtained with nanoparticles coated with FA. Moreover in the latter case the stiffness is slightly influenced by the filler content reaching the maximum improvement by addition of 1% (wt/wt) of C-FA nanoparticles; while in the case of C-PPMA a dependence of modulus value up on the filler content was observed.

These results allow assessing that there is a significant influence of the interaction nature between filler particles and polymeric matrix on the iPP-based nanocomposites tensile properties. In fact, it is well known that in a multiphase system such as nanocomposites an external applied load can be transferred from the polymer to the reinforcement phase through the interphase region obtaining mechanical improvements as a function of the reached polymer/nanoparticles adhesion level. As above described, C-PPMA nanoparticles permit to obtain a better interfacial adhesion between the components (fig. 1) due to a strong surface modifiers-polymer interaction, assuring, in this way, a larger increase of the Young's modulus.

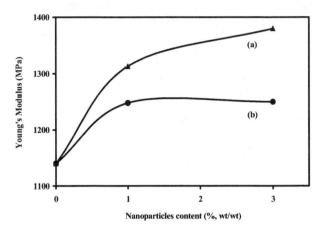

Figure 3: Young's Modulus values of iPP and iPP based nanocomposites filled with: a) C-FA nanoparticles; b) C-PPMA nanoparticles.

3.3   Barrier Properties

Generally speaking for an impermeable filler of volume fraction ϕ, the composite permeability (Pc) is directly linked to the tortuosity factor (τ) of nanoparticles, according to the following relationships [14]:

$$Pc = Pm \ (1- \phi)/\tau \qquad (1)$$

$$\tau = 1 + (L/2W)\phi \qquad (2)$$

where L and W denote the length and the thickness of the fillers and the ratio L/2W represents the aspect ratio.

In Figs. 4-5 oxygen and carbon dioxide permeability coefficient values versus nanoparticles content are plotted. As reported, C-FA and C-PPMA nanoparticles are responsible for a decrease of iPP permeability either to oxygen or to carbon dioxide. Moreover, the surface

modifier nature seems to play a key role in the extent of the iPP permeability improvement. In fact, while the C-FA nanoparticles lower the iPP permeability to oxygen up to 50% (fig. 4b), in the case of C-PPMA nanofillers this decrease is around 35% (fig. 4a). A similar trend was observed for the permeability to carbon dioxide as summarized in fig 5a-b. This result could be explained by considering that additional interactions generating between surface modifier (PP-g-MA or FA) and either oxygen or carbon dioxide occur. Although the C-PPMA nanoparticles induce a stronger C-PPMA/iPP interfacial adhesion, the nature of the surface modifier is similar to that of the polymeric matrix such as their permeability properties to oxygen and carbon dioxide. As a matter of fact, barrier properties are influenced only by the tortuous path that diffusing molecules must bypass with a consequent improvement of polymer permeability .

As far as C-FA nanoparticles, fatty acids unsaturations can be considered as potential sites of interactions with both oxygen and carbon dioxide molecules, thus inducing the additional effect of hindering and slowing the diffusion of the gases through the nanocomposites.

In a previous work, it was demonstrated that spherical $CaCO_3$ nanoparticles also induce an improvement of iPP barrier properties, due to the high nanoparticles volume fraction occupied, the homogeneous $CaCO_3$ dispersion and the strong matrix-nanofiller interfacial adhesion reached [11]. However, the relevance of this phenomenon is less pronounced with respect to that discussed in this work.

As a matter of fact, it can be assessed that elongated nanoparticles permit to magnify the well known effect of $CaCO_3$ spherical nanoparticles on the permeability properties, in agreement with the equations (1) and (2) above reported.

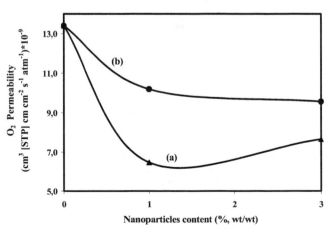

Figure 4: Oxygen permeability of iPP and iPP based nanocomposites filled with: a) C-FA nanoparticles; b) C-PPMA nanoparticles.

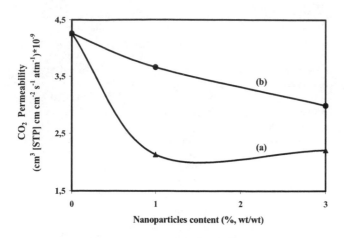

Figure 5: Carbon dioxide permeability of iPP and iPP based nanocomposites filled with: a) C-FA nanoparticles; b) C-PPMA nanoparticles

4 CONCLUSION

iPP based nanocomposites filled with innovative calcium carbonate nanoparticles were prepared and characterized. $CaCO_3$ nanoparticles were coated with two different surface modifiers, polypropylene-maleic anhydride graft copolymer (iPP-g-MA) or fatty acids (FA). The presence of the surface modifier prevents nanoparticles agglomeration phenomena allowing homogeneous dispersion of the nanophase into polymeric matrix. A close relationship between the surface modifier nature and final properties of the material was evidenced. Young's modulus increases in presence of nanoparticles. C-PPMA are responsible for a more pronounced improvement of iPP stiffness with respect to that achieved by addition of C-FA, due to stronger physical interactions between surface modifier and polymer via entanglements. $CaCO_3$ drastically reduces the iPP permeability to both oxygen and carbon dioxide because of nanoparticles elongated shape and good iPP/$CaCO_3$ interfacial adhesion. C-FA nanoparticles permit to obtain a larger effect on the barrier properties due probably to the presence of unsaturations that can be considered as potential sites of interactions with tested gases. These interactions induce the additional effect of hindering and slowing the diffusion of the gases through the nanocomposites.

REFERENCES

[1] J Karger Kocsis, "Polypropylene: structure, blends and composites" Chapman and Hall, 1995.

[2] H.G. Karian, "Handbook of polypropylene and polypropylene composites" Marcel Dekker, 1999.

[3] R.D. Deanin, M.A. Manion, "Handbook of Polyolefins" 2nd edition, Marcel Dekker, 633, 2000.

[4] M. Avella, P. Laurienzo, M. Malinconico, E. Martuscelli, M.G. Volpe, "Handbook of Polyolefins" 2nd edition, Marcel Dekker, 723, 2000.

[5] N. Hasegawa, H. Okamoto, M. Kato, A. Usuki, J. Appl. Polym. Sci 78 (11), 1918, 2000.

[6] E. Manias, A. Touny, L. Wu, K. Strawhecker, B. Lu, T.C. Chung, Chem. Mater. 13 (10), 3516, 2001.

[7] W.C.J. Zuiderluin, C. Westzaan, J. Huetink, R. J. Gaymans, Polymer 44(1), 261, 2003.

[8] M. Avella, M.E. Errico, E. Martuscelli, Nanoletters, 1(4), 213, 2001.

[9] M.L. Di Lorenzo, M.E. Errico, M. Avella, J Mater Sci 37(11), 2351, 2002.

[10] C.M. Chan, J. Wu, X. Li, Y.K. Cheung, Polymer 43(10), 2981-2992, 2002.

[11] M. Avella, S. Cosco, M.L. Di Lorenzo, E. Di Pace, M.E. Errico, G. Gentile European Polymer Journal in press.

[12] M.A. Del Nobile, G. Mensitieri , L. Nicolais, Polymer International 41, 73, 1996.

[13] L. Nicodemo, A. Marcone , T. Monetta, G. Mensitieri, F. Bellucci, J. Membrane Sci. 70, 207, 1, 1992.

[14] T.S. Ellis, J.S. D'Angelo, J. Appl. Polym. Sci. 90, 1639, 2003.

Enlarge the distance of water molecules to incise microorganism cell----the development and application of nano-grade microorganism cell crushing machine

Yao Hongwen[*] and   Guo Suge[**]

*College of Bioengineering , Hebei University of Science
and Technology, Shijiazhuang, 050018, China, yaohw163@sina.com
**Weisheng Pharmaceutical (Shijiazhuang)
Co., Ltd., Shijiazhuang, 050035, China, gsg1027@sohu.com

ABSSTRACT

This article elaborates the principle and method which microorganism cell is broken by nano-grade microorganism cell crushing machine. The machine has tailor-made nozzles which send out mightiness sector jet flow enlarged space between of molecules of water to 1~100nm within target distance of jet flow, and incise microorganism cell to nanometer microorganism granule between 10nm and 100nm.

Key words: space between of molecules of water, microorganism cell, nanometer microorganism granule, jet flow of water molecule

1   ENLARGE THE DISTANCE OF WATER MOLECULES TO INCISE MICROORGANISM CELL

Nano-grade microorganism cell crushing machine(patent number: 03143321.9)can break the microbial cell to nanometer level. It established the machine rationale to study out the enlargement water intermolecular distance formula from the design special-purpose spray nozzle, the enlargement water intermolecular distance, jetting flow of water molecule, cutting the microorganism cell. The birth of nano-grade microorganism cell crushing machine ended the disintegrator family cannot crush microscopic material the history.

1.1   Sending jet flow of water molecule and the structure of spray nozzle[1] [2] [3] [4] [5]

Microorganism cell is equably distributed in water and its density is between 10%~12%. The superhigh pressure jet flow carries secretly microorganism cell that sending by the jet flow generator comes into two pipelines which their diameters are between 0.8nm and 1.2mm, then into special-purpose spray nozzles, and enlarges the space between of molecules of water to 4-100 times. Two jet flows collide each other in collided room and jet to microorganism cell, then cut it into nano-grade biology granule. The pressure of the jet flow by generator send is different with microorganism cell. The structre of the spray nozzle see fig 1. Its broken pressure and effect can be seen in the table 1.

From fig1 we can see that the superhigh pressure jet flow carries secretly microorganism cell that sending by the jet flow generator comes into 5 left spray pipe and 6 right spray pipe, then into 3 left nozzle and 4 right nozzle, and collides each other in jet flow correlation and incision room, then cutting microorganism cell later. Its collided each other power is between 500kg/cm^2 and 1400 kg/cm^2. Its particular process is as following fig 2.

Table 1 Various industry microorganism broken pressure and effect:

Name	Crushing pressure(Mpa)	Crushing rate(%)	Microognizizm cell original size(μ)	Nanometer microorganism granule size(nm)
Penicillium	75	>90	3x50	10-80
Pseudomonas	75	>90	0.5x0.5	25-50
Streptomyces rimosus	85	>90	1x30	10-100
Bacillus cereus	70	>90	3x10	25-50
Streptomyces griseus	85	>90	1.2x1.2	----
Acremonium chrysogenum	85	>90	1.5x1.5	----
Yeast	110	>90	3x6	20-60

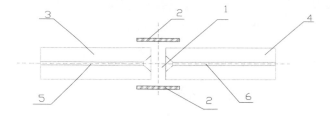

Fig 1 The structure of spray nozzle of nano-grade microorganism cell crushing machine

In fig 1, 1. jet flow ccorrelation and incision room; 2. shaken slice; 3. left spray nozzle; 4. right spray nozzle; 5. left pipe; 6. right pipe.

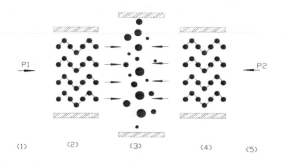

Fig 2 a schematic diagram of supervoltage flow of molecule of water to cut microorganism cell to nanometer biology grain

In fig2, (1)P1 is the directional left stock superhigh pressure jet flow by generator sending out; (2)left stock superhigh pressure jet flow schematic drawing, namely the generator sends out the high-pressured water jet flow which enlarging the molecules spacing between 1nm and 100nm and directional motion to incise the microorganism cell in collided room to the nanometer level granule; (3) Microorganism cell which is crushed is made the biology granule between 1nm and 100nm; (4)The right stock superhigh pressure jet flow; (5)P2 is the directional right stock superhigh pressure jet flow by generator sending out.

1.2 Water intermolecular distance enlargement and example[3] [4] [7] [8] [9]

It is well known that the liquid water has the certain volume not to be compressed easily. The very strong action is being in the water intermolecular because of the water intermolecular distance is very   small. Only molecule who has enough kinetic energy can overcome other molecular action and enters the free space forms the gas hydrone however the general member is not easy to be separated from the water liquid into free gas.

The water has fluidity not solid form because affects between the liquid molecule does not strong like the solid, the water molecule time of day changes its position of equilibrium, each of them does not fixed position, arrangement rule between them is the temporary combination again as necessary decomposes. Bestowed the formidable pressure, the water forced into the high-pressured nozzle and enlarged in the nozzle place according to its characteristics. We can see figure 3:

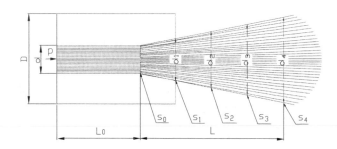

Figure 3 a schematic diagram of water intermolecular distance enlargement

(1)D---spray pipe outer diameter, unit: mm; (2)P---jet flow pressure, unit: Mpa; (3)d---spray pipe inside diameter, unit: mm; (4)S_0---spray pipe inside cross-sectional area, unit: mm^2; (5)S_1, S_2, S_3, S_4---the cross-sectional area of the place where the jet flow is in the target distance, unit: mm^2; (6)d_1, d_2, d_3, d_4---the diameter of place where the taper jet flow pass through,passes through the place diameter, unit: mm; (7)L_0---pipe length, unit: mm; (8)L---the jet flow length in target distance, unit: mm.

It is known that the space between molecules of water was enlarged in the d tube in the fig 3. The space between molecules of water is very small in the normal, only 0.4nm, r_0 = 0.4nm, and was enlarged in the place where d_1, d_2, d_3, d_4 is when the liquid water sprayed out after added high-pressured P. The calculation formulae of enlargement factor η is as follows:

In the place d_n, enlargement factor η_n :

$$\eta_n = \frac{S_n}{S_0} = \frac{\pi \left(\dfrac{d_n}{2}\right)^2}{\pi \left(\dfrac{d}{2}\right)^2} = \frac{d_n^{\,2}}{d^{\,2}} \qquad n=1,2,3\ldots$$

It shows spray nozzle structure in chart 1, the thickness is between 10%and 12%, microorganism cell is equably distributed in the water in the left nozzle 3 and right nozzle 4, in the place r_0 = 0.4nm. We can give enove strong pressure P to jet flow in the left and right pipe to crush the microorganism cell inside target distance. The enlarged

Cleantech 2007, ISBN 1-4200-6382-0

distance of water molecules calculation formulae is as follows:

$$r_F = \eta \times r_0$$

In nature ,d is measured:
d=1.2mm, d_1=2.4mm, d_2=4.8mm, d_3=9.6mm, d_4=19.2mm.

In the place d distance of water molecules r can be calculated, for example, r_3,

$$r_3 = \eta_3 \times r_0 = \frac{d_3{}^2}{d^2} \times r_0 = \frac{9.6^2}{1.2^2} \times 0.4 = 10.24\text{nm}$$

After crushed by nano-grade microorganism cell crusher, the size of microorganism granule is seen in table 2 from the Beijing physics and chemistry center examination result.

Serial number	Name	S.
1	The fresh yeast	20 nm - 60nm
2	The bacterium original fluid	25 nm - 50nm
3	The penicillin fungus	10 nm - 80nm
4	The streptomyces rimosus	40 nm - 100nm
5	The corn flour	50 nm - 100nm

Table 2 the size of microorganism granule examination result

S* After crushed by nano-grade microorganismcell crusher examination result

The machine crushing effect is as follows fig 4 and fig 5.

Fig4 the phototelegraph that pseudomonas and bacillus cereus was broken and then enlarged 48000 times

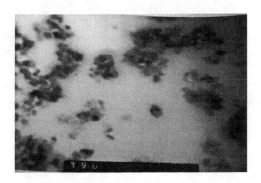

Fig5 the phototelegraph that the yeast was broken and then enlarged 100000 times

2 DEVELOPMENT AND CONCLUSION OF NANO-GRADE MICROORGANISM CELL CRUSHING MACHINE[7] [8] [9] [10]

2.1 Nano-grade microorganism cell breaker machine sculpt

Nano-grade microorganism cell breaker machine can be seen in fig 6.

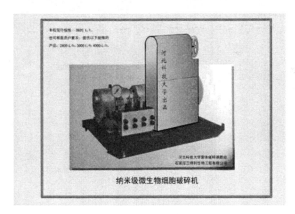

fig 6 nano-grade microorganism cell breaker machine sculpt

2.2 Nano-grade microorganism cell crusher can be developped and manufactured according to model machine

Nano-grade microorganism cell crusher that is used in production is taken to research and produce on the basis of model machine. !The mathematics model of systematic is educed on the basis of doniel bermoulli equation and relevant mathematics. See it as follows:

$$D-1.233 \sqrt{\frac{Q}{\sqrt{P} \times n}} \times \omega \times \psi \qquad (1)$$

D---spray nozzle diameter, unit: mm; p---jet pressure, unit: MPa; Q---spraying flux, unit: L/min; n---spray nozzle number, unit: quantity; ω ---water flux quotiety, ω =0.85~1; ψ ---spray nozzle efficiency quotiety.

Velocity of jet flow V calculation formulae of nano-grade microorganism cell crushing machine is as follows

$$V=\frac{Q}{\pi\gamma^2} \qquad (2)$$

V---velocity of jet flow, unit: m/s; Q---jet flow flux, unit: m^3/s; r---spray nozzle semidiameter, unit: m.
Jet flow impulsive force F calculation formulae effecting 1g particle of Nano-grade microorganism cell crusher is as follows

$$F=\frac{m(v-v_0)}{t} \qquad (3)$$

F---impulsive force of 1g particle received, unit: kgm/s^2; m----1g particle weight, unit: g; v---the velocity of jet flow striking 1g particle, unit: m/s; v_0---muzzle velocity of jet flow, unit: m/s, v_0=0; t---time of jet flow striking 1g particle, unit: s, t=6.0 x 10^{-3}s.

Spray pipe sectional area S calculation formulae of nano-grade microorganism cell crushing machine is as follows

$$S=\pi r^2 \qquad (4)$$

S---spray pipe sectional area, unit: m^2; π---3.14; r---spray pipe semidiameter, unit: m.

Jet flow flux Q calculation formulae of nano-grade microorganism cell crushing machine is as follows

$$Q=S \times V \qquad (5)$$

Q---jet flow flux, unit: m^3/s; S---spray pipe sectional area, unit; m^2; V---velocity of jet flow, unit: m/s.

Electromotor power N calculation formulae of nano-grade microorganism cell crushing machine is as follows

$$N=\frac{\psi P_N Q_N}{600\eta_P} \qquad (6)$$

N---engine power, unit: Kw; Ψ---transform quotiety

$\frac{p_{max}}{P_N}$, $\frac{p_{max}}{P_N}$ =0.52~0.54; P_N---rating working pressure, unit: MPa; Q_N---rating working flux, unit: L/min; η_p ---engine efficiency , 80%~90%.

2.3 Conclusion

1. The special-purpose spray nozzle in this disquisition enlarge triumphantly the distance of water molecules, calculation formulae is as follows: $r_F=\eta x r_0$. Nano-grade microorganism cell crusher based on this theory.

2. Nano-grade biology granule by this machine produced can be used in the beer, the soy sauce, the antibiotic and the microorganism drugs manufacture industry as the nitrogen raw material, and realizes the waste mycelium circulation economy in fermentation industry.

3. Nano-grade microorganism cell crusher end the disintegrator family cannot crush microscopic material the history at present and realizes nano-grade biology crushing. This machine become one of the disintegrator family.

REFERENCES

[1] Dong Zhiyong, etc..Jet flow mechanics, Beijing: Science publishing house, 2005,3.

[2] Wang Huimin, etc..Project hydromechanics, Nanjing: The river and sea university publishing house, 2005,1.

[3] Wang Huajiu, etc..Technical mathematics, Beijing: Engineering industry publishing house, 2003,8.

[4] Huang Weimin,etc..Technical physics foundation (volume one),2002,9.

[5] Yao Hongwen, etc..Study on the Technique of yeast cell Disruption , Beijing: China brews, 2005, (4).32-34.

[6] Yao Hongwen, etc..Nanometer grade of microorganism cell crushing machine and applies in the beer industry, Beijing: Beer science and technology, 2004 (7), 48-50.

[7] Liu Guoquan. The bio-engineering downriver technology, Beijing: the chemical industry publishing house, 2003. 65-69.

[8] Mao Zhonggui. Bio-engineering downriver technology, Beijing, Chinese light industry publishing house, 2000, 61-79.

[9] H.Schlichting, K.Gersten. Boundary Layer Theoey (8thRevisedandEnlaerged Edition), Springer, 2000.

[10] A.J.Smits.A. Physical Introduction To Fluid Mechanics. John Wiley & Sons. Inc.,2000.

Systematic Approach on Modeling and Identification for Nanobattery Prototyping

Pradeep Bhattacharya, Zhengmao Ye[*], Ernest Walker, Fred Lacy, Madhusmita Banerjee

Department of Electrical Engineering
Southern University
Baton Rouge, LA 70813, USA
zhengmaoye@engr.subr.edu[*]

ABSTRACT

This article is concerned with the systematic design of nanobattery prototyping. Miniaturization of power sources is a challenging area of nanotechnology research. There are four major parts in miniaturized Li-Ion nanobattery: anode, cathode, electrode and separator. Correspondingly, some appropriate material must be distinguished. The multi-walled carbon nanotube array electrode is used as anode, which exhibits high current density. $LiMn_2O_4$ spinel oxide is used as the cathode. Nanoporous dielectric membrane is selected for mixture storage of gel electrolyte. Ni is chosen as a suitable current collector. The separator and electrolyte container accounts for the reduction of dendrites and compatibility increment of electrode-electrolyte. According to nanobattery physical mechanism, mathematical model has been identified. At last, some numerical simulations of nanobattery characteristics have been conducted.

Keywords: modeling, identification, miniaturization, nano-technology, nanobattery

1 INTRODUCTION

Nanobattery technology is crucial to active nano-devices which has numerous applications on miniature cameras, miniature cell phones, micro-sensors and nano-sensors, micro-chips and nano-chips, and so on. Scaling laws are applicable to miniaturized nano-system design. Geometrical scaling is strictly dependent on the size of physical objects (volume and surface). Phenomenological scaling reflects phenomenological behaviors of miniaturized power source [Hsu, 2002]. The simulation of a Li-Ion battery with phase change material thermal management system has been studied. Mathematical model is applied to Li-Ion battery design in electric scooter. Li-Ion discharge curves under various thermal conditions are clearly plotted [Khateeb, 2004]. A low temperature approach has been proposed for fabricating nanostructure of $LiMn_2O_4$ spinel oxide instead of high temperature solid-state reaction. Electrochemical impedance spectrum demonstrates that reaction activity and diffusion property of electrode material are significantly improved at nano-scale due to the stronger interaction of lithium ion with nano-material at the surface than inside the bulk [Ye, 2004]. Carbon nanotube array electrodes are prepared in another case using alumina template. Lithium ions are shown to diffuse through outer surface and upper tips into the inner graphene without any interference of binder and conductor in an ordered carbon nanotube. The electrodes could tolerate high current density of a micro-battery [Zhao, 2002]. It is concluded that the Li-Ion system provides the highest energy density among rechargeable battery systems available so far. It is thus necessary to investigate mathematical models of Li-Ion batteries in order to simulate the behavior of single electrode particles, single electrodes and full cells under a variety of operating conditions like constant current discharge, pulse discharge, impedance, cyclic voltammetry. In this paper, nano-scale characteristics of electrode and electrolyte material are identified and different components of nanobattary are determined. Suitable fabricating method is also presented on a basis of anode, cathode, electrolyte and separator being identified [1-16].

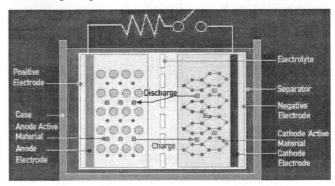

Fig. 1 Operating Mechanism of Li-Ion Battery

2 NANOBATTARY DESIGN AND EXPERIMENT PROCEDURE

Nanobattary design problem should be accompanied by suitable fabricating methods for anode, cathode, electrolyte, and miniaturized separator. Firstly mathematical modeling and optimization must be conducted theoretically. Then numerical simulations have to be applied to inspect possible properties out of the current design. Experimental methods will be the next step for preparing nanostructures for all components and procedures during fabricating should also be identified. Eventually, all prototypes must be tested.

2.1 Design of Anode

Carbon nanotube array electrode can provide high Li-ion intercalation in the hexagonal structure. It is selected as a suitable anode in a miniature Li-Ion battery, which shows rechargeability with almost 100% couloumbic efficiency, high current density, high capacity and light weight. Multi-walled carbon nanotube can be conducted using Chemical Vapor Deposition method. Uniform and straight pores can be formed in alumina substrate. The template method has the advantage of simple reproduction. It is predictable in the sense that both the diameter and length of the pores can be adjusted in order to obtain nanotubes of any desired size. Another advantage is that all pores are vertically oriented in the same direction. The minute defects existing in the array electrode after dissolution of alumina template have a very little effect on the regular array of carbon nanotubes. These nanotube attributes have a remarkable impact on making genuine anode material.

2.2 Design of Cathode

Rechargeable Li-Ion batteries require broad environment acceptability, low cost and preparation simplicity. $LiMn_2O_4$ spinel oxide is obviously much better a cathode candidate than those of nickel and cobalt oxides. Lithium insertion and extraction can be easily conducted on nano-material with the enhanced electrochemical reaction activity and diffusion capacity. It acts as the best potential candidate for miniaturization presently. It is known that the particle size, surface morphology and homogeneity are difficult to adjust precisely. The simple and adjustable method is ball-milling approach, where particles of sample will be provided with more homogeneous crystal grains. It is less complex than template method and soft chemistry method. From Thermo-Gravity simulation differential curves, temperature of raw material treated with ball milling decreased substantially compared to raw material untreated. The low synthetic temperature gives rise to a decrease of particle size and an increase of the oxidation state of manganese. It takes a more important role on surface than inside bulk.

2.3 Design of Electrolyte

Conventional battery contains aqueous acidic or alkaline electrolyte. Non-aqueous electrolyte possesses the quality of zero electrolyte leakage when conversion of electrolytes in electrochemical cell occurs. It is adhesiveness to active electrode material, which shows high ionic conductivity in a wide temperature range, provides low vapor pressure, prevents electrolyte leakage, enhances thermal stability, reduces solvent decomposition and improves life cycle. Therefore, it is possible for polymer electrolyte to absorb solvent up to three times its weight. The phase separation method can be used for preparations of gel electrolyte membranes. The copolymer is dissolved and dispersed in acetone via stirring inside container. Then ethanol is slowly added when the gel solution is stirred. Resulting slurry is cast for evaporation at ambient temperature. After vacuum drying, the liquid electrolyte can be absorbed by thick and stable membranes obtained.

2.4 Design of Separator

In the one-step membrane, electrolyte wet pores are more effective for reasons of high porosity, low pore tortuosity, big pore density, extraordinary hardness, uniform thickness. At the nano-scale range, oxidized anodical aluminum is capable of keeping this mechanical and dielectric property over vast temperature conditions. Anodizing of aluminum is used for fabricating nanoporous alumina membranes. The specimen is detached from the substrate and treated in the sulphuric acid solution. Then it is washed and dried so as to immerse in a hot colloid solution at once. The gelatin powder is added to water under room temperature, when solid gelatin particles imbibe water as discrete swollen particles. The mixture is then heated gently until all swollen gelatin particles collapse and hydrated gelatin goes into solution. The solution is heated and spread to surface of the porous film. Now its surface tension has reduced which allows the gelatin sol to penetrate into pores by capillary action. The uniform pore size, high pore density, pore thinness and vertical pore structure of porous films result from anodic oxidation of aluminum possess.

In summary, Li-Ion nanobattery has been proposed. Its anode is designed as multi-walled carbon nanotube array electrode and cathode is designed as $LiMn_2O_4$ spinel oxide. Nanoporous dielectric membranes made from anodically oxidized aluminum foil is used for storing $LiPF_6$ in gel electrolyte, vinylidene fluoride polymer and hexafluoro-propylene copolymer. A separator and electrolyte container is required to reduce formation of dendrites and to increase electrode-electrolyte compatibility.

3 MATHEMATICAL MODELING AND SIMULATION

Coupled nonlinear differential equations are used for the modeling of Li-Ion nanobattery. The equations contain six dependent variables and two independent variables (length x and time t), representing both galvanostatic charge and discharge processes using concentrated solution theory instead of dilute solution theory for accuracy consideration. As a matter of fact, interactions occur among all species in solution in concentrated solution theory. Interactions occur between ion and solvent while nothing happens among ions in dilute solution theory. Composite electrodes consist of inert conducting material, electrolyte and active insertion particle. One dimensional transport of those lithium ions is taken into account for mathematical models from negative electrode through separator into positive electrode.

Cleantech 2007, ISBN 1-4200-6382-0

To reduce computational complexity, some assumptions are made to solve the complex and coupled mathematical equations. Firstly, transport of lithium ions from negative electrode through separator to positive electrode is regarded as one dimensional. Secondly, conditions of binary solvent and electrolyte are assumed whose transport model follows the concentrated solution theory. Thirdly, the transference number of lithium ion, electrical conductivity, the diffusion coefficient of lithium ion are fixed. In addition, Al and Cu are taken as collectors for negative electrode and positive electrode, respectively.

Composite cathode consists of inert conducting material, polymer/salt electrolyte and solid active insertion particles in these equations. The material balance on the lithium at polymer/salt phase gives rise to (1).

$$\varepsilon(\partial c/\partial t) = \nabla.(D \nabla c) - (i_2. \nabla t_+^0)/F + aj_n(1- t_+^0) \qquad (1)$$

The variation in potential of separator is calculated by:

$$i_2 = -\kappa \nabla \Phi_2 + 2\kappa RT/F (1 + \partial \ln f_+/\partial \ln c)(1+t_+^0) \nabla \ln c \qquad (2)$$

The pore wall flux of lithium ions across the interface is averaged over the interfacial area between solid matrix and electrolyte, which is related to divergence of current flow or current density at electrolyte phase by eq. (3).

$$aj_n = (1/F) \nabla i_2 \qquad (3)$$

The current flowing in the matrix follows the Ohm's law.

$$i_1 = -\sigma \nabla \Phi_1 \qquad (4)$$

where, active cathode material is assumed to be made up of spherical particles with diffusion considering mechanisms of the lithium transport.

$$\partial c_s/\partial t = D_s [\partial^2 c_s/\partial r^2 + (2/r) \partial c_s/\partial r] \qquad (5)$$

Its direction is to the surface of particles. Eqs (4-5) are electrode phase equations. Its boundary condition is given by the relationship between the pore wall flux across the interface and the rate of diffusion of Li-Ion into the surface of the insertion material:

$$j_n = -D_s \partial c_s/\partial r \text{ at } r = R_s \qquad (6)$$

Butler-Volmer kinetics model shows the cell kinetics.

$$j_n = kc^{1/2}(c_t - c_s)^{1/2}(c_s)^{1/2} \{e^{(F/2RT(\eta - U))} -e^{(-F/2RT(\eta - U))}\} \qquad (7)$$

The over-potential is modeled by eq. (8).

$$\eta = \Phi_1 - \Phi_2 \qquad (8)$$

Eqs (1-8) formulate a set of coupled nonlinear differential equations which need solving simultaneously, where ε is Bruggeman's exponent, D is diffusion coefficient without turtuosity correction, σ is electronic conductivity without turtuosity correction, T is temperature, t_+^0 is initial time and F is Faraday's constant (96,487 C/mol). There are six dependent variables all together: solid phase concentration of lithium c_s, electrode phase potential Φ_1, electrolyte phase potential Φ_2, concentration c, reaction rate j_n and current density at electrolyte phase i_2.

4   SOME SIMULATION RESULTS

A practicable Li-Ion battery is designed via a thorough study of battery behavior. Fig. 2 shows the relationship between cell potential and utilization. Discharge curves are shown as a function of manganese electrode stoichiometric parameter in $LiMn_2O_4$. The cell can achieve a maximum stoichiometry in a positive electrode of 0.76, demonstrating that this cell is negative-electrode limited where the carbon runs out of lithium ahead of manganese oxide.

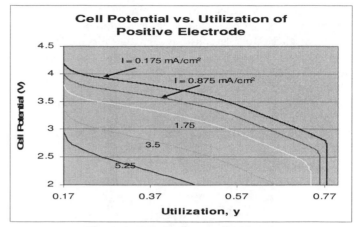

Fig. 2 Cell Potential v.s. Utilization

The current density at solution phase is plotted across the cell with 1.75mA/cm^2 discharge. Current density increases in negative electrode, remains constant in separator region and gradually decreases in positive electrode. This current density increases dissimilarly along different time instants at the discharge phase, while it reaches the maximum in the separator showing electrolytic current density occurs in the separator exclusively since it contains only the electrolyte. After certain durations, gradient of current density decays with time increment since a low salt concentration appears after a certain period of discharge (Fig. 3).

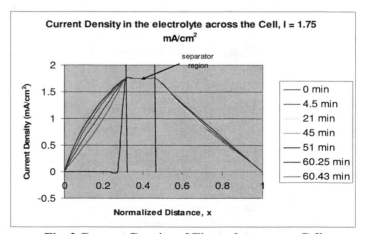

Fig. 3 Current Density of Electrolyte across Cell

Li-Ion battery simulations of macro scale have been conducted above using mathematical models with satisfied performance characteristics. Scaling law of dimensionality states if certain parameter is proportional to a dimension, then a change in that dimension by a certain factor changes the parameter by the same factor. By applying scaling law, values of all relevant parameters at nano-scale are obtained.

For nanobattery simulations, the unchanged mathematical model is assumed and values are scaled down to nano-scale. Each of four graphs in Fig. 4 has shown the cell potential against discharge time curves when the battery thickness keeps diminishing. Initially, it is of an order of microns. Then it is scaled down by certain factor of ten. Other parameters can be scaled down in a similar way. An analysis of these four graphs reveals that for the same current density, discharge time of nanobattery increases significantly. It has shown that the discharge time is inversely proportional to current density for any fixed thickness. From Fig. 4, it is concluded that the discharge time increases with the decrement in dimension under the condition that the current density remains a constant.

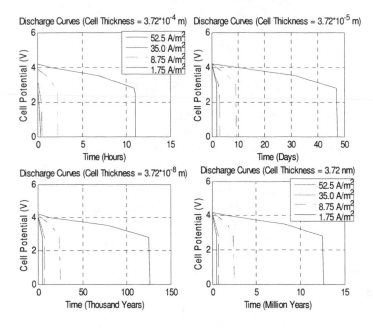

Fig. 4 Cell Potential v.s. Discharge Time Curves for Simulated Nano-battery

5 CONCLUSIONS

In this preliminary research, Li-Ion is picked up owing to its competitive properties over other rechargeable batteries. Then the mathematical model is proposed for nanobattery. For numerical simulations, scaling law has been considered, where exactly the same mathematical model is assumed whose values are scaled down to nano-scale. The result indicates the relationship of cell potential against discharge time along with the reduction of battery thickness. For other parameter scaling cases, similar numerical simulations are conducted. It gives an insight into extensive investigation of the nanobattery performance in future, where more precise modeling and optimization should be involved and further clean room fabrication should be followed.

REFERENCES

[1] Tai-Ran Hsu, MEMS & Microsystems: Design and Manufacture, McGraw-Hill, New York, NY, 2002.

[2] Khateeb, S.A., Farid, M.M., et al, "Design and simulation of a lithium-ion battery with a phase change material thermal management system for an electric scooter", Journal of Power Sources, v 128, n.2, 5 April 2004, p 292-307.

[3] S. Ye, J. Y. Lv, X. P. Gao, F. Wu and D. Y. Song, Synthesis and electrochemical properties of $LiMn_2O_4$ spinel phase with nanostructure, Electrochimica Acta 49 (2004) 1623–1628.

[4] J. Zhao, Q. Gao, C. Gu, Y. Yang, Preparation of multi-walled carbon nanotube array electrodes and its electrochemical intercalation behavior of Li ions, Chemical Physics Letters 358 (2002) 77–82.

[5] A. Mozalev, S. Magaino, H. Imai, The formation of nanoporous membranes from anodically oxidized aluminium and their application to Li rechargeable batteries, Electrochimica Acta 46 (2001) 2825–2834.

[6] Z. Cai and C. R. Martin, Electronically conductive polymer fibers with mesoscopic diameters show enhanced electronic conductivities, Journal Chem. Society, 111 (1989) 4138.

[7] J. Billingsley, Mechatronics and Machine Vision 2003: Future Trends, Research Studies Press LTD, England.

[8] J. Barisci, G. Wallace, R. Baughman, Electrochemical quartz crystal microbalance studies of single-wall carbon nanotubes in aqueous and non-aqueous solutions, Electrochim. Acta 46 (2000) 509.

[9] J. Zhao, A. Buldum, J. Han, First-Principles Study of Li-Intercalated Carbon Nanotube Ropes, Phys. Rev. Lett. 85 (2000) 1706.

[10] G. Wu, C. Wang, X. Zhang, H. Yang, Z. Qi, W. Li, Lithium insertion into CuO/carbon nanotubes, J. Power Sources 75 (1998) 175

[11] C. J. Brumlik and C. R. Martin, Anal. Chem. 59 (1992) 2625.

[12] B. Gao, T. Cagin, W. Goddard, Position of K Atoms in Doped Single-Walled Carbon Nanotube Crystals, Phys. Rev. Lett. 80 (1998) 5556.

[13] B. Gao, A. Kleinhammes, X. Tang, C. Bower, L. Fleming, Y. Wu, Electrochemical intercalation of single-walled carbon nanotubes with lithium, Chem. Phys. Lett. 307 (1999) 153.

[14] E. Frackowiak, S. Gautier, H. Gaucher, S. Bonnamy, F. Beguin, Electrochemical storage of lithium multiwalled carbon nanotubes, Carbon 37 (1999) 61.

[15] T. Kuzumaki, Y. Takamura, H. Ichinose and Y. Horiike, Structural change at the carbon-nanotube tip by field emission, Appl. Phys. Letter. 78 (2001) 3699.

[16] Y. Yang, D. Shu, J. You, Z. Lin, "Performance and characterization of lithium–manganese-oxide cathode material with large tunnel structure for lithium batteries", J. Power Sources, 81–82 (1999) 637.

Biosensor for heavy metals using hydrothermally grown ZnO nanorods and metal-binding peptides

W.Z. Jia, E.T. Reitz and Y.Lei

Department of Chemical, Materials and Biomolecular Engineering
University of Connecticut, 191 Auditorium Road, Unit 3222
Storrs, CT 06226
ylei@engr.uconn.edu

ABSTRACT

We report herein a biosensor for the determination of heavy metals based on hydrothermally grown ZnO nanorods and metal-binding peptides. Metal binding peptide is immobilized on ZnO nanorods through non-specific binding. Peptide binding with heavy metal ion causes the electrical signal change, which measured and correlated to the concentration of heavy metals. The sensor performance will be optimized with respect to the operating conditions. The new biosensor format will offer great promise for real-time environmental monitoring of heavy metals.

Keywords: biosensor, zinc oxide, heavy metal binding peptide

1 INTRODUCTION

Every year, around 2.4 million tons of metal wastes from industrial sources and 2 million tons from agriculture are generated in US. Pollution caused by heavy metals poses a great danger to humans and the environment which has led to stringent regulations over the allowable limits of heavy metals in drinking water. Heavy metals such as Pb^{2+}, Hg^{2+}, and Cd^{2+} are currently ranked second, third, and seventh, respectively, on the EPA's priority list of metals that are of major environmental concern. The heavy metals mercury, lead and cadmium can be released into the environment through industrial use and leakage from dump sites, and once released can contaminate water, soil and air[1]. These heavy metals can not be broken down in the human body and their accumulation over time can result in serious health risks. Studies have indicated that long term exposure to mercury can cause permanent damage the brain, kidneys, and the developing fetus[2]. Many studies demonstrate lead's toxicity in the nervous system and its association with increased blood pressure in adults[1]. Long term exposure to cadmium can include nephropathies, emphysematous alterations in the lung, and cardiovascular diseases, whereas acute exposure is usually restricted to the lungs[3]. Because of their extreme toxicity, there are growing needs for rapid and sensitive detection of the heavy metals.

Many spectroscopic and electrochemical techniques have been developed thus far for detection of heavy metals, such as atomic absorption spectrophotometry (AAS), Auger electron spectroscopy (AES), inductively coupled plasma – mass spectrometry (ICP-MS), ICP-optical emission spectrometer (ICP-OES), ion-selective electrode (ISE), and polarography. However, their applications to the detection of environmental samples are limited due to high cost and, sometimes, to lack of real-time. Therefore, there is an urgent need to develop more sensitive and cost-effective methods to detect heavy metal concentrations in real-time format. Biosensors are good choices in this respect. A biosensor is an analytical device that combines a biological sensing element with a transducer to produce a signal proportional to the concentration of analyte. Affinity peptides, such as hexahistidines, glutathione S-transferase, maltose-binding proteins and synthetic phytochelatins $(EC)_n$ have been demonstrated to bind heavy metals even at dilute concentrations, and can be used as biological sensing elements.

Since the discovery and characterization of carbon nanostructures, extensive research has been conducted to characterize other nanostructures. ZnO is of particular interest because of its wide band gap (3.37 eV), high exciton binding energy (60 meV), high stability and high melting point. These physical properties make ZnO an attractive material for use in short-wavelength optoelectronic devices[4], such as LED, ultra-violet lasers[5], chemical and biological sensors[6,7], and solar cells[8]. Various synthesis methods are used to produce the ZnO nanostructures. Although fewer morphologies have been discovered by hydrothermal synthesis than by a solid-vapor synthesis processes, the simplicity of equipment and low temperatures required for synthesis makes the hydrothermal method a better candidate for real application. Using the hydrothermal method, techniques have been developed to produce quasi-one-dimensional ZnO structures including nanorods[9,10] and nanotubes[11]. Three-dimensional structures have also been constructed, including nanoflowers[12-14] and nanospheres[15,16]. While the morphology of ZnO nanostructures has been extensively studied, applications of ZnO in biosensors are rare[17].

We propose herein a biosensor for determination of heavy metals based on hydrothermally grown ZnO nanorods and metal-binding peptides. Heavy metal binding peptide is immobilized on ZnO nanorods. Its binding with heavy metals causes the electrical signal change which can be measured and correlated to the concentration of heavy metals. The sensor performance will be optimized with respect to the operating conditions. The new biosensor offers great promise for rapid environmental monitoring of heavy metals.

2 EXPERIMENTS

2.1 ZnO nanorods synthesis

ZnO nanorods were directly grown on a glass by hydrothermal decomposition described elsewhere[7]. Briefly, the reaction solution was prepared by mixing proper quantity of ammonia (25%) and 0.1 M zinc chloride solution in a bottle with autoclavable screw cap. The glass substrate was vertically immersed in the reaction solution. A layer of product was deposited on the substrate after the sealed bottle was heated at 95 °C for 2 hour in an oven. Then, the sample was thoroughly washed with DI water and dried in air. Scanning electron microscopy (SEM) was employed to examine the morphology of the hydrothermal product. Figures 1 and 2 show a typical SEM image of the as-grown product by hydrothermal decomposition. It can be seen that, the nanostructure presents a rod-like shape with a hexagonal cross-section, a typical morphology of wurtzite ZnO. It is also noted that the nanorods are primarily aligned along the perpendicular direction of the substrate. The nanorods are uniform in size with a diameter of about 300 nm and a length of 4 μm. The size uniformity is a unique feature of hydrothermally grown nanostructures compared to vapor phase transport method.

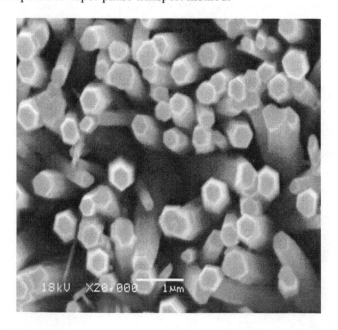

Figure 1. SEM image of the as-grown ZnO nanorods on glass substrate

Figure 2. Large area of ZnO nanorods.

2.2 Electrode fabrication

The interdigitated electrodes (IDE) configuration (Figure 3) is chosen to enable effective electronic contact between hydrothermally grown ZnO and the electrodes over large areas. The interdigitated electrode will be fabricated using a conventional photolithographic method with a finger width of 10 μm and a gap size of 8 μm. The IDE fingers are made by thermally evaporating 20-nm Ti and 40-nm Au either on ZnO nanorods.

Figure 3. Schematic illustration of interdigitated electrodes on ZnO nanorods film

2.3 Metal binding peptide based biosensor

Hexahistidines, glutathione S-transferase, maltose-binding proteins and synthetic phytochelatins $(EC)_n$ can bind different heavy metal ions with different affinity. In this research, $(EC)_{10}$, which can retrieve Cd^{2+} at sub-parts per million levels, is used to develop the biosensor.

Metal binding peptide $(EC)_{10}$, purchased as a customized product, will be used to functionalize the ZnO nanorods through non-specific binding. Briefly, a small volume of metal binding peptide solution is dropped onto the ZnO for 1 hour, and then the sample is rinsed with deionized water to remove unbound peptides and dried in air. The developed biosensor is stored at 4 °C for future use. The biosensor will be challenged with simulated Cd^{2+} solution. The binding of target heavy metal with metal binding peptide will change the conductance of ZnO, which can be sensitively measured and correlated to the concentration of heavy metals.

The biosensor is still under investigation and it will be optimized and the analytical characteristics such as calibration curve, the limit of detection, selectivity, accuracy, and stability will be investigated in detail.

3 ACKNOWLEDGEMENT

We greatly appreciate UConn for supporting studies on this project.

REFERENCES

[1] J. Bernier, P. Brousseau, K. Krzystyniak, H. Tryphonas and M. Fournier, Environ Health Perspect, 103(Suppl 9), 23-34, 1995.

[2] WHO, Environmental Health Criteria 101: Methylmercury. Geneva:World Health Organization, 1976.

[3] ATSDR, Toxicological Profile for Cadmium. Rpt TP-92/06. Atlanta, GA:Agency for Toxic Substances and Disease Registry, 1993.

[4] M. Toneta, K. Yoshino, M. Ohishi and H. Haito, Physica B, 376-377, 745-748, 2006.

[5] L. N. Dem'yanets, L. E. Li and T. G. Uvarova, J. Mater. Sci., 41, 1439-1444, 2006.

[6] Z. Fan and J. G. Lu, IEE Transactions on Nanotechnology, 5, 4, 2006.

[7] A. Wei, X. W. Sun, J. X. Wang, Y. Lei, X. P. Cai, C. M. Li, Z. L. Dong and W. Huang, Applied Physics Letters, 89, 1, 2006.

[8] J. B. Baxter, A. M. Walker, K. van Ommering and E. S. Aydil, Nanotechnology, 17, S304-S312, 2006.

[9] B. Lie and H. C. Zeng, J. Am. Chem. Soc., 125, 4430-4431, 2003.

[10] C. H. Lu and C. H. Yeh, Ceramics International, 26, 351-357, 2000.

[11] Y. Tong, Y. Liu, C. Shao, Y. Liu, C. Xu, J. Zhang, Y. Lu, D. Shen and X. Fan, J. Phys. Chem. B, 110, 14714-14718, 2006.

[12] Z. Li, Y. Xiong and Y. Xie, Nanotechnology, 16, 2303-2308, 2005.

[13] Z. Fang, K. Tang, G. Shen, D. Chen, R. Kong and S. Lei, Materials Letters, 60, 2530-2533, 2006.

[14] Y. L. Liu, Y. H. Yang, H. F. Yan, Z. M. Liu, G. L. Shen and R. Q. Yu, J. of Inorganic Biochemistry, 99, 2046-2053, 2005.

[15] S. Gao, H. Zhang, X. Wang, R. Deng, D. Sun and G. Zheng, J. Phys. Chem. B., 110, 15847-15852, 2006.

[16] M. Mo, J. C. Yu, L. Zhang and S. A. Li, Adv. Mater., 17, 2005.

[17] F.F. Zhang, X. L. Wang, S. Y. Ai, Z. D. Sun, Q. Wan, Z. Q. Zhu, Y, Z, Xian, L. T. Jin and K. Yamamoto, Anal. Chim. Acta. 519, 155, 2004.

Multiplexed immunoassays by flow cytometry for detection of clenbuterol, chloramphenicol and sulfadimidine with high sensitivity and selectivity

Yuan Zhao[1], Mingqiang Zou[1*], Haixia Gao[1], Qiang Xue[1], Peng Zhou[2]

[1]Chinese Academy of Inspection & Quarantine Science; [2] Kionix Inc., NY 14850, USA

An accurate, rapid and cost-effective detection on the environmental hazardous chemicals is the cornerstone of efficient food safe management. Here we discuss the relevance of an emerging technology, multiplexed competitive immunoassays read by flow cytometry, for the detection of Clenbuterol, Chloramphenicol and sulfadimidine. In these assays, multiple fluorescent microspheres, conjugated to different test antigens, constitute the solid phase for detecting antigens in biological samples based on the competitive ELISA principle. These assays are more sensitive than traditional immunoassays , have a high throughput capacity provide a wide analytical dynamic range and are powerful tools for exposure analysis and assessment offering low-cost screening with minimal sample pretreatment requirements combined and served a better alternative for the instrumental detection on the derivatives of those metabolites that often require expensive instrumentation. The sensitivity for the detection limit of the simultaneous identification of clenbuterol, chloramphenicol and sulfadimidine can reach 0.5ng, 2.0ng and 0.5ng/mL, which shows the promising multiplexing ability. Therefore, we predict a widespread application for a new breed of small, affordable, practical flow cytometrics as field instruments for replacing conventional ELISA and sophisticated GC or HPLC analysis.

Key Words: multiplexed immunoassays, flow cytometrics, clenbuterol, chloramphenicol , sulfadimidine, competitive ELISA

Clenbuterol (2-[*tert*-butylamino]- 1-[4-amino-3,5-dichloro-phenyl]-ethanol hydrochloride) is a major β-agonist drug reported to be used for illegal purposes in man and animals. The use of clenbuterol (CL) to improve athletic performanceis banned by the sport authorities [1]. CL is also fraudulently used to promote growth in meat-producing animals [2-5]. The most serious adverse effect of chloramphenicol is bone marrow depression. Serious and fatal blood dyscrasias (aplastic anemia, hypoplastic anemia, thrombocytopenia and granulocytopenia) are known to occur after the administration of chloramphenicol. An irreversible type of marrow depression leading to aplastic anemia with a high rate of mortality is characterized by the appearance weeks or months after therapy of bone marrow aplastia or hypoplasia. Peripherally, pancytopenia is most often observed, a reversible type of bone marrow depression, which is dose related, may occur. This type of marrow depression is characterized by vacuolization of the erythroid cells, reduction of reticulocytes and leukopenia, and responds promptly to the withdrawal of chloramphenicol. Chloramphenicol is categorised by the IARC (International Agency for Research on Cancer) as probably carcinogenic in humans.

Measurement of those illicit drug levels has in recent years been dominated by instrumental and ELISA methods. While ELISA offers a sensitive approach with detection in the low pg/mL range, the technique is hampered by its heterogenerous format with multiple addition and wash steps and an absorbance readout with substantial variability and interference. Complex pretreatments of derivation did not allow for the popularity of GC and HPLC methods to determine those molecules. Uniquely, this method provides the ability to combine different immunoassay formats in a single detection system: a competitive-inhibition format for the simultaneous measurement of the three representative drugs.

Therefore, an alternative microsphere-based immunoassay by flow cytometrics that offers comparable or higher sensitivity, better reproducibility, greater dynamic range, shorter preparation time, higher throughput capacity and the capability of simultaneous measurement of analytes in biological matrices was initially put forward in this study [6].

MATERIALS

Carboxyl SPHERO™ microspheres (Φ 4.0μm, Spherotech, Libertyville, IL) were applied as solid carriers.. The McAb, hapten of the three analytes (clenbuterol, chloramphenicol and sulfadimidine) and bovine serum albumin (BSA) were purchased from Wanger Biotechnology Inc. FITC goat anti-mouse IgG antibody conjugate was obtained from USA. N-Hydroxysuccinimide(NHS) and 1-ethy-3-(3- dimethyaminopropyl) carmodiimide hydrochloride (EDC) were purchased from Fluka(switzerland). Activation buffer (0.1 mol L^{-1} NaH_2PO_4, pH 6.2), storage/blocking buffer (PBS, 1%BSA, 0.05% NaN_3 (*Sigma*), pH 7.4), wash buffer (PBS containing 0.05% Tween20; agdia), dilution buffer (PBS, pH 7.4) and coupling buffer (0.05 mol L^{-1} 2-(*N* morpholinoethanesulfonic acid, MES; *Sigma*; pH 5.0). Labeling result is determined by measuring the absorbances at the related wavelengths on a spectrophotometer (Beckman ALTRA HyPerSort System) according to EXPO™ 32 MultiCOMP software.

METHODS
PREPARATION OF MICROSPHERE

All prepared capture antigens diluted in PBS buffer were at a level of 5–10µg/mL. After the sonication for 30s, every 10^4 microsphere was immediately added 200µL PBS, and centrifuged (*Sigma*,3k30 super-speed bench ,UK) at 5,700 rpm for 20 min. The microspheres were washed twice with PBS to secure the pellets were not adhered to the wall of tubes. The supernatant was removed and the rest microspheres were suspended in 80 µL PBS. The level of EDC and sulfo-NHS was adjusted to prepare a final concentration of 5mg/mL. A solution of hapten–BSA(5µg/mL) was added to each aliquot and they were incubated by vortex shaker for 3 h in the dark and centrifuged for 20 min to remove the supernatant, incubated in a blocking buffer for 2h and washed twice and then resuspended in 500µL PNT(PBS+1%BSA), stored at 4! overnight.

MICROSPHER-BASED COMPETITIVE FLUORENSCENT IMMUNOASSAY

Aliquots of the suspension of functionalized microspheres were recentrifuged, the supernatant was removed. Added 50µL analyte, then added 20µL 4.23µg/ml antibody to the respective microspheres tube, vortexed immediately for 10 s. The mixture was incubated on a rocker at room temperature in the dark for 2 h. The microspheres were centrifuged for 20 min, and then the separated microsphere were washed with wash buffer twice. Next, microspheres were removed supernatants and resuspended in 100µL of 10µg/ml FITC-goat anti-mouse IgG to each tube. The microspheres was shaken in the dark for 30 min and centrifuged for 20 min, washed with wash buffer twice and then diluted in 600µL wash buffer. The tube was shaken vigorously for approximately 1 min to disperse the microspheres.

Flow Cytometric Analysis

A COULTER EPICS ALTRA HYPerSort™ system (Beckman Coulter, Inc. Fullerton,CA) equipped with an air-cooled argon ion laser (488 nm) and HeNe ion laser(633nm) was used for this study. Fluorescence of thousands of beads was reported by EXPO™ 32 MultiCOMP software. To design a protocol for immunophenityping dual-signals, fluorescent light intensities of individual microspheres were measured and four-parameter histograms showed adhesion ratio. To determine the fluorescence intensity of a microsphere, a region of interest (ROI) was drawn around the microsphere and the fluorescence intensity within the ROI measured according four-parameter logistic log fits.

RESULTS
DETERMINATION OF MICROSPHERE-BASED COMPETITIVE FLUORESCENT IMMUNOASSAY PARAMETERS

Flow cytometry displays the intensity of the fluorescent microspheres bound with Antibody-FITC with the most intense signal present in the absence of competing free clenbuterol (see Fig. 2).

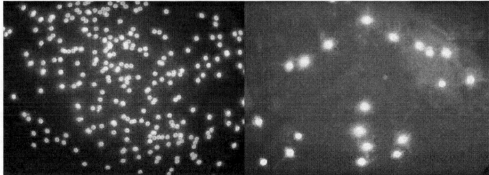

Fig. 2 The fluorescence of microsphere excited by green and red laser

Effect of the Mode of Microsphere-based competitive fluorescent immunoassay

Images showing the analytical and encoding signals were captured by a flow cytometry interfaced with a PMT detector. Highly significant linear relationships were observed.

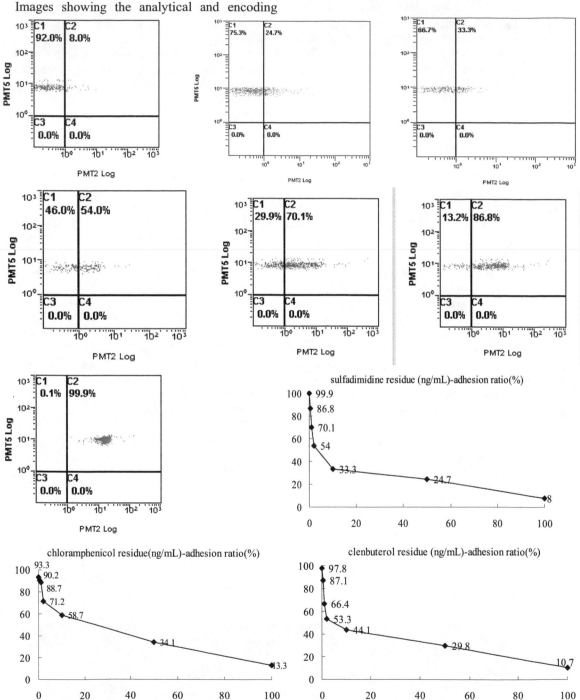

sulfadimidine residue (ng/mL)-adhesion ratio(%)

chloramphenicol residue(ng/mL)-adhesion ratio(%)

clenbuterol residue (ng/mL)-adhesion ratio(%)

Comparison of Microsphere-based competitive fluorescent immunoassay with ELISA

(ELISA MDC =0. 5 ng/mL) .This novel method will be exploited to detect immune responses from multiplex veterinary residuals.

Discussion

First, we devised and optimized competitive immunoassays for the individual analyte using labeled BSA immobilized on microspheres. Furthermore, we are contuine studying multianalyte based on this method. To determine the fluorescence intensity of a microsphere, a region of interest (ROI) was drawn around the microsphere and the fluorescence intensity within the ROI measured according four-parameter logistic log fits. The specificity of this gating is such that two different-sized or different fluorescent particles, each with a different antigen used to bind a different fluorescent ligand, can be used to simultaneously measure multiple analytes.

In the early stage of our investigation, antibodies labeled with a fluorescent dye were used in relatively high concentrations to generate appropriate fluorescent signals. Therefore, the sensitivity of these assays was only modest. To achieve high sensitivity with heterogeneous competitive immunoassays, it is usually necessary to keep the concentrations of the immunoreagents low. Furthermore, low analyte and labeled antigen concentrations are needed to minimize nonspecific effects and interferences between the individual immunoassays of multiplexed assays. Nonspecific binding to the solid surface, cross-reactivity, and aggregation of the individual tracer molecules can decrease the performance of multiplexed immunoanalyses[7]. It appears that it is more difficult to meet these requirements for competitive than for noncompetitive immunoassays because the latter assays are generally more sensitive [8]. Most reported multianalyte immunosensing methods are in the noncompetitive (sandwich) format.

In our experiment, we are previously attributed to decrease microsphere, FITC and monoclonal antibody nonspecific reaction. Coupling BSA to active microspheres, Incubate microspheres for 2 h on a rocker in dark, then centrifugation at 5,700 rpm for 20 min and wash twice with 150 µL wash buffer. Direct coupling monoclonal antibody or FITC goat anti-mouse IgG antibody can perform FITC or monoclonal antibody between microsphere nonspecific connection. For best assay response and sensitivity, we must be optimized to provide maximal labeling efficiency of bound antigen accompanied by minimal nonspecific binding with each other.

The sensitivity, dynamic range, and robustness of the Microsphere-Based Flow Cytometric Immunoassay are comparable to those of ELISA because ELISA is traditional method to determination residues. ELISA, although quantitative, precise, sensitive, and accurate, are designed to measure only one analyte per assay, for example,clenbuterol residual immunoassay kit, yielding 96 data points per plate. The Microsphere-Based Flow Cytometric Immunoassay give accurate, specific and reproducible results while saving time, sample and money, also small simple, multiplexing capabilities, more sensitivity and specificity. As the level of multiplexing is increased, the throughput is increasingly elevated. We will design the simultaneous measurement of several veterinary medicine using microsphere-based competitive fluorescent immunoassay for a total throughput of many analytes per sample.

In future work, the precision of our microsphere-based assays could be increased by using arrays with higher microsphere density, improved immobilization chemistry, and optical imaging fibers and an imaging system better suited for these analyses. Furthermore, it appears that the sensitivity of the analyses could be increased and the interferences between the assays could be minimized by employing even more efficient signal generation material such as quantum dot, a novel and superior fluorescent material.

In conclusion, we devised and optimized Microsphere-based competitive fluorescent immunoassay for the individual veterinary medicine(clenbuterol) residueusing labeled antigen immobilized on fluorescence microspheres. The result proved to be very sensitive for the determination of clenbuterol. Lower limit of detection and quantification was obtained by the new method. Whereas, there are few methods at present for the examination of veterinary residual. This method can be used to monitor clenbuterol residue in edible tissues, pork and dairy products.

References

[1] *M. J. Sauer, R. J. H. Pickett, A. L. McKenzie,* Anal. Chim. Acta 275, 195 (1993).

[2] *J. P Hanrahan,* "13-Agonists and their Effects on Animal Growth and Carcass Quality", Elsevier, London, 1987.

[3] *P. K. Baker, R. H. Dalrymple, D. L. Ingle, C. A. Ricks,* J. Anim. Sci. 59, 1256 (1984).

[4] *G.A. Qureshi, A. Eriksson,* J. Chromatogr. 441, 197 (1988).

[5] *M. M. Jimdnez Carmona, M. T Tena, M. D. Luque de Castro,* J. Chromatogr. 711! 269 (1995).

[6] *H. H. Meyer, L. Rinke, L Diirsch,* J. Chromatogr., Biomed. Appl. 564, 551 (1991).

[7] *L Yakamoto, K. Iwata,* J. Immunoassay 3, 155 (1982).

[8] *114. A. Bacigalupo, A. Ius, G. Meroni, M. Doris, E. Petruzzelli,* Analyst 120, 2269 (1995)..

The Chemical Nano-Sensor Development and Characterization

A.J. Jin

Hoda Globe Company, 800 El Camino Real, Mountain View, CA 94042, aj.jin@ieee.org

ABSTRACT

The nanotechnology will be critical for the near future success of the US economy. In this paper, I will present study of the chemical nanosensors for the space and environmental applications, the safety alert devices, etc. I will demonstrate a high-resolution CNS that is applied to the rocket fuel hydrazine leak detection. The CNS detects the changes in the electrical conductivity response during the chemical species presence. When the hydrazine is leaked into air, it immediately dissociates into NO_2. Detailed works in this paper are as follows. I will discuss the sensor chips preparation and process control. Furthermore, detailed studies of the CNS show responses to varying NO_2 concentration and nanomaterials.

Keywords: chemical nano sensor, hydrazine, nano technology, nanosensor stability

1   BACKGROUND

Nanoscience and nanotechnology, through the exploration and control of the nanomaterials at the nanometer scale, is considered as one of the key research areas for the future growth of US economy. Many sensor devices are part of our everyday life. More sensor improvements are needed for small size, great sensitivity and selectivity, fast response, minimal power consumption, and reliability demands, etc. Due to the well organized structure in atomic level of nanomaterials and their large surface-to-volume ratio, nanosensors are becoming very attractive for the next-generation of the sensing devices. Chemical nanosensors (CNS) are fabricated for the space and environmental applications. For example, we can apply the CNS that detects the electrical signal during the chemical species presence.

Figure 1 is a conceptual diagram where the sensor is placed so that the physical and chemical environment can be monitored and controlled. These conditions include the total gas flow rate, chemical concentration, humidity, chemical interface, temperature, pressure, etc. When chemical gases pass by CNS, the nanomaterials in the sensor platform respond correspondingly. The sensor response by electrical conductivity change is a result of the chemical sensing. Each sensor response is monitored electronically and is recorded in the computer as a sensor signal and for the further data processing.

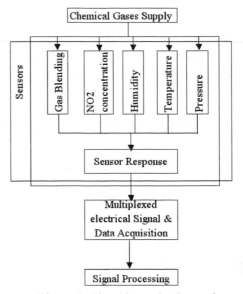

Figure 1: The schematic shows the conceptual CNS NO_2 experiments.

The purpose of our CNS project is to monitor the trace amount of NO2 composed from the leakage of one fuel component, hydranzaine. The liquid hydrazine (N2H4) is an efficient rocket propellant. When the N2H4 is leaked into air, it immediately dissociates and produces NO2.

As published in the previous literatures [1], carbon nanotubes (CNT) is very sensitive to NO_2 and it is therefore a very promising CNS to be employed as a commercial sensor product. The sensor development in this study will focus on the CNS and its NO_2 response in relationship to various NO_2 concentrations. In terms of the dry NO_2 analyte response, we will investigate the

CNS on the effects of various variables such as nanomaterials and gap size, etc.

We surveyed many sensing nanomaterials that include: 1) CGNT, 2) CGNT+MPC, 3) CGNT+polymer. The nanomaterials are as follows. The CGNT is the CVD grown nanotubes [2]. The MPC is monolayer-protected gold clusters (MPC) [3]. The polymer is cellulose hydroxypropyl [1,3].

We employ CGNT as the base nanomaterial in the form of the CVD growth. Illustration of Figure 2 procedures shows the nano coating to prepare a sensor chip before its sensor application.

2 IV CURVE, BIAS OPTIMIZATION

2.1 IV characterization

The IV characteristics of several nanomaterials are studied in the voltage windows and in the reversal voltage as well. The typical

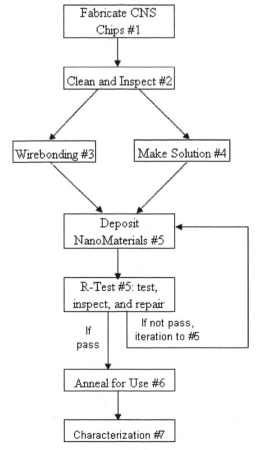

Figure 2: The flow chart shows preparation procedures of the clean CNS chip.

IV curves are shown in Figure 3. For example, we investigate the IV sweep curves from typical sensors with a CNT/MPC nanomaterial (on the upper left), a CNT/cellulose nanomaterial (on the upper right), and the CNT nanosensors (on the lower left). We have plotted a variety of IV curves in Fig. 3a), 3b), and 3c) for the 4 m feature gap of three typical inter-digitated electrode sensors. The non-linearity of the IV curves is also very interesting in order to identify the optimal sensors operating regime.

2.2 Bias voltage optimization

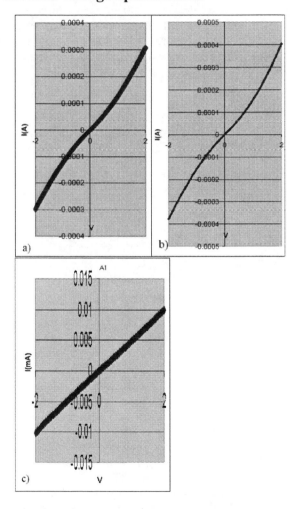

Figure 3: The IV curves of the nano-sensors are shown for a) CNT/MPC, b) CNT/polymer, c) CNT.

Moreover, we study the bias effects by applying different DC bias values. As stated at above, the electrical resistance of the CNS may be nonlinear. Therefore, we choose several different dc-bias voltages to measure the sensor response curve at various NO_2 concentration.

3. SENSOR STUDIES

We detect the baseline for the gas flow with pure air that shows no signal. Then we expose a sensor by applying on a sensor the chemical/gas flow with a concentration programmed in the same total flow. The change in the electrical signal is measured and the response is extracted from the sensor. Following this step, the sensor is purged to recover. After having enough purge time, go back and iterate the exposure-purge steps until the sensing process finish.

Figure 4 is a typical KAC31 run, where the chip KAC31 is deposited with nanomaterials of CGNT-only, CGNT plus MPC composite, and CGNT plus polymer composite, respectively. The sensors show strong response.

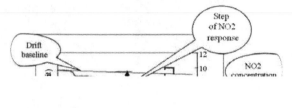

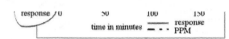

Figure 4: typical data trace of CNS in response to NO2 variation

We analyze the data and the relationship between the resistance change and the chemical flow as follows:
1. Make a linear fit to the drift baseline (on the initial 15' conditioning);
2. The baseline resistance is taken near the end of the recipe step-1, that is the end of the initial conditioning;
3. The baseline is extrapolated as a function of time by using a linear regression method;
4. The response dR is calculated as the difference between the resistance signal and the baseline at the time immediately after the exposure step. As shown in Figure 4, the CNS response steps are extracted for every concentration.

Furthermore, the analysis yields the dR and dR/R_0 dependence upon NO_2 concentration. By extracting dR/noise ratio for every sensor at all concentrations, we calculate the sensitivity function and plot this function in Figure 5 versus

the sensors. As a remark, the sensor's resolution in terms of the sensitivity limit, S/N, can also be derived by an extrapolation method.

Table 2: The table shows a typical recipe of the NO_2 sampling with total flow rate 400CCM.

Step Number	1	2	3	4	5	6	7	8	9	10	11
Time (minutes)	15'	10'	15'	10'	15'	10'	15'	10'	15'	10'	15'
NO2 (ppm)	0	0.5	0	1	0	2	0	5	0	10	0

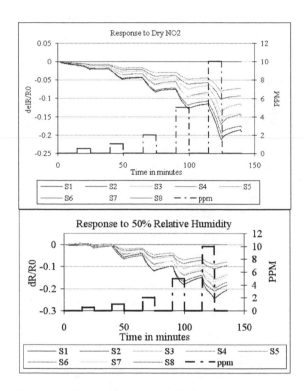

Figure 5: These charts show the relative response traces of the KAC31 chip with two RH levels: a) 0% or dry, b) 50%. The NO_2 concentration varies as shown at 0.5-, 1-, 2-, 5-, and 10-ppm.

We studied the sensor's responses at various humidity values. We observed that, for the response at low humidity range of 0% to 30% RH, the relative sensor response shows that the humidity in this range has quite little effects on the nanosensor response. The NO_2 response increases with the increasing humidity at a value of 50% and greater.

4. SUMMARY

In summary, We have studied the trace concentration of NO_2 below 1-ppm regime. Further studies are in the progress to characterize the life-expectance of CNS and the effect of temperature, pressure, etc.

References:
[1]: J. Li, Y. Lu, Q. Ye, M. Cinke, J. Han, and M. Meyyapan, NanoLett. 3, 929 (2003).

[2]: Jing Li, Yijiang Lu, Qi Ye, Lance Delzeit, and M. Meyyappan, Electrochemical and Solid-State Letters, 8 _11_ H100-H102 _2005.
[3]: P. Young, Y. Lu, R. Terrill, and J. Li, Journal of Nanoscience and Nanotechnology 5, 1 (2005).

Composite Carbon Nanostructures as Promising Carriers for Gaz Analysis

Laetitia Majoli[*], Alexis Evstratov[*], Jean-Michel Guillot[*], Aurélie Rouvière[**], Patrick Baussand[**]

[*] Ales High School of Mining, 6 Avenue de Clavières, Alès, F-30319 Cedex, France
alexis.evstratov@ema.fr
[**] Research Group on Environment and Atmospheric Chemistry (GRECA), University Joseph Fourier, 39 Boulevard Gambetta, Grenoble, F-38000, France

ABSTRACT

A lot of applications of carbon nanostructures (CNS) have been developed for the last fifteen years. Besides some particular electronic properties, such structures possess considerable adsorption capacities and so could be proposed both for gas treatment and for gas sampling and analysis. In this work, nanostructured carbon based composite samples were elaborated by the Catalytic Pyrolysis Method over the catalyst beds containing different transition metals supported by activated alumina (γ-Al$_2$O$_3$). Their efficiencies for volatile organic compounds (VOC) trapping, compared with the ones of Tenax-TA as the reference material, were evaluated in dynamic conditions. The obtained results show a real interest for new active material application for VOC adsorption procedures.

Keywords: carbon nanostructures, volatile organic compounds, gas sampling, analytical techniques.

1 INTRODUCTION

Carbon nanostructures (CNS) have been widely studied for a number of advanced technical applications [1, 2], from gas storage [3, 4] to electronics [5] and bio-detection [6]. However, their important adsorption capacities allow also a probable using of the CNS based composites as alternative adsorbents for gas treatment and / or sampling [7, 8]. Unfortunately, this application axis is not yet significantly explored.

In the framework of the project ACI NPD-37 "NSC-Environment" supported by the French Ministry of Research and New Technologies, a series of composite nano- and micrometric carbon layered materials was developed in order to test them for air pollution sampling, namely for trapping of volatile organic compounds (VOC). One of widely spread commercial products for VOC sampling, Tenax-TA, had been selected as the reference material.

A development of new composite adsorbents was started from an elaboration of the material's conception. The objectives were to confer to the material prototype some special properties in order to assure its technical ability for adsorption applications.

These properties are the following:
- elevated specific surface,
- great hydrophobic capacity,
- perfect crystalline structure,
- high thermal and mechanical resistance,
- high adsorption capacity for VOC,
- easy thermal regeneration of the surface enriched with adsorption products.

It was supposed that these properties achieved in their ensemble for the same elaborated species, have to assure a good VOC sampling material quality allowing its high efficiency and intensive multi-usage for an important service period duration. In fact, an *elevated specific surface* coupled with a *great hydrophobic capacity* promotes a *high adsorption capacity* for the hydrophobic organic compounds such as many VOCs, a *perfect crystalline structure* of the nano- and micrometric carbon aggregates minimizes the surface polar site population avoiding so a negative water adsorption impact and also favoring the sample *thermal regeneration* in moderated conditions. Suitable *mechanical properties* are necessary in order both to allow material's handling and favor its lifetime.

2 EXPERIMENTAL

Nano-structured carbon based composite samples were elaborated by the Catalytic Pyrolysis Method over the catalyst beds being constituted from activated alumina supported transition metals (pure Fe, Ni, Co or their binary mixtures).

2.1 Synthesis of CNS

The catalyst samples were obtained by an impregnation of γ-Al$_2$O$_3$ support (specific surface – 250–269 m^2/g, mercury detected total pore volume – 0.683–0.684 cm^3/g, fabrication – AXENS IFP Group Technologies) with nitrate salt solutions of either Ni or Fe, acetate salt solutions of Co and mixture acetate salt solutions of Ni and Fe or Co and Fe. After the impregnation step followed by the drying and calcination procedures, the samples were fulfilled in 1.5–

2.5 % ms. of pure or mixture metal oxides. The CNS elaboration was carried out at 550–850°C using a flow tubular laboratory reactor. The synthesis duration varied between 15 and 90 minutes. A synthesis gas mixture – N_2, H_2 and C_2H_2 – was used as the reaction middle. Due to the hydrogen presence, this mixture was also called to reduce, in situ, the oxide catalyst precursor to the metal one.

2.2 Characterization of CNS

The CNS species morphology, crystalline structure and thermo-chemical properties were investigated by SEM, TEM, XRD, Raman spectroscopy and DTA.

The sample affinities to water (or, contra versa, hydrophobic properties) were tested by so called "drying losses method". According to this method, the mass *Ms* of a sample saturated in water is compared with the one the same sample *Md* after its drying in air at 120°C during 3 hours. The mass difference related to the dry sample mass and expressed in percent scale corresponds to its water affinity. Subtracted from 100 %, the water affinity value yields the sample hydrophobic capacity (HC) given by the equation 1:

$$HC = 100 - \left[(Ms - Md)/Md\right] * 100, \% \qquad (1)$$

2.3 CNS adsorption test

The VOC adsorption capacities of elaborated composite samples were evaluated in dynamic conditions at constant temperature and relative humidity (20–21°C and ≤ 20 %, respectively). The sampling flow was fixed at 0.1 L/min through the stainless steel cartridges (i.d.-5 mm) packed with 200 mg of the tested adsorbents. The selected VOCs for the test were: Toluene, Ethyl-Benzene, Xylene, Hexane, Octane, Octene, Trichloroethylene, Tetrachloroethylene, α-Pinene and Limonene. The VOC levels of a glass chamber (200 L) in the gas phase were about 1 ppm for each compound.

3 CHARACTERIZATION RESULTS

The specific surfaces and the pore structures of the initial alumina support and of the CNS composite samples based on 2.5 % ms. cobalt oxide / 97.5 % ms. γ-Al₂O₃ precursor (products 1–3) are shown in the Table 1.

Sample	Specific surface, m^2/g	Pore volume **V**, cm^3/g			
		V≥37 Å	V≥100 Å	V≥0,1μ	V≥1,0 μ
Initial γ-Al₂O₃	250	0.684	0.341	0.009	0.003
1	170	0.263	0.084	0.024	0.007
2	205	0.444	0.247	0.009	0.002
3	211	0.596	0.309	0.016	0.003

Table 1: Structure characteristics of the alumina support and of the CNS cobalt oxide based composite samples.

The samples 2 and 3, elaborated using the optimal parameter set, manifest good structure characteristics whereas their CNS contents exceed 20 % ms.

The Table 2 contains the data on the sample hydrophobic capacities related with their CNS contents and the catalytic precursor chemical nature (samples 4, 5 – 2.5 % ms. iron oxide / 97.5 % ms. γ-Al₂O₃; sample 6 – 2.5 % ms. nickel oxide / 97.5 % ms. γ-Al₂O₃; sample 7 – 2.5 % ms. iron and cobalt oxide mixture / 97.5 % ms. γ-Al₂O₃).

Sample	4	5	6	7
HC, % ms.	100.00	99.99	99.96	99.97
CNS content, % ms.	29.8	25.0	34.2	33.0

Table 2: Sample hydrophobic capacities via their CNS contents and the catalytic precursor nature.

The iron based CNS samples containing less than 30 % ms. of carbon seem to have the best hydrophobic capacities which are near closed or even equal to the "ideal" (100 %) values.

Raman spectroscopy study of elaborated CNS structures clearly shows three bands from 1200 cm⁻¹ to 3500 cm⁻¹ and no signal from 100 to 500 cm⁻¹ (Figure 1). This lack of signal indicates that single wall structures are not present. In the range of 1300 to 1400 cm⁻¹, the deformation D-band indicates carbon microcrystalline structures like disordered graphite [9, 10]. Around 1600 cm⁻¹, the G-band confirms the presence of graphite plans.

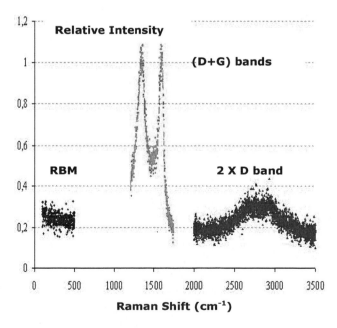

Figure 1: Raman spectra of CNS.

With XRD analyses, graphite structure development is also confirmed (Figure 2). The examined CNS structures were obtained both with single metallic and bi-metallic catalytic supports.

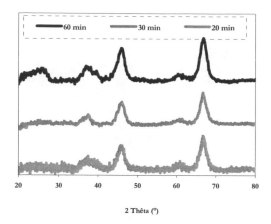

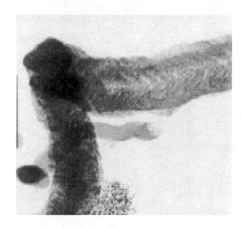

Figure 4: Internal structure with TEM analysis.

Figure 2: Synthesis duration influence on the crystalline state of CNS structures (XRD data are obtained for the samples elaborated with 2.5 % ms. iron and cobalt oxide mixture / 97.5 % ms. γ-Al$_2$O$_3$ precursor).

The CNS thermal stability was investigated by Differential Thermal Analysis (DTA) under air atmosphere. Results with bi-metallic catalyst supports are presented in Figure 3. For these CNS, no oxidative thermal degradation is observed until ~ 300°C. Under reduction atmosphere, the CNS composite thermal stability will be conditioned by the support phase thermal resistance (600–650°C for activated alumina).

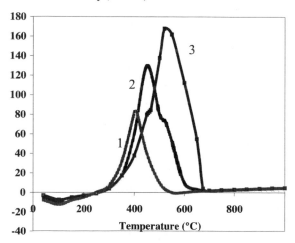

Figure 3: DTA results for 3 adsorbent with bimetallic catalysts (1-Oxides of Ni and Co; 2-Oxides of Fe and Co; 3-Oxides of Fe and Ni).

The internal CNS structure is visualized by means of Transmission Electronic Microscopy (TEM). Figure 4 shows that carbon filaments possess an internal channel and external walls constructed by superposed graphite plans.

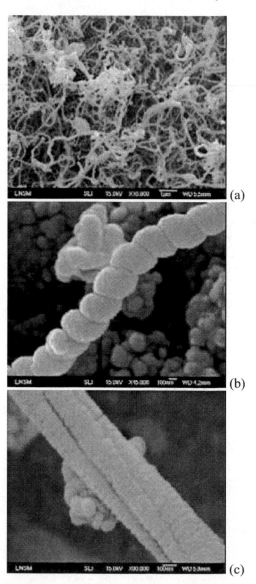

Figure 5: General view of CNS (a) and details of a nanocoil (b) and a nanofiber (c) (SEM analysis).

Cleantech 2007, ISBN 1-4200-6382-0

The CNS morphology and its details (nanocoils or nanotubes) were investigated by Scanning Electronic Microscopy (SEM). Some results are shown on the Figure 5.

All characterization results indicate that the sample properties correspond to the main objectives listed above. The elaborated CNS composites are so considered to be ready for testing as VOC trapping agents.

4 ADSORPTION TEST RESULTS

The adsorption test was carried out on with a mixture of selected VOCs into a glass chamber. For both cartridges packed containing Tenax-Ta and CNS, the capacities of the adsorbents were determined for each compound. An example of results obtained for a CNS sample elaborated with iron oxide – alumina catalyst is given in Table 3. The breaktrough volume (in L), as well as the adsorption capacity (mass ratio), are given. The capacity of CNS is always at least 9 times higher than for Tenax-TA and can reach to more than 20 times (cases of octane and octane). So CNS structures present a strong adsorption capacity for VOC trapping and can be used for analytical application. CNS composites with bi-metallic catalysts lead also to greater adsorption capacities than reference adsorbent. An important CNS density and their high adsorption capacity allow the packing of cartridges with very high efficiency where breakthrough probability is considerably decreased.

Selected VOCs For the test	Tenax TA (reference)		CNS (Oxides of Fe)	
	V (L)	m_c/m_a (µg/g)	V (L)	m_c/m_a (µg/g)
Toluene	0.85	27.6	11.2	364.1
Ethylbenzene	1.00	32.6	11.2	365.4
o-Xylene	1.20	39.5	11	362.2
Hexane	0.30	7.4	4.9	121.3
Octane	0.50	13.3	10.5	278.8
Trichloroethylene	0.50	27.4	4.6	252.2
Tetrachloroethylene	0.60	36.5	7	425.3
α-Pinene	0.80	25.7	10.9	350.5
D-Limonene	> 20	n.m.	> 17	n.m
Octene	0.50	13.4	10.6	284.2

Table 3: Comparison of adsorption capacities between Tenax-TA and CNS adsorbent for several VOCs (m_c: compound mass, m_a: adsorbent mass, n.m.: not measured).

5 CONCLUSION

This study has shown the characterization results and the application field for new composite adsorbents based on carbon nanostructures. These supports present the specific surface similar to the initial support (alumina) with values close to 200 m^2/g. The carbon level is in the range 25–35 % ms. The structure is perfectly crystalline (graphite type). This fact permits to confer to elaborated composites not only suitable mechanical hardness and thermal resistance (until 280-300°c under air atmosphere) but also hydrophobic behaviour. A comparative test, based on breakthrough volume, between CNS and Tenax-TA has shown than CNS composites manifest high adsorption efficiency. Comparatively to Tenax-TA, adsorption capacities of CNS are 10 times higher than capacities of the reference adsorbent. This fact opens a real application field for CNS as potential adsorbent in analytical chemistry. The adsorption capacity gives also the possibility to use these adsorbent for some treatment steps (VOC removal) especially for confined atmospheres or all applications where material resistance criteria are determinant for the adsorbent choice.

6 AKNOWLEDGEMENTS

The authors thank the French Ministry of Research and New Technologies for financial support of this project called NPD-37 "NSC-Environment" in ACI program. Some characterization have been carried out with the help of other French research teams: J.F. Chapat and J.L. Le Loarer (Axens-IFP Group Technologies), B. Ducourant and J.L. Sauvajol (University of Montpellier II) and A. Frazckiewicz and P. Jouffrey (Saint-Etienne High School of Mining).

REFERENCES

[1] J.C. Charlier and J. P. Issy, J. Phys. Chem. Solids, 57, 957-965, 1996.

[2] L.P. Biro, G.I. Mark, J. Gyulai and P.A. Thiry, Mater. Struct., 6, 104-108, 1999.

[3] M. Yumura, S. Ohshima, K. Uchida, Y. Tasaka, Y. Kuriki, F. Ikazaki, Y. Saito and S. Uemura, Diam. Relat. Mater., 8, 785-791, 1999.

[4] E. Frackowiak and F. Béguin, Carbon, 40, 1775-1787, 2002.

[5] A. Minett, J. Fraysse, G. Gang, G.-T. Kim and S. Roth, Curr. Appl. Phys., 2, 61-64, 2002.

[6] S. Carrara, V. Bavastrello, D. Ricci, E. Stura, C. Nicolini, Sensor.Actuat. B-Chem., 109, 221-226, 2005.

[7] L. Majoli, A. Evstratov and J.M. Guillot, 6th International Conférence on Nanotechnology in Carbon, Nanotec'04, Batz-sur-Mer, France, 2004.

[8] L. Majoli, A. Evstratov, and J.M. Guillot, Colloque du Groupe Français d'Etudes des Carbones, Saint Agnan-en-Morvan, France, 2005.

[9] T. Luo, J. Liu, L. Chen, S. Zeng and Y. Qian L., Carbon, 43, 755-759 , 2005.

[10] A. Maroto Valiente, P. Navarro López, I. Rodríguez Ramos, A. Guerrero Ruiz, Can Lin and Qin Xin, Carbon, 38, 2003-2006, 2000.

Molecular Simulation Studies on the Adsorption of Mercuric Chloride

R.R. Kotdawala, Nikolaos Kazantzis* and
Robert W. Thompson
Department of Chemical Engineering
Worcester Polytechnic Institute
Worcester, MA 01609, USA
*E-mail address: nikolas@wpi.edu

In the present research study, we make an attempt to understand the physical adsorption of mercuric chloride in zeolite NaX and activated carbon through Monte Carlo simulations in the temperature range of 400-500 K. In particular, we consider zeolite NaX with spherical cavities and sodium cations, as well as activated carbon with slit carbon pores and hydroxyl, carboxyl, and carbonyl sites. The capacity and affinity of zeolite NaX are compared with activated carbon with different acid sites by explicitly assessing the impact on mercuric chloride adsorption of a range of magnitudes of the electrostatic interactions considered, namely charge–induced dipole and charge–quadrupole interactions, as well as dispersion interactions.

1. INTRODUCTION

The Clean Air Act amendments identify a number of hazardous air pollutants (HAPs) of particular concern to human health and the environment. Data suggest that coal-fired power plants and municipal solid waste (MSW) incinerators are a significant source of some of these compounds, particularly elemental mercury and mercuric chloride. In the combustion zone, all mercury in coal is vaporized yielding vapor concentrations of mercury in the range of 1-20 $\mu g/m^3$ (0.1-2 ppbv). At flame temperatures (1700 K), all of the mercury is expected to be as elemental mercury in the gas phase. In the post flame region (1700-400 K), equilibrium predicts the oxidation of Hg^0 to $HgCl_2$ in the gas phase. Measurements in pilot and full-scale systems show that 10-80 % of the vapor phase mercury is likely to be $HgCl_2$ which is more easily removed from flue gas streams than elemental mercury [1,2,3].

These emissions of mercuric chloride from MSW incinerators and coal burning power plants can be reduced by adsorption on dry sorbents, which can be carried out either by injecting the sorbents into the exhaust gases, or by using multistage fixed beds for selective adsorption of acid gases, mercuric chloride and dioxins [1,2]. Processes which use adsorption on dry sorbents do not pose the problem of the treatment and stabilization of waste liquid streams, and therefore seem very attractive for both small and large combustors such as those used for incineration of hospital wastes for example [1,2,3]. The need to develop technologies capable of achieving high removal efficiencies for mercuric chloride emission control led many researchers to focus their attention on the evaluation of the adsorption capacity and selectivity shown by different solids. Selections of appropriate adsorbent compel researchers to understand the adsorption behavior of mercuric chloride at the molecular level.

The next section provides the requisite information on the particular simulation models proposed for zeolite NaX and activated carbon, followed by a description of the Grand Canonical Monte Carlo (GCMC) simulation scheme employed in the present research study. Finally, Section 4 discusses the simulation results derived.

2. MODEL DESCRIPTIONS
2.1 Zeolite NaX Model

Zeolite NaX was modeled by considering the zeolite cavity as spherically shaped with sodium cations located uniformly in the cavity [4]. The locations of sodium cations were taken from Karavias et al. [4]. The interactions with spherical cavity were calculated using a spherically–averaged potential for the dispersion and repulsion given by [4]:

$$\psi^{disp} = 4C\varepsilon_{1s}\left[\left(\frac{\sigma_{1s}}{R}\right)^{12} L\left\{\frac{r^2}{R^2}\right\} - \left(\frac{\sigma_{1s}}{R}\right)^6 M\left\{\frac{r^2}{R^2}\right\}\right]$$

(1)

where,

$$L\{x\} = (1 + 12x + 25.2x^2 + 12x^3 + x^4)/(1-x)^{10}$$

(2)

$$M\{x\} = (1+x)/(1-x)^4$$

(3)

with, $x = \dfrac{r}{R}$

and ψ^{disp} being a function of r, the radial distance of the adsorbed molecule from the center of the cavity. The cavity radius R was chosen as the distance from the center to the nearest oxygen atom. For zeolite NaX, the cavity radius(R) is 7.057 Å [5]. In the above expression, $C\varepsilon_{1s}$ and σ_{1s} are the Lennard-Jones energy and collision diameter of molecule i with the solid wall respectively. The interactions of mercuric chloride with sodium cations are mainly cation-induced dipole and cation-quadrupole moment interactions which can be calculated using equations (4) and (5) respectively, as shown below [6]:

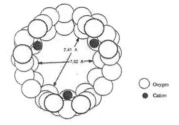

Fig. 1 Cross-section of a zeolite NaX adsorption cavity [4]

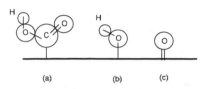

Fig. 2 Schematic representation of the surface sites on carbon: (a) a carboxyl group, (b) a hydroxyl group, and (c) a carbonyl group

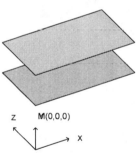

Fig. 3 A schematic diagram of a slit pore with origin at M(0,0,0)

$$\psi^{ind} = -\frac{\alpha q^2}{2r^4(4\pi\varepsilon_o)^2} \qquad (4)$$

where α is the polarizability of the molecule, q is the electronic charge of the ion on the surface, ε_o is the permittivity of vacuum. R is the distance between the centers of the interacting pair [6]. Notice, that interactions between the ion field and the point dipole is given by [6]:

$$\psi^{quadrupole} = -\frac{Qq(3\cos^2\theta - 1)}{4r^3(4\pi\varepsilon_o)} \qquad (5)$$

where, Q is the linear quadrupole moment of the molecule, and θ is the angle between the direction of the field and the axis of the quadrupole.

It should be pointed out, that two types of interactions are considered in the present study, namely dispersion interactions and quadrupole-quadrupole interactions given by Eqns. (4) and (5). The following adsorbate (i)-adsorbate (j) interactions also were considered [7,8]:

$$U_{ij}^{QQ} = \frac{3Q_iQ_j}{4r^5}\begin{pmatrix} 1 - 5\cos^2\theta_i - 5\cos^2\theta_j + 17\cos^2\theta_i\cos^2\theta_j + 2\sin^2\theta_i\sin^2\theta_j\cos^2\phi_{ij} \\ -16\cos\theta_i\cos\theta_j\sin\theta_i\sin\theta_j\cos\phi_{ij} \end{pmatrix}$$

$$(6)$$

$$U_{ij}^{disp}(r) = \left[-\frac{1}{(4\pi\varepsilon_0)^2 r_{ij}^6}\left(\frac{3\alpha_i\alpha_j}{I_iI_j 4}\right)\right] \qquad (7)$$

where I_i and I_j are the first ionization potentials for molecules i and j, respectively.

2.2 Activated Carbon Model

The pores of activated carbon were modeled by considering two parallel walls, each of which comprises an infinite number of layers of graphite. The graphite layers are composed of Lennard- Jones sites, but these are smeared out uniformly over each layer. The interaction between an adsorbate molecule and this smooth carbon surface is represented by the 10-4-3 potential of Steele [9]. Three types of polar surface sites: hydroxyl, carboxyl, and carbonyl groups, were considered in the simulations, and are represented schematically in Fig. 2. The carboxyl, hydroxyl, and carbonyl sites can be considered as a collection of five, two and three point charges. The parameters (size of charge atom and partial charge) of point charges and their positions were taken from Jorge et al. [10]. Cross–species parameters were calculated using the Lorentz-Berthelot combining rule [10]. As mentioned earlier, the electrostatic interactions namely charge-induced dipole moment and charge-quadrupole moment were calculated using equations (4) and (5), respectively.

3. SIMULATION METHODS

We calculated adsorption isotherms using grand canonical Monte Carlo (GCMC) simulations, in which the temperature (T), volume (V), and chemical potentials (μ) were kept constant. The algorithm for GCMC simulations is well documented [11,12], and we used the general methodology. The pressure (P) was calculated from the chemical potential and the equation of state for an ideal gas, and all simulations were performed in equilibrium bulk gas temperature in the range of 400-500 K. Each type of Monte Carlo trial (creation, destruction, and displacement/rotation) was chosen randomly with the same probability. The number of equilibrium steps in the simulations varied according to the operating conditions, in the range of 30 to 70 millions. During the sampling period, typical configurations for each run were stored in files and then converted into images. In the case of zeolite NaX, the interactions with molecules in four neighboring cavities also were considered. The molecules in the neighboring cavities were the images of molecules in the central cavity. In the case of activated carbon slit pores, we used rectangular simulation cells, bound in the Y direction by pore walls and replicated in X and Z directions using periodic boundary conditions. In order to account for the long range interactions, especially charge–quadrupole, charge-induced dipole and quadrupole–quadrupole interactions, the interactions of molecules in the central cavity with neighboring cavities were considered. The size of the simulation box was 3.0x1.5x3nm.

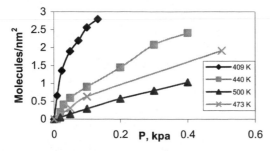

Fig. 4 Adsorption isotherm from GCMC simulation for HgCl$_2$ in zeolite NaX

4. RESULTS AND DISCUSSION
4.1 Adsorption in zeolite NaX

GCMC calculations of mercuric chloride adsorption in zeolite NaX in the temperature range of 409-500 K are shown in Fig. 4, up to an operating pressure of 0.6 kPa. The sorption capacity was predicted to decrease with temperature, as expected. In order to understand the phenomena, the interaction energies of the molecules with the zeolite cavity, sodium cations, and with other mercuric chloride molecules were studied,

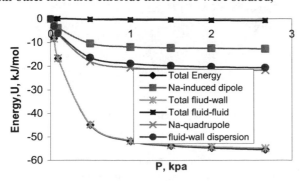

Fig. 5 Relative energy contributions to adsorption for zeolite NaX at 473 K

and are shown in Fig. 5. The figure indicates that the Na–quadrupole moment and fluid-wall dispersion interactions with the spherical cavity dominate over the Na-induced dipole and other interactions among mercuric chloride molecules. In fact, the Total Energy is almost completely due to the Total fluid-wall energy, comprised of the Na-quadrupole, the fluid-wall dispersion, and the Na-induced dipole. However, the Na-induced dipole only plays a minor role in this total.

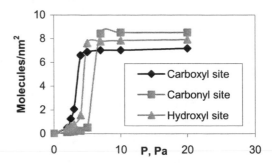

Fig. 6 Adsorption isotherms for activated carbon with acid site concentration of 2.2 sites/nm^2 at 500 K

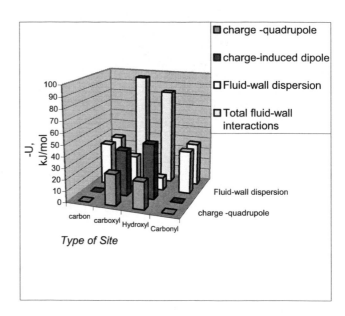

Fig. 7 Relative interaction energy for HgCl$_2$ – activated carbon as a function of acid site at 2 Pa and 450 K

The higher quadrupole interaction is attributed to a high quadrupole moment [13] (-1.48 x 10^{-39} C.m^2) of the molecule, and dispersion interactions might be attributed to the high polarizability (9 Å^3) of the molecule [14].

4.2 Adsorption in Activated Carbon

The simulations were carried out by introducing two carboxyl, hydroxyl, and carbonyl functional sites in the slit pore as well as in the absence of functional sites. The two sites are located at (0.5,1.5,0.5 nm) and (0.5,1.5,1 nm), with the origin M as shown in Fig.3, and correspond to a carbon atom of the functional site that is also a part of the carbon base of the slit pore. The coordinates of the other atoms in the functional sites can be obtained from the bond length and bond angle among them as shown by Jorge et al. [10]. The isotherms for each functional site at 500 K are depicted in Fig. (6). The pore with the carboxyl sites has the highest Henry's Law constant (slope of isotherm in the Zero pressure limit) followed by those with hydroxyl and carbonyl sites. The isotherms also give an indication of a sudden pore filling, implying that the strength of the interactions among mercuric chloride molecules and interactions with the functional sites and the carbon base are comparable. However, the densities of the molecules were low in the case of carboxyl and hydroxyl sites as compared to carbonyl groups. This might be attributed to less available space for the molecule in the pore, due to the greater size of the carboxyl site, indicating the trade off between affinity for the surface and capacity of the adsorbent.

Fig. 7 gives insights of the relative strengths of the interactions with different functional sites and without any functional site. It indicates that the presence of a carbonyl site doesn't increase the affinity to the adsorbate molecules, since the charge-quadropole and charge-induced dipole interactions (electrostatic interactions) are negligible. However, in the case of carboxyl and hydroxyl sites, the electrostatic interactions are greater than dispersion interactions, which would lead to higher affinity for mercuric chloride.

5. CONCLUSIONS

We carried out GCMC simulations to study adsorption phenomena of mercuric chloride molecules in zeolite NaX and activated carbon with and without functional groups. The results from zeolite NaX implies that the Na-quadrupole and dispersion interactions dominate over other types of interactions, suggesting that by manipulating cation density, or types, one can enhance the sorption of the molecule. The evaluation of the sorption of the molecule in activated carbon showed that hydroxyl and carboxyl sites are more important than carbonyl sites. The carboxyl sites give superior affinity than the other two types of functional sites at the expense of the adsorbent's capacity.

References

[1] B. Hall, O.Lindquvist, and E.Ljungstrom, *Env. Sci. Techn.*, **24**, 108 (1990)

[2] S.V. Krishnan, B.K. Gullett and W. Jozevicz, *Env. Sci. Techn.*, **28**, 1506 (1994)

[3] D. Karatza, A. Lancia, D. Musmarra, F.Pepe and G.Volpicelli, *Comb. Sci. Tech.*, **112**, 163 (1996)

[4] F. Karavias, *Ph.D Thesis*, Univ. of Pennsylvania (1992)

[5] J.L. Soto,and A.L.Myers, *Mol.Phys.*, **42**, 971(1981)

[6] R.T .Yang , *Adsorbents: Fundamentals and applications*, Wiley – Interscience (2003)

[7] T.M. Reed, and K.E. Gubbins, *Applied Statistical Mechanics*, Mc-Graw Hill, NewYork, (1973)

[8] J.M.Prausnitz, R.N. Lichtenthaler and E. Gomes de Azevedo, *Molecular Thermodynamics of Fluid-Phase Equilibria*, 3rd ed, Prentice-Hall, NJ (1999)

[9] W A Steele, *The Interaction of gases with Solid surfaces*, Pergamon Press, Oxford (1974)

[10] M. Jorge, C. Schumacher and N..Λ. Seaton, *Langmuir*,**18**, 9296 (2002)

[11] D. Frenkel and B. Smit, *Understanding Molecular simulation*, Academic Press, London (1996)

[12] M P Allen and D J Tildesley , *Computer simulation of Liquids*, Clarendon Press: Oxford (1989)

[13] A.N. Pandey, A. Bigotto and .K.Gulati, *Acta Phys. Polon. A*, **80**, 503 (1991)

[14] K.Watanabe, *J. Chem. Phys.* **26**, 542 (1957)

Layer by Layer Fabrication of an Amperometric Nanocomposite Biosensor for Sulfite

S.B. Adeloju[*] and A. Ohanessian

Water and Sensor Research Group
School of Applied Sciences and Engineering, Monash University
Churchill, Victoria 3842, Australia, Sam.Adeloju@sci.monash.edu.au

ABSTRACT

A layer by layer strategy is described for the galvanostatic fabrication of a polypyrrole nanocomposite amperometric biosensor for sulfite. The strategy can be used to fabricate bilayer and trilayer sulfite biosensors, consisting of nanolayers of polypyrrole-sulfite oxidase, polypyrrole-dextran-sulfite oxidase, polypyrrole-chloride and/or polypyrrole-nitrate films. It has been demonstrated that the nature of the outer nanolayer has a significant influence on the selectivity and sensitivity of the amperometric nanocomposite biosensor. The presence of interferants, such as oxalic and tartaric acids at levels usually present in wine and beer did not interfere with the nanocomposite biosensor. The successful application of the nanolayer biosensor to sulfite determination in some wine and beer samples without sample pre-treatment is demonstrated.

Keywords: Nanocomposite, biosensor, polypyrrole, sulfite, layer-by-layer

1 INTRODUCTION

Traditionally, layer-by-layer assembly (LBL) is used for fabrication of films with molecular order and stability, based on alternating adsorption of oppositely charged macromolecules, such as biomacromolecules and polymers.[1-3] The resulting assembly of these ultrathin films (5-500 nm in thickness) is usually held together by electrostatic forces, covalent bonding, and, to a lesser extent, hydrogen bonding and hydrophobic interactions. This approach has gained considerable interest and has been used to prepare layered nanocomposites with high degree of organization from polymers and different nanocolloids such as nanoparticles, nanowires, nanotubes, clay platelets, and proteins. Recently, the use of conducting polymer polypyrrole (PPy) for layer-by-layer nano-assembly has been reported for the fabrication of highly sensitive and fast response humidity sensors. Also, layer-by-layer nano-assembly of polystyrenesulfonate (PSS) and PPy, as polyanion and polycation, respectively, on glass substrate has been reported. In general, most of the reported LBL methods to date rely on alternating adsorption of oppositely charged macromolecules or similar processes. However, where the use of such an adsorption process is not possible, electrochemical deposition can provide an alternate effective strategy for achieving LBL assemblies at the nanoscale level.

This paper presents the results of a layer by layer strategy that we have developed for the fabrication of a polypyrrole nanocomposite amperometric biosensor for sulfite. The use of this strategy to fabricate bilayer and trilayer sulfite biosensors, consisting of nanolayers of polypyrrole-sulfite oxidase, polypyrrole-dextran-sulfite oxidase, polypyrrole-chloride and/or polypyrrole-nitrate films will be discussed. In particular, the influence of the outer nanolayer on the selectivity and sensitivity of the amperometric nanocomposite biosensor will be highlighted. The performance of sulfite nanocomposite biosensor in the presence and absence of interferants, such as oxalic and tartaric acids will also be discussed. Furthermore, the successful application of the nanolayer biosensor to sulfite determination in some wine and beer samples without sample pre-treatment will be demonstrated.

2 EXPERIMENTAL

Reagents

All reagents used were AR grade unless otherwise stated. Pyrrole was obtained from Merck-Shuchardt (Sydney, Australia) and was distilled prior to use and stored covered with aluminium foil in the refrigerator to prevent UV degradation. Potassium chloride and potassium nitrate were from BDH Laboratory Supplies and sodium sulfite was obtained from Ajax Chemicals. Sulfite solution was prepared fresh prior to use. Dextran (average MW 40,000 g/mol was obtained from ICN Biomedical and sulfite oxidase (SOx) was obtained from Sigma-Aldrich. All solutions were prepared with Milli-Q water.

Instrumentation

The conducting polymer nanolayers were prepared galvanostatically with a potentiostat/galvanostat. Voltammetric and amperometric measurements were performed with a MacLab 4s (ADInstruments Pty Ltd) connected to a computer and a printer.

Layer-by-Layer Galvanostatic Assembly

Nanolayers of polypyrrole-dextran-sulfite oxidase (PPy-Dex-SOx) was grown as previously described.[4] The desired outer nanolayers were deposited galvanostatically onto PPy-Dex-SOx inner layer, as either bilayer or trilayer. Prior to the deposition of each layer, the monomer solutions were purged with nitrogen for 10 mins to remove oxygen. A three-electrode cell which consists of a platinum working electrode, Ag/AgCl reference electrode and a platinum auxiliary electrode was used for all electropolymerisation.

Characterisation of Nanolayered Films

After galvanostatic deposition of the nanolayers, the electrodes were rinsed several times with Milli-Q water to remove loosely bound molecules. Cyclic voltammetry and amperometry were then performed in phosphate buffer (pH 7.2) with a conventional three-electrode cell.

3   RESULTS AND DISCUSSION

The layer-by-layer galvanostatic assembly of the nanolayers, as bilayers or trilayers, is shown in Figure 1. The use of three different outer nanolayers for the fabrication of bilayer biosensors, based on galvanostatic assembly in layer-by-layer modes resulted in substantial improvement in the amperometric response for sulfite. As shown in Figure 2, the inclusion of all the outer nanolayers resulted in improvement of the sulfite response beyond that obtained with the single (mono-) layer. This is due, in part, to the benefit of the outer layer in retaining the enzyme in the inner layer and enabling better containment of the enzymatic product for more improved detection. It is particularly interesting to note that the sulfite response with the use of PPy-Cl and PPy-NO$_3$ outer nanolayers were less than half of that obtained with the outer PPy-SOx nanolayer. It is therefore apparent that the presence of the required enzyme in two or more subsequent nanolayers enabled maximum catalytic reaction and, hence, generated the most catalytic product for amperometric detection at the platinum electrode. It was also found that the influence of the outer PPy-SOx nanolayer is dependent on the galvanostatic polymerization period and, hence, on the thickness of the layer. A film thickness of 50-60 nm for the outer PPy-SOx nanolayer gave optimum amperometric response for sulfite.

The galvanostatic assembly, based on layer-by-layer, of various combination of triple layers also resulted in the improvement of the sulfite amperometric response beyond that obtained with the single (mono-) layer. However, the addition of the third nanolayer did not result in more improvement in sulfite response beyond those obtained with the bilayers.

Both the bilayer and trilayer assemblies limited the interference of oxalic acid and tartaric acid on the amperometric response of sulfite. In fact, at the concentrations usually present in wine, < 20 mg/l, no interference was observed for sulfite determination with both types of layer-by-layer assemblies. The bilayer assembly was successfully applied to a range of beverages, with a recovery efficiency of 101±1 %.

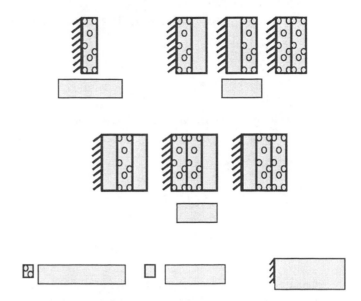

Figure 1: Different configurations for galvanostatic assembly of nanolayers based on layer-by-layer.

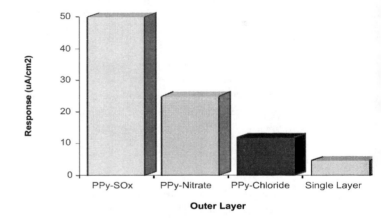

Figure 2: Influence of outer nanolayer composition on amperometric response of sulfite on a bilayer electrode. Inner layer is PPy-Dex-SOx. E_{app} is -0.1 V.

4   CONCLUSION

Galvanostatic polymerisation, based on layer-by-layer assembly, of bilayers and trilayers has been successfully used to fabricate an ultra-sensitive biosensor for sulfite, and to enable its reliable determination in wine and beer. Oxalic and tartaric acids did not interfere with the biosensor.

REFERENCES
[1] Kim, T. W.; Lee, D. U.; Yoon, Y. S. *J. Appl. Phys. 88*, 3759-3761, 2000.
[2] Chan, W. C.; Nie, S. *Science 281*, 2016-2018, 1998.
[3] Lvov, Y.; Decher, G.; Mohwald, H. *Langmuir 9*, 481-486, 1993.
[4] Adeloju, S.B.; Ohanessian, A.; Duc, N. Synthetic Metals 153, 17-20, 2005.

Electrochemical 'Lab on a chip' for Toxicity Detection in Water

Rachela Popovtzer[*], Tova Neufeld[**], Eliora z. Ron[**], Judith Rishpon[**] and Yosi Shacham-Diamand[*]

[*]Department of Electrical Engineering – Physical Electronics and the TAU Research Institute for Nano Science and Nano-technologies, Tel-Aviv University.
[**]Department of Molecular Microbiology and Biotechnology. Faculty of Life Sciences, Tel Aviv University.

ABSTRACT

An electrochemical 'Lab-on-a-chip' for water toxicity detection is presented. This miniaturized device containing an array of nano liter electrochemical cells, which integrates bacteria and can emulate physiological reactions in response to different chemicals. Bacteria, which have been genetically engineered to respond to environmental stress, act as a sensor element and trigger a sequence of processes, which leads to generation of electrical current.

The silicon chip contains an array of nano-volume electrochemical cells that house the bacteria, connected to a sensing and data analysis unit. Each of the electrochemical cells in the array can be monitored independently and simultaneously with the others.

A measurable current signal, well above the noise level, was produced within less than 10 minutes of exposure to representative toxicants. This miniature device provides high throughput rapid and sensitive real-time detection of acute toxicity in water.

Keywords: Electochemical sensor, bio-chip, whole cell biosensors, lab on a chip, BioMEMS.

1. INTRODUCTION

The use of living organisms as active components in electronic devices is an innovative and challenging area combining recent progress in molecular biology and micro technology. The aim of this study is to develop a system, composed of new design and process of MEMS that integrates genetically engineered living organisms serving as physiological sensors, and an electrochemical system functioning as an electrical sensor.

Nowadays, there is a tendency to limit the use of animals in laboratory experiments and to prefer technologies that uses microorganisms as experimental models. Although microorganisms differ from highly developed organism, at the molecular and physiological levels there are many shared features by both species including response to toxicants, drugs and stressful conditions[1].

The vast development in genetic engineering of live cells, enables the use of recombinant cells as cell-based sensing systems [2, 3]. The cascade of mechanisms by which *E.coli* bacterial reactions to toxic chemicals or to stressful condition are electrochemically converted into electronic signals have been previously reported[4-6].

The potential uses for microorganisms-based systems could have important applications in pharmacology, medicine, cosmetics, and environmental monitoring.

2. CHIP DESIGN AND FABRICATION

The chip has been designed and fabricated using micro-system-technology (MST) methods. The materials for chip construction have been selected with special considerations on their biocompatibility characterization, since they are aimed to be in direct contact with living cells. The chip was produced on silicon wafers and includes an array of eight independent electrochemical cells, which are temperature-controlled. Each electrochemical cell can hold 100 nL of solution and consists of three embedded electrodes: 1) Gold working electrode, 2) Gold counter electrode and 3) Ag/AgCl reference electrode. The wall of the chambers constructed from photopolymerized polyimide (SU-8). The shape and size of the nano chambers as well as of the microelectrodes were designed so it could be easily modified, and their size could be scaled up and down for any specific applications. The device was manufactured as two parts: the first part is a disposable silicon chip - with an array of electrochemical cells. The silicon chip was wire bonded to a special printed circuit board (PCB) platform, which was connected to the data processing units. The PCB manufacturing process consists of selective deposition of gold which creates conducting bands. The bands width in the center were designed to be the same width as the external gold pads of the silicon chip in order to wire bond between them, while the bands width in the external region of the PCB fitted precisely to the socket of the electronic sensing unit (Figure 1). The second part of the device is reusable, which includes a multiplexer, potentiostat, temperature control and a pocket PC for sensing and data analysis. This setting allows continuous reusing for multiple measurements.

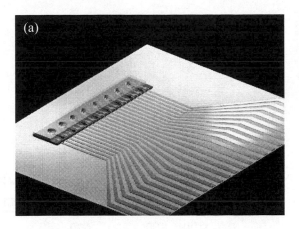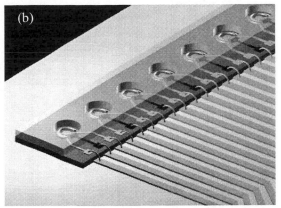

Figure 1: Image of the electrochemical silicon chip wire bonded to the PCB platform. (a) Array of eight 100nL electrochemical cells on a silicon chip. The chip is glued to the tailored PCB platform, and the chip's gold pads are wire bonded to the gold PCB's electrodes. The PCB board enters directly to the socket of an external sensing circuit. (b) Electrochemical-cells on chip consist of three embedded electrodes: gold working electrode, gold counter electrode, and Ag/AgCl reference electrodes.

3. RECOMBINANT BACTERIA

Genetically engineered bacteria where used as whole cell sensors for acute toxicity in water. The recombinant bacteria react to the presence of toxin by activating specific promoter (regulatory DNA sequence). This promoter induces the production of the reporter enzyme β-galactosidase. This enzyme reacts with the PAPG substrates (molecules that where initial placed inside the chambers) to produce two different products: electrochemical active product *p*-aminophenol (PAP), and inactive product β-d-galactopyranoside. The PAP molecules are oxidized on the working electrode at 220mV. This oxidation is converted to a current signal using an Amperometric technique.

4. EXPERIMENTAL

In the present work, we used recombinant E.coli bacteria bearing plasmid with one of the following promoters: *Dnak*, *grpE* or *fabA*. These promoters were fused to the reporter enzyme β-galactosidase[5, 7]. Eethanol [1%] or phenol [1.6ppm] were introduced to the bacterial samples in the presence of the substrate PAPG. Immediately after (~1second), the suspensions were placed in the electrochemical cells. The response of the bacteria to the toxic chemicals was measured on-line by applying a potential of 220 mV. The substrate, PAPG, was added to a final concentration of 0.8 mg/ml (100 nL total volume). The product of the enzymatic reaction (PAP) was monitored by its oxidation current. Additional measurements in the absence of the bacteria were performed to exclude the possibility of electroactive species in the LB medium, in the substrate, or in the substrate and the LB medium mixture, which can contributes to the current response.

Bacteria at 3×10^7 cells/ml were used for all experiments.

5. RESULTS AND DISCUSSION

Real time detection of the response of recombinant E. coli bacteria, with one of the promoters *Dnak*, *grpE* or *fabA*, to ethanol and phenol are shown in figure 2 and 3 respectively. The results show that concentration of 1% ethanol could be detected within less than 10 minutes, and concentration as low as 1.6ppm phenol could be detected within less than 6 minutes. Different intensity response of the various bacterial sensors, *dnaK*, *grpE* and *fabA*, to ethanol and phenol is due to the specific activation of each promoter to the type of the toxicant. The promoters *dnaK* and *grpE* are sensitive to protein damage (SOS system), thus, they were induced in response to ethanol which is known as protein damage agent [2] (figure 2). *grpE* showed high induction activity in response to ethanol, *dnaK* showed reduced enzyme activity and *fabA* was only slightly induced above the background level. *fabA* promoter is sensitive to membrane damage, and thus, reacts to phenol exposure, which is known membrane damage chemical[]. As expected, *grpE* and *dnaK* promoters were less activated by phenol (figure 3).

In comparison to equivalent optical detection methods using whole cell biosensors for water toxicity detection, these results proved to be more sensitive and produce faster response time. Concentrations as low as 1% of ethanol and 1.6 ppm of phenol could be detected in less than 10 minutes of exposure to the toxic chemical, whilst a recent study[8], which utilized bioluminescent E.coli sensor cells, detected

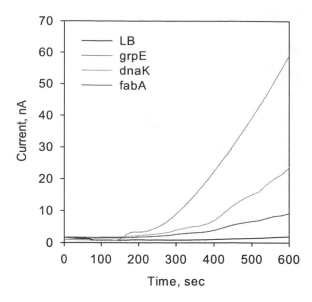

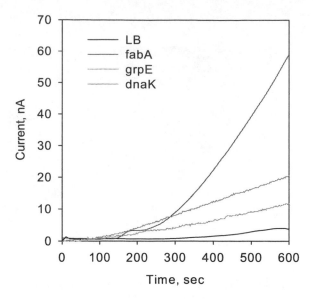

Figure 2: Amperometric response curves for on line monitoring of different *E. coli* reporters in response to the addition of 1% ethanol, using the nano-bio-chip. The different *E. coli* reporters are *fab*A *dna*K and *grp*E. Measurement performed immediately after the ethanol addition (~1min) at 220mV working potential vs Ag/AgCl reference electrode. The LB curve represents the bacterial response to the LB medium with the substrate PAPG without ethanol.

Figure 3: Amperometric response curves for on line monitoring of different *E. coli* reporters in response to the addition of 1.6ppm phenol, using the nano-bio-chip. The different *E. coli* reporters are *fab*A *dna*K and *grp*E. Measurement performed immediately after the ethanol addition (~1min) at 220mV working potential vs Ag/AgCl reference electrode. The LB curve represents the bacterial response to the LB medium with the substrate PAPG without phenol.

0.4M (2.35%) ethanol after 220 minutes. An additional study[9] based on fluorescent reporter system (GFP), enabled detection of 6% ethanol and 295 ppm phenol after more than one hour. Cha et al [10] used optical detection methods of fluorescent GFP proteins, detected 1g of phenol per liter (1000ppm) and 2% ethanol, after 6 hours. Other studies[11], could not be directly compared due to different material used, however their time scale for chemicals identification is hours.

These results emphasize the advantages of merging electrochemical detection methods with adjusted design and process of MEMS, which resulting in fast response time and low detection limit. Enhanced sensitivity and high signal to noise ratio is achieved by optimizing the ratio between working electrode area and cell volume. The larger the ratio the higher the signal.

In order to prevent false alarms, all arrays include positive and negative controls chambers. In the positive control chamber other than adding the tested sample with the unknown chemicals to the bacterial solution, pure water was added. In case a current signal was generated, it is a false alarm. A negative control chamber includes w.t. (MG1655) *E.coli* bacteria that constitutively expresses β-galactosidase, thus, current should be generated in all cases. When no current generated, measurement is incorrect, due

to bacterial death from highly toxic chemicals added or from other unknown reason. However, chemicals can produce only constant DC current signal, while the enzymatic reaction act as an intrinsic amplifier, and generates increasing current signal.

Biochemical process, which intends to produce a measurable signal, has a great benefit while utilizing enzymatic activity. Since enzymes form continuously, and each enzyme reacts with many substrate molecules successively, this enzymatic mechanism serves as an intrinsic amplifier; consequently the signal produces faster, more sensitively and increasing with time [12]. Combining enzymatic system with electrochemical detection methods enables measurements in turbid solutions and under anaerobic conditions[5].

In addition to the aforementioned capabilities, this 'lab on a chip' system could be easily adapted to different applications, including specific identification of chemicals by using binding techniques, i.e., each electrochemical cell in the array can incorporate different biosensor, Thus, large amount of analytes can be detected simultaneously and independently. Similarly, in experiments aiming to analysis physiological reactions, bacteria harboring different types of promoters can be introduced to the chambers, and thus, this lab on a chip can detect simultaneously a variety of toxicant types.

Cleantech 2007, ISBN 1-4200-6382-0

6. CONCLUSIONS

Integrating recombinant microorganisms tailored to respond to specific biochemical events, together with unique and adjusted nanoscale electronic systems, provides exciting opportunities in the biosensors field as well as in the basic research on microorganisms. In this study, a miniaturized and portable electrochemical analytical device was fabricated, studied and characterized. The benefit of this new geometry, in which electrochemical cell dimensions were reduced to nano-scale, and the ratio between the working electrode area and cell volume was optimized, is demonstrated by the results presented here, which showed high sensitivity and extremely fast response time. The construction of an array of nano chambers on one silicon chip leads to high throughput in addition to the capability of performing multi experiment simultaneously and independently.

This design enables performance of multi experiments simultaneously and each electrochemical cell can be measured independently. The total weight of the entire system is ~900 grams, making it ideal for field environmental monitoring and for medical applications. Further size reduction is achievable using identical process and can provide MEMS for single cell measurements as well as for ensemble, but should be considered by the specific application requirements. In addition, better packaging can reduce the size and weight, keeping all the electrical parameters intact. Miniaturization of the electrochemical cell and favourable area to volume ratio, leads to short diffusion distance of the analytes towards electrode surface, therefore, providing improved signal to noise ratio, faster response time, enhanced analytical performance and increased sensitivity

The main features of the proposed "lab on chip" include small sampling requirements, high throughput, low false positive and false negative rates, high robustness and reliability and most probably very low cost in mass production. The proposed platform can be used for various applications that require 'electrochemical cell-based lab on a chip'.

REFERENCES

[1] L. H. Hansen and S. J. Sorensen, "The use of whole-cell biosensors to detect and quantify compounds or conditions affecting biological systems," *Microbial Ecology*, vol. 42, pp. 483-494, 2001.

[2] S. Daunert, G. Barrett, J. S. Feliciano, R. S. Shetty, S. Shrestha, and W. Smith-Spencer, "Genetically engineered whale-cell sensing systems: Coupling biological recognition with reporter genes," *Chemical Reviews*, vol. 100, pp. 2705-2738, 2000.

[3] S. Kohler, T. T. Bachmann, J. Schmitt, S. Belkin, and R. D. Schmid, "Detection of 4-chlorobenzoate using immobilized recombinant Escherichia coli reporter strains," *Sensors and Actuators B-Chemical*, vol. 70, pp. 139-144, 2000.

[4] I. Biran, L. Klimentiy, R. Hengge-Aronis, E. Z. Ron, and J. Rishpon, "On-line monitoring of gene expression," *Microbiology-Sgm*, vol. 145, pp. 2129-2133, 1999.

[5] Y. Paitan, D. Biran, I. Biran, N. Shechter, R. Babai, J. Rishpon, and E. Z. Ron, "On-line and in situ biosensors for monitoring environmental pollution," *Biotechnology Advances*, vol. 22, pp. 27-33, 2003.

[6] Y. Paitan, I. Biran, N. Shechter, D. Biran, J. Rishpon, and E. Z. Ron, "Monitoring aromatic hydrocarbons by whole cell electrochemical biosensors," *Analytical Biochemistry*, vol. 335, pp. 175-183, 2004.

[7] R. Babai and E. Z. Ron, "An Escherichia coli gene responsive to heavy metals," *Fems Microbiology Letters*, vol. 167, pp. 107-111, 1998.

[8] J. R. Premkumar, O. Lev, R. S. Marks, B. Polyak, R. Rosen, and S. Belkin, "Antibody-based immobilization of bioluminescent bacterial sensor cells," *Talanta*, vol. 55, pp. 1029-1038, 2001.

[9] S. Belkin, "Microbial whole-cell sensing systems of environmental pollutants," *Current Opinion in Microbiology*, vol. 6, pp. 206-212, 2003.

[10] H. J. Cha, R. Srivastava, V. M. Vakharia, G. Rao, and W. E. Bentley, "Green fluorescent protein as a noninvasive stress probe in resting Escherichia coli cells," *Applied and Environmental Microbiology*, vol. 65, pp. 409-414, 1999.

[11] D. E. Nivens, T. E. McKnight, S. A. Moser, S. J. Osbourn, M. L. Simpson, and G. S. Sayler, "Bioluminescent bioreporter integrated circuits: potentially small, rugged and inexpensive whole-cell biosensors for remote environmental monitoring," *Journal of Applied Microbiology*, vol. 96, pp. 33-46, 2004.

[12] E. Sagi, N. Hever, R. Rosen, A. J. Bartolome, J. R. Premkumar, R. Ulber, O. Lev, T. Scheper, and S. Belkin, "Fluorescence and bioluminescence reporter functions in genetically modified bacterial sensor strains," *Sensors and Actuators B-Chemical*, vol. 90, pp. 2-8, 2003.

Investigating the Benefit-Cost of MEMS Application for Structural Health Monitoring of Transportation Infrastructure

M. K. Jha[*] and A. B. Davy[**]

[*]Morgan State University, Dept. of Civil Engineering
Baltimore, Maryland 21251, USA, mkjha@eng.morgan.edu
[**]Morgan State University, Dept. of Civil Engineering
Baltimore, Maryland 21251, USA, davy@eng.morgan.edu

ABSTRACT

In recent years MEMS has been widely recognized as an effective device for structural health monitoring of transportation structures, such as bridges and tunnels. It is unclear, however, whether the benefits of MEMS application far outweigh the associated cost. A quantitative approach for benefit and cost calculation of MEMS application for structural health monitoring of transportation infrastructure will be a major step forward to provide guidance to potential MEMS users. In this paper, we develop a fuzzy logic-based approach (since MEMS benefits are generally fuzzy in nature and at best, they can be quantified using fuzzy-logic) for benefit-cost calculation associated with MEMS application. Real-world case studies will be presented in future works using the proposed fuzzy-logic approach.

Keywords: MEMS, Benefit-Cost, Fuzzy Logic, Transportation Infrastructure

1 INTRODUCTION

Tools available to conduct structural health monitoring (SHM) of transportation structures have increased in number and sophistication over the past twenty years. These structures include subway and roadway tunnels, highway and railroad bridges and other above ground structures that carry our transportation network. The Transportation Equity Act for the 21st century (TEA-21) enacted for the years 1998 to 2003 and reauthorized for the years through 2009, provides a vast infusion of funds to repair and for improvement of the nation's transportation infrastructure [1]. However the majority of existing structures will not be included in the repair and rehabilitation program. Clearly, the transportation infrastructure is aging faster than it can be repaired or replaced.

Structures begin to deteriorate through wear and tear, corrosion, ever-increasing traffic and overloads, and fatigue soon after being placed into service. SHM is a strategy to consistently detect damage through cost-effective monitoring of the structure's condition early enough to schedule preventative maintenance and repair.

Three strategies have been employed in the assessment of the infrastructure condition. They are scheduled visual inspection, scheduled non-destructive evaluation (NDE) and continuous health monitoring. Visual inspection is still predominant in most jurisdictions and will be for some time to come. It cannot however detect damage until it is visible, it is time-consuming and results are very dependent on the experience of the inspector. NDE is widely supported by the Federal Highway Administration (FHWA) through its Validation Center and is mainly used on highway bridges. Ground penetrating radar, x-rays, ultrasonic testing and acoustic emission (AE) monitoring are employed to detect and monitor structural cracks and delaminations. AE monitoring utilizes detection sensors, signal transmitters and computers to process the data. NDE systems tend to be expensive, cumbersome and are generally only brought in after a problem has been determined to exist.

Health monitoring utilizes a variety of sensors, some with built in measurement, interpretation and actuation capability to provide more continuous real-time data aimed at detecting the onset of serious deterioration. Initially the tools employed in SHM were expensive, difficult to set up and maintain, and yet deterioration could still go undetected. The promise of MEMS technology in this application is that it will provide an inexpensive, durable, compact information collection, processing and storage system for use in SHM.

Many applications of MEMS applied to the transportation infrastructure have been cited in recent publications and the most efficient ones appear to be those using intelligent sensors and wireless technology. Robust, cost-effective SHM solutions are best achieved by integrating and extending technologies from various engineering disciplines [2]. The challenge of developing networking algorithms for reliable communications in a wireless data acquisition network was described in [2] and [3]. Wireless active sensors with the capability to question response data was described in [4]. Some articles ([5-6]) describe monitoring techniques that have been used in wireless sensor networks and the cost of a wireless modular monitoring system was discussed in [6].

2 BENEFIT-COST OF MEMS APPLICATION

Typically, any new application that is likely to have promise is widely embraced by people in the beginning. However, in the absence of having a clear idea about the benefits of any new technology over its cost, its long-term usability and success may be doubtful. Research suggests that while cost can be easily quantifiable, benefit quantification is often a challenging task [7]. Benefit is generally the "perception" of individuals about a commodity, which can be captured and quantified through fuzzy-logic. The first author of this article was motivated to apply fuzzy-logic to quantify the benefit of Computer Visualization (CV) in highway development since just as MEMS holds greater promise in structural health monitoring CV holds grater promise in highway development; but whether the benefits of CV outweighs its cost is generally known. In this paper we provide the theoretical framework for applying fuzzy-logic in quantifying Benefit and cost of MEMS application in structural health monitoring.

2.1 Quantifying Benefits and Costs with Fuzzy-Logic

Fuzzy logic is particularly suited for benefit and cost estimations of MEMS application for two reasons: (1) as noted earlier, while cost of MEMS application may be quantifiable it may be difficult to quantify MEMS benefits due to their fuzzy nature. At best one can describe MEMS benefit in linguistic terms (commonly known as "linguistic hedges" in fuzzy logic), such as "high," "medium," or "low." (2) by converting both benefits and costs in fuzzy form we can be assured that we are comparing apples to apples; thus, the resulting B/C ratio might make more sense since fuzzy representation of a quantity always lies between 0 and 1. A review of literature [7-10] suggests that there are very limited applications of fuzzy-logic in benefit cost assessment. Neitzel and Hoffman [8] described benefits and costs in linguistic rather than numerical terms. Our procedure of fuzzy representation of MEMS benefit and cost is similar to that developed by Neitzel and Hoffman [8] in that we also use linguistic terms in describing benefit and cost and then convert them to numerical values using triangular fuzzy numbers (described later).

Bailey et al. [11] applied fuzzy logic to quantify public perception towards visualization in highway development. However, they did not quantify visualization benefits and costs. For highway economic analysis, Benefit-Cost ratio (B/C) is generally calculated [12] to assess if a proposed development is worth undertaking. The FHWA procedure uses actual numerical values using life-cycle costs and does not consider fuzzy characteristics of benefits. Next, we derive a general fuzzy B/C calculation procedure (similar to that available in [7]), which can be applied to investigate the cost effectiveness of MEMS application in the structural health monitoring of transportation infrastructure.

3 FUZZY-LOGIC APPROACH TO B/C CALCULATION

Fuzzy logic is a superset of conventional (Boolean) logic that has been extended to handle the concept of partial truth, i.e., truth values between "completely true" and "completely false." It was introduced by Zadeh [13] who defined fuzzification as a methodology to generalize any specific theory from a crisp (discrete) to a continuous (fuzzy) form. Since its inception in 1965 there has been numerous applications of fuzzy-logic in many fields where linguistic hedges often prohibit quantification. This makes MEMS benefit an excellent candidate for fuzzy-logic application.

3.1 Fuzzy Benefit of MEMS Application

In linguistic terms, benefit of MEMS application in structural health monitoring of transportation infrastructures can be characterized as "improved precision in identifying structural deterioration over time" and "reduced manual time/labor requirements for structural health monitoring." Let's introduce two sets $M(x)$ and $N(y)$ which represent "improved precision in identifying deterioration" and "reduced manual time/labor requirements" measured on a scale of 1-10. Intuitively, a higher number in both categories will imply "higher" increase in precision identification and "higher" reduction in time/labor requirements. The corresponding fuzzy membership functions can be represented as $\mu_M(x)$ and $\mu_N(y)$, respectively whose values range from 0 and 1. Thus, expected benefit, B_E of MEMS application can be expressed as:

$$B_E = 1 - \min\{\mu_M(x), \mu_N(y)\} \qquad (1)$$

Equation (1) implies that expected fuzzy benefit of MEMS application is a combined measure of increased precision identification AND reduced manual time/labor requirements. $M(x)$ and $N(y)$ can be described as linguistic variables, such as "low," "medium", or "high." For consistency in numerical analysis those linguistics variables can be represented on a scale of 1-10. For example, on a 1-10 scale, "low," "medium," and "high" can be described as 2, 5, and 7, respectively. Furthermore, $M(x)$ and $N(y)$ can be represented as triangular fuzzy numbers with different confidence intervals [14]. A typical triangular fuzzy number is most often presented in the form:

$$A = (a_1, a_2, a_3) \tag{2}$$

where a_1 is lower (left) boundary of the triangular fuzzy number, a_2 is number corresponding to the highest level of presumption, and a_3 is upper (right) boundary of the fuzzy number. The membership function of the fuzzy number $\mathbf{A}$ is:

$$\mu_A(x) = \begin{cases} 0, & x \le a_1 \\ \left(\dfrac{x - a_1}{a_2 - a_1}\right), & a_1 \le x \le a_2 \\ \left(\dfrac{a_3 - x}{a_3 - a_2}\right), & a_2 \le x \le a_3 \\ 0, & x \ge a_3 \end{cases} \tag{3}$$

By the similar notion assuming that $\mu_M(x)$ and $\mu_N(y)$ can be represented as triangular fuzzy numbers with $\mathbf{M} = (m_1, m_2, m_3)$ and $\mathbf{N} = (n_1, n_2, n_3)$, respectively. For example, if $\mathbf{M}$=(2, 6, 10) and $\mathbf{N}$=(2, 4, 6), then $\mu_M(x)$ and $\mu_N(y)$ can be expressed as:

$$\mu_M(x) = \begin{cases} 0, & x \le 2 \\ \dfrac{x}{4} - 0.5, & 2 \le x \le 6 \\ -\dfrac{x}{4} + 2.5, & 6 \le x \le 10 \\ 0, & x \ge 10 \end{cases} \tag{4}$$

$$\mu_N(y) = \begin{cases} 0, & y \le 2 \\ \dfrac{y}{2} - 1, & 2 \le y \le 4 \\ -\dfrac{y}{2} + 3, & 4 \le y \le 6 \\ 0, & y \ge 6 \end{cases} \tag{5}$$

The confidence intervals for membership functions $\mu_M(x)$ and $\mu_N(y)$ are user-specified; this has no effect on the general methodology developed here since the proposed methodology assumes a standard triangular fuzzy forms for $\mu_M(x)$ and $\mu_N(y)$. The confidence intervals are interpreted as being "low", "medium," and "high" and a quantitative measures of these linguistic variables are to be determined based on the characteristics of to the project to which the analysis is being applied. The assumption of triangular fuzzy numbers in similar applications can be found in Teodorovic and Vukadinovic [14] and Dompere [10].

3.2 Fuzzy Cost of MEMS Application

The different components of MEMS implementation cost can be classified as: (1) cost of hardware/software; (2) installation and setup cost; and (3) monitoring cost (which will be a recurring cost). Let $\mathbf{V}(z_1)$, $\mathbf{A}(z_2)$, and $\mathbf{R}(z_3)$ be these fuzzy costs, respectively. $\mu_V(z_1)$, $\mu_A(z_2)$, and $\mu_R(z_3)$ represent corresponding membership functions having triangular fuzzy forms with confidence intervals $\mathbf{V} = (v_1, v_2, v_3)$, $\mathbf{A} = (a_1, a_2, a_3)$, and $R = (r_1, r_2, r_3)$, respectively. Using the similar analysis presented earlier about the "OR" operator the fuzzy MEMS application cost, C_M can be expressed as:

$$C_M = \max[\mu_V(z_1), \mu_A(z_2), \mu_R(z_3)] \tag{6}$$

Since C_M is a combined measure of the three MEMS implementation costs noted above the use of "OR" operator is justifiable.

3.3 Fuzzy B/C Ratio

Using Eqs. (1) and (6) the fuzzy benefit-cost ratio of MEMS application can be expressed as:

$$\frac{B_E}{C_M} = \frac{1 - \min\{\mu_M(x), \mu_N(y)\}}{\max[\mu_V(z_1), \mu_A(z_2), \mu_R(z_3)]} \tag{7}$$

4 CONCLUSIONS AND FUTURE WORK

In this paper we presented an overview of emerging MEMS application in structural health monitoring of transportation infrastructure. We developed a fuzzy benefit-cost procedure for MEMS application, which seems quite promising. It will help MEMS users in correctly assessing if MEMS application outweighs its cost. In our future works, case studies will be presented using real-world examples, cost data, and a survey of the degree of satisfaction of MEMS users.

REFERENCES
[1] Clarke, E., "TEA-21, something for everyone," ASCE Civil Engineering Vol. 68, pp. 52-55, Oct. 1998.
[2] C. Farrar, H. Sohn, M. Fugate, J. Czarnecki. "Integrated Structural Health Monitoring", SPIE, 8th Annual International Symposium on Smart Structures and Materials, Newport Beach, CA, March 4-8, 2001.
[3] J. Caffrey, R. Govindan, E. Johnson, B. Krishnamachari, S. Masri, G. Sukhatme, K. Chintalapudi, K. Dabtu, S. Rangwala, A. Sridharan, N. Xu, M. Zuniga, "Networked Sensing for Structural Health Monitoring", 4th International Workshop on Structural Control, Columbia University, Ney York, June 2004.

[4] J. Lynch, A. Sundararajan, K. Law, A. Kiremidjian, T. Kenny, E. Carryer, "Embedment of Structural Monitoring Algorithms in a Wireless Sensing Unit", Structural Engineering and Mechanics, Vol. 15, No.3, 2003.

[5] J. Lynch, A. Sundararajan, K. Law, H. Sohn, C. Farrar, "Design of a Wireless Active Sensing Unit for Structural Health Monitoring", SPIE, 11th Annual International Symposium on Smart Structures and Materials, San Diego, CA, March 14-18, 2004.

[6] J. Lynch, K. Law, E. Straser, A. Kiremidjian, T. Kenny, "The Development of a wireless Modular Health Monitoring System for Civil Structures", Proc. Of the MCEER Mitigation of Earthquake Disaster by Advanced Technologies (MEDAT-2) Workshop, Las Vegas, NV, USA, November, 30-31, 2000.

[7] Jha, M.K., "Feasibility of Computer Visualization in Highway Development: A Fuzzy Logic-Based Approach," Computer-Aided Civil and Infrastructure Engineering, Vol. 21, No. 2, pp. 136-147, 2006.

[8] L.A. Neitzel, and L.J. Hoffman, "Fuzzy Cost/Benefit Analysis." In Fuzzy Sets: Theory and Applications to Policy Analysis and Information Systems (eds: Wang, P. P. and Chang, S.K.), pp. 275-290, Plenum Press, New York, 1980.

[9] Kahraman, C., "Fuzzy Versus Probabilistic Benefit/Cost Ratio Analysis for Public Work Projects," International Journal of Applied Mathematics and Computer Science, Vol. 11, No. 3, pp. 705-718, 2001.

[10] Dompere, K.K., "Cost-Benefit Analysis and the Theory of Fuzzy Decisions," Springer-Verlag, New York, 2004.

[11] K. Bailey, T. Grossardt, and J. Brumm, "Enhancing Public Involvement Through High Technology," TR news, pp. 16-17, May-June 2002.

[12] FHWA, "Economic Analysis Primer," U.S. Department of Transportation, Federal Highway Administration, Office of Asset Management, Washington, D.C., 2003.

[13] Zadeh, L. "Fuzzy Sets." Information and Control, Vol. 8, pp. 338-353, 1965.

[14] D. Teodorovic, and K. Vukadinovic, "Traffic Control and Transportation Planning: A Fuzzy Sets and Neural Networks Approach," Kluwer Academic Publishers, Norwell, MA, 1998.

Wireless Micro and Nano Sensors for Physiological and Environmental Monitoring

Bozena Kaminska

Simon Fraser University and Adigy Corporation

kaminska@sfu.ca

Abstract: A practical implementation of a wearable physiological and environmental monitoring system is presented. The technical requirements for wearable electronics and sensors are analyzed. A proposed system includes micro and nano device design, wireless network based on TCP/IP protocol and software application. The result samples from monitoring in ambulatory environment are discussed.

Keywords: wearable computing, medical sensors, wireless network, monitoring, SoC

1. Introduction

Recent advances in miniature devices as well as mobile communication and ubiquitous computing have fostered interest in wearable technology [1, 2, 3]. Radio telemetry of human and animal vital functions, first introduced in the 1950's, has today evolved into microelectronics for remote sensing of patients' motion and location, heart (ECG) and brain (EEG) electrical activity, arterial pulse, blood pressure, and oxygen saturation, intestinal motility and acidity, internal tissue chemistry and gas pressures, as well as orthopedic and dental measurements.

Wearable systems (sometimes incorporated into garments, shoes, costume jewellery, or "bandaids") facilitate noninvasive and unobtrusive monitoring of individuals over extended periods of time. Such systems generally rely on wireless, miniature transmitters with adequate memory capacity to temporarily store data from sensors, than upload/transmit that data to a database server via a secure high reliability receiver link (radio, optical, induction) often through a LAN or Internet connection. New techniques for short range radio communication unencumbered by regulatory restrictions (viz, wideband spread spectrum) enable wireless monitoring of ambulatory subjects both in home care and hospital that is relatively immune to interference.

Wireless sensor networks and biosensors are both subjects of intense current research. Employed together they permit uninterrupted physiological monitoring across broad geography in daily routine as well as emergency medical situations. Significant benefits can be realized in population disease screening, individual diagnosis (especially of unpredictable pathological events with ephemeral symptoms), evaluation of treatment efficacy, and broad delivery of individualized preventative care.

For health care to effectively employ wearable technology, several system criteria all need to be satisfied [1, 2. 3]. Hardware must be sufficiently robust to make measurements reliable during all activities of daily living, including demanding athletics, fitness training and heavy physical work. Data processing and decision-making algorithms need to provide timely communication and trustworthy interpretation, particularly for life-threatening events. Bidirectional communication and interactive control/test is necessary to assess and optimize measurement accuracy to improve user outcomes and care-provider confidence. Finally, electronic technology for widespread deployment must be clearly cost effective compared to more primitive alternatives or to simply ignoring problems.

These criteria create multiple design requirements: compactness and light weight, stability of signal during user motion or location change, tolerance to electrical interference and other environmental disturbances, durability for long life, data storage to allow opportunistic radio communication, and low "just enough" power consumption. Additionally, wearable instruments need to be easy to apply and adjust without assistance, and comfortable enough to wear for extended periods of time. A particular design challenge unique to wearable monitors is the tradeoff between long-term comfort and reliable sensor attachment. Our ultimate objective is to select, refine, design, develop, test and clinically validate those technologies and solutions best suited in this unique environment for reliable, miniature wearable biosensors and behavioral/environmental monitors easily addressable by wireless area networks connected to the Internet.

The overarching goal of this *Wearable Biomonitors* project is to create solutions that will *optimize lifetime wellness of a person*, thereby enhancing quality of life, permit independent living as long as possible, provide real-time support and advice as an electronic "health companion", and (eventually) reduce the overall life cycle cost of health and medical care.

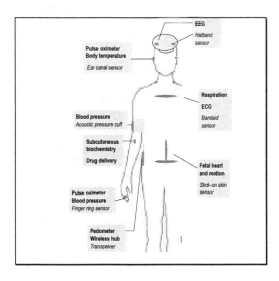

Figure 1. Wearable Body Network with Biosensors and Transceiver.

The proposed wearable biomonitoring system performs the following tasks:

- Reliably measure important indicators of an individual's state of health;

- Process/combine data into time-weighted transmission protocols;

- Provide fault-tolerance, self-test, and self-correction to assure data validity and engender user confidence;

- Develop protocols for intermittent opportunistic transmission of stored data to local receivers networked to the Internet;

- Provide appropriate feedback for care recipients/providers to enable responsive intervention for treatable disorders.

The variety of biomonitors and monitors that are candidates for wearable, long-term use are summarized in Figure 1. Not all of these monitoring devices or ideas have been reduced to practical devices, but those that are not available are either currently being researched or have promise from an engineering and/or medical perspective. Here the focus will be on a most popular ECG sensor and its practical wearable implementation.

2. Proposed Wearable System

Our wireless medical monitors send non-invasive physiologic monitoring signals (e.g. ECG and pulse oximetry) directly from the patient's skin via digital radio to nearby vital signs monitors and/or straight to the hospital network. There are several advantages to our approach. First, providing a means for inexpensively un-cabling the patient will allow the majority of patients, not just a few percent, to be un-tethered from their monitoring equipment. The second advantage of the proposed approach is that no additional equipment will be required; in the long run specialty monitoring equipment can be completely eliminated from the hospital. Third, the system allows much more flexibility for signal collection, transmission, display, printing, and data storage using commercially-available computer equipment. Fourth, where biosensors are designed with digital processing and storage at the sensor site, significantly better physiologic information can be gathered and delivered to caregivers. Finally, since our approach allows patients to be monitored without expensive monitoring equipment, effective monitoring can be implemented in many more instances.

The biosensors and digital radio components replace the patient monitor with a simple network receiver device and thereafter use existing elements of the hospital's IT system to display and record this important patient data, thus eliminating the expensive specialty monitoring equipment required today. Beyond the hospital application additional use include: long-term care (chronic disease) monitoring; in-home and ambulatory monitoring; and specialty screening, monitors, and measurement such as sleep physiology, pediatric apnea, pregnant mom/fetus heart rate monitor, and sports rehab and training.

The multiple benefits result from the developed solution. This includes patients, nurses and other hospital and care providing personnel, and hospital.

- **Patient Benefits:** Minimize slips, trips and falls; Eliminate lead wire entanglement; Provides monitored mobility; Enables easier and earlier ambulation; Increases patient comfort; Reduces risk of hospital acquired infections.

- **Nurse Benefits:** Facilitates ease of patient transport; Improves job satisfaction; Saves nursing time – fewer lead-off events.
- **Hospital Benefits:** Eliminates large cost of specialty monitors; Enhances nurse productivity; Fewer slips, trips and falls reduce adverse events; Monitored mobility reduces adverse events; Earlier ambulation leads to shorter hospital stays.

3. Technology Development

There are several examples where RF technology is being applied to medical devices. Baxter and others are using RFID tags for hospital equipment management and Precision Dynamics and others have developed and are selling an RFID identity bracelet for hospitals. Medtronic has the "Bravo" swallowed diagnostic capsule and a number of companies are developing implanted devices that can communicate with ex-body receivers. In patient monitoring area, such companies as Philips Medical and others are applying the newly-defined, medical-device-only frequencies to their telemetry monitoring systems. At the intersection of biosensors for medical applications and the new developments in active RFID chip-level technology, however, there is little action. Synergistic designs that fully leverage the latest-available chip-level technology will dramatically change, not incrementally improve, the products to which they are applied. Adigy's newly developed products fit into this category.

Active RF systems which are used here include, Figure 2: 1) a "chip" element that itself accomplishes RF sending and receiving, analog processing, digital processing, and biological, chemical, or physical sensor elements; 2) remote sender/receiver hardware; and 3) a software application for communications and data collection and display, as shown below.

Active RF "chips" (microelectronic devices) have a long read range for remote read-out, can provide real time status and telemetry, can be integrated with many available sensors, have available memory and may be used as data-loggers, can perform measurements even in the absence of a reader, and are reprogrammable. Their disadvantage is that they require a battery or direct source for power.

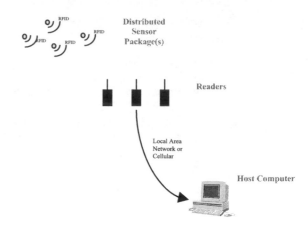

Figure 2. 3-tier architecture for an RF-enabled system

A schematic of the microelectronic device, application specific integrated circuits (ASIC's), developed for this project is shown below, Figure 3. The device has a sensor, analog circuitry, an analog to digital converter, amplifiers, and references. It has an on-chip clock with 1sec/week precision, and has on-board memory. It has bi-directional RF communication with encryption, anti-collision and error correction. The design has been done in a low power technology resulting in the extremely low standby current of 3 micro Amperes, emission current of 3 mili Amperes, and operating voltage of 3V. The currently used operating frequency is 433.92 MHz and 920 MHz. The transmission mode is programmable with periodic transmission and sensor activated transmission. There is a battery low alarm and easy connection to additional sensors. The transmission range is up to 100 m and life expectation over five years.

Figure 3. ASIC-based active RF device, a part of wearable system solution.

The developed RF-enabled system requires the readers. The reader is designed to receive, store and analyze data. A memory buffer is available to store captured data, including flash memory, so it will not lose information at power failure. The flash memory is 40 kB. The anti-collision and error detection/correction mechanism have been implemented allowing the reliable communication for up to 16 mln unique codes. Other features include: firmware update through TCP/IP, remote test functions, optional power supply for readers through the RJ-45 connector.

The reader is designed to communicate bi-directionally via TCP/IP interface or cellular wireless band. The reader dimensions are approximately 10x5 cm. The throughput is 50 sensors per second with data rate up to 20 Kbps and sensitivity -103 dBm.

Many system applications also require software database for collection and analysis of system data. The Adigy team has developed these database applications with appropriate user interface. The software establishes active connection with the readers, modifies network configuration, performs network tests, collects data from readers, and performs reader firmware upgrade and configuration.

4. Experimental Results

Specifically, the following innovative solutions are developed: a) low-noise, pre-processing ECG electrode with motion cancellation (Patent Pending); b) self-learning physiologic signal acquisition and transmission (Patent Pending); c) transmission codes for low volume and high reliability; d) new-generation ECG and pulse oximetry sensor; e) pattern recognition technique.

Figures 4 and 5 show the monitoring results in home environment. The variety of places and circumstances are monitored and compared with traditional stationary equipment. The correlation is excellent and there was no faulty monitoring observed.

Figure 4 shows the monitored differences comparing to the established normal patterns for the ECG and oxygen meter monitoring between 9 am and 3 pm from 5 sensors placed on 5 persons representing various health conditions. Figure 6 shows daily summaries from the sensors which represent the differences for the same person health monitoring (ECG).

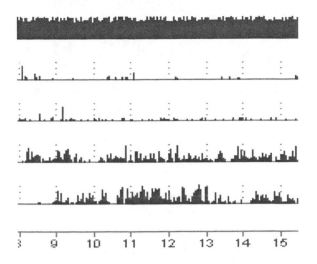

Figure 4. Hourly monitoring example

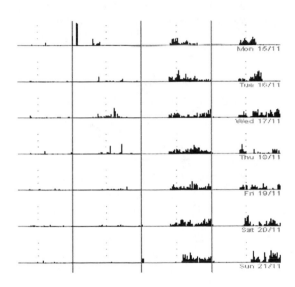

Figure 5. Daily summary monitoring

5. Conclusion

We have defined a set of requirements for a practical wearable and wireless monitoring system. The practical and commercial implementation has been described and the results shown.

References

1. Virone, G., Noury, N. Demongeot, J. 2002. IEEE Trans Biomed Eng 49:1463-1469.
2. Mendeza,P.et al."Web-based Vital Sign Monitoring", *IEEE EMBS*, San Fran., 2196-2199,2004.
3. Vehkaoja,A.,Lekkala,J. "Wearable Wireless Biopot."*IEEE EMBS*, S.F.,2177-2179, 2004.

Electrospun TiO$_2$ nanofibers for gas sensing applications

Il-Doo Kim[*], Avner Rothschild[**], Harry L. Tuller[***], Dong Young Kim[****], and Seong Mu Jo[*****]

[*]Optoelectronic Materials Research Center, KIST, Seoul, Republic of Korea, idkim@kist.re.kr
[**]Department of Materials Science and Engineering, MIT, Cambridge, MA, USA, avner@mit.edu
[***]Department of Materials Science and Engineering, MIT, Cambridge, MA, USA, tuller@mit.edu
[****]Optoelectronic Materials Research Center, KIST, Seoul, Republic of Korea,dykim@kist.re.kr
[*****]Optoelectronic Materials Research Center, KIST, Seoul, Republic of Korea, smjo@kist.re.kr

ABSTRACT

Nanostructured TiO$_2$ has attracted much attention for a variety of applications including photocatalysts, electrodes for water photolysis, dye-sensitized solar cells, and gas sensors. In this work we report on TiO$_2$ fiber mats for use in gas sensors demonstrating exceptionally high sensitivity to NO$_2$, a toxic gas responsible for acid rain and other air pollution effects, and high sensitivity to H$_2$, a potentially explosive gas.

Keywords: TiO$_2$, nanofiber, electrospinning, sensor, NO$_2$, H$_2$

1 INTRODUCTION

Recently, significant progress had been made in developing chemical sensing architectures based on various one-dimensional nanostructures by using semiconducting oxides [1-3]. The use of nanofibers in gas sensing provides unique structural features and high surface areas that are expected to promote the sensitivity of the oxide materials to the gaseous components as well as affecting the temperature dependence on sensing [4]. One-dimensional (1D) structures can also provide the lowest dimensionality for effective transport. Therefore they can be useful as building blocks in bottom-up assembly in many areas including nano-electronics and photonics [5]. In particular, increasing worldwide concerns regarding environmental degradation and health hazards have stimulated growing interest in means for detecting and monitoring potentially toxic chemicals.

A commonly applied gas sensing mechanism involves chemically induced resistivity changes in semiconducting materials. For example, the conductivity of TiO$_2$, a semiconducting oxide, can be modulated by the chemisorbed species on its surface. Chemisorbed oxygen species (O$_{2\text{(ads)}}^{-}$ or O$_{\text{(ads)}}^{-}$), accept electrons from the conduction band, thereby inducing an electron depletion layer which in turn results in increased band bending. In polycrystalline semiconductors and, as discussed in this work, in fiber mats, this results in the development of potential barriers between the grains or fibers. As a result, the conductivity of TiO$_2$ specimens is expected to decrease when exposed to an oxidizing gas and inversely decrease when exposed to a reducing gas. Although the use of one-dimensional nanostructures provides the prospect of high sensitivity and fast detection due to high surface-to-volume ratios, the incorporation of nanowires or nanofibers into sensing device systems is complicated by difficulties associated with selection of appropriate processes and achieving reproducibility. A variety of methods have been suggested for the preparation of nanofiber or nanowire, macroporous structures [6-8]. Among them, electrospinning is one of the most simple, versatile and cost-effective approaches. In a typical process for inorganic fibers, a sol-gel precursor solution with polymeric binder is extruded from the orifice of a needle under a high electric field. The polymeric binders are subsequently easily removed during calcination of the electrospun fibers. Indeed, there are many reports related to the preparation of semiconducting oxide nanofibers by electrospinning. However, to date, there are no reports about electrospun TiO$_2$ nanofiber sensors.

In this work we report the fabrication of electrospun TiO$_2$ fibers obtained through the phase separation of TiO$_2$ gel and poly(vinylacetate) during solidification. We investigate the response of TiO$_2$ nanofiber gas sensors to a number of gases and demonstrate exceptionally high sensitivity to NO$_2$, an atmospheric pollutant and H$_2$, a potentially explosive gas.

2 EXPERIMENTAL

TiO$_2$ fibers were electrospun from a solution of dimethyl formamide (DMF) (37.5 ml) of 3 g of poly(vinyl acetate) (PVAc, Mw=850000 g/mol), which was synthesized using bulk radical polymerization, 6 g of titanium(IV) propoxide (Aldrich), and 2.4 g of acetic acid as a catalyst. As in typical electrospinning procedures, the precursor solution was loaded into a syringe and connected to a high-voltage power supply. An electric field of 15 kV was applied between the orifice and the ground at a distance of 10 cm. TiO$_2$ fiber mats (Fig. 1(a)) were directly electrospun on Al$_2$O$_3$ substrates prepared with interdigitated Pt electrodes arrays (200 μm Pt fingers spaced 200 μm apart). The as-spun TiO$_2$ fiber mats were then pressed using preheated plates at 120°C for 10 min. Subsequently, the samples were calcined at 450°C for 30 min in air to remove the organic constituents and crystallized the TiO$_2$ fiber mats into the anatase phase. X-ray diffraction (XRD) and Scanning Electron Microscopy (SEM) were used to examine the phase composition and microstructure of the films, respectively. X-ray diffraction patterns showed anatase phase of TiO$_2$ structure with (101) and (200) peaks.

The sensitivity of the TiO$_2$ nanofibers towards H$_2$ and NO$_2$ gases was tested at temperatures between 300 and 350°C. The electrospun TiO$_2$ nanofibers were mounted on Al$_2$O$_3$ sample holders and contacted by Pt wires which were attached to the interdigitated electrode arrays on the Al$_2$O$_3$ substrates using silver paste (SPI Silver Paste Plus, SPI Supplies, Chester, PA, USA). The sample holders were then inserted inside a quartz tube placed within a tube furnace (Lindberg/Blue Model M 0.8 KW Tube Furnace). Pt/Pt-Rh (type S) thermocouples were used to measure temperature *in-situ*. The resistance was measured under a DC bias voltage of 0.1 V using a 4-channel DC power supply and ammeter (HP 6626A and 4349B, respectively). The TiO$_2$ nanofiber resistances were measured during exposure to different gas compositions using dry air as a carrier gas and pre-mixed bottled gases of 1000 ppm H$_2$ in air, 100 ppm NO$_2$ in air, 1000 ppm CO in air, and 1% CH$_4$ in air (BOC Gases, Riverton, New Jersey), together with mass flow controllers (MKS 1359C mass flow controllers and an MKS 647A controller). To eliminate interfering effects due to changes in gas flow rate, the gas sensing tests were carried out at a constant flow rate of 200 sccm. The flow rates of the carrier and test gases were varied between 200:0 sccm to 100:100 sccm in order to modulate the concentration of the test gas between 0 to 50% of the gas concentration in the pre-mixed gas bottle. The gas sensitivity of the TiO$_2$ fiber mats is defined as the ratio R$_0$/R or R/R$_0$ for reducing (H$_2$, CO, CH$_4$) or oxidizing (NO$_2$) gases, respectively, where R$_0$ is the baseline resistance in air and R is the resistance during the test under the presence of the test gas.

3 RESULTS

Figure 1 shows SEM images of electrospun TiO$_2$ nanofibers. The as-spun PVAc/TiO$_2$ composite fibers exhibit a range of diameters from 200 to 600 nm. After calcination in air for 30 min to remove the PVAc, TiO$_2$ fiber mats were observed as shown in Fig. 2 (a). As previously reported in Ref [9], in the initial stages of electrospinning, the TiO$_2$ sol precursor converts to a TiO$_2$ gel when the fibers are exposed to moisture. Liquid-liquid phase separation results in TiO$_2$-rich and PVAc-rich phases due to the concentration instability that arises after solvent evaporation. The separated phases are elongated during the spinning step, resulting in an aligned fibrillar structure in the fiber-axis direction. This unique structure could further enhance surface activity compared to normal TiO$_2$ nanofibers without fibrillar structure.

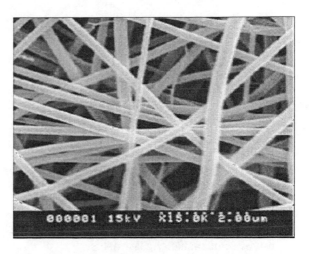

Figure 1. SEM image of as-spun TiO$_2$/PVAc composite fibers fabricated by electrospinning from a DMF solution.

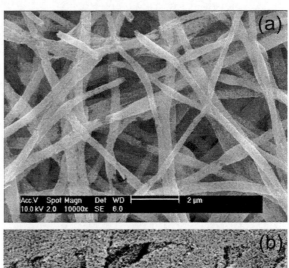

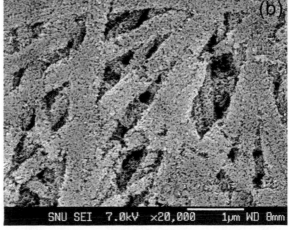

Figure 2. (a) SEM image of TiO$_2$ fiber mats after 450°C calcination. (b) High resolution SEM image of TiO$_2$ nanorod structure which was pretreated with mechanical press at 120°C for 10 min before calcination.

The resultant fibers exhibited a bundle structure consisting of ~ 20 nm thick fibrils as shown in Fig. 2 (b). This unique morphology results in an exceptionally high surface-to-volume ratio, highly advantageous for gas sensors. The specific area measured by BET was 90 m^2/g.

In order to investigate the potential advantages of the enhanced surface activity of the TiO_2 fiber mats and short diffusion lengths associated with the 20nm fibrils, prototype gas sensors, using Pt interdigital electrode structures, were fabricated. The electrical response of several gas sensor prototypes was measured during exposure to traces of reducing (H_2) and oxidizing (NO_2) gases mixed in air, at operating temperatures of 300°C and 350°C. The response was reversible and reasonably fast as shown in Figs. 3 and 4, with response times of the order of min. The resistance decreased upon exposure to reducing gases and increased during exposure to oxidizing gases, typical of n-type semiconductor gas sensors.

These sensor prototypes demonstrated exceptional sensitivity to NO_2, with response magnitudes, R/R_0, as high as ~100 upon exposure to some tens of ppm of NO_2 in air and response to levels as low as 500 ppb (Fig. 3).

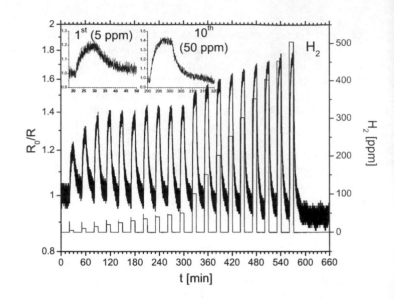

Figure 4. The resistance response during exposure to successive pulses with increasing concentrations of H_2 in air, operating at a temperature of 300°C.

Thus, these sensors demonstrated preferential selectivity towards H_2 amongst the reducing gases tested. We demonstrate detection of H_2 down to 5 ppm level using TiO_2 fiber mats as shown in inset of Fig. 4. Note that the response magnitude (y-axis) in Figs. 3 and 4 is opposite, R/R_0 and R_0/R, respectively.

4 SUMMARY AND CONCLUSIONS

In summary, TiO_2 fiber mats were prepared for use in gas sensors by the electrospinning method. These demonstrated high sensitivity to NO_2 (oxidizing gas) and H_2 (reducing gas). The response was reversible and reasonably fast with response times of the order of 1 min. Sensitivities as high as ~100 were measured upon expose to 50 ppm NO_2 in air. This work demonstrates that TiO_2 fibers are promising candidates for ultra-sensitive gas sensors capable of detection of various gases down to sub ppm levels.

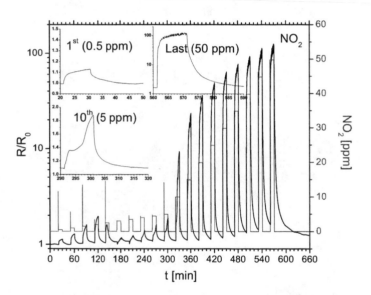

Figure 3. The resistance response during exposure to successive pulses with increasing concentrations of NO_2 in air at an operating temperature of 350°C.

These results point to TiO_2 nanofibers as promising candidates for gas sensors in environmental and medical applications in which low ppm levels of NO_2 need be detected. As shown in inset of Fig. 3, such ultra-sensitive levels of detection can be obtained by use of nanofiber TiO_2 sensor protoypes.

The response to reducing gases, while not as high as for NO_2, was still significant. Amongst the gases tested in this study, the sensitivities to H_2 where significantly higher than for CO and CH_4.

ACKNOWLEDGEMENTS

This work was supported, in part, by the National Science Foundation under contract #ECS-0428696.

REFERENCES

[1] D. Zhang, Z. Liu, C. Li, T. Tang, X. Liu, S. Han, B. Lei, and C. Zho, Nano Lett., 4, 1919, 2004.

[2] J. Chen, L. Xu, and W. Li, and X. Gou, Adv. Mater., 17, 582, 2005.

[3] D. Li, Y. Wang, and Y. Xia, Nano Lett., 3, 1167, 2003.

[4] P. I. Gouma, Rev. Adv. Mater. Sci. 5, 147, 2003.

[5] N. Dharmaraj, H. C. Park, C. K. Kim, H. Y. Kim, and D. R. Lee, Mater. Chem. Phys., 87, 5, 2004.

[6] I. D. Kim, A. Rothschild, T. Hyodo, and H. L. Tuller, Nano Lett. 6, 193, 2006.

[7] D. Li and Y. Xia, Adv. Mater., 16, 1151, 2004.

[8] Y. Im, C. Lee, R. P. Vasquez, M. A. Bangar, N. V. Myung, E. J. Menke, R. M. Penner, and M. H. Yun, Small, 2, 356, 2006.

[9] M. Y. Song, Y. R. Ahn, S. M. Jo, D. Y. Kim, and J. P. Ahn, Appl. Phys. Lett., 87, 113113, 2005.

Cleantech 2007, ISBN 1-4200-6382-0

Index of Authors

Cleantech 2007, www.techconnect.org, ISBN 1-4200-6382-0

Index of Keywords

Cleantech 2007, www.techconnect.org, ISBN 1-4200-6382-0